Experimental Brain Research Series 24

Professor Dr. med. Otto Detlev Creutzfeldt
1. 4. 1927–23. 1. 1992

B. Albowitz K. Albus U. Kuhnt
H.-Ch. Nothdurft P. Wahle (Eds.)

Structural and Functional Organization of the Neocortex

Proceedings of a Symposium in the Memory of Otto D. Creutzfeldt, May 1993

With 250 Figures

Springer-Verlag
Berlin Heidelberg New York
London Paris Tokyo
Hong Kong Barcelona
Budapest

Dr. Birgit Albowitz
Prof. Dr. Klaus Albus
Dr. Ulrich Kuhnt
PD Dr. Hans-Christoph Nothdurft
Dr. Petra Wahle

Max-Planck-Institut für Biophysikalische Chemie
Postfach 2841
37018 Göttingen
FRG

ISBN-13:978-3-642-78501-6 e-ISBN-13:978-3-642-78499-6
DOI: 10.1007/978-3-642-78499-6

Softcover reprint of the hardcover 1st edition 1994

Cover Illustration by P. Wahle
Production Editor: Renate Münzenmayer
Typesetting: Camera-ready by authors
25/3130-5 4 3 2 1 0 – Printed on acid-free paper

Contents

Preface

On April 1, 1992, Otto Creutzfeldt would have celebrated his 65th birthday. About a year before, those of us who worked with him in the Department of Neurobiology decided to organize a meeting to honour Otto and his contributions to neuroscience. Since Otto Creutzfeldt had no intention of retiring before his 67th birthday, this was not to be a "retirement conference" but a celebration on the occasion of a major anniversary.

However, events turned out differently. In the spring of 1991, Otto Creutzfeldt became seriously ill, before we could consult with him on his suggestions for such a meeting, and we decided to postpone further discussion until his health improved. In the summer, when we talked to him about the meeting, he basically approved of the idea but immediately suggested our turning it into a proper scientific congress, rather than limiting ourselves to inviting his friends and former co-workers. He had very good friends in many parts of the world and would certainly have liked to see many of them again. However, his interest in science remained the dominant factor. He proposed the main topics for the conference and suggested several speakers to be invited. Although he knew that his health was critical and that his condition could rapidly worsen, we had the feeling that the planning gave him hope and perhaps an opportunity to look forward. At one stage, when he saw a chance of being able to go abroad for a few months around the time of his 65th birthday, he even suggested postponing the conference until the fall of 1992. Sadly, his plans could not be fulfilled and he died early in 1992.

After a time of mourning, we decided to go ahead with the conference in his memory. The meeting took place in May 1993, at the Max Planck Institute for Biophysical Chemistry in Göttingen, where Otto Creutzfeldt worked for the last 20 years of his life. It turned out to be full of interesting presentations and stimulating discussion, and quite in a style he would have approved; only his questions, statements and critical comments were missed.

This book gives an almost complete collection of the contributions to the conference. The topics cover different aspects of structure and function of the neocortex and provide a survey of the areas of greatest interest to Otto Creutzfeldt. It can be considered as the end of a path which he had taken more than 20 years ago, and in comparison with a meeting he had organized a few years after his arrival in Göttingen[1], this book also documents the development of neuroscience during his tenure in Göttingen as head of the Department of Neurobiology.

[1] Creutzfeldt O (ed) Afferent and intrinsic organization of laminated structures in the brain. Exp Brain Res Suppl 1, Springer, Berlin Heidelberg New York, 1976

Despite the many different areas this book presents, we must apologize for its incompleteness in certain aspects. Many more neuroscientists — all very good friends of Otto Creutzfeldt — would have liked to have contributed to the meeting but could not because of the limited time available. Otto Creutzfeldt would not have wanted to exclude anybody from his conference, but we had to be guided by practical considerations. It was physically impossible to include scientific presentations from all his friends.

Now that the book has been edited to a final version and only needs to be printed and bound (and sold), we would like to thank everyone who contributed to our lively meeting, the speakers especially, but also the members of the audience, all those who helped to organize it (mainly the former secretary of Otto Creutzfeldt, Eva-Maria Hölscher) and the Deutsche Forschungsgemeinschaft for financial support. Finally, we would like to thank, in particular, Jutta Tolksdorf who worked for several months editing all the manuscripts, many of them several times over. Without her continued help this book would not have been possible.

Göttingen, November 1993

On behalf of the Organizing Committee,
Christoph Nothdurft

Otto Creutzfeldt 1927–1992

W. SINGER
Max Planck Institute for Brain Research, Franfurt am Main, Germany

Adapted from Experimental Brain Research (1992) 88:463–465

Otto Creutzfeldt was born on April 1, 1927, in Berlin, the son of the great neurologist Hans-Gerhard Creutzfeldt, whose name has become associated with Jacob-Creutzfeldt disease. After receiving his school education in a humanistic gymnasium, Otto Creutzfeldt started studying theology, history and philosophy in 1945 at the universities of Kiel and Tübingen, subjects which continued to fascinate him throughout his life and which provide a key to understanding his motivations and aspirations. Otto Creutzfeldt was a humanist in the full sense of the word, passionately searching for ways to penetrate the mysteries of life and above all the *conditio humana*. We will ignore the reasons he decided in 1948 not to continue his studies of the humanities but to enter the faculties of medicine at the universities in Kiel, Heidelberg, and finally Freiburg, where he obtained his approbation as a physician in 1953. Maybe he succumbed to the attraction of a discipline that promises a holistic approach to the understanding of human beings, allowing for the coexistence of rigorous reductionism and intuitive comprehension.

During the years between 1953 and 1959 Otto Creutzfeldt served as an assistant and trainee in various medical disciplines: in physiology with Prof. Hoffmann in Freiburg, in psychiatry with Prof. Müller in Bern, and in neurophysiology and neurology with Prof. Jung, again in Freiburg. At the end of these *Lehr- und Wanderjahre* Otto Creutzfeldt was a fully trained neurologist, familiar with neurological and psychiatric diseases, the domains where the specific limitations and dangers of human existence are particularly clearly exposed. In addition, he was already a member of the scientific community, having authored a number of important papers in the field of electrophysiology. The years with Richard Jung, the eminent neurologist and pioneer of single neuron recordings in mammalian brains, were decisive for Creutzfeldt's future orientation. Richard Jung's main interest was to relate electrophysiological events in the brain to behavior, to clinical phenomena such as motor disorders and epilepsy, to mental states such as sleep, wakefulness and attention, and, above all, to perception. And Jung insisted that it is the search for such relations which matters, not the analysis of the neuronal substrate *per se*. Two congenial minds had found one another; Richard Jung and Otto Creutzfeldt formed a close and wonderful friendship until Richard Jung died in 1986.

During his stay in Freiburg Otto Creutzfeldt laid the foundations upon which he built the admirable range of his interests and the fields of research to which he would later make so many seminal contributions. He adapted the single unit recording technique — at that time still a daring and highly sophisticated endeavour — and applied it to the most complex structure there is, the cerebral cortex, the structure which subsequently never ceased to exert a deep fascination on him and became his major scientific challenge. Otto Creutzfeldt studied neuronal activity during epileptic seizures, described changes in firing patterns associated with changes in arousal and investigated responses to sensory stimulation in the somatosensory and visual cortex. He also produced seminal papers on interactions at the level of single cortical neurons of nonspecific modulatory afferents from central core structures and specific sensory inputs. He was one of the first to emphasize the important influence of central states on sensory responses and to point out that inhibition plays an important role in sculpting the response properties of neurons in the central nervous system. O.J. Grüsser, K. Lehmann, H. Kornhuber and G. Baumgartner are among the colleagues who witnessed Otto Creutzfeldt's years in Freiburg and remained his friends since then. The spirit of these years, which were not only decisive for individual careers but also for the development of basic and clinical neurophysiology in Germany, can be appreciated by reading the obituaries which Otto Creutzfeldt wrote in 1986 for Richard Jung and then in 1991 for his close friend Günther Baumgartner (see the appropriate issues of *Experimental Brain Research*).

Otto Creutzfeldt understood his scientific efforts as an integral part of his life. In both domains priorities were therefore the same. He was driven by genuine curiosity and a striving for harmony, esthetic solutions and synthesis. He had a fine mastery of the flute and it was one of his greatest pleasures to join with his friends, and over the last years also with his children, to play chamber music, preferably from composers of the renaissance and the baroque periods. Competition and priority did not figure in his vocabulary other than as attributes of fair sportsmanship. This is why his colleagues, coauthors and later his pupils were always at ease with him, grew besides him, and if time allowed became his friends. Otto Creutzfeldt was strict and categorical when consistency, logical stringency and validation of results were at stake, but he was inexhaustibly generous and flexible when it came to personal matters, curricular worries, or the very private problems of those for whom he had assumed responsibility. Those who had the priviledge to get close to him, know about his touching inability to say "no" if a "yes" was going to relieve someone's grief even if there were good reasons at an impersonal, political level to remain strict.

As most German postdocs who wished to begin an independent scientific career at the time, Otto Creutzfeldt went abroad. He worked for two years as research anatomist at the medical school of UCLA before moving to the Max Planck Institute for Psychiatry in Munich where he was to stay for nearly 10 years, from 1962 to 1971. These were decisive years both for this institute and for all of us who had the chance to share part of the Munich era with him. In 1963 he obtained the aggregation for clinical neurophysiology from the University of Munich, and in 1965 he was nominated scientific member of the Max Planck Society, head of the department of neurophysiology and consultant neurologist at the MPI for Psychiatry. He was now in a position to formulate and realize his scientific concepts with great freedom, to create a permissive environment for crossfertilization between basic and clinical research and to form his own school of pupils. The then revolutionary intracellular recordings from neurons in the central nervous system were pioneered in Creutzfeldt's department. These were seminal studies on the relation between postsynaptic potentials in cortical neurons and EEG activity, on synaptic interactions between specific and non-specific thalamic inputs in the motor and the visual cortex, on synaptic transmission in the lateral geniculate nucleus and visual cortex and on the role of postsynaptic inhibition in the formation of visual receptive fields. The names of H.D. Lux, A.C. Nacimiento, J.M. Fuster, S. Watanabe, M. Konichi, J. McIllwain, M. Ito and U. Kuhnt will remain associated with these pioneering studies and illustrate the then unusual international breadth of Creutzfeldt's lab. It is still highly rewarding to consult these now classical papers, especially since there has been a renaissance of intracellular studies in vivo during the last decades. However, these investigations, which were at the time still extremely demanding, were only one, albeit an important facet, of Creutzfeldt's approach. In parallel, he initiated a systematic search for neurophysiological correlates of visual perception, starting with a detailed analysis of receptive field properties in all visual centers known at that time and extending these studies to recordings from awake animals. H. Noda, H. Scheich, A. Herz, M. Straschill, B. Sakmann, R. Freeman, and H. Wässle were among the many colleagues and students who figured repeatedly as authors on papers in this field of visual neurophysiology. But despite these numerous, extremely successful and rewarding advances in basic research, Otto Creutzfeldt had not forgotten what Richard Jung had imprinted in all of his pupils, that basic and clinical research should and can go hand in hand. The conditions in Munich were ideally suited because the clinic was under the same roof as the departments of basic research. Creutzfeldt, thus, developed a research unit for clinical neurophysiology and thereby allowed for an optimal exchange of concepts and techniques between clinicians and neurobiologists. This proved to be an exceptionally fertile marriage, and the offspring was a respectable number of important studies on the applicability of computer-based analyses of EEG patterns. These approaches also prepared the ground for the development of algorithms to locate the dipoles responsible for evoked potentials, work that would be pursued after Creutzfeldt left the institute in 1971.

The time in Munich is already sufficiently remote to be seen as a historical epoch. It has been extraordinarily influential and stands for an unmistakeable synthesis of fundamental and clinical research in the field of neurophysiology. It saw the initiation to neuroscience of an impressive number of pupils who now populate more than a dozen chairs in German universities and Max Planck Institutes. Creutzfeldt formed a school with a profound and lasting impact on German neuroscience.

Despite, or perhaps even because of this splendid career, Otto Creutzfeldt, then 44 years old accepted the challenge to move to Göttingen and to create a new department of Neurobiology in the newly founded Max Planck Institute for biophysical Chemistry. Creutzfeldt followed the call of Manfred Eigen, the eminent biophysicist, who then was famous not only for his measurements of ultrafast chemical reactions but already for his theories on the molecular processes at the transition from matter to life. Here was an institute devoted to the analysis of elementary mechanisms at a level which, as far as living structures were concerned, was maximally distant from the level of complexity of higher nervous systems. The question why Otto Creutzfeldt opted for this new challenge is intimately related to the question why he had chosen to devote his life to the exploration of higher brains. Answers to these questions are bound to remain fragmentary. But Otto Creutzfeldt's *oevre*, especially where it goes beyond the publication of data in specialized journals, his famous monograph on the "cortex cerebri", an impressive collection of book chapters and his conferences to interdisciplinary audiences are revealing. There is a passionate interest in philosophical problems associated with the subdivision of the world into levels of inanimate matter, living organisms and mental phenomena. Thoroughly acquainted with the history of philosophical and theological concepts, endowed with a rich clinical experience about the close relation between mental processes and neuronal substrates and finally with his "hands on" knowledge about brain mechanisms, he was in the rare position of being able to treat the mind-body problem competently at both levels. And the more he knew, the more he felt attracted and fascinated by the question, to what extent mental phenomena, perception of qualia, consciousness and emotions could be considered as emergent properties of an immensely complex aggregation of matter. A variant of this question poses itself also at the transition from inanimate to living structures: In both cases a continuous increase of complexity of aggregated matter appears to support the emergence of new qualities. Here, then, is perhaps one of the motivations for Creutzfeldt's move to Göttingen, a vision of continuity, a longing for synthesis, a fascinating ability for abstraction which allowed him to view together what seemed to others far apart.

The conditions in Göttingen permitted a substantial expansion of the department and Creutzfeldt used this opportunity to develop a permissive environment for a broad range of activities. Work in the acoustic and somatosensory system complemented the continuing studies of the visual system, the latter focusing more and more on mechanisms of colour coding in the primate lateral geniculate and on principles of cortical organization. Particularly revealing for Creutzfeldt's ability to sense quality and to grant freedom for individual developments is the fact that it was in his department that B. Sakmann and E. Neher had performed the experiments in the early eighties which were honoured with the Nobel Prize for Medicine in 1991.

Despite his many responsibilities as head of a large department, as acting director for the whole institute, as editor in chief of "Experimental Brain Research", as author of multiple contributions to the discussion on "animal rights", Otto Creutzfeldt continued to have his own lab and to perform experiments. He had to collect the data himself to satisfy his curiosity. This is also why he chose to spend a sabbatical with the neurosurgeon G. Ojemann in Seattle and to realize his dream of recording from neurons in the cerebral cortex of humans and to establish direct relations between cellular responses and cognitive behavior. This collaboration led to an impressive series of papers, the last ones still being in press.

It is touching to read through the last titles of his impressive publication list which altogether comprises more than 200 experimental reports and more than 20 major philosophical contributions on the mind-matter problem and on ethical implications of the neurosciences. One realizes Otto Creutzfeldt's fidelity to the projects that he had developed as a young scientist: there is the renaissance of work with patients, there are intracellular studies in the cortex in vivo, this time with the whole cell patch technique, there are single unit studies in awake monkeys in search for a correlation between single unit activity and perception and there is his last book, edited together with Sir John Eccles, on *The Principles of Design and Operation of the Brain.*[1] This book summarizes a symposium at the Pontificia Academia Scientiarium of which Otto Creutzfeldt had become a member in 1990.

Throughout his life Otto Creutzfeldt generously and without hesitation gave the most precious of the goods he could share, his time: time to listen, time to discuss, time to help, and time to take over yet another obligation. In the years to come he would have had the chance to use some of this time for himself, to look back on his rich life, to enjoy the recognition of his achievements, the careers of his pupils and the friends all over the world

[1] Eccles JC, Creutzfeldt OD (eds) The principles of design and operation of the brain. Study week Pontificia Academia Scientiarium, Scripta Varia. Exp Brain Res Series, vol 21, Springer, Berlin Heidelberg New York, 1990

who were waiting for him. And above all he was longing to find more time for his wonderful family, his wife Mary, his son Benjamin, and the three daughters Corinna, Ruth and Claire. Otto Creutzfeldt was a warm-hearted husband and father, a pioneer of modern neurophysiology and to many of us a dear friend.

List of Speaker

Birgit Albowitz, Max-Planck-Institut für biophysikalische Chemie, Abteilung Neurobiologie, Postfach 2841, D-37018 Göttingen, Germany

Klaus Albus, Max-Planck-Institut für biophysikalische Chemie, Abteilung Neurobiologie, Postfach 2841, D-37018 Göttingen, Germany

Richard A. Andersen, Division of Biology, California Institute of Technology, Pasadena CA 91125, USA

Alain Artola, Max-Planck-Institut für Hirnforschung, Abteilung Neurophysiologie, Deutschordenstraße 46, D-60528 Frankfurt am Main, Germany

Giovanni Berlucchi, Istituto di Fisiologia umana, Strada Le Grazie, I-37134 Verona, Italy

Colin Blakemore, University Laboratory of Physiology, Parks Road, Oxford OX1 3PT, United Kingdom

Jürgen Bolz, Cerveau et Vision, INSERM Unité 371, 18, Av du doyen Lépine, F-69500 Bron/Lyon, France

Valentino Braitenberg, Max-Planck-Institut für biologische Kybernetik, Spemannstraße 38, D-72076 Tübingen, Germany

Jean Bullier, Cerveau et Vision, INSERM Unité 371, 18, Av du doyen Lépine, F-69500 Bron/Lyon, France

Verne S. Caviness Jr., Department of Neurology, Harvard Medical School, Massachusetts General Hospital, 25 Fruit Street, Boston MA 02114, USA

Barry W. Connors, Department of Neuroscience, Division of Biology and Medicine, Brown University, Providence RI 02912, USA

Robert Desimone, Laboratory of Neuropsychology, National Institute of Mental Health, Building 9, Room 1N107, Bethesda MD 20205, USA

Ursula Dräger, EK Shriver Center, Division Developmental Neuroscience, 200 Trapelo Road, Waltham MA 02254, USA

Ulf Eysel, Ruhr-Universität Bochum, Medizinische Fakultät, Abteilung für Neurophysiologie, D-44780 Bochum, Germany

David Ferster, Department Neurobiology & Physiology, Northwestern University, 2153 North Campus Drive, Evanston IL 60208, USA

Burkhart Fischer, Neurologische Universitätsklinik, Abteilung für Neurophysiologie, Hansastraße 9a, D-79104 Freiburg im Breisgau, Germany

Christoph Fromm, Max-Planck-Institut für biophysikalische Chemie, Abteilung Neurobiologie, Postfach 2841, D-37018 Göttingen, Germany

Charles D. Gilbert, Rockefeller University, 1230 York Avenue, New York NY 10021, USA

Michael E. Goldberg, Laboratory of Sensorimotor Research, Building 49, Room 2A-50, NIH, 9000 Rockville Pike, Bethesda MD 20892, USA

Amiram Grinvald, Department of Neurobiology, The Weizmann Institute of Science, 76100 Rehovot, Israel

Otto-Joachim Grüsser, Physiologisches Institut, Freie Universität, Arnimallee 22, D-14195 Berlin (Dahlem), Germany

Paul Heggelund, University of Oslo, Department of Neurophysiology, Institute of Basic Medical Sciences, PB 1104 Blindern, N-0317 Oslo, Norway

Stewart H. C. Hendry, Zanvyl Krieger Mind/Brain Institute, 338 Krieger Hall, John Hopkins University, 3400 N. Charles St., Baltimore MD 21218, USA

Klaus-Peter Hoffmann, Ruhr-Universität Bochum, Lehrstuhl für Allgemeine Zoologie und Neurobiologie, D-44780 Bochum, Germany

Manfried Hoke, Institut für Experimentelle Audiologie, Westfälische Wilhelms-Universität Münster, Kardinal-von-Galen-Ring 10, D-48129 Münster, Germany

Giorgio Innocenti, Institut d'Anatomie, Université de Lausanne, Rue du Bugnon 9, CH-1005 Lausanne, Switzerland

Jon H. Kaas, Department of Psychology, Vanderbilt University, 301 Wilson Hall, Nashville TN 37240, USA

Takuji Kasamatsu, The Smith-Kettlewell Eye Research Institute, 2232 Webster Street, San Francisco CA 94115, USA

Hans H. Kornhuber, Abteilung für Neurologie, Universität Ulm, Steinhövelstraße 9, D-89073 Ulm/Donau, Germany

Alan U. Larkmann, Laboratory of Physiology, University of Oxford, Parks Road, Oxford OX1 3PT, United Kingdom

Barry B. Lee, Max-Planck-Institut für biophysikalische Chemie, Abteilung Neurobiologie, Postfach 2841, D-37018 Göttingen, Germany

Marla B. Luskin, Department of Anatomy and Cell Biology, Emory University - School of Medicine, Atlanta GA 30322, USA

Christoph von der Malsburg, Lehrstuhl für Biophysik, Institut für Neuroinformatik, Ruhr-Universität Bochum, D-44780 Bochum, Germany

Kevan A. C. Martin, Medical Research Council Anatomical Neuropharmacology Unit, Department of Pharmacology, Mansfield Road, Oxford OX1 3TH, United Kingdom

Konrad Maurer, Klinik für Psychiatrie I, Zentrum für Psychiatrie, Universität Frankfurt, Heinrich-Hoffmann-Strasse 10 D-60428 Frankfurt am Main, Germany

David A. McCormick, Yale University, Section of Neurobiology, School of Medicine, C303 SHM, P.O.Box 3333, New Haven CT 06510-8001, USA

Helen J. Neville, Department of Cognitive Science, Mail Code 0515, Universtity of California, San Diego, La Jolla CA 92093, USA

Hans-Christoph Nothdurft, Max-Planck-Institut für biophysikalische Chemie, Abteilung Neurobiologie, Postfach 2841, D-37018 Göttingen, Germany

George A. Ojemann, Department of Neurological Surgery, RI 20, University of Washington, School of Medicine, Seattle WA 98195, USA

Hellmuth Petsche, Institut für Neurophysiologie der Universität, Währingerstraße 17, A-1090 Wien, Austria

Bert Sakmann, Abt. Zellphysiologie, Max-Planck-Institut für Medizinische Forschung, Jahnstraße 29, D-89518 Heidelberg, Germany

Henning Scheich, Institut für Neurobiologie, Abteilung Neurophysiologie, Brenneckestraße 6, Postfach 1860, D-39110 Magdeburg, Germany

Almut Schüz, Max-Planck-Institut für biologische Kybernetik, Spemannstraße 38, D-72076 Tübingen, Germany

Wolf Singer, Max-Planck-Institut für Hirnforschung, Abteilung für Neurophysiologie, Deutschordenstraße 46, D-60528 Frankfurt am Main, Germany

Peter Somogyi, Medical Research Council Anatomical Neuropharmacology Unit, Department of Pharmacology, Oxford University, Mansfield Road, Oxford OX1 3TH, United Kingdom

Daniel Y. Ts'o, Division of Neuroscience, Baylor College of Medicine, 1 Baylor Plaza S603, Houston TX 77030–3411, USA

Tadaharu Tsumoto, Department of Neurophysiology, Biomedical Research Center, Osaka University Medical School, Suita, 565 Japan

Hendrik Van der Loos, Institut d'Anatomie, Université de Lausanne, Rue du Bugnon 9, CH-1005 Lausanne, Switzerland

David C. Van Essen, Department of Anatomy and Neurobiology, Washington University Medical School, Box 8108, 660 South Euclid Avenue, St. Louis MO 63110, USA

Trichur R. Vidyasagar, Visual Neuroscience Group, John Curtin School of Medical Research, Australian National University, G.P.O. Box 334, Canberra ACT 2601, Australia

Heinz Wässle, Max-Planck-Institut für Hirnforschung, Abteilung Neuroanatomie, Deutschordenstraße 46, D-60528 Frankfurt am Main, Germany

Petra Wahle, Fakultät für Biologie, Allgemeine Zoologie und Neurobiologie, Ruhr-Universität Bochum, ND7/31, D-44780 Bochum, Germany

Claudia Walther, Max-Planck-Institut für biophysikalische Chemie, Abteilung Molekulare Zellbiologie, Postfach 2841, D-37018 Göttingen, Germany

1 David A. McCormick, **2** Ulrich Kuhnt, **3** Barry W. Connors, **4** Ulf Eysel, **5** Alan U. Larkman, **6** Klaus-Peter Hoffmann, **7** Takuji T. Kasamatsu, **8** Alain Artola, **9** Heinz Wässle, **10** Christoph Nothdurft, **11** Petra Wahle, **12** David VanEssen, **13** Robert Desimone, 14 Burkhart Fischer, **15** Birgit Albowitz, **16** Otto-Joachim Grüsser, **17** Henning Scheich, **18** Jon H. Kaas, **19** Christoph Fromm, **20** David L. Ferster, **21** Trichur R. Vidyasagar, **22** Helen J. Neville, **23** Wolf Singer, **24** Jean Bullier, **25** Michael E. Goldberg, **26** Charles D. Gilbert, **27** Klaus Albus, **28** Barry B. Lee, **29** Amiram Grinvald, **30** Christoph von der Malsburg, **31** Richard A. Andersen, **32** George A. Ojemann, **33** Mary Creutzfeldt, **34** Benny Creutzfeldt, **35** Hans H. Kornhuber, **36** Ruth Creutzfeldt, **37** Tadahuru Tsumoto, **38** Giorgio Innocenti, 39 Marla B. Luskin, **40** Ursula Dräger, **41** Jürgen Bolz, **42** Daniel Y. Ts'o, **43** Giovanni Berlucchi, **44** Verne S. Caviness, **45** Almut Schüz, **46** Kevan A.C. Martin, **47** Stewart H.C. Hendry, **48** Colin Blakemore, **49** Peter Somogyi, **50** Paul Heggelund, **51** Valentino Braitenberg, **52** Hendrik Van der Loos, **53** Claudia Walther

missing: Manfried Hoke, Konrad Maurer, Hellmuth Petsche, Bert Sakmann

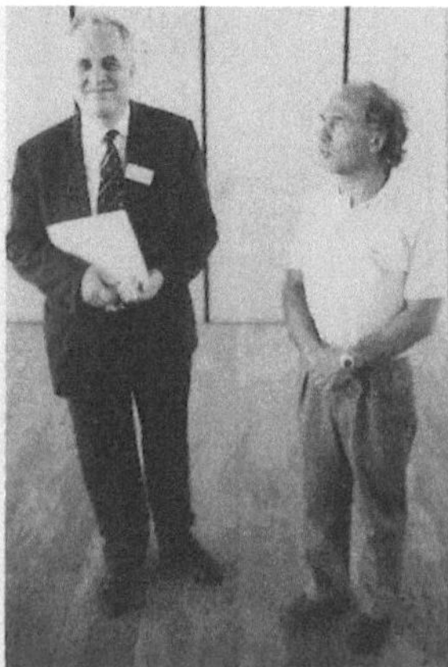
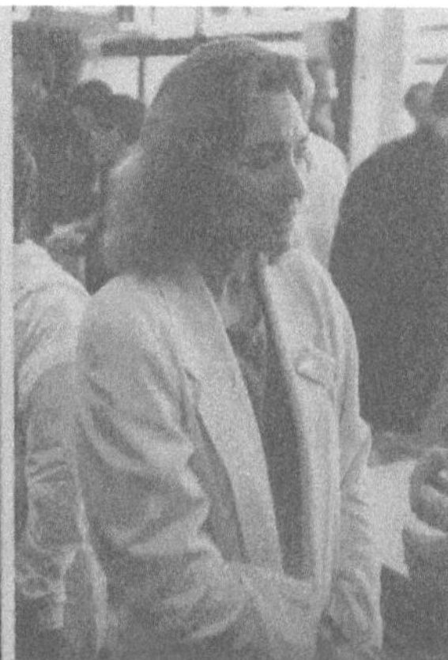

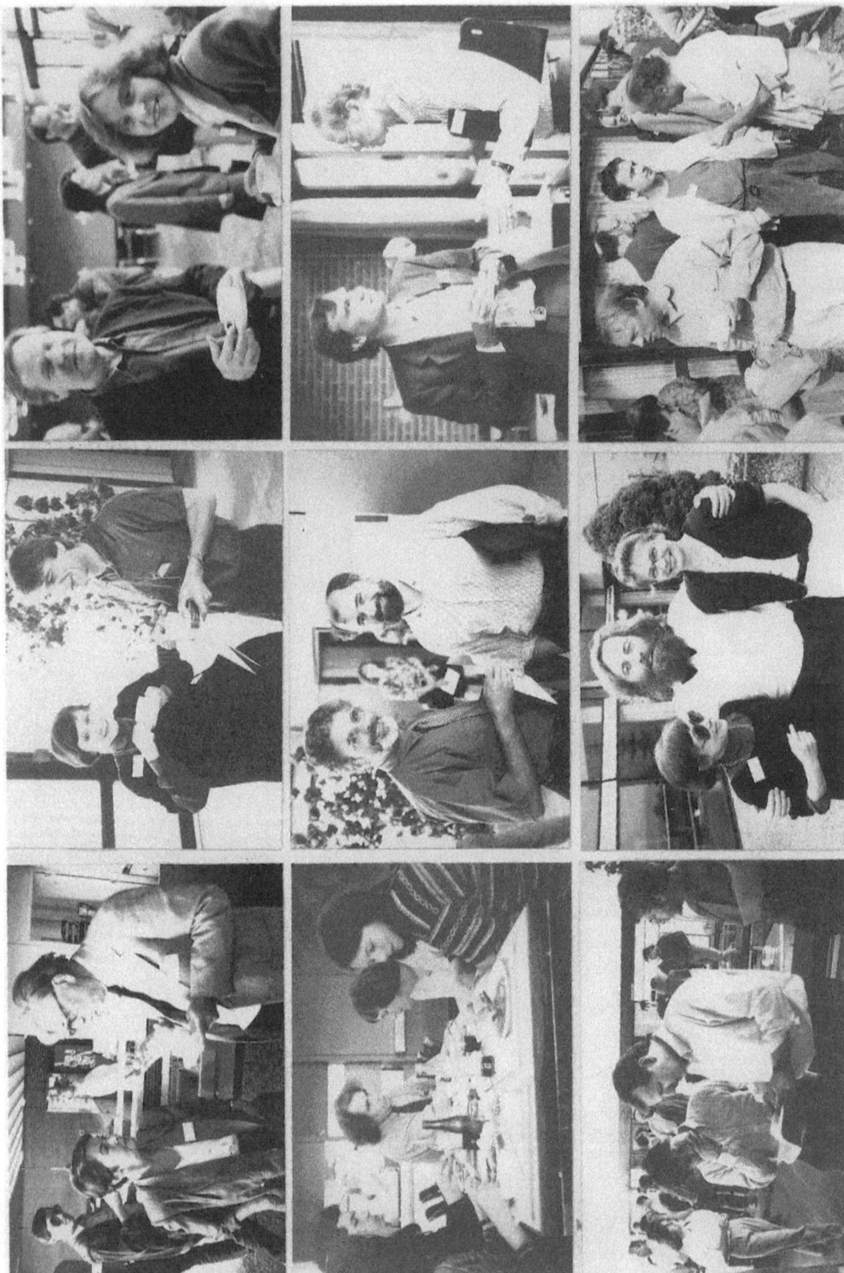

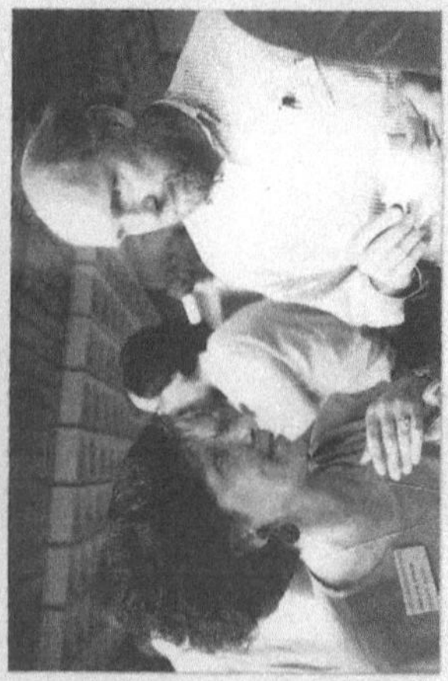

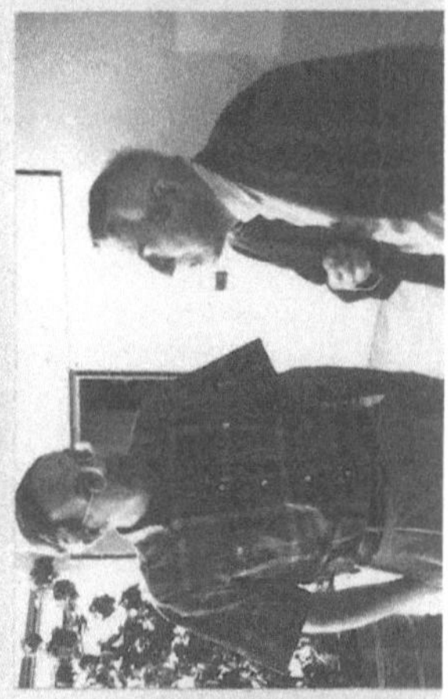

Chapter 1

Development

1.1 Introduction

U. Dräger
Division Developmental Neuroscience, E.K. Shriver Center, Waltham, Massachusetts

Any comprehensive treatise on the cerebral cortex has to contain a section on development like the following one. Nonetheless, while quite a few developmental studies were done in Otto Creutzfeldt's institute, his own research was relatively little concerned with such questions. The reason for this was that he was a neurophysiologist by inclination and was thus interested in mechanisms and complex functional relationships. At the early stage of his career, when he established the directions of his future research, very little experimental or functional research could be done on mammalian development; purely descriptive anatomical studies were much too confining for his expansive temperament and all-round interests. At that time, neurophysiologists had turned mainly to psychophysics, psychology, and physics for the exchange of ideas and expansion of concepts. There was hardly any vocabulary or techniques in common with the wet sciences such as biochemistry and molecular biology. In recent years, however, a revolution of techniques has been taking place which pervades all of biology and is increasingly invading neuroscience. These techniques now make it possible to address functional questions on developmental mechanisms, and they firmly integrate neurobiology into the rest of modern biology.

The first contribution from Claudia Walther and the laboratory of Peter Gruss is an excellent example of the revolution of techniques. The Pax genes belong to a class of factors that have the greatest potential as determinants for the functional parcellation of the brain. Despite the fundamental significance for neurobiology, these and related studies did not originate in neurobiology at all. Rather, they are rooted in a wealth of genetic and molecular insights on *Drosophila* development which have been accumulated in recent years.

Similar to the approach chosen by Walther and Gruss, the retroviral transfer method, a technique essential to the next two contributions by Verne Caviness and by Marla Luskin, was developed outside of neurobiology in the pursuit of virological and molecular biological goals. The contribution by Caviness provides for the first time a theoretical and comprehensive framework for the wealth of existing lineage studies on the cerebral cortex. Luskin's work reveals unexpected complexities in the generation of the developing olfactory bulb.

The fact that Otto Creutzfeldt was very interested in using innovations from other fields in neurobiological problems is evident from his encouragement of Petra Wahle's studies, which were largely done in his institute. Her work demonstrates how longstanding developmental questions can be addressed with the combined technology of neuroanatomical, cell biological, and molecular biological methods.

Given the diversity of developmental studies and the rapid speed of progress at present, it is clear that the four contributions in this section cannot cover this field, but they provide an exciting introduction to one of the most fascinating processes in neurobiology at this time. If Otto Creutzfeldt were to start his career now, he would probably try to link higher cortical functions and molecular developmental mechanisms.

1.2 The Role of Pax Genes in Mammalian Development

C. WALTHER
Department of Molecular Cell Biology, Max Planck Institute of biophysical Chemistry, Göttingen, Germany

One of the most fascinating questions in biology has for centuries been how a single cell, the zygote, can give rise to such a complex organism as a vertebrate. Due to the rapidly developing techniques in molecular biology, it is now feasible to decipher the molecular mechanisms underlying vertebrate development in general and CNS development in particular. Invaluable for the identification of genes involved in the regulation of embryonal development has been the discovery of conserved sequence motifs in developmental control genes of *Drosophila melanogaster*, for example, homeobox (McGinnis et al. 1984). Different sequence motifs have been used to isolate vertebrate genes involved in different developmental processes (Kessel and Gruss 1990).

One such sequence motif, the paired box, encodes a DNA binding domain (the paired domain) which was originally identified in segmentation genes of *Drosophila* (Bopp et al. 1986). Subsequently, paired box containing genes have been identified in mouse, human, frog, turtle, chick, and zebrafish (Deutsch and Gruss 1991; Walther et al. 1991; Gruss and Walther 1992; Chalepakis et al. 1992). To date, eight murine genes, called Pax (paired-box) genes, have been described. Some members of this gene family encode, in addition to the paired domain, a paired-type homeodomain and a conserved motif encoding an octapeptide. Sequence comparison using statistical geometry has permitted the allocation of Pax genes into at least four different classes (Fig. 1). Genes of a particular class show extensive sequence homology not only in the paired box and, if present, the homeobox but also in the flanking coding regions. In addition, they have similar genomic organizations, suggesting that they arose during evolution via duplication (Walther et al. 1991).

With the exception of *Pax-1*, all Pax genes analyzed to date are expressed in the developing CNS in a distinct pattern along the dorsoventral and the anterior-posterior axes. The primitive neural tube consists of a layer of mitotically active cells from which cells that have undergone final mitosis migrate radially to specific positions in the intermediate zone, where they differentiate. During this process a functional polarity along the dorsoventral axis of the developing spinal cord becomes evident. Motor neurons, for

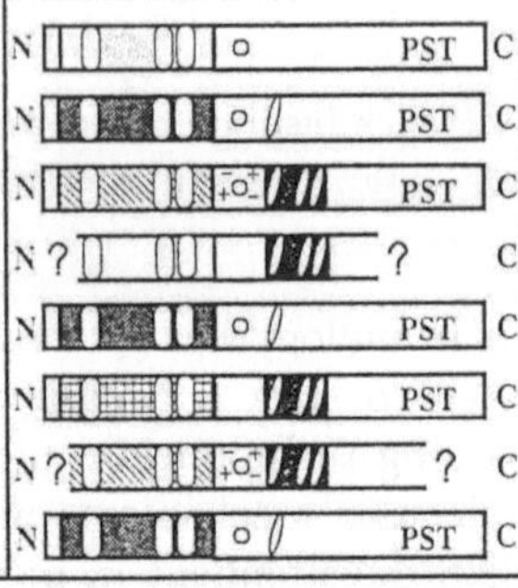

GENE	CHROMO-SOME	STRUCTURE (PD HD)	PROTEIN	MUTANTS MOUSE	MUTANTS HUMAN
Pax-1	2	N PST C	361 aa	*undulated (un)*	
Pax-2	19	N PST C	392 / 415 aa		
Pax-3	1	N PST C	479 aa	*splotch (Sp)*	*Waardenburg Syndrome I*
Pax-4	6	N ? ? C	?		
Pax-5	4	N PST C	391 aa		
Pax-6	2	N PST C	422 / 436 aa	*small eye (Sey)*	*Aniridia*
Pax-7	4	N ? ? C	>300 aa		
Pax-8	2	N PST C	457 aa		

Fig. 1.2–1. The Pax gene family. The paired domain and homeodomain are shown as *boxes* in the linear representation of the Pax protein structure. α-Helices included in these domains are indicated by *white ovals*; the octapeptide is indicated by a *circle* C-terminal to the paired domain. The chromosomal localization, the size of the protein, and the mouse mutant and human diseases associated with certain Pax genes are shown. *aa*, Amino acids; *PD*, paired domain; *HD*, homeodomain; *?*, not determined; *PST*, proline-, serine-, threonine-rich region

example, are located exclusively in the ventral portion of the neural tube, whereas neurons of the sensory pathway are located in the dorsal portion. The notochord and the floor plate have been shown to play a crucial role in this patterning of the developing spinal cord (van Straaten et al. 1988; Placzek et al. 1990).

The Pax genes are expressed in distinct regions along the ventrodorsal axis of the neural tube. *Pax-3*, *Pax-6*, *Pax-7* are expressed mainly in the mitotically active cells of the ventricular zone. *Pax-3* expression is detected dorsally in the alar plate and in the roof plate (Goulding et al. 1991), *Pax-6* more ventrally in the basal and intermediate plate (Walther and Gruss 1991), and *Pax-7* in the alar plate (Jostes et al. 1991). In contrast, *Pax-2*, *Pax-5*, and *Pax-8* are detected in subgroups of cells which have already stopped dividing and migrated into the intermediate zone (Nornes et al. 1990; Asano and Gruss 1992; Plachov et al. 1990).

Since *Pax-3*, *Pax-6*, and *Pax-7* show a defined spatial pattern in the neuroepithelium stem cells before neural differentiation starts, it appears that these genes play a role in the formation of the organization of the neural tube. For this reason it was tested whether Pax gene expression pattern is regulated by induction events directed by the notochord. In collaboration with Prof. Andrew Lumsden, notochord transplantation experiments were carried out in the chicken embryo. An additional notochord was placed laterally to the neural tube, and the effect of this supernumerary notochord on the expression pattern of *Pax-3* and *Pax-6* was studied (Goulding et al. 1993). It was demonstrated that the graft of an additional notochord leads to the turning off of the dorsal Pax gene (*Pax-3*), whereas the boundary of expression of the ventral Pax gene (*Pax-6*) in the basal plate is moved more dorsally. Moreover, this effect depends upon the position of the notochord: the more dorsally the notochord is implanted, the stronger is the inhibitory effect on *Pax-3* and *Pax-6*. As a supplementary experiment, an extirpation of the notochord was undertaken. This resulted in the *Pax-6* pattern being almost completely switched off and the *Pax-3* pattern being moved ventrally. These experiments showed that the defined expression pattern of the Pax genes depends on the induction cascade originating from the notochord. Therefore, these genes represent excellent candidates for the control of the development of the neural tube.

Pax genes are also expressed in the developing brain, and their expression domains coincide with established anatomical boundaries. Thus, they might also be involved in the establishment of different regions in the anterior neural tube along the anterior-posterior axis. *Pax-2, Pax-5*, and *Pax-8* show an anterior limit of expression at the midbrain-hindbrain boundary (Nornes et al. 1990; Asano and Gruss 1992; Plachov et al. 1990). In early embryonic stages, *Pax-3* and *Pax-7* are expressed in all regions of the brain. In midgestation embryos the anterior boundary of expression is retracted to the posterior commissure at the border between diencephalon and mesencephalon (Stoykova and Gruss 1993). The posterior commissure delineates also the boundary of *Pax-6* expression in the brain. *Pax-6* is expressed in the prosencephalon and myelencephalon but not in the roof of the mesencephalon, and this pattern is maintained until late in development. Moreover, in the case of the zebrafish counterpart of *Pax-2*, pax[zf-b], injection of antibodies against pax[zf-b] resulted in the loss of the hindbrain-midbrain boundary (Krauss et al. 1992). The activity of *Pax-3* extends to the neural crest and its derivates, particularly the spinal ganglia, the nasal process and structures derived from the first and second branchial arches. *Pax-6* is expressed in the developing eye: in the retina, lens, cornea, and notably in early stages in the inducing (optic vesicle) as well as in the induced (overlying surface ectoderm) tissues. In addition, *Pax-6* expression is detected in the olfactory epithelium and in the olfactory bulbs.

The functional significance of these expression patterns can now be analyzed in mutant mice carrying mutations in particular Pax genes. Interestingly, three Pax genes, *Pax-1, Pax-3*, and *Pax-6*, map near known mouse mutants that show malformations of structures in which these genes are expressed. Analysis of the particular Pax gene in these mice substantiated this correlation. The mutant *undulated* which shows skeletal malformations carries a mutation in the *Pax-1* gene (Balling et al. 1988). In two alleles of

Splotch (Sp^{2H} and Sp) the *Pax-3* gene is affected (Epstein et al. 1991, 1993). Mutants show exencephaly, spina bifida as well as dysgenesis of Schwann cells and the spinal ganglia, and defects in heart structures. Heterozygotes show aberrations in pigmentation. This phenotype underlines the requirement of *Pax-3* for the proper development of the CNS and neural crest derived structures.

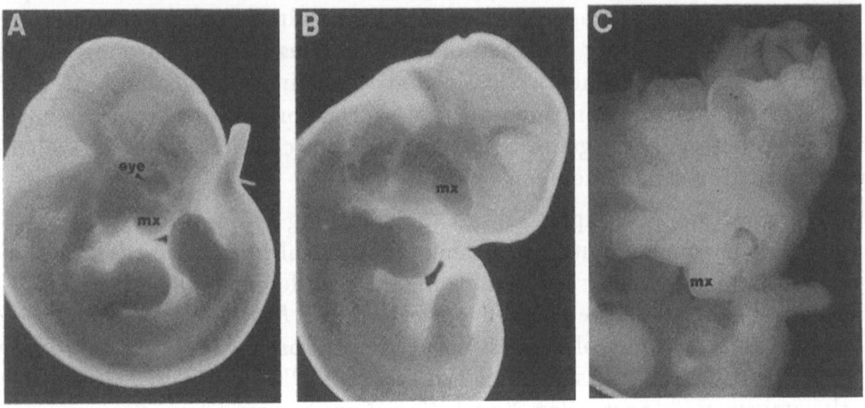

Fig. 1.2–2. Phenotype of *Splotch* and *Small eye*. **A** Day 11.5 postcoitus wild-type mouse embryo. **B** Day 12 postcoitus *Sey/Sey* embryo. **C** Day 11.5 postcoitus *Sp/Sp* embryo. Note the absence of the eye in the *Sey/Sey* embryo and the exencephalus and partial spina bifida in the *Sp/Sp* embryo. *mx*, Maxillary process. (From Gruss and Walther 1992)

Mutations in the *Pax-6* gene have been found in several alleles of *Small eye (Sey, Sey^H*; Hill et al. 1991). Homozygous *Small eye* mice do not develop eyes and nose and lack olfactory bulbs. Heterogygotes develop small eyes, cataracts, hypoplasia of the optic nerve and iris defects (Fig. 2). In contrast to the *Pax-3* mutant *Splotch* whose phenotype correlates well with the expression of *Pax-3*, *Small eye* mutants show defects in only a subset of structures in which *Pax-6* is expressed. Possible explanations are that more subtle defects in the neural tube have been missed due to the lack of appropriate markers, or that there are functional redundancies.

At least two human syndromes can be correlated with the human homologues of Pax genes and are counterparts of the mouse mutants *Splotch* and *Small eye*. Waardenburg's syndrome I is an autosomal dominant combination of pigmentary disturbances, lateral displacement of the inner canthus of each eye, occasional deafness of each ear, and mental retardation. Waardenburg's syndrome I has been genetically mapped to a region of human chromosome 2 that is synthenic to the proximal part of mouse chromosome 1, where the *Sp/Pax-3* is located. Based on the chromosomal mapping and the phenotypic features between Waardenburg's syndrome and

Splotch, subsequent inspection of *Pax–3* coding sequences from Waardenburg's syndrome patients has led to the identification of several mutations in the *Pax–3* gene (Baldwin et al. 1992; Tassabehji et al. 1992). The human counterpart of the *Small eye* phenotype is the aniridia syndrome, which is characterized by complete or partial absence of the iris (Ton et al. 1991; Jordan et al. 1992; Glaser et al. 1992)

In summary, by providing an entry point to approach an intricate network of events, Pax genes serve as a tool to study the molecular basis of developmental processes, particularly that of the nervous system.

References

Asano M, Gruss P (1992) Pax–5 is expressed at the midbrain-hindbrain boundary during mouse development. Mech Dev 39: 29–39

Baldwin CT, Hoth CF, Amos JA, Da-Silva EO, Milunsky A (1992) An exonic mutation in the HuP2 paired domain gene causes Waardenburg's syndrome. Nature 355: 637–638

Balling R, Deutsch U, Gruss P (1988) Undulated, a mutation affecting the development of the mouse skeleton, has a point mutation in the paired box of Pax1. Cell 55: 531–535

Bopp D, Burri M, Baumgartner S, Frigerio G, Noll M (1986) Conservation of a large protein domain in the segmentation gene paired and in functionally related genes of Drosophila. Cell 47: 1033–1040

Chalepakis G, Tremblay P, Gruss P (1992) Pax genes, mutants and molecular function. J Cell Sci 16 (Suppl.): 61–67

Deutsch U, Gruss P (1991) Murine paired domain proteins as regulatory factors of embryonic development. Sem Dev Biol 2: 413–424

Epstein DJ, Vekemans M, Gros P (1991) Splotch (Sp^{2H}), a mutation affecting development of Pax–3. Cell 67: 767–774

Epstein DJ, Vogan KJ, Trasler DG, Gros P (1993) A mutation within intron 3 of the Pax–3 gene produces aberrantly spliced mRNA transcripts in the splotch (Sp) mouse mutant. Proc Natl Acad Sci USA 90: 532–536

Glaser T, Walton DS, Maas RL (1992) Genomic structure, evolutionary conversation and aniridia mutations in the human Pax-6 gene. Nat Genet 2: 232–239

Goulding MD, Chalepakis G, Duetsch U, Erselius J, Gruss P (1991) Pax–3, a novel murine DNA binding protein expressed during early neurogenesis. EMBO J 10: 1135–1147

Goulding MD, Lumsden A, Gruss P (1993) Signals from the notochord and floor plate regulate the region specific expression of two Pax genes in the developing spinal cord. Development 117: 1001–1016

Gruss P, Walther C (1992) Pax in development. Cell 69: 719–722

Hill RE, Favor J, Hogan BLM, Ton CCT, Saunders GF, Hanson JM, Prosser J, Jordan T, Hastie ND, van Heyningen V (1991) Mouse Small eye results from mutations in a paired-like homeobox-containing gene. Nature 354: 522–525

Jordan T, Hanson I, Zalatayev D, Hodgson S, Prosser J, Seawright A, Hastie ND, van Heyningen V (1992) The human PAX6 gene is mutated in two patients with aniridia. Nat Genet 1: 328–332

Jostes B, Walther C, Gruss P (1991) The murine paired box gene, Pax-7, is expressed specifically during the development of the nervous and muscular system. Mech Dev 33: 27–38

Kessel M, Gruss P (1990) Murine developmental control genes. Science (Washington) 249: 374–379

Krauss S, Maden M, Holder N, Wilson SW (1992) Zebrafish pax[b] is involved in the formation of the midbrain-hindbrain boundary. Nature 360: 87–89

McGinnis W, Garber RL, Wirz J, Kuriowa A, Gehring WJ (1984) A homologous protein-coding sequence in Drosophila homeotic genes and its conservation in other metazoans. Cell 37: 403–408

Nornes HO, Dressler GR, Knapik EW, Deutsch U, Gruss P (1990) Spatially and temporally restricted expression of Pax–2 during murine embryogenesis. Development 109: 797–809

Plachov D, Chowdhury K, Walther C, Simon D, Guenet J-L, Gruss P (1990) Pax-8, a murine paired box gene expressed in the developing excretory system and thyroid gland. Development 110: 643–651

Placzek M, Tessier-Lavigne M, Yamada T, Jessell T, Dodd J (1990) Mesodermal control of neural cell identitiy: floor plate induction by notochord. Cell 64: 635–647
Stoykova A, Gruss P (1993) Roles of Pax-genes in developing and adult brain as suggested by expression patterns. J Neurosci (in press)
Tassabehji M, Read AP, Newton VE, Harris R, Balling R, Gruss P, Strachan T (1992) Waardenburg's syndrome patients have mutations in the human homologue of the Pax–3 paired box gene. Nature 355: 635–636
Ton CCT, Hirvonen H, Miwa H, Weil MM, Monaghan P, Jordan T, van Heyningen V, Hastie ND, Meijers-Heijboer H, Drechsler M, Royer-Pokora B, Collins F, Swaroop A, Strong LC, Saunders GF (1991) Positional cloning of a paired box- and homeobox-containing gene from the Aniridia region. Cell 67: 1059–1074
Van Straaten HWM, Hekking JWM, Wiertz-Hoessels EJLM, Thors F, Drukker J (1988) Effect of the notochord on the differentiation of a floor plate area in the neural tube of the chick embryo. Anat Embryol 177: 317–324
Walther C, Gruss P (1991) Pax-6, a murine paired box gene, is expressed in the developing CNS. Development 113: 1435–1449
Walther C, Guenet J-L, Simon D, Deutsch U, Jostes B, Goulding MD, Plachov D, Balling R, Gruss P (1991) Pax: a murine multigene family of paired box containing genes. Genomics 11: 424–434

1.3 Regulated and Nonregulated Parameters of Neocortical Cytogenesis

V.S. Caviness Jr.[1], T. Takahashi[1], R.S. Nowakowski[2], and L.-H. Tsai[3]
[1]Department of Neurology, Massachusetts General Hospital, Harvard Medical School, Boston, Massachusetts
[2]Department of Neuroscience and Cell Biology, UMDNJ-Robert Wood Johnson Medical School, Piscataway, New Jersey
[3]Massachusetts General Hospital Cancer Center, Charlestown, Massachusetts

We consider here quantitative behavior of the proliferative population that gives rise to neurons of the cerebral neocortex. Specifically, we characterize this behavior in terms of the rate of cell production and the total number of cells generated through the neocortical cytogenetic epoch in a defined and representative region of the mouse cerebral wall. Certain of the parameters that characterize this proliferative behavior vary while others are invariant as cytogenesis proceeds. The variable parameters are given particular emphasis; these may be "regulated" by cellular interactions. Their identification is seen as the entry step into a molecular neurobiology of neocortical cytogenesis.

1.3.1 Neocortical Histogenesis: An Overview

The neurons and neuroglial cells which inhabit the neocortex arise from proliferative populations lying deep within the embryonic cerebral wall (His 1904; Schaper 1897; Sidman and Rakic 1982). Neurons enter the neocortex in advance of the glial populations (Imamoto et al. 1978; Levinson and Goldman 1993; Luskin et al. 1988; Misson et al. 1991b; Takahashi et al. 1991a). The earliest neurons to be formed come to occupy the deepest neocortical layers; progressively later arising neurons are directed to successively more superficial layers (Angevine and Sidman 1961; Caviness 1982; Rakic 1974). The majority are thought to ascend from the proliferative zone to the cortex by migrating along the surfaces of radial glial fibers (Misson et al. 1991a; Rakic 1971; Rakic et al. 1974). As a result, those neurons arising from a "point source" within the proliferative population tend to occupy closely adjacent loci within the developing cortical layers (Misson et al. 1991a; Rakic 1988).

Two distinct proliferative populations contribute to neocortical cytogenesis (Boulder Committee 1970; Sauer 1935; Takahashi et al. 1993). A pseudostratified ventricular epithelium (PVE), the proliferative epithelium lying at the margin of the ventricular cavity and approximately coextensive with the architectonically defined ventricular zone (VZ; Boulder Committee 1970), is the source of most, if not all, neocortical neurons. Radial glial cells and, perhaps, other glial precursors also proliferate in this epithelium (Levitt et al. 1981; Misson et al. 1988). Cells of the PVE undergo mitosis at the ventricular margin (Sauer 1935; Sauer and Walker 1959; Sidman et al. 1959; Takahashi et al. 1992, 1993). The nucleus and soma are progressively displaced from the ventricular margin in the course of the G1 phase, although the cell maintains contact via a descending process. The S phase is initiated with the nucleus maximally displaced from the ventricular margin. The nucleus reapproximates the ventricle as it proceeds through the S phase and then the G2 phase.

A secondary proliferative population (SPP), which emerges in the course of neocortical cytogenesis, partially overlaps the PVE within the VZ (Takahashi et al. 1992, 1993). The greater number of its cells spread outward through the subventricular (SVZ) and intermediate (IZ) zones of the overlying cerebral wall (Boulder Committee 1970). Cells of the SPP, in contrast to those of the PVE, undergo no such interkinetic nuclear movement in the course of the cell cycle (Takahashi et al. 1993).

Retroviral transfection of the *Escherichia coli* galactosidase (Cepko 1988; Gray and Sanes 1991; Walsh and Cepko 1988) and other reporter genes (Halliday and Cepko 1992) into individual proliferating cells of these two populations has clarified lineage relationships among different neuronal and glial elements. Evidence accumulating from applications of this method indicates that neurons of different classes may arise across multiple cell cycles from the same lineage, whereas those arising at the same time are of the same class (Austin and Cepko 1990; Luskin et al. 1988; Mione et al. 1992; Parnavelas et al. 1992; Price and Thurlow 1988; Walsh and Cepko 1988; 1990). Quite startling complexities have been revealed where this method has been applied to the analysis of the relationship of clonal origin to intracortical patterns of neuron distribution following migration. Only a portion, perhaps 50% or less, of neurons of clonal origin initiate their migrations from the same "point" in the proliferative epithelium (Mione et al. 1992; Parnavelas et al. 1992; Tan and Breen 1993; Walsh and Cepko 1992a,b). It appears that the proliferative cells themselves have substantial mobility within the plane of the proliferative epithelium, with the result that the full complement of a clone may initiate their migrations at greatly separated points in the cerebral wall and come to rest after migration at greatly separated points within the neocortex (Fishell et al. 1993; Tan and Breen 1993; Walsh and Cepko 1993).

1.3.2 Quantitative Cytogenesis: Parameters and Experimental Approach

The cell division cycle may be viewed as beginning with the G1 phase, continuing with the S and the G2 phases, and ending with M phase or mitosis. Cells arise from the proliferative population by exiting the cell division cycle following mitosis. Over time, the rate of cell output from a proliferative population of defined size will be directly proportional to the growth fraction (GF = the fraction of the population that is cycling) and inversely proportional to the length of the cell division cycle (Tc). It will be directly proportional to the fraction Q which is that proportion of daughter cells which leaves the PVE with each cycle (Takahashi et al. 1991b). 1 – Q (= P fraction) corresponds to the fraction of daughter cells which reenters the S phase to remain in the cell cycle as a part of the proliferative population. The fractions P and Q, essentially ignored as variable parameters in previous analyses and theoretical formulations of cerebral histogenesis, are the most forceful determinants of proliferative behavior (Nowakowski 1991; Takahashi et al. 1991b). We will consider the theoretical reasons for the importance of these two parameters.

1.3.3 The Q and P Fractions

For the purposes of formulating or interpreting models of neocortical histogenesis, including those based upon observations with retroviral experiments, the stem cell model developed for steady state, lifelong populations such as the hematopoietic system and the gut epithelium (Hall and Watt

1989; Potten and Loeffler 1990) has been assumed, without experimental support, to operate in the developing vertebrate central nervous system. This model, as currently applied to studies of central nervous system histogenesis, specifies asymmetric cell divisions. What is meant by this is that the two daughter cells arising from each cell division will partition differently: one will exit the cycle and one will reenter the proliferative pool.

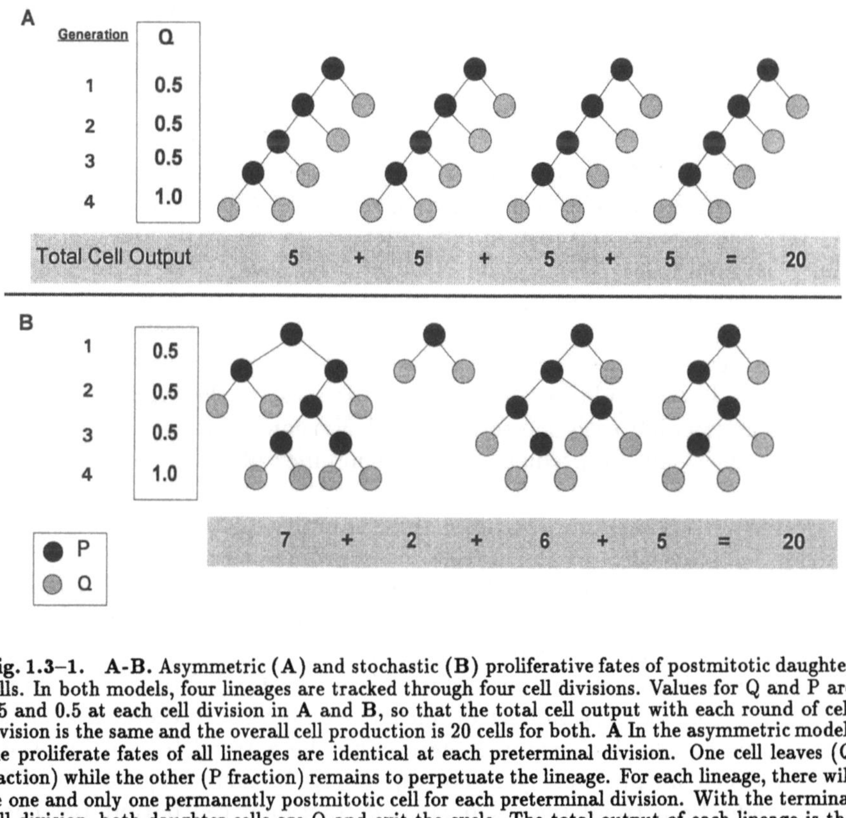

Fig. 1.3–1. A-B. Asymmetric (**A**) and stochastic (**B**) proliferative fates of postmitotic daughter cells. In both models, four lineages are tracked through four cell divisions. Values for Q and P are 0.5 and 0.5 at each cell division in **A** and **B**, so that the total cell output with each round of cell division is the same and the overall cell production is 20 cells for both. **A** In the asymmetric model, the proliferate fates of all lineages are identical at each preterminal division. One cell leaves (Q fraction) while the other (P fraction) remains to perpetuate the lineage. For each lineage, there will be one and only one permanently postmitotic cell for each preterminal division. With the terminal cell division, both daughter cells are Q and exit the cycle. The total output of each lineage is the total number of divisions +1. **B** In the stochastic model, a daughter cell may be either Q or P, allowing the possibility that a lineage may terminate after a single division if both are Q (*second lineage from left*). Both daughter cells may be P or the daughter cells may have Q/P fates

In theory, a daughter cell might instead have two possible "proliferative fates," either to exit (Q fraction) or to remain with (P fraction) the proliferative pool (Nowakowski 1991; Takahashi et al. 1991b). (In the cerebral PVE, we do not encounter and will ignore cell death as a proliferative fate.) Thus, Q and P fates might be assigned to the two daughter cells of a given mitotic division by two fundamentally different mechanisms (Fig. 1). There might be, as has been generally assumed, obligatory different fates (Fig. 1A), where one daughter cell is assigned a Q and the other a P fate. Such a partition, invariant over serial cell cycles, would, one assumes, be determined for each daughter cell by premitotic events and mechanisms. Alternatively, the assignment might be stochastic for the individual daughter cell (Fig. 1B). Here, the probability of Q or P assignment would be independent for each cell, in some way a consequence of interactions of daughter cells with components of the milieu. We suggest that daughter cells are competitive with respect to these interactions. The probability of Q or P assignment would vary over serial cell cycles with variations in the force of these determinative interactions.

Cell lineages governed by either the predetermined or the competitive-stochastic modes of fate determination could result in an identical accumulation of postmitotic cells over a series of cell cycles. However, the proliferative fate paths could be substantially different. In the predetermined mode, where asymmetric Q/P partition is obligatory, each cycle would produce one and only one postmitotic cell until the terminal division of the cycle, which would produce two postmitotic cells. The rate of cell output from each lineage would be identical and would not vary through the series of cell cycles. There would be no variation in the size of the proliferative pool until its exhaustion with the terminal cell division. Total cell output from each lineage would be predictable as the number of cell cycles +1 (Rakic 1988).

Proliferative lineages operating in the competitive-stochastic mode would provide substantially greater variation in proliferative fates among their progeny. The two daughter cells arising from a given mitosis might adopt the Q/P, P/P, or Q/Q fates. Individual lineages would achieve less than or greater than the aggregate average output, and some lineages would abruptly terminate, should daughters adopt the Q/Q fate. The average output per cycle for the full set of lineages could vary in the course of the cytogenetic epoch. For example there might be a trend toward increased fractional cell output per cycle corresponding to increasing values of Q; there might be a trend toward decreased fractional cell output per cycle corresponding to decreasing values of Q (Nowakowski 1991; Takahashi et al. 1991b). We suggest that these variations would reflect variations in the milieu with which the proliferative cells are interactive.

1.3.4 Experimental Designs

We have undertaken estimates of these parameters of cytogenesis through experiments of two separate designs. The first design provides measures of the cytokinetic parameters (lengths of the overall cell cycle, Tc, and the separate phases of the cycle, TG_1, Ts, and TG_{2+M}) and the growth fraction. The second design, dependent upon the first, provides measures of the Q and P fractions. We have undertaken the experiments in the mouse embryo from the 12th (E12) through the 17th (E17) days of gestation, corresponding to all but the earliest hours of the cytogenetic epoch in this species. The experiments are conducted in a coronal sector of the dorsomedial cerebral wall approximately corresponding to the future S1 representation. At the ventricular margin, this sector is 100 μm in its medial — lateral dimension and 4 μm in its rostral — caudal dimension (corresponding to section thickness). The radial dimension increases progressively over the E12–E17 interval of neocortical histogenesis.

1.3.5 Cytokinetic Parameters

Cells of the PVE in the S phase are labeled cumulatively by repeated exposures to bromodeoxyuridine (BUdR), a thymidine analog (Miller and Nowakowski 1988), beginning at 9:00 A.M. on each of the days E12–E17. The injections are given frequently enough so that all cells will be labeled upon entering the S phase. The exposure is continued long enough so that all proliferating cells will enter the S phase and become labeled (Nowakowski et al. 1989; Sidman 1970). The proportion of labeled cell nuclei with respect to all nuclei (LI = labeling index) is plotted for each age as a function of cumulative labeling time (Fig. 2). At each age the LI reaches 1.0, indicating that all cells in the population are proliferating. Because the population is proliferating asynchronously (Nowakowski et al. 1989; Takahashi et al. 1993)

$$\mathrm{LI}_{(at\,Y\,intercept)} = \mathrm{Ts/Tc} \text{ and}$$
$$\text{Time to } \mathrm{LI}_{max} = \mathrm{Tc{-}Ts}$$

The computed values for Tc increase from the value of approximately 10 h at E12 to 20 h in the course of E16, with this measurement completed only a matter of hours before the cytogenetic activity of the PVE is exhausted. The values for Ts are invariant at 4 h, while those for TG_{2+M} (the time required for 100% labeling of mitotic figures at the ventricular margin) is invariant at approximately 2.0 h throughout the E12–E16 interval. Thus, the increments in Tc in the interval E12–E16 are due entirely to increments in the developmentally regulated parameter, TG_1.

1.3.6 The Q and P Fractions

The cytokinetic parameters Ts and Tc are critical to the design of experiments which estimate the values of Q and P. The experiments are initiated

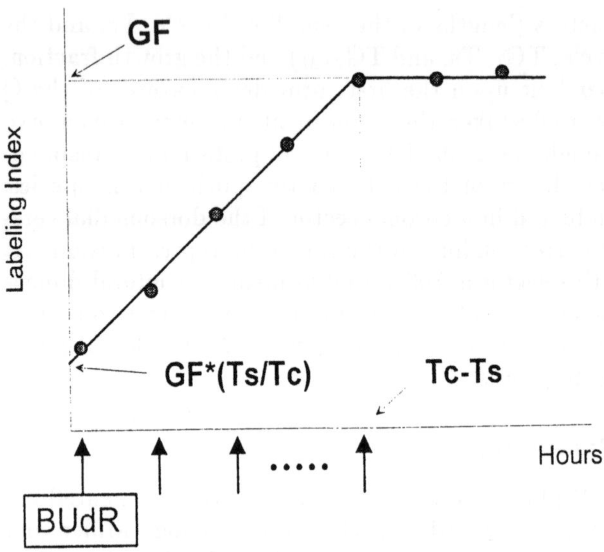

Fig. 1.3–2. Cumulative bromodeoxyuridine (*BUdR*) labeling of the pseudostratified ventricular epithelium (PVE). The labeling index (LI) gives the proportion of labeled nuclei. The maximum LI, corresponding to the growth fraction (*GF*, the proportion of cells which are mitotically cycling), is reached after a period of cumulative labeling which is the length of the overall cell cycle (*Tc*) minus the length of the S phase (*Ts*). The labeling index at the Y intercept corresponds to GF × Ts/Tc

on each of the embryonic days E12–E16, by a single exposure to tritiated thymidine (^{3}H–TdR). Exposure to ^{3}H–TdR is followed after 2 h by exposure to BUdR. This sequence has the effect of defining (cells labeled only with ^{3}H–TdR) a synchronous ("2 h") cohort of immediately post-S phase cells (Takahashi et al. 1991b). For one set of embryos, the initial exposure to BUdR is repeated through an interval greater than Tc–Ts. At that time, the cells of the Q and P fractions will have undertaken their separate fates. Those of the P fraction will have entered the S phase and will have become labeled with BUdR. The number of residual cells marked only with ^{3}H–TdR will correspond to the number of cells of the Q fraction (NQ) which arise in 2 h, on the respective day of the experiment. In a second paradigm, the initial BUdR exposure is not repeated. At Tc–Ts, there will be no BUdR available to label the cells of the P fraction as they enter the S phase. Here, the number of cells labeled with 3H-TdR will correspond to the number of cells of the Q+P fractions (NQ+P) which arise in 2 h, on the respective day of the experiment. From these numbers

$$Q = NQ/NQ{+}P \text{ and}$$
$$P = 1{-}Q$$

The value of Q, from these experiments, changes in a nonlinear fashion throughout the cytogenetic epoch. This finding stands in sharp contrast to the generally accepted view that daughter cell fates are "asymmetric" with the implication that Q and P are held constant at 0.5 throughout neocortical cytogenesis. Q ascends slowly from an initial value near 0.1 on E12 to a value which is still less than 0.4 at E14. After E14, its ascent accelerates, with the steady state value of 0.5 reached at some time between E14 and E15 and a value of 1.0 reached early on E17. The terminal value of 1.0 is assigned early on E17 because the PVE is replaced with nonproliferative ependymal cell layer by E18.

1.3.7 Cell Output from the Dorsomedial Cerebral PVE

Eight cycles will be executed between 9:00 A.M. on E12 and the end of the period of neuronogenesis on E17. Because the outset of production of postmitotic cells in the dorsomedial murine cerebral wall occurs between E10 and E11 (Caviness 1982) and because there are approximately two cell cycles in 24 h, there are approximately ten cell division cycles throughout the full cytogenetic period in this region of the cerebrum. The values of Tc for each of the eight cycles beginning on E12 were estimated by interpolation from a best fit linear regression based on the values of Tc determined experimentally on E12–E16. The cell output from the 400 μm^2 ventricular surface for each of the integral cycles may be estimated as NQ $\times$ Tc/2h. This value is relatively low at 10–25 cells/cycle over the initial four cycles, when the cell sparse infragranular layers are being formed. The value climbs to 100 cells/cycle for the cycles occurring around E15, when the cell rich supragranular layers are being formed (Caviness 1982). The total output for the sector through the complete series of nine cycles during which the neocortex is formed will be the sum of the output for each cycle (= 334 cells).

1.3.8 Values for Q and P and Clone Size

We see from these experiments that the values of Q and P change continuously with progression through the series of cell cycles. The values of Q and P determined here reflect the average proliferative fate assignments of the postmitotic cells of the dorsomedial PVE. We assume that these values reflect the probability of a Q or P fate for each daughter cell and that each daughter cell will adopt its fate independently of the fate of its sister cell.

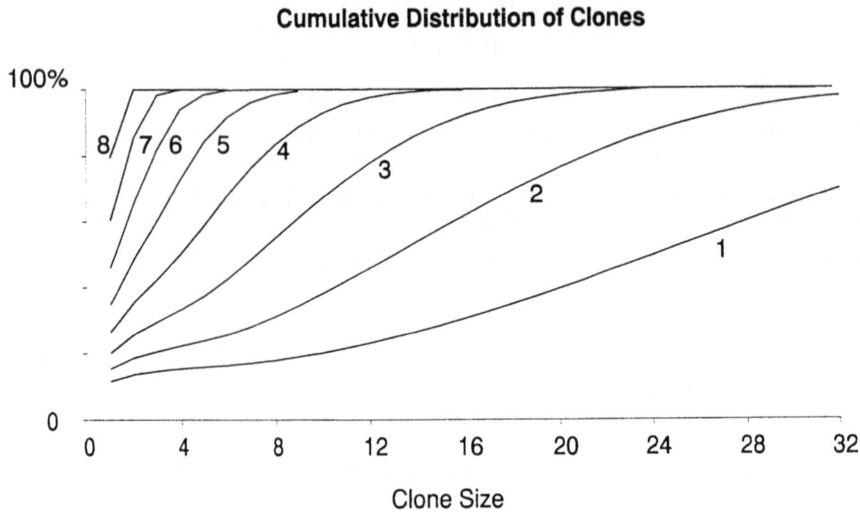

Fig. 1.3–3. Theoretical distribution (percentages) of clone sizes for lineages initiated with cell cycles 1 (E12) – 8 (E16–E17) computed from experimentally determined values of Q. For each lineage, the most probable clone size is one cell, the probability corresponding to the value of Q for the first mitotic division. For lineages initiated with early cycles where the initial values are less than 0.5, more than 50% of clones will be greater than ten cells. For lineages initiated with late cycles where the initial value of Q is near or greater than 0.5, maximum clone sizes are less than ten cells and the average clone size will be less than five cells

These experimentally determined values of Q and P allow estimates of the probability of encountering any given clonal size from a lineage initiated with a proliferative progenitor cell at any of the sequential integral cell cycles in the dorsomedial murine cerebral wall (Fig. 3). A clone of a specific size will arise through the same dichotomous branching patterns, independent of the cycle of initiation. The probability of each of the branching options (i.e., Q versus P) corresponds to the values of Q and P for the respective cycle. The area under each of the curves in Fig. 3 is 100%; that is, the plot for each cycle of lineage initiation depicts all possible clonal size outcomes and the probabilities for each.

For each of the lineages, the most probable clone size is one cell. This probability will correspond to the value of Q for the cell cycle initiating the lineage, as this is simply the probability that the single founder cell will leave the cycle without progeny. For any lineage, the probability of any clone size greater than one cell is very much smaller than that for clone sizes of one cell. For lineages initiated early in the sequence of cell cycles where values of Q are less than 0.5, no clone size has a probability greater than 10%. Extinction of the lineage initiated early is improbable, and it is to be expected that many clones will be constituted of many cells. For example, where a lineage is initiated at cycle 1, there is a 30% chance of clone sizes greater than 32 cells. By contrast, when the lineage is initiated late, for example with one of the cycles 6–8, where Q is greater than 0.5 and early extinction is certain, the probability of encountering clones of only a few cells is much larger, and no clones as large as ten cells are to be expected.

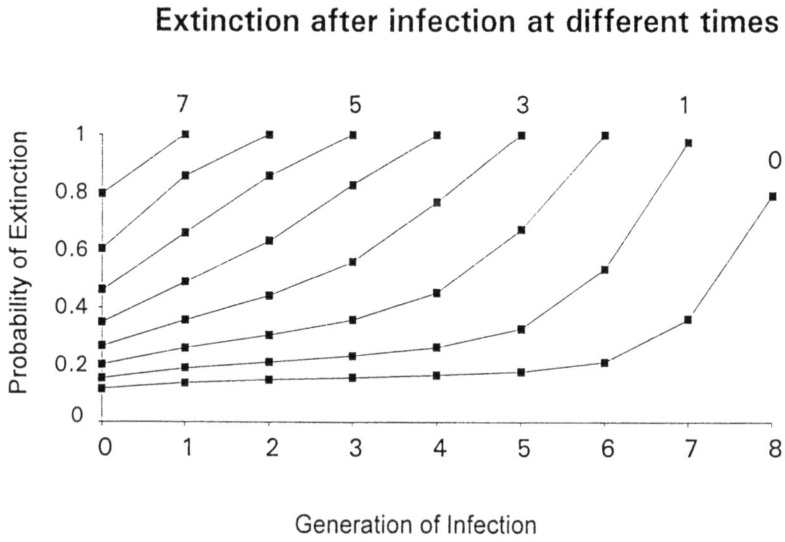

Fig. 1.3–4. Theoretical generational life expectancy for lineages initiated with cell cycles 1 (E12) – 8 (E16–E17) computed from experimentally determined values of Q. For lineages initiated with the final cycles 6–8, where the initial value of Q is near or greater than 0.5, lineage extinction is highly probable after only one to three generations, depending upon the initial value of Q. For lineages initiated with early cycles 1–4, where the initial values are less than 0.5, lineage extinction early in the cytogenetic period is improbable

In Fig. 4, the opposite perspective of the probability of extinction for lineages initiated through the series of cycles is plotted; the operation of the value for Q determined for each of the successive integral cycles of the dorsomedial PVE is reflected. This plot re-enforces the inferences from the plot in Fig. 3. Thus, it is highly probable that lineages initiated early will proliferate without extinction through most or all of the subsequent cycle sequence. Such lineages will give rise to many cells throughout the full cytogenetic epoch. Such lineages may be expected to give rise to progeny destined for virtually every neocortical layer with multiple progeny destined for the granular and supragranular layers. This is certainly the expectation for clones initiated for the first 4–5 integral cycles underway in the E12–E14 interval where values of Q are lower than the steady state probability of 0.5. For the last 3–4 cycles, by contrast, where values of Q mount quickly above the steady state value of 0.5, extinction is highly probable and the expected contribution of the lineage to more than one cortical layer is low.

1.3.9 Clone Size: Correlations with Retroviral Experiments

The most frequent clone size estimated from retroviral experiments is one cell; the average size of multicellular clones encountered with this method is less than five cells, and only infrequently are multicellular clones encountered which are greater than ten cells (Austin and Cepko 1990; Luskin et al. 1988; Price and Thurlow 1988; Walsh and Cepko 1988, 1990, 1992a,b). The average size of clones has been little different with viral injections early or relatively late in the cytogenetic period. For example, with experiments done in the mouse by Luskin et al. (1988), the average clones resulting from injections at E12 was 4.5 cells, and for E14, 3.4 cells. Infrequently encountered clones larger than ten cells have been observed with injections at E12 in mice (Austin and Cepko 1990).

How do these estimates based upon retroviral tracking of neuronal lineages compare with the theoretical distribution of clone sizes, computed here from values of Q? They are in reasonably good accord with the expected distribution of single cell clones. Where viral insertion is achieved during the middle period of the cytogenetic epoch, that is, around E14 in mice and near E16 in rats (Bayer and Altman 1991), the frequency of one-cell clones has tended to be at least 40%–60% of total clones recovered. This is to be expected, because this is precisely the time when Q approaches and then exceeds 0.5.

The similarity of clone sizes for early and late injections does not, however, accord with the expected theoretical distributions of lineages initiated with widely separated cycles. An observed clone size of 2–5 cells is exactly what would be expected with viral injections late in the period of cytogenesis when Q is approaching or has exceeded 0.5. Thus, for a lineage initiated between E14 and E15 with cycle 5, 98% of clones would be eight cells or fewer and 50% would be no larger than 2–3 cells (Fig. 3). However, for lineages initiated as early as E12, corresponding to cycle 1, 80% of clones would be expected to be 10–12 cells and 50% at least 20 cells in size and not 4–5 cells as observed in the retroviral experiments of Luskin et al. (1988). Presumptive clones of this much larger magnitude are evident in the preparations of Tan and Breen (1993), where the reporter gene is present in founder cells of lineages initiated at the outset of neocortical cytogenesis.

This disparity between the observed and predicted clone sizes for lineages marked by early viral injections is not unexpected, because common clonal origin of retrovirally marked cells is inferred in these analyses only from proximity of postmigratory cells. In the later experiments of Walsh and Cepko (1992b), where clonal relationships are defined instead by the specific restriction fragment length polymorphism (RFLP) fingerprints of early inserted retroviral vectors, only approximately 50% of cells of a clone were clustered after migration into the cortex. Proliferating cells have been directly demonstrated to move tangentially in the PVE (Fishell et al. 1993). Substantial dispersion of proliferating cells and their progeny might, therefore, occur if viral insertion is achieved in the early cycles of the cytogenetic period (Fishell et al. 1993; Walsh and Cepko 1993). Because the disparities between the observed and predicted distributions hold only for the retrovirally defined lineages of early initiation, they need not be ascribed to low sensitivity of the labeling method itself (Halliday and Cepko 1992) or to nonadherence of postmitotic neurons to the radial glial fiber system in the course of migratory ascent (O'Rourke et al. 1992)

1.3.10 Clone Size and Histogenesis

We assume that the progressive and systematic changes in the probabilities of the Q and P fates reflect progressive and systematic changes in Q/P determinants that impinge upon the cell as the cytogenetic epoch advances. Further these systematic and progressive changes in the determinants with their effect upon the Q/P partition among daughter cells are consonant with histogenetic needs, reflecting the regulation of fundamental events of histogenesis at the level of regulation of proliferate fate determination. The most salient of the consequences of the Q fraction crescendo is the late crescendo of cell output, indispensable to assembly of the amplified cell populations of granular and supragranular neuronal populations relative to the low-density neuronal populations of the infragranular cortex. The obvious consequence of the high values for P early in the cycle is the augmentation of the proliferative pool, matched by expansion of the PVE, in preparation for the later crescendo of cell output.

1.3.11 Mechanisms of Cell Cycle Regulation

Large-scale variations in the rate of neuron output from the PVE reflect the combined effects of three parameters: TC, GF, and Q. Only two of these, TC and Q, are variable throughout the cytogenetic epoch. Variations in TC are due entirely to variations in TG_1. In other words, TG_1 and Q are the regulated parameters, while the lengths of the other cell cycle phases and GF are unregulated parameters of the cytogenetic behavior of the PVE.

Progression of the cell cycle of eukaryotic cells is regulated by cyclic synthesis and degradation of a class of proteins referred to as cyclins. The principal transitions of the cycle, G_2 to M and G_1 to S, are triggered by kinases which are activated by binding to the cyclins (Draetta 1990; Murray and Kirschner 1989; Nurse 1990; Pines and Hunter 1990). In mammalian cells, there are multiple cyclins and kinases, closely identical in their structure but probably having critically complementary roles, which act to implement the cell cycle progressions. As a general phenomenon, the synthesis of cyclins essential to the G_1 to S transition is substantially modulated by environmental substances, in particular by substances recognized as mitogens (Dou et al. 1993; Matsushime et al. 1991). Variation in cyclin synthesis in G_1, modulated by the effect of factors in the environment of the proliferating cell, may correlate with and possibly be a determinant of the variations in TG_1 and Q characteristic of cell cycle progression in the course of neocortical cytogenesis.

We introduce here the first investigations into the patterns of activity of these molecular operators in a proliferating cell population of a vertebrate central nervous system. We have considered specifically the patterns of protein expression as well as kinase activity of two cyclin-dependent kinases, cdk2 and cdk5. The former is known from in vitro experiments in non-neural transformed and immortalized cell lines to be indispensable to the G_1/S transition (Pagano et al. 1993; Tsai et al. 1993). This kinase is activated by binding to cyclins E and A (Dulic et al. 1992; Koff et al. 1992; Rosenblatt et al. 1992; Tsai et al. 1991). The expression and activity of this kinase in the forebrain of the embryonic mouse are observed to be maximum at the outset of cytogenesis and decline progressively to essentially nondetectable levels by the end of the cytogenetic epoch on E16. Cdk5, by contrast, has been found to be expressed only in testes and brain and to have kinase activity only in brain. The cyclin with which it is associated is not known. The complexity of the evolutionary history of the family of cyclin dependent kinases is underscored by the observation that expression and activity of cdk5, in contrast to cdk2, are essentially nondetectable before E11 but are increased sharply thereafter as the output of postmitotic neurons proceeds. Expression of cdk5 is virtually limited to axons of postmitotic, differentiating neurons. Paradoxically, it is not expressed in proliferating cells. Future experiments will be concerned with the role of these and related elements in the regulation of neocortical neuronal cytogenesis.

References

Angevine JB, Sidman RL (1961) Autoradiographic study of cell migration during histogenesis of the cerebral cortex in the mouse. Nature 192: 766–768

Austin CP, Cepko CL (1990) Cellular migration patterns in the developing mouse cerebral cortex. Development 110: 713–732

Bayer SA, Altman J (1991) Neocortical development. Raven, New York

Boulder Committee (1970) Embryonic vertebrate nervous system: revised terminology. Anat Rec 166: 257–262

Caviness VS Jr (1982) Neocortical histogenesis in normal and reeler mice: a developmental study based upon [3H]thymidine autoradiography. Dev Brain Res 4: 293–302

Cepko CL (1988) Retrovirus vectors and their applications in neurobiology. Neuron 1: 345–353

Dou O-P, Levin AH, Zhao S, Pardee AB (1993) Cyclin E and cyclin A as candidates for the restriction point protein. Cancer Res 53: 1493–1497

Draetta G (1990) Cell cycle control in eukaryotes: molecular mechanisms of cdc2 activation. Trends Biochem Sci 15: 378–383

Dulic V, Lees E, Reed SI (1992) Association of human cyclin E with a periodic G_1-S phase proteinkinase. Science 257: 553–557

Fishell G, Mason CA, Hatten ME (1993) Dispersion of neural progenitors within the germinal zones of the forebrain. Nature 362: 636–638

Gray GE, Sanes JR (1991) Migratory paths and phenotypic choices of clonally related cells in the avian optic tectum. Neuron 6: 211–225

Hall PA, Watt FM (1989) Stem cells: the generation and maintenance of cellular diversity. Development 106: 619–633

Halliday AL, Cepko CL (1992) Generation anad migration of cells in the developing striatum. Neuron 9: 15–26

His W (1904) Die Entwicklung des menschlichen Gehirns während der ersten Monate. Hirzel, Leipzig

Imamoto K, Peterson JA, Leblond CP (1978) Radioautographic investigation of gliogenesis in the corpus callosum of young rats. I. Sequential changes in oligodendrocytes. J Comp Neurol 180: 115–138

Koff A, Giordano A, Desai D, Yamashita K, Harper W, Zlledge S, Nishimoto T, Morgan D, Franza R, Roberts J (1992) Formation and activation of a cyclin E–cdk2 complex during the G_1 phase of the human cell cycle. Science 257: 1689–1693

Levinson SW, Goldman JE (1993) Both oligodendrocytes and astrocytes develop from progenitors in the subventricular zone of the postnatal rat forebrain. Neuron 10: 201–212

Levitt P, Cooper ML, Rakic P (1981) Coexistence of neuronal and glial precursor cells in the cerebral ventricular zone of the fetal monkey: an ultrastructural immunoperoxidase analysis. J. Neurosci. 1: 27–39

Luskin MB, Pearlman AL, Sanes JR (1988) Cell lineage in the cerebral cortex of the mouse studied in vivo and in vitro with a recombinant retrovirus. Neuron 1: 635–647

Matsushime H, Roussel MF, Ashmun RA, Sherr CJ (1991) Colony-stimulating factor 1 regulates novel cyclins during the G_1 phase of the cell cycle. Cell 65: 701–713

Miller MW, Nowakowski RS (1988) Use of bromo-deoxyuridine-immunohistochemistry to examine the proliferation, migration and time of origin of cells in the central nervous system. Brain Res 457: 44–52

Mione MC, Danevic C, Cavanagh MA, Boardman P, Paranavelas JG (1992) Neurotransmitter expression in clonally related neurons in the rat cerebral cortex. Soc Neurosci Abst 18: 1111

Misson J-P, Edwards MA, Yamamoto M, Caviness jr VC (1988) Mitotic cycling of radial glial cells of the fetal murine cerebral wall: A combined autographic and immunohistochemical study. Dev Brain Res 38: 183–190

Misson J-P, Austin CP, Takahashi T, Cepko CL, Caviness VS Jr (1991a) The alignment of migrating neural cells in relation to the murine neopallial radial glial fiber system. Cerebral Cortex 1: 221–229

Misson J-P, Takahashi T, Caviness VS Jr (1991b) Ontogeny of radial and other astroglial cells in murine cerebral cortex. Glia 4: 138–148

Murray AW, Kirschner MW (1989) Dominoes and clocks: the union of two views of the cell cycle. Science 246: 614–621

Nowakowski RS (1991) The contribution of competition during G_1 and reentry into the proliferative population to variation in clone size. Soc Neurosci Abst 17: 30

Nowakowski RS, Lewin SB, Miller MW (1989) Bromodeoxyuridine immunohistochemical determination of the lengths of the cell cycle and the DNA-synthetic phase for an anatomically defined population. J Neurocytol 18: 311–318

Nurse P (1990) Universal control mechanism regulating onset of M phase. Nature 344: 503–508

O'Rourke NA, Dailey ME, Smith SJ, McConnell KS (1992) Diverse migratory pathways in the developing cerebral cortex. Science 258: 299–302

Pagano M, Pepperkok R, Lukas J, Baldin V, Ansorge W, Bartek J, Draetta G (1993) Regulation of the cell cycle by the cdk2 protein kinase in cultured human fibroblasts. J Cell Biol 121: 101–111

Parnavelas JG, Danevic C, Boardman P, Mione MC, Cavanagh ME (1992) Separate progenitor cells in the ventricular zone give rise to cortical pyramidal and nonpyramidal neurons throughout the period of neurogenesis. Soc Neurosci Abst 18: 1111

Pines J, Hunter T (1990) p34cdc2: the S and M kinase? New Biologist 2: 389–401

Potten SC, Loeffler ML (1990) Stem cells: attributes, cycles, spirals, pitfalls and uncertainties. Lessons for and from the crypt. Development 110: 1001–1020

Price J, Thurlow L (1988) Cell lineage in the rat cerebral cortex: a study using retrovirus-mediated gene transfer. Development 104: 473–482

Rakic P (1971) Neuron-glia relationship during granule cell migration in developing cerebellar cortex: a Golgi and electron microscopic study in Macaca rhesus. J Comp Neurol 141: 283–312

Rakic P (1974) Neurons in rhesus monkey visual cortex: systematic relation between time of origin and eventual disposition. Science 183: 425–427

Rakic P (1988) Specification of cerebral cortical areas. Science 241: 170–176

Rakic P, Stensaas LJ, Sayer EP, Sidman RL (1974) Computer aided three-dimensional reconstruction and quantitative analysis of cells from serial electron microscopic montage of foetal monkey brain. Nature 250: 31–34

Rosenblatt J, Gu Y, Morgan DO (1992) Human cyclin dependent kinase 2 (cdk2) is active during the S and G_2 phases of the cell cycle and associates with cyclin A. Proc Natl Acad Sci USA 89: 2824–2828

Sauer FC (1935) Mitosis in the neural tube. J Comp Neurol 62: 377–405

Sauer ME, Walker BE (1959) Radioautographic study of interkinetic nuclear migration in the neural tube. Proc Soc Ex Biol 101: 557–560

Schaper A (1897) Die frühesten Differenzierungsvorgänge im Zentralnervensystem. Arch Entwicklungsmech Organ 5: 81–132

Sidman RL (1970) Autoradiographic methods and principles for study of the nervous system with thymidine-H3. In: Nauta WJH, Ebbesson SOE (eds) Contemporary research methods in neuroanatomy. Springer, Berlin Heidelberg New York, pp 252–274

Sidman RL, Rakic P (1982) Development of the human central nervous system. In: Haymaker W, Adams RD (eds) Histology and histopathology of the nervous system. Thomas CC, Springfield, pp 3–145

Sidman RL, Miale IL, Feder N (1959) Cell proliferation and migration in the primitive ependymal zone: an autoradiographic study of histogenesis in the nervous system. Exp Neurol 1: 322–333

Takahashi T, Misson JP, Caviness VS Jr (1991a) Glial process elongation and branching in the developing murine neocortex: a qualitative and quantitative immunohistochemical analysis. J Comp Neurol 302: 15–28

Takahashi T, Nowakowski RS, Caviness VS Jr (1991b) Cell output of the ventricular zone of the E14 mouse neocortex. Soc Neurosci Abst 17: 29

Takahashi T, Nowakowski RS, Caviness VS Jr (1992) Cytogenesis in the secondary proliferative population of the murine cerebral wall. Soc Neurosci Abst 18: 30

Takahashi T, Nowakowski RS, Caviness VS Jr (1993) Cell cycle parameters and patterns of nuclear movement in the neocortical proliferative zone of the fetal mouse. J Neuroscience 13: 820–833

Tan S-S, Breen S (1993) Radial mosaicism and tangential cell dispersion both contribute to mouse neocortical development. Nature 362: 638–640

Tsai L-H, Harlow E, Meyerson M (1991) Isolation of the human cdk2 gene that encodes the cyclin A- and adenovirus E1A-associated p33 kinase. Nature 353: 174–177

Tsai L-H, Lees E, Faha B, Harlow E, Riabowol K (1993) The cdk2 kinase is required for the G1-to-S transition in mammalian cells. Oncogene 8: 1593–1602

Walsh C, Cepko CL (1988) Clonally related cortical cells show several migration patterns. Science 242: 1342–1345

Walsh C, Cepko CL (1990) Cell lineage and cell migration in the developing cerebral cortex. Experientia 46: 940–947

Walsh C, Cepko C (1992a) Generation of widespread cerebral cortical clones. Soc Neurosci Abst 18: 925

Walsh C, Cepko CL (1992b) Widespread dispersion of neuronal clones across functional regions of the cerebral cortex. Science 255: 434–440

Walsh C, Cepko CL (1993) Clonal dispersion in proliferative layers of developing cerebral cortex. Nature 362: 632–635

1.4 Cell Fate Determination and Cell Migration in the Developing Forebrain

M.B. LUSKIN
Department of Anatomy and Cell Biology, Emory University School of Medicine, Atlanta, Georgia

The development of the neocortex and olfactory bulb, two highly organized and laminated structures of the forebrain, have been studied to great advantage in order to gain an understanding about the rules governing the formation of the mammalian central nervous system. Just as in the mature neocortex and olfactory bulb, where individual layers can be characterized by their connectivity patterns and constellation of cell types, the immature neocortex and olfactory bulb can also be subdivided into distinct zones. These zones contain predominantly dividing cells, migrating cells, or post-migratory cells (Fig. 1). The dividing or progenitor cells which give rise to the assortment of cells destined for the cerebral cortex and olfactory bulb compose a layer of cells known as the ventricular zone or ependymal zone surrounding the lateral ventricles of the forebrain (Boulder Committee 1970). Encircling and immediately superficial to the ventricular zone is another layer of progenitor cells, the subventricular zone or subependymal layer (Fig. 1). It also contributes cells to the cerebral cortex and olfactory bulb (reviewed by Farbman 1991). Hence, during the formation of the neocortex and olfactory bulb, cells must migrate from their site of origin to their final laminar position in a highly coordinated manner and differentiate into distinct cell types. The main types to be accounted for are the glia (astrocytes and oligodendrocytes) and neurons (projection neurons and interneurons). In my laboratory we have been investigating how cell proliferation, cell migration, and cell identity are governed. Discussed in this article are our findings and hypotheses concerning the composition of cells in the ventricular and subventricular zones, how the progeny of progenitor cells are distributed, and how cells acquire their phenotypes. Using retroviral lineage tracers to introduce a heritable marker into progenitor cells of the prenatal ventricular zone and postnatal subventricular zone, we have analyzed the lineage relationships of the different cell types within the neocortex and the course of migration and destination of neurons derived from a discrete part of the subventricular zone.

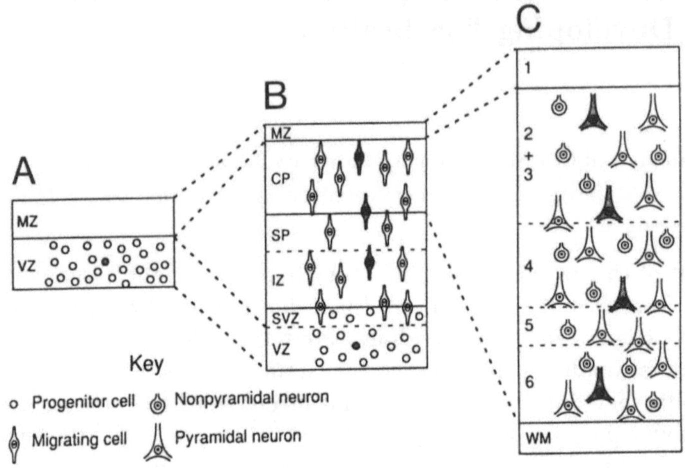

Fig. 1.4–1. Sequential stages in the development of the laminae of the cerebral cortex and hypothetical formation of a neocortical clone. **A** Schematic representation of the developing telencephalon before the generation of postmitotic neurons. The ventricular zone consists of a cell-dense layer of proliferating cells surrounding the lateral ventricles of each cerebral hemisphere. The marginal zone, predominantly acellular, is mainly composed of processes of the ventricular zone cells extending to the pial surface. **B** Schematic representation of the cerebral wall during neurogenesis. The ventricular zone and overlying subventricular zone, another layer of dividing cells, give rise to the neurons and glia comprising the cerebral cortex. Most newly generated postmitotic neurons (*elongated profiles*) arrive at the cortical plate, where they differentiate, after migrating radially through the intermediate zone and subplate, a transient layer of early-born neurons. **C** Schematic representation of the mature cerebral cortex illustrating the final position of postmigratory cortical neurons occupying layers II–VI. The *shaded* cells in each panel represent the hypothetical arrangement of lineally related progenitor, migrating and postmigratory cells during the formation of a clone of neurons. *CP*, cortical plate; *IZ*, intermediate zone; MZ, marginal zone; *SP*, subplate; *SVZ*, subventricular zone; *VZ*, ventricular zone; *WM*, white matter. (Modified, with permission, from Luskin 1993a)

1.4.1 Organization of the Ventricular Zone and Subventricular Zone of the Developing Cerebral Cortex

Early in the embryonic development of the telencephalon, the ventricular zone constitutes the sole progenitor cell population. Shortly after the onset of cortical neurogenesis (the production of postmitotic neurons), the subventricular zone emerges overlying the ventricular zone. Next, coinciding with the cessation of cortical neurogenesis, the ventricular zone is exhausted and only the subventricular zone remains. Subsequently, there is a postnatal diminution of the subventricular zone. Thus, the relative thicknesses of the ventricular zone and subventricular zone vary at different times.

Besides the difference in the position of ventricular and subventricular zone cells relative to the ventricular surface, different modes of cell division are characteristic of each zone. For the ventricular zone, cytokinesis occurs along the ventricular surface and DNA synthesis takes place farthest from the ventricular surface. By a process referred to as interkinetic nuclear migration, the nuclei of ventricular zone cells undergo a movement towards and away from the ventricular surface (Sauer 1935). This cyclical behavior of the nuclei of dividing cells is not executed by subventricular zone cells. Rather, there is no spatial segregation between the cells synthesizing DNA and those undergoing cytokinesis (Boulder Committee 1970). Thus, the differential distribution and kinetic behavior of progenitor cells in the ventricular and subventricular zones has made it possible, with the use of retroviral lineage tracers, to track and identify the progeny of progenitor cells at any stage of development and to define their respective cellular contributions to the cerebral cortex (Austin and Cepko 1990; Luskin et al. 1988, 1993a; Parnavelas et al. 1991; Price and Thurlow 1988; Walsh and Cepko 1988, 1992) and olfactory bulb (Luskin 1993a,b; Luskin et al. 1993b).

Another important difference between the ventricular and subventricular zone is the contribution each makes to the neuronal and glial populations of the forebrain. Tritiated thymidine birthdating studies have demonstrated that the ventricular zone is the main source of neurons for the mammalian cerebral cortex and that these cells arise prenatally during a particular period of time for each species (e.g., primate: Rakic 1974; cat: Luskin and Shatz 1985; rat: Raedler and Raedler 1978). The cortical plate — the forerunner of the future cortical layers — is elaborated by the sequential addition and differentiation of postmitotic neurons that have left the ventricular zone and have basically undergone a radialward migration through the intermediate zone (Rakic 1972; Tan and Breen 1993). After the ventricular zone is depleted, the subventricular zone continues to generate glial cells postnatally (Patterson et al. 1973; Privat 1975; Skoff 1980). Furthermore, while neurons are postmitotic after leaving the ventricular zone, glial cells (astrocytes and oligodendrocytes) or glial progenitor cells are thought to retain the capacity to divide after leaving the subventricular zone (Ling et al. 1973; Mares and Bruckner 1978).

The seminal question that needed to be addressed in order to understand how the variety of neurons (pyramidal cells and nonpyramidal cells) and glia (astrocytes and oligodendrocytes) of the cerebral cortex arise was this: are the progenitor cells multipotential or do individual progenitor cells generate cells of one type only? This question presupposes that early in development all dividing cells of the ventricular zone are multipotential, but that over time a series of decisions may be made to restrict the range of possibilities of a progenitor cell or its offspring. We tested whether there were separate progenitor cells for the four major cell types by the onset of cortical neurogenesis in the developing rodent cerebral cortex. The existence of restricted progenitor cells would mean that certain aspects of a cell's identity or phenotype would be determined by its lineage and not exclusively by environmental factors.

1.4.2 Types of Progenitor Cells in the Prenatal Ventricular Zone of the Developing Cerebral Cortex

To determine if and when there are separate progenitor cells in the ventricular zone for the major cell types constituting the cerebral cortex, we and others have used the approach of retrovirus-mediated gene transfer (for review see Sanes 1989). The replication-defective recombinant retroviruses adapted for use as lineage tracers have had the reporter gene *Escherichia coli* β-galactosidase added to their genome and the viral genes *gag, pol*, and *env* removed. In essence, this renders the recombinant retrovirus suitable as a lineage tracer, because it is competent to infect a host cell, integrate into its DNA and be passed on to all its progeny, but it is unable to replicate on its own and infect unrelated cells. (The introduced reporter gene *E. coli* β-galactosidase is referred to as the *lacZ* gene, and its gene product lacZ.) The expression of lacZ can be readily detected histochemically or immunohistochemically in the progeny of an infected cell.

Using the method of retrovirus-mediated gene transfer to infect dividing cells of the ventricular zone, the first question addressed was whether there are separate progenitor cells for the major cortical cell types by the time postmitotic cortical neurons begin to be generated. For the rat, the animal we used in these studies, cortical neurogenesis starts on embryonic day 15 (E15) or E16. At various times after delivering retrovirus to the prenatal ventricular zone, the phenotypes of lacZ-positive cells forming discrete clusters of labeled cells, considered to be clonally related, were compared. We found that virtually all clones, so defined, were comprised of a homogeneous population of cells (e.g., Figs. 2,3). They were composed wholly of astrocytes, oligodendrocytes, pyramidal neurons or nonpyramidal neurons, (Parnavelas et al. 1991; Luskin et al. 1993a). In many important ways these results extend the findings of our earlier study (Luskin et al. 1988). Furthermore, they are compelling because the phenotype of lacZ-positive cells was assessed at the ultrastructural level using well-established criteria (Jones and Powell 1970; Parnavelas et al. 1977; Colonnier 1981; Peters 1985), rather than simply at the light microscopic level. The problem with relying on the histochemical staining achieved at the light microscopic level is that it is not adequate to definitively distinguish the phenotype of all lacZ-positive cells (see Figs. 2A,3B). At the ultrastructural level, the histochemical reaction product not only stains predominantly the nuclear membrane

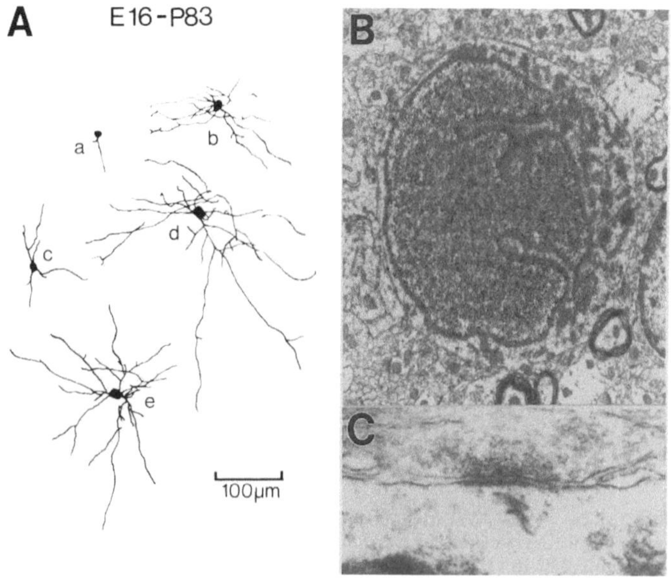

Fig. 1.4–2. Representative example of a homogeneous clone of nonpyramidal neurons in the mature visual cortex from a rat brain injected with retrovirus on embryonic day 16 (E16) and perfused at postnatal day 83 (*P83*). **A** Camera lucida drawings illustrating the similarities at the light microscopic level of the five clonally related cells. Notice that except for *cell a*, the least well stained member of the clone, they all exhibit characteristics of multipolar nonpyramidal neurons. (**B**) However, at the ultrastructural level, *cell a* also displayed the features of a nonpyramidal neuron, including an invaginated nucleus of irregular shape and darkly stained nucleoplasm and nuclear membrane. In addition, *cell a*, as well as all cells classified as nonpyramidal neurons, had symmetrical and asymmetrical synapses on its cell soma, a feature specific to nonpyramidal neurons. **C** An example of an asymmetrical synapse on a lacZ-positive nonpyramidal cell soma. **B** ×6370; **C** ×11 000. (Modified, with permission, from Luskin et al. 1993a)

and endoplasmic reticulum, but it is cell-type specific, which facilitates phenotype classification. In addition, the staining does not obscure the ultrastructure of the synaptic contacts associated with the lacZ-positive cells, which are indicative of neuronal cell phenotype (Jones and Powell 1970; Collonier 1981; Luskin 1992). Pyramidal cells receive exclusively symmetrical synapses on their cell soma, and nonpyramidal cells receive symmetrical and asymmetrical synapses. Furthermore, clonally related cells were shown to have similar patterns and intensity of histochemical staining. We previously suggested that the level of β-galactosidase expression is a function of the integration site of the viral DNA (Luskin et al. 1993a), which ought to be the same in all lineally related cells.

Some of the remaining questions that require future attention before we can understand how cell identity is established are: (1) How early do the lineage restrictions revealed in our study arise? Compatible with our re-

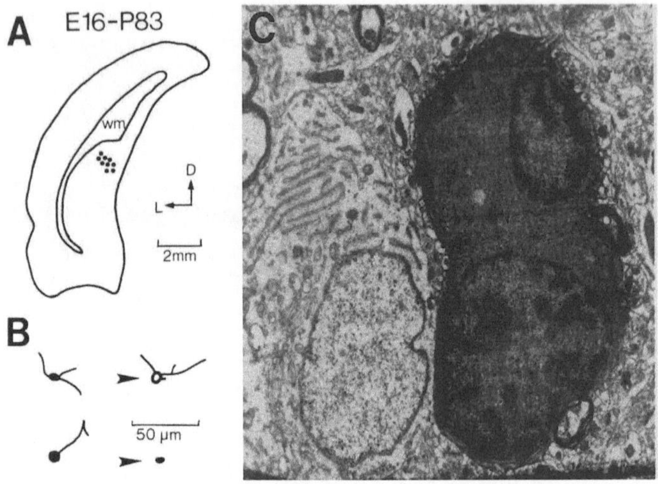

Fig. 1.4–3. Representative example of a clone in the mature neocortex containing exclusively oligodendrocytes from a rat brain which received an intraventricular injection of retrovirus on embryonic day 16 (E16) and perfused at postnatal day 83 (P83). **A** A reconstruction of the distribution of the nine clonally related oligodendrocytes (*black dots*) obtained by superimposing two adjacent 100 μm sections. **B** Camera lucida drawings of four of the nine lacZ-positive oligodendrocytes. The identity of many of the lacZ-positive cells cannot be determined at the light microscopic level because of their meager process staining. However, at the ultrastructural level, lacZ-positive oligodendrocytes can be recognized by their cell-type specific features, such as clumped heterochromatin, and by their cell-type specific staining pattern; lacZ-positive oligodendrocytes exhibit heavy histochemical staining throughout their nucleus and cytoplasm. The *arrowheads* in **B** point to the two lacZ-positive oligodendrocytes shown in the electron micrograph in **C**. **C** As shown, clonally related lacZ-positive oligodendrocytes are usually tightly clustered and commonly juxtaposed, which is another indication of their common origin. The cell on the *left* is an unlabeled neuron. **C** ×2635 *wm*, white matter; *D*, dorsal; *L*, lateral. (Modified, with permission, from Luskin et al. 1993a)

sults, Bernard et al. (1992) demonstrated, on the basis of the differential expression of the different forms of the *myc* oncogene, that the lineages for neurons and glia diverge at the onset of cortical neurogenesis. Although a molecular correlate or explanation has not been revealed, our studies indicate that lineages for the two main types of neurons and glia have also diverged. (2) Is there an even greater range of neuronal progenitor cell types than demonstrated by our studies? For example, do bipolar nonpyramidal cells originate from different progenitor cells than multipolar nonpyramidal cells? This is suggested by the light and electron microscopic appearance of the clones we have studied. (3) How much can the "preordained" phenotype of a cell be altered by its environment at its site of genesis, destination, or somewhere in transit? The laminar fate of a cell can be regulated by the environment in which it becomes postmitotic (McConnell and Kaznowski 1991). Are other features also malleable? (4) What are the properties of the small proportion of widespread clones described by Walsh and Cepko (1992)? Are they fundamentally different from the discrete clusters of clonally related cells we examined with respect to the phenotype of their component cells?

1.4.3 Types of Progenitor Cells in the Postnatal Subventricular Zone of the Developing Cerebral Cortex

Although there is increasing evidence that the astrocytes and oligodendrocytes of the telencephalon come from separate progenitor cells (LeVine and Goldman 1988; Luskin et al. 1988, 1993a; Grove et al. 1993), we investigated this more directly by infecting the subventricular zone postnatally, by which time all the cortical neurons had been generated. When, 2 weeks later, the clones of cells in the cerebral cortex were examined ultrastructurally, a result consistent with the prenatal series (see above) was obtained. The clones were composed of a homogeneous population of cells. They contained either all subependymal cells situated in the subventricular zone, all astrocytes, or all oligodendrocytes (McDermott et al. 1991). Some of the glial clones were located in the gray matter and some in the white matter, and some spanned the two compartments. This suggests that there is one type of astrocyte progenitor cell and one type of oligodendrocyte progenitor cell and that the progeny of each may reside in the gray matter or white matter and undergo further differentiation appropriate for their specific location.

Most glial clones contained two to nine cells, although larger clones arise when the time between injection and fixation is increased (M.B. Luskin, unpublished results). Furthermore, individual clones could contain immature and mature cells types, indicating that the constituent cells were generated at different times. Accordingly, as expected, if there is ongoing gliogenesis, cells with the characteristics of astroblasts and oligodendroblasts (Peters et al. 1991; Parnavelas et al. 1983) were often present in clones analyzed at 2 weeks postinjection. In addition, the cells within glial clones derived from both neonatal subventricular zone injections as well as prenatal intraventricular injections were usually closely spaced and even frequently juxtaposed. A likely possibility based on the evidence for persistent gliogenesis within the first month or two after birth and the close proximity of labeled cells is that following the infection of a subventricular zone cell, its progeny migrate away from the subventricular zone and continue to divide locally, where they produce a cluster of clonally related cells of the same phenotype. Even owing for the possibility, for which presently there is little evidence, that the originally infected cell is bipotential and its progeny disperse, all subsequent divisions would produce cells of an identical phenotype, arguing for lineage restrictions.

1.4.4 Course of Migration and Destination of Olfactory Bulb Neurons Derived from Progenitor Cells in the Anterior Part of the Postnatal Subventricular Zone

Recent reports from our laboratory (Luskin 1993a,b) and others (Reynolds and Weiss 1992) require a revision of the premise that the telencephalic postnatal subventricular zone produces only glial cells. Reynolds and Weiss cultured dissociated cells from the adult striatum and showed that in the presence of epidermal growth factor, a fraction of the cells under favorable conditions would divide and differentiate into neurons. The exact source of the dividing cells is not known, though conceivably they were derived from cells lining the lateral ventricle that retain the capacity for mitosis (Morshead and van der Kooy 1992). We recently demonstrated that progenitor cells situated in a discrete part of the neonatal subventricular zone, distinct from the region isolated by Reynolds and Weiss, generate an immense number of neurons destined for the olfactory bulb (see Fig. 4). The particular part of the subventricular zone containing neuronal progenitor cells was deduced by the outcome of injections of retrovirus into different locations along the anterior-posterior axis of the subventricular zone surrounding the lateral ventricle. After various postinjection survival times, the distribution and phenotype of the labeled lacZ-positive cells was analyzed. While most injections produced small numbers of glial clones, the more anterior placed injections virtually always yielded vast number of cells en route to or residing in the olfactory bulb (Luskin 1993b).

An analysis of the laminar distribution and phenotype of the labeled cells in the mature olfactory bulb indicates that the subventricular zone-derived cells differentiate into olfactory bulb interneurons. They become either periglomerular cells of the glomerular layer or granule cells of the granule cell layer (Fig. 4B,C). We also established that the postnatally generated interneurons reach the olfactory bulb by migrating along a highly restricted pathway (Fig. 4A), which can be considered a continuation of the subventricular zone extending from the lateral ventricle to the middle of the olfactory bulb (Hinds 1968; Altman 1969). Moreover, unlike cortical neurons (see Rakic 1971, 1972; O'Rourke et al. 1992), interneurons derived from the subventricular zone approach the olfactory bulb in a direction orthogonal to the orientation of radial glial fibers (Kishi et al. 1990) and migrate at approximately twice the rate of cortical neurons (Luskin et al. 1993b). These observations suggest that cells bound for the olfactory bulb advance to their destination by a mechanism different from the one used by the cortical neurons in their ascent from the ventricular zone to the cortical plate. One of our current aims is to determine the means of migration and

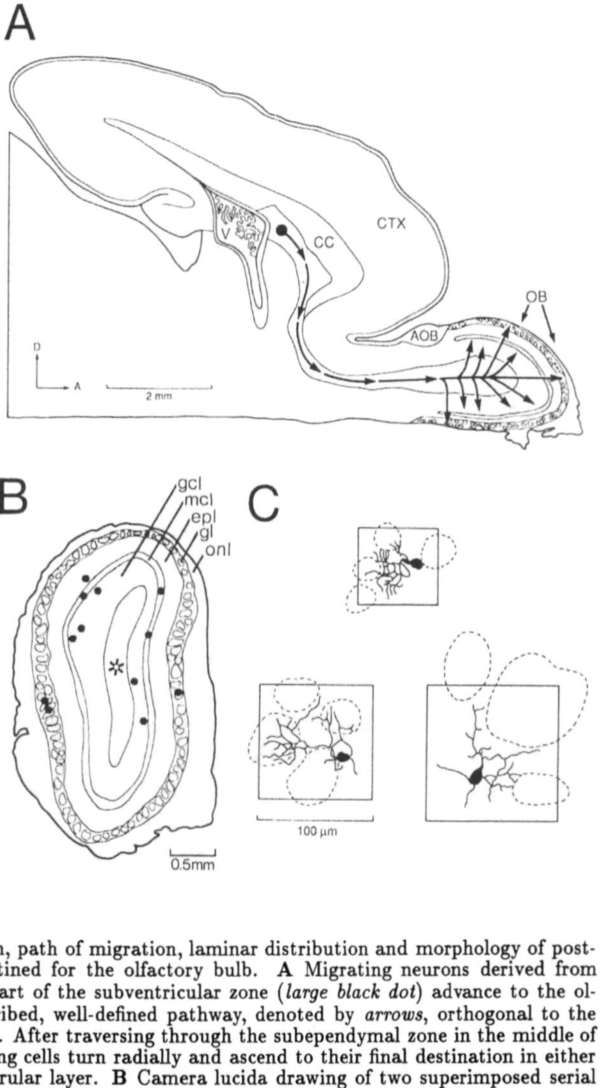

Fig. 1.4–4. Site of generation, path of migration, laminar distribution and morphology of postnatally generated neurons destined for the olfactory bulb. **A** Migrating neurons derived from progenitor cells in a discrete part of the subventricular zone (*large black dot*) advance to the olfactory bulb along a circumscribed, well-defined pathway, denoted by *arrows*, orthogonal to the orientation of radial glial fibers. After traversing through the subependymal zone in the middle of the olfactory bulb, the migrating cells turn radially and ascend to their final destination in either the granule cell layer or glomerular layer. **B** Camera lucida drawing of two superimposed serial 100 μm coronal sections showing the laminar distribution of lacZ-positive cells in the postnatal day 28 (P28) olfactory bulb resulting from an injection of a retroviral lineage tracer into the anterior part of the subventricular zone at P1. The labeled cells (*black dots*), though widely distributed, are concentrated in the granule cell layer and glomerular layer. Labeled cells in the mitral cell layer have the appearance of displaced granule cells. The subependymal zone in the center of the olfactory bulb (*asterisk*) is devoid of labeled cells. **C** Morphological appearance of lacZ-positive periglomerular cells in the olfactory bulb elucidated by immunohistochemically staining the P28 brain with an antibody to β-galactosidase following an injection of retrovirus into the anterior part of the subventricular zone at P1. The soma of each lacZ-positive periglomerular cell is situated next to a glomerulus and the ramified lacZ-positive dendrites extend into nearby glomeruli demarcated by *dashed lines*. *A*, anterior; *AOB*, accessory olfactory bulb; *CC*, corpus callosum; *CTX*, cerebral cortex; *D*, dorsal; epl, external plexiform layer; *gl*, glomerular layer; *gcl*, granule cell layer; *mcl*, mitral cell layer; *OB*, olfactory bulb; *onl*, olfactory nerve layer; *V*, lateral ventricle

molecular cues guiding the postnatally generated neurons to the olfactory bulb.

1.4.5 Conclusions

Increased understanding of the strategies used to assemble diverse arrays of characteristic cell types in the developing telencephalon have been revealed in a number of recent studies. These involve tracking the position and phenotype of cells which originate from either individual progenitor cells or distinct pools of progenitor cells. Our studies, reported here, suggest that the scheme devised by the developing cerebral cortex to obtain phenotypic diversity is to separate the lineages for each of the major cell types (astrocytes, oligodendrocytes, pyramidal neurons, and nonpyramidal neurons) by the onset of cortical neurogenesis. This conclusion rests on the definitive identification at the ultrastructural level of clonally related cells derived from progenitor cells of the prenatal ventricular zone or postnatal subventricular zone. Although the degree of dispersion among the members of a cortical clone is still not resolved (Kirkwood et al. 1992), there is plenty of support for a radial mode of migration for most postmitotic neurons migrating from the ventricular zone to the cortical plate (O'Rourke et al. 1992; Nakatsuji et al. 1991). On the other hand, evidence is presented here that many interneurons of the olfactory bulb, which are derived from a discrete part of the postnatal subventricular zone, traverse a highly restricted, well-defined path that courses perpendicular to the direction of migrating cortical neurons (Luskin 1993b). How and at what stage of development progenitor cells of the ventricular zone and subventricular zone become specialized, and the fate of their progeny determined, remains to be established.

References

Altman J (1969) Autoradiographic and histological studies of postnatal neurogenesis. IV. Cell proliferation and migration in the anterior forebrain, with special reference to persisting neurogenesis in the olfactory bulb. J Comp Neurol 137: 433–458

Austin CP, Cepko CL (1990) Cellular migration patterns in the developing mouse cerebral cortex. Development 110: 713–732

Bernard O, Drago J, Sheng, H (1992) L-myc and N-myc influence lineage determination in the central nervous system. Neuron 9: 1217-1224

Boulder Committee (1970) Embryonic vertebrate central nervous system: revised terminology. Anat Rec 166: 257–261

Colonnier M (1981) The electron-microscopic analysis of the neuronal organization of the cerebral cortex. In: Schmitt FO, Worden FG, Adelman G, Dennis SG (eds) The organization of the cerebral cortex. MIT Press, Cambridge, Maine, pp 125–152

Farbman AI (1991) Developmental neurobiology of the olfactory system. In: Getchel TV et al. (eds) Smell and taste in health and disease. Raven, New York

Grove EA, Williams BP, Li D-Q, Hajihosseini M, Friedrich A, Price J (1993) Multiple restricted lineages in the embryonic rat cerebral cortex. Development 117: 553–561

Hinds JW (1968) Autoradiographic study of histogenesis in the mouse olfactory bulb. II. Cell proliferation and migration. J Comp Neurol 134: 305–322

Jones EG, Powell TPS (1970) Electron microscopy of the somatic sensory cortex of the cat. I. Cell types and synaptic organization. Phil Trans R Soc Lond B 257: 1–11

Kirkwood BL, Price J, Grove EA (1992) Technical comment. Science 258: 317–318

Kishi K, Peng JY, Kakuta S, Murakami K, Kuroda M, Yokota S, Hayakawa S, Kuge T, Asayama T (1990) Migration of bipolar subependymal cells, precursors of the granule cells of the rat olfactory bulb, with reference to the arrangement of the radial glial fibers. Arch Histol Cytol 53: 219–226

LeVine SM, Goldman JE (1988) Embryonic divergence of oligodendrocyte and astrocyte lineages in developing rat cerebrum. J Neurosci 8: 3992–4006

Ling, EA, Paterson, JA, Privat, A, Mori S, Leblond, CP (1973) Investigation of glial cells in semithin sections. I. Identification of glial cells in the brain of young rats. J Comp Neurol 149: 43–72

Luskin MB (1992) Generation of cell diversity in the mammalian visual cortex: lineage relationship. In: Lent R (ed) The visual system for genesis to maturity. Birkhäuser, Cambridge, Maine, pp 111–130

Luskin MB (1993a) Mammalian forebrain development: how cells find a permanent home and establish their identity. J NIH Res 5: 60–64

Luskin MB (1993b) Restricted proliferation and migration of postnatally generated neurons derived from the forebrain subventricular zone. Neuron 11: 173–189

Luskin MB, Shatz CJ (1985) Neurogenesis of the cat's primary visual cortex. J Comp Neurol 216: 611–631

Luskin MB, Pearlman AL, Sanes JR (1988) Cell lineage in the cerebral cortex of the mouse studied in vivo and in vitro with a recombinant retrovirus. Neuron 1: 635–647

Luskin MB, Breding E, Miller KE, Maultsby M (1992) Cells destined for the olfactory bulb originate in the telencephalic subventricular zone and reach their target by migrating along a restricted longitudinal pathway. Soc Neurosci Abstr 18: 924

Luskin MB, Parnavelas JG, Barfield JA (1993a) Neurons, astrocytes, and oligodendrocytes of the rat cerebral cortex originate from separate progenitor cells: an ultrastructural analysis of clonally related cells. J Neurosci 13: 1730–1750

Luskin MB, Boone MS, Breding E (1993b) Rate and pattern of migration of olfactory bulb interneurons generated postnatally in the subventricular zone. Soc Neurosci Abstr 19: 873

Mares V, Bruckner JG (1978) Postnatal formation of non-neuronal cells in the rat occipital cerebrum; an autoradiographic study of the time and space pattern of cell division. J Comp Neurol 177: 519–528

McConnell SK, Kaznowski CE (1991) Cell cycle dependence of laminar determination in developing neocortex. Science 254: 282–285

Morshead CM, van der Kooy D (1992) Postmitotic death is the fate of constitutively proliferating cells in the subependymal layer of the adult mouse brain. J Neurosci 12: 249–256

McDermott KW, Breding E, Luskin MB (1991) Clonal analysis of gliogenesis in the neonatal rat cerebral cortex. Soc Neurosci Abst 17: 733

Nakatsuji N, Kadokawa Y, Suemori H (1991) Radial columnar patches in the chimeric cerebral cortex visualized by use of mouse embryonic stem cells expressing β-galactosidase. Develop Growth Differ 33: 571–578

O'Rourke NA, Dailey ME, Smith SJ, McConnell SK (1992) Diverse migratory pathways in the developing cerebral cortex. Science 258: 299–302

Parnavelas JG, Sullivan K, Lieberman AR, Webster KE (1977) Neurons and their synaptic organization in the visual cortex of the rat. Electron microscopy of Golgi preparations. Cell Tissue Res 183: 499–517

Parnavelas JG, Luder R, Pollard SG, Sullivan K, Lieberman AR (1983) A qualitative and quantitative ultrastructural study of glial cells in the developing visual cortex of the rat. Phil Trans R Soc Lond B 301: 55–84

Parnavelas JG, Barfield JA, Franke E, Luskin MB (1991) Separate progenitor cells give rise to pyramidal and nonpyramidal neurons in the rat telencephalon. Cerebral Cortex 1: 463–468

Patterson JA, Privat A, Ling EA, Leblond CP (1973) Investigation of glial cells in semithin sections. III. Transformation of subependymal cells into glial cells, as shown by radioautography after ^{3}H-thymidine injection into the lateral ventricle of the brain of young rats. J Comp Neurol 149: 83–102

Privat A (1975) Postnatal gliogenesis in the mammalian brain. Int Rev Cytol 40: 281-323

Peters A (1985) The visual cortex of the rat. In: Peters A, Jones EG (eds) Cerebral cortex. Visual Cortex, vol 3. Plenum, New York, pp 19–80

Peters A, Palay SL, Webster H deF (1991) The fine structure of the nervous system — neurons and their supporting cells, 3rd edn. Oxford University Press, New York
Price J, Thurlow L (1988) Cell lineage in the rat cerebral cortex: a study using retroviral-mediated gene transfer. Development 104: 473–482
Raedler E, Raedler A (1978) Autoradiographic study of early neurogenesis in rat cerebral cortex. Anat Embryol 154: 267–312
Rakic P (1971) Guidance of neurons migrating to the fetal monkey neocortex. Brain Res 33: 471–476
Rakic P (1972) Mode of cell migration to the superficial layers of fetal monkey neocortex. J Comp Neurol 145: 61–84
Rakic P (1974) Neurons in the rhesus monkey visual cortex: systematic relation between time of origin and eventual disposition. Science 183: 425–427
Reynolds BA, Weiss S (1992) Generation of neurons and astrocytes from isolated cells of the mammalian central nervous system. Science 255: 1707–1710
Sanes JR (1989) Analyzing cell lineage with a recombinant retrovirus. Trends Neurosci 12: 21–28
Sauer FC (1935) Mitosis in the neural tube. J Comp Neurol 62: 377–405
Skoff RP (1980) Neuroglia: a reevaluation of their origin and development. Pathol Res Pract 168: 279–300
Tan S-S, Breen S (1993) Radial mosaicism and tangential cell dispersion both contribute to mouse neocortical development. Nature 362: 638–640
Walsh C, Cepko CL (1988) Clonally related cortical cells show several migration patterns. Science 241: 1342–1345
Walsh C, Cepko CL (1992) Widespread dispersion of neuronal clones across functional regions of the cerebral cortex. Science 255: 434–440

1.5 Transient Neuronal Systems and Expression of Marker Molecules in the Developing Cerebral Cortex

P. Wahle
Lehrstuhl Allg. Zoologie und Neurobiologie, Fakultät für Biologie, Ruhr-Universität, Bochum, Germany

Cortical maturation is characterized by a sequence of progressive and regressive events. The neurons are generated and migrate away from the ventricular zone (see Caviness Chap. 1.3). Postmigratory neurons of cortical plate layers VI–II are successively sandwiched in an inside first – outside last fashion (Rakic 1982) between the earliest generated cortical cell populations: the neurons of the marginal zone and neurons of the upper intermediate zone (Boulder Committee 1970). The marginal zone transforms into layer I. The intermediate zone transforms into white matter. The upper intermediate zone has been termed the subplate zone (Rakic 1982) and has no counterpart in the adult cortex. It is a transient embryonic zone, and the subplate neurons, and also certain layer I neurons, have been regarded as transient neurons which largely become eliminated by cell death. Only very few survive (Luskin and Shatz 1985; Chun and Shatz 1989a).

In a general view, the neurons in the transient cortical zones are thought to be involved in the regulation of developmental processes (Shatz et al. 1988; McConnell et al. 1989; but see Molnár and Blakemore 1991), and they die after the functions are no longer needed. More recent studies have shown, however, that quite a percentage of subplate neurons in fact do survive through adulthood (Woo et al. 1991), as does the majority of layer I neurons. Moreover, transient neuronal types occur not only in the transient zones but also in the grey matter. In contrast, many individual neurons of persisting types in the grey matter are "transient" because they undergo cell death. The problem is the precise identification of the various cell types with specific structural and chemical markers. A transient cell type of the marginal zone with a unique morphology is the Cajal-Retzius cell. It is prominent in all mammals during the embryonic phase, dying perinatally. Rather few markers have been available for subplate neurons. Most also characterize the persisting neurons of the adult cortical grey matter, e.g., the ubiquitous neuronal protein MAP2 (Chun and Shatz 1989b), γ-amino butyrate (GABA), neuropeptide Y (NPY), and somatostatin (SOM; Chun et al. 1987).

The points to be addressed here are that "transient neurons" cannot be defined solely by time of birth and position in the cortex. Rather, the "transient neuronal types" should be defined by morphology, target specificity and neurochemical or molecular phenotypes. In the *first part* I describe the maturation of transient and persisting neuropeptidergic cell types and the structural reorganization of the peptidergic innervation patterns during development. In the *second part* I present evidence that intracortical cell types dynamically regulate the expression of neuropeptides. In the *third part* I describe monoclonal antibody SUBPLATE-1, a molecular marker for the inverted pyramidal neurons of the subplate and the cloning of a putative SP-1 encoding cDNA.

1.5.1 Transient Neuronal Types are Present in Developing Grey and White Matter

The NPY-ir Neuron System. A particular NPY-ir cell population in the perinatal cortex is a heterogeneous group of neurons with small somata and a rather primitive morphology, situated in grey and white matter. The neurons resemble types described in the reptilean pallium and tectum, and have been termed "archaic" neuronal types (Wahle and Meyer 1987). Archaic types occur more frequently at the late embryonic stage (Fig. 1, E57) especially within the most immature cortical plate. At this age the cortical plate has only a low density of fibers. Most axons accumulate in the white matter, few in layer I. Archaic neurons are completely eliminated pre- and perinatally. The cell death hypothesis for these and other types is supported by immunohistochemical (Wahle and Meyer 1987) and light and electron microscopical observations (Golgi method; Valverde and Facal-Valverde 1988).

Starting perinatally, another cell type differentiates in the white matter and later also in deeper layers of the grey matter of all cortical areas. It has a larger soma size, elaborate dendritic fields, and a peculiar axonal morphology with the main axon forming hairpin loops. This "axonal loop" cell (Wahle and Meyer 1987) establishes long-range intracortical projections with axons running through the white matter. Fibers then ascend vertically through the cortical plate layers into layer I where they follow a horizontal course before ending obliquely in superficial layers (Fig.1, P10). The projection does not respect areal boundaries. The fibers connect cortices of different modalities and even cross the corpus callosum toward the other hemisphere. Its dense axonal plexus in the white matter and the vertical innervation of the grey matter remain prominent for the first 3 postnatal weeks. During this time we observe differentiating and degenerating neurons of the same type side by side, indicating a neuronal turnover. Thereafter, the "axonal loop" cells successively degenerate and are gradually 'replaced' by newly differentiating NPY-ir neurons in layers V and VI and white matter having more restricted axonal arbors. These basket-type cells form the adult innervation with terminal strata in grey matter layers V/VI and II/III and connecting collaterals passing vertically through layer IV (Fig. 1, adult). Not before the second postnatal month are the typical NPY-ir terminal elements detectable: basket-type endings on somata, en passant boutons and axodendritic boutons onto proximal dendrites of pyramidal cells (Wahle et al. 1986). NPY-ir "axonal loop" cells, as the adult NPY-ir neurons, are GABAergic (Wahle et al. 1987). They represent a long range, white matter based, inhibitory neuron system that reaches the entire developing cortex. Its function may be to deliver inhibition, thus balancing excitation, to the very immature superficial layers, the interneurons of which have not yet completed differentiation.

The VIP/PHI Neuron System. Transient interneuron systems are not confined to the transient subplate zone but also occur in the developing cortical plate. The morphology of perinatally differentiating neurons containing vasoactive intestinal peptide (VIP) and the closely related peptide histidine isoleucine (PHI) differs dramatically from the adult cell types (Wahle and Meyer 1989). The dominant types were "transient bipolar"

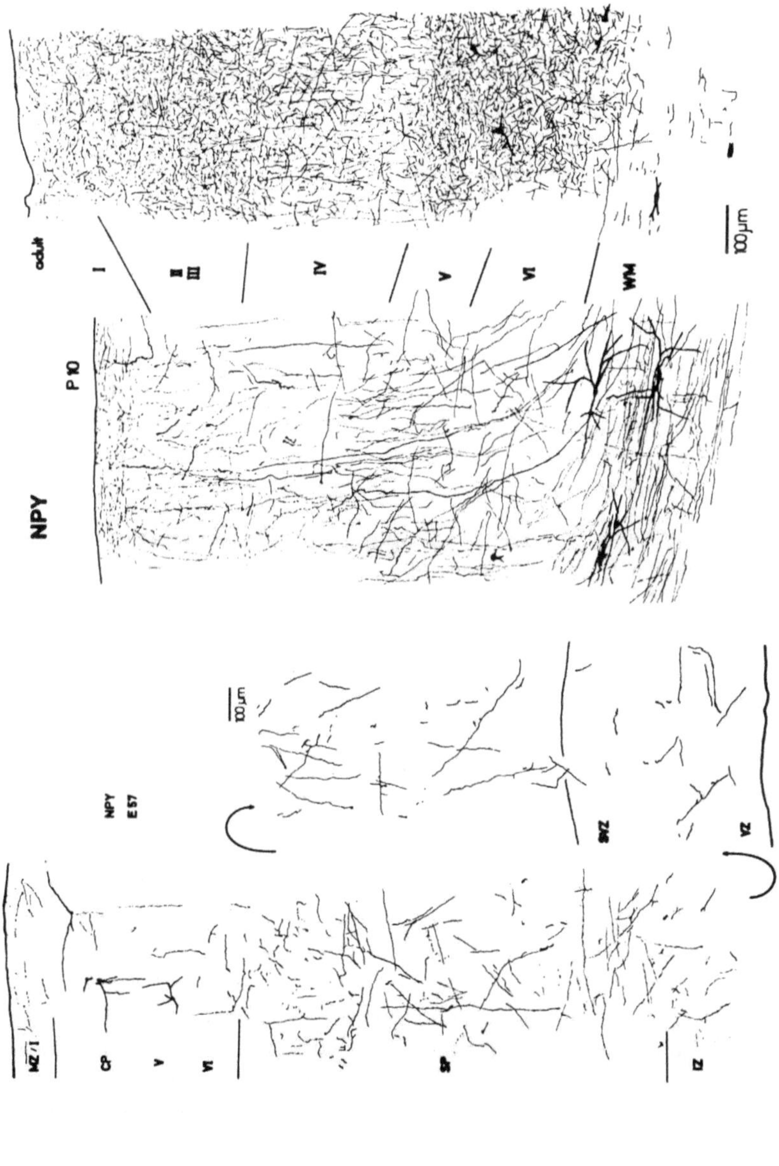

Fig.1.5–1. NPY-ir innervation in the occipital (visual) cortex at embryonic day (*E*) 57, postnatal (*P*) day 10 and the adult (> 1 year). Mostly "archaic" neurons are present at E57. The "axonal loop" cell innervation is dominant at P10, and is replaced by the adult local axon cells of layers V/VI. *WM*, White matter; *MZ*, marginal zone; *CP*, cortical plate; *SP*, subplate; *IZ*, intermediate zone; *SVZ*, subventricular zone, *VZ*, ventricular zone

and "pseudo-horsetail" neurons. The former type resides in infragranular layers and projects into the subjacent white matter, possibly targeting subplate neurons. The latter type resides within the most immature superficial layers and in layer I and forms, with bundled axon collaterals, a vertically oriented innervation terminating in layer VI. Both types and their prominent innervation patterns persist through the first 3 postnatal weeks, gradually disappearing by degeneration during the fourth and fifth weeks. Concurrently, a morphologically different population of neurons matures in the middle cortical layer first and then also in supragranular layers. These neurons form a dense innervation which, around the end of the second postnatal month, is reduced in density particularly in layers V/VI. Thus, in the adult cortex a small basket neuron remains as the most prominent cell type in upper layers. Its axons form round to columnar arbors and basket endings possibly targeting small pyramidal neurons.

In other words, a vertically projecting interlaminar neuron system of the grey matter, which also reaches the subplate neurons is gradually replaced by a more or less intralaminar small basket cell neuron system. Further, the persisting cell types differentiate initially in a higher cell density, which is later reduced to the adult level. The consequence is a dramatic reorganization of the cortical innervation starting around the peak of the critical period and proceeding through the second postnatal month.

The CCK-ir Neuron System. Compared to the dramatic changes in the NPY-ir and VIP/PHI-ir neuron systems, the ontogenesis of the cholecystokinin-ir (CCK-ir) system appears as a rather continuous process (Meyer and Wahle 1988). Neuronal types differentiate sequentially as cortical layers mature, and their innervation patterns superimpose onto the existing ones. As a rule, types with long-range connections develop first. Thus, neurons with long horizontal axons in layer VI were among the first CCK-ir neurons, followed by neurons in layers V, IV, and III with vertically descending axons which also terminate in layer VI. The neurons of these types mature during the first 2 weeks, and many individual cells are "transient" because they undergo cell death during the second and third postnatal weeks (again a neuronal turnover to a certain degree). However, the types and their innervation pattern persists through adult life. Further, the local axon cells of layer I (possibly early generated, but persisting cells of the marginal zone) differentiate during the first postnatal week. The next to differentiate, as soon as "their" layers become postmigratory, are cells in supragranular layers, many of which have short descending axons with horizontal collaterals terminating with en passant boutons. Superimposed is then the patch like termination of small basket-type neurons in layer II. The last to appear during the second month is a type of small basket neurons in layer IV, seen mainly in area 17. A peculiar feature of area 17 is a two-tiered fine punctate terminal plexus in upper and lower layer IV. This plexus originates partly from the layer IV basket neurons. Besides, CCK peptide (Wahle and Albus 1985) and CCK mRNA (Burgunder and Young 1988) is expressed in the thalamus. Possibly, the puncta represent the termination of thalamic afferents, corresponding to the finding that CCK receptors are enriched in thalamic input layers (Kritzer et al. 1987). The findings indicate that peptidergic cotransmitters possibly modulate thalamocortical transmission.

1.5.2 Neuropeptides Are Transiently Expressed in Developing Neocortex

Regressive events may also occur at the molecular level (Wahle 1993). Martinotti cells of layers V and VI are characterized by ascending axonal bundles which pass layer IV and terminate with horizontal collaterals in layers II/III and layer I. Local collaterals terminate in layer V and may descend into layer VI and into the white matter. The fibers terminate with en passant boutons and seem to contact dendrites rather than somata. The Martinotti neurons differentiate early postnatally. They are GABAergic and initially coexpress two peptides: SOM and substance P. The substance P expression reaches a peak in the second postnatal week; then it gradually declines to virtually undetectable levels in area 17. Not only is the peptide undetectable, also the substance P encoding mRNA is no longer expressed. In contrast, SOM and GABA expression remain unchanged. SOM is the fate marker. It reveals that during the second postnatal month many individual Martinotti neurons disappear. About 30% remain in adult layers V and VI. Transmitters may thus be expressed only for a limited developmental period and are downregulated at the transcriptional level in a distinct cortical cell type. Further, cortical neurons apparently regulate the expression of neuropeptides differentially. The role of transiently expressed substance P is far from clear. Besides direct physiological effects, it mediates metabotrophic effects (induction of DNA synthesis and cell division, neuronal differentiation, and protein phosphorylation). It is possible that it influences on differentiation and axon collateral formation of supragranular neurons as soon as they become postmigratory during the second and third postnatal week (Shatz and Luskin 1986).

To summarize, the peptidergic neuron systems undergo specific structural and molecular changes before an "adult" innervation finally becomes established. Transient cell types occur not only in the white matter (being probably subplate derived) but also in the grey matter (thus being cortical plate-derived). The different peptidergic types mature sequentially, some not before the second postnatal month. The adult pattern may be established by the addition of more and more locally projecting cells (e.g., CCK-ir system). Even in the course of such a rather "progressive" process, individual neurons belonging to persisting types are lost by degeneration. Our impression is that the earlier a neuron matures, the higher is its risk of becoming eliminated. Environmental factors or the problem of finding sufficiently mature synaptic partners may reduce its "fitness", inducing degeneration. In other cases the maturation of persisting types concurs with regressive events, the elimination of transient types. This results in a dramatic structural reorganization of the cortical innervation and thus of cortical circuitry (e.g., the replacement of a NPY-ir white matter based, long-range neuron system by a more locally projecting basket cell system in the grey matter). The peptidergic neuronal subsystems are all GABAergic. The inhibitory cortical interneuron population thus displays a remarkable structural and also molecular plasticity.

1.5.3 Cloning Molecular Markers for Subplate Neurons

In collaboration with Dr. Janice R. Naegele, monoclonal antibody technology has been applied to identify markers for subplate cells (Naegele et al. 1991). Monoclonal antibody (Mab) SUBPLATE-1 (SP-1) labels MAP2-positive neurons in the postnatal kitten cortical white matter (Fig. 2A), only few of which coexpress GABA, SOM, and/or NPY. The vast majority of the SP-1-ir neurons are non-GABAergic. We expected pyramidal features, although the dendritic pattern was only incompletely resolved by immunohistochemistry. We therefore combined, in collaboration with Dr. J. Lübke, the method of intracellular injections of Lucifer yellow (LY) in fixed sections with SP-1 immunostaining (Wahle et al., 1990 and submitted). The majority of double-labeled (LY fill + SP-1-ir) neurons are inverted pyramids, with a major descending dendrite and a bunch of finer dendrites arising from the upper somatic pole (Fig. 2F). All dendrites are covered with spines. The dendritic fields remain confined to the white matter. The axon arises from the descending dendrite or from the basal somatic pole. It descends into the white matter but gives off recurrent collaterals oriented toward the grey matter.

Studies using retrograde tracers have revealed the projection targets of the inverted pyramidal neurons (Wahle et al. 1991). In collaboration with Dr. K. Albus, rhodamine conjugated latex beads were injected into grey matter, thalamus, and midbrain of young and adult cats. For the subcortical injections one hemisphere was carefully retracted, and the injection pipet was inserted obliquely into the thalamus or midbrain of the opposite hemisphere. In this way we avoided local uptake in the hemisphere being analyzed. As in a series of pilot experiments which employed 1,1'-dioctadecyl-

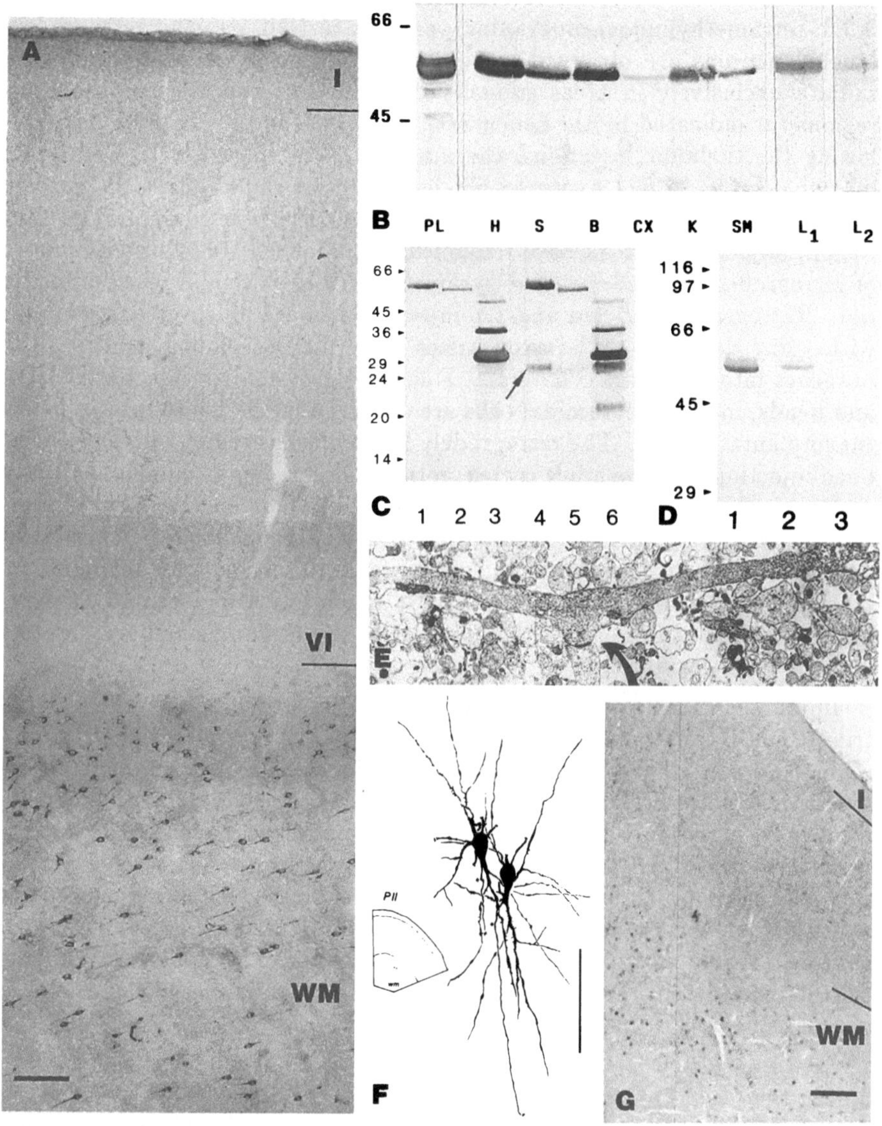

Fig. 1.5–2. **A** Mab SP-1-ir staining (P12 kitten visual cortex). Many neurons are labeled in the white matter (*WM*), besides a diffuse staining of the extracellular matrix. **B** The 56 kDa SP-1-ir band detected in blood plasma (*PL*), heart (*H*), spleen (*S*), whole blood (*B*), adult neocortex (*CX*), kidney (*K*), skeletal muscle (*SM*), and liver (L_1). After perfusion, staining of the band is dramatically reduced (L_2). **C** The 56 kDa SP-1 antigen detected with Mab SP-1 (*lanes 1–3*) and with the polyclonal antiserum against cat IgG (*lanes 4–6*). *Lanes 1,4*, untreated antigen. The polyclonal antiserum recognizes heavy and light chains (*arrow*). *Lanes 2,5*, the SP-1 antigen is a glycoprotein, and the band shifts after deglycosylation with *N*-glycanase but remains immunoreactive. *Lanes 3,6*, the same set of bromocyan fragments is stained by both antibodies. **D** Mab SP-1 staining (*lane 1*) is quenched after preabsorbing SP-1 with increasing amounts of cat IgG (*lanes 2,3*). **E** SP-1 positive dendrite receives an asymmetrical synapse on a spine (*curved arrow*). **F** SP-1 expressing, LY injected inverted pyramidal neurons. **G** In situ hybridization detects cell bodies in the white matter of P6 neocortex expressing an mRNA complementary to the putative SP-1 encoding cDNA

3,3,3'-tetramethylindocarbocyanine perchlorate (DiI) tracing retrogradely labeled neurons are observed in the white matter as deep as the corona radiata exclusively in areas axonally connected to the injected thalamus regions (as indicated by the concurrent labeling of layer VI pyramids). Following the thalamic injections, the majority of retrogradely labeled white matter neurons in fact expresses SP-1-ir material and displays the characteristic inverted pyramidal shape. The retrogradely labeled pyramidal neurons in adjacent layer VI have remained immunonegative. Quite a number of retrogradely labeled inverted pyramidal cells display glutamate-ir material. Retrogradely labeled and LY-injected cells are all inverted pyramids with spiny dendrites. The axon arises from the descending dendrite and branches into recurrent collaterals. Following grey matter injections of DiI and beads, inverted pyramidal cells are observed in the white matter below the implantation site. The retrogradely labeled cells are SP-1-ir. Even after bead injection into the adult cortex, retrograde labeling occurs in the white matter, and although the white matter cells have long lost the SP-1 expression, they display the typical inverted pyramidal morphology when filled with LY. Following midbrain injections (superior colliculus) retrogradely labeled cells are only observed in layer V, and they are glutamate-ir pyramids. White matter neurons thus project to the thalamus and grey matter but not to the midbrain.

Immunoelectron microscopy reveals that SP-1-ir material is diffusely distributed in the cytoplasm. Many cells have a dilated endoplasmatic reticulum, and some have large vacuolar inclusions, while others display normal ultrastructural features. The cell bodies receive symmetrical synapses. Dendrites, particularly the spines, receive asymmetrical synapses (Fig. 2E.; Nicksch and Wahle, in preparation). SP-1-ir neurons thus receive excitatory and inhibitory afferent input. The latter finding confirms earlier observations of white matter basket neurons ending on large non-GABAergic neurons (Wahle et al. 1987). It thus appears that the inhibitory and excitatory neurons in the white matter are wired locally in the same way as neurons in the grey matter.

SP-1 expression in white matter neurons starts around birth and peaks during the second postnatal week (Fig. 2A). Thereafter, the SP-1-ir staining intensity declines and disappears during the fourth week. The SP-1 antigen is thus transiently expressed in a developmentally regulated cell type, and SP-1 does not reveal its late postnatal fate. We do not have the impression that not all neurons degenerate and die. This has been confirmed using specific antisera against glutamate, the permanent marker for the inverted pyramidal cells (Wahle et al. 1990). In the glutamate-stained material, the white matter cell density, being already high in the newborn kitten, increases further to peak during the second and third weeks. Then follows a progressive reduction in density through the second postnatal month, presumably caused by degeneration and cell death. About 30% of the GLUT-ir population survive into adult life, and, as described above, the cells retain their typical morphology and form part of the adult circuitry.

To summarize, Mab SP-1 is a molecular marker for inverted pyramidal, probably glutamate-ir neurons in the subplate zone. The subplate zone is thus not solely inhibitory, as was previously thought (Chun et al. 1987). Rather, the proportion of interneurons to projection neurons resembles that found in the grey matter layers. SP-1 further distinguishes between early generated neurons, in that it does not stain the marginal zone-derived, cogenerated neurons (Luskin and Shatz 1985). The SP-1-ir pyramids project specifically to the thalamus and overlaying grey matter. The projection seems to be topographically ordered. In these aspects the inverted pyramidal neurons of the subplate zone strongly resemble layer VI pyramids. Further, at least the ascending projection persists into adult life. This finding indicates that the inverted pyramidal cell type, although located in the same compartment, is not transient as is the *axonal loop* cell type. Rather, it undergoes a dramatic reduction in density, but about 30% persist through adulthood. This is quite similar to the reduction of other persisting cell types in adjacent layers VI and V (e.g., SOM-ir Martinotti neurons). It is assumed that in most species early born neurons settle deep to the cortical plate. Layer VIb (subplate) in the rodent cortex shows a remarkable persistence of neurons (Woo et al. 1991). Even the aged human cortical white matter displays a high density of mostly pyramidal cells (Meyer et al. 1992).

The SP-1 antigen is a transiently expressed marker highly enriched in a putatively subplate-derived cell type. It is not a marker for all transient neurons since those in the grey matter are not labeled. What is the SP-1 antigen? We decided to isolate the SP-1 encoding cDNA from expression libraries which we made in lambda-ZAP^{TM} with mRNA from P1, P11, P30, and adult cat cortex. The P11 library was screened with Mab SP-1, but this first approach was not successful.

R. Henschel, a graduate student in my laboratory, then found that the 56 kDa SP-1-ir band (reported initially in kitten cortex; Naegele et al. 1991) can also be detected in other organs, even in adult cortex, where SP-1-ir structures can no longer be stained with immunohistochemistry (Fig. 2B). Perfusing the organs prior to protein extraction, however, dramatically reduces the 56 kDa-ir band (Fig. 2B). We suspected the antigen to be bloodborne. Affinity chromatography enriches the 56 kDa antigen from cat liver and blood plasma. The protein was purified and sequenced. The 20 N-terminal amino acids have an 80% homology to the heavy chain of human immunoglobulin. If the antigen is an immunoglobulin or an Ig-like molecule, antibodies against cat IgG should, in tissue sections and in Western blots, identify the SP-1 positive structures and bands. We obtained a Mab and a polyclonal serum, and indeed both probes selectively labeled subplate cells and identified the same 56 kDa band (Fig. 2C) on Western blots in tissue sections. Similar findings were reported by Dunn et al. (1993). The staining with the two probes and Mab SP-1 can be quenched and abolished after preabsorption of the antibodies with a commercial cat IgG fraction (Fig. 2D). We then subjected the SP-1 antigen to affinity purification, using P15 cat brain exsanguinated by perfusion with Ringer solution prior to protein extraction. This leaves only the intracellularly stored SP-1 antigen in the tissue, and removing the plasma-borne Ig's. Again, the 56 kDa band occurs, and on the Western blots it is immunoreactive with all three antibodies. Having established the SP-1 antigen as an immunoglobulin that is stored intracellularly in neurons, we rescreened the expression libraries with the polyclonal serum. We have identified a 1.8 kb cDNA putatively encoding the SP-1 antigen. cRNA probes were transcribed from the clone in the presence of digoxigenin-uridine triphosphate and were hybridized in situ (Wahle and Beckh 1992) to tissue sections prepared from newborn and P6 cat cortex. The antisense probe hybridized to an mRNA in white matter cells (Fig. 2G; Henschel and Wahle 1993). These results suggest that cells (inverted pyramids?) in the postnatal cat cortical white matter express and/or accumulate Ig-like molecules. Are we dealing with specific Ig's expressed and secreted by subplate neurons? They may contribute to the execution of molecular changes (Sheppard et al. 1991) in the extracellular matrix (e.g., masking guidance molecules?) which could transform white matter into a substrate which no longer permits migration and axon growth, or to an immune-mediated subplate cell death.

Acknowlegdement. Supported by the Deutsche Forschungsgemeinschaft.

References

Boulder Committee (1970) Embryonic vertebrate central nervous system: revised terminology. Anat Rec 166: 267–272

Burgunder JM, Young WS III (1988) The distribution of thalamic projection neurons containing cholecystokinin mRNA using in situ hybridization histochemistry and retrograde transport. Mol Brain Res 4: 179–189

Chun JJM, Shatz CJ (1989a) Interstitial cells of the adult neocortical white matter are the remnant of the early generated subplate neuron population. J Comp Neurol 282: 555–569

Chun JJM, Shatz CJ (1989b) The earliest generated neurons of the cat neocortex: characterization by MAP2 and transmitter immunohistochemistry during fetal life. J Neurosci 9: 1648–1667

Chun JJM, Nakamura MJ, CJ Shatz (1987) Transient cells of the developing mammalian telencephalon are peptide immunoreactive. Nature 325: 617–620

Dunn JA, Kirsch JD, Naegele JR (1993) Dying neurons in developing and adult cerebral cortex are detected with IgG-specific antibodies. Soc Neurosci Abstr 19: 1504

Henschel R, Wahle P (1993) Identification of the SP-1 antigen as an immunoglobulin-like molecule. Soc Neurosci Abstr 19: 627

Kritzer MF, Innis RB, Goldman-Rakic PS (1987) Regional distribution of cholecystokinin receptors in primate cerebral cortex determined by in vitro receptor autoradiography. J Comp Neurol 263: 418–435

Luskin MB, Shatz CJ (1985a) Studies of the earliest generated cells of the cat visual cortex: cogeneration of subplate and marginal zone. J Neurosci 5: 1062–1075

McConnell SK, Gosh A, Shatz CJ (1989) Subplate neurons pioneer the first axon pathway from the cerebral cortex.Science 245: 978–982

Meyer G, Wahle P (1988) Early postnatal development of cholecystokinin ir structures in the visual cortex of the cat. J Comp Neurol 276: 360–386

Meyer G, Wahle P, Castaneyra-Perdomo A, Ferres-Torres R (1992) Morphology of neurons in the white matter of the adult human neocortex. Exp Brain Res 88: 204–212

Molnár Z, Blakemore C (1991) Lack of regional specificity for connections between thalamus and cortex in culture. Nature 351: 475–477

Naegele JR, Barnstable CJ, Wahle P (1991) Expression of a unique 56 kDa polypeptide by neurons in the subplate zone of the developing cerebral cortex. Proc Natl Acad Sci USA 88: 330–334

Rakic P (1982) Early developmental events: cell lineages, acquisition of neuronal positions and areal laminar development. In: Rakic P, Goldman-Rakic P (eds) Development and modifiability of the cerebral cortex. MIT Press, Cambridge, USA, p 440

Shatz CJ, Luskin MB (1986) The relationship between the geniculocortical afferents and their cortical target cells during development of the cat's primary visual cortex. J Neurosci 6: 3655–3668

Shatz CJ, Chun JJM, Luskin MB (1988) The role of the subplate on the development of the mammalian telencephalon. In: Peters A, Jones EG (eds) Cerebral cortex. Plenum, Vol 7: 35-58

Sheppard AM, Hamilton SK, Pearlman AL (1991) Changes in the distribution of extracellular matrix components accompany early morphogenetic events of mammalian cortical development. J Neurosci 11: 3928–42

Valverde F, Facal-Valverde MV (1988) Postnatal development of interstitial (subplate) cells in the white matter of the temporal cortex of kittens: a correlative Golgi and electron microscopic study. J Comp Neurol 269: 168–192

Wahle P (1993) Differential regulation of substance P and somatostatin in Martinotti cells of the developing cat visual cortex. J Comp Neurol 329: 519–538

Wahle P, Albus K (1985) CCK-octapeptide like ir material in neurons of the intralaminar nuclei of the cat thalamus. Brain Res 327: 348–353

Wahle P, Beckh S (1992) A method of in situ hybridization combined with immunocytochemistry, histochemistry and tract tracing to characterize the mRNA expressing cell types in heterogeneous neuronal populations. J Neurosci Meth 41: 153–166

Wahle P, Meyer G (1987) Morphology and quantitative changes of transient NPY-ir neuronal populations during early postnatal development of the cat visual cortex. J Comp Neurol 261: 165–192

Wahle P, Meyer G (1989) The early postnatal development of VIP and PHI - immunoreactive structures in the cat visual cortex. J Comp Neurol 282: 215–248

Wahle P, Meyer G, Albus K (1986) Localization of NPY-ir in the cat visual cortex. Exp Brain Res 61: 364–374

Wahle P, Meyer G, Wu J-Y, Albus K (1987) Morphology and axon terminal pattern of glutamic acid decarboxylase – immunoreactive cell types in the white matter of cat occipital cortex during development. Dev Brain Res 36: 53–61

Wahle P, Lübke J, Naegele JR, Albus K (1990) SUBPLATE-1: a molecular marker for excitatory neurons in the subplate zone of developing cat neocortex? Soc Neurosci Abstr 16: 987

Wahle P, Albus K, Naegele JR (1991) Thalamic and intracortical projections originate from SUBPLATE-1-ir and glutamate-ir inverted pyramids in the white matter of developing and adult cat neocortex. Soc Neurosci Abstr 17: 899

Woo TU, Beale JM, Finlay BL (1991) Dual fate of subplate neurons in a rodent. Cerebral Cortex 1: 433–443

Chapter 2

Plasticity

2.1 Introduction

H. VAN DER LOOS[1]
Institute of Anatomy, University of Lausanne, Switzerland

"Der Mensch muss bei dem Glauben verharren dass das Unbegreifliche begreiflich sei. Er würde sonst nicht forschen." This is not my proposition, but Goethe's, made in his *"Maximen und Reflexionen."* In the spirit of this symposium, which is becoming a mixture of serious and light, much as Otto Creutzfeldt would have wanted it and certainly as it would have been under his aegis and in his presence, I would like to paraphrase this statement as follows: "You should stick to the belief that what's not understandable is in fact understandable; if you don't, you may as well give up research."

[1]Hendrik Van der Loos passed away on the 11th of October 1993

Why all this Goethe, who was never really my love? First, this is not a bad maxim and, second, it is the motto Otto chose for his *"Philosophische Probleme der Neurophysiologie,"* published 12 years ago in a collection acrobatically entitled *"Rückblick in die Zukunft"* — let us call it this afternoon *"Looking back into the future"* (Creutzfeldt 1981). I believe that the motive expressed in Goethe's maxim guided much of Otto's thoughts and scientific enterprise, and I predict that, if we force ourselves through the motions that Roessner's title suggests, we will judge (in a future I cannot define) that this session was good, that this symposium was good, and that the man and the scientist whom we honor today was very good.

According to the organizers, this meeting was planned with Otto, that is, the meeting was scheduled to take place one year ago, as a festive accompaniment to his 65th birthday. Through one of the grossly unreasonable moves of nature — that make us angry and triste at the same time and with which I shall never learn to have peace — Otto died; and so it happened that we are having his symposium now. But we may take it that, in a rather direct way, we are Otto's guests.

I did not know Otto well, but I knew him in many guises, and I liked him. I believe there was a reciprocal friendship. I say "I believe," for I don't use the word friendship lightly, and we saw much too little of each other. I see him as a man who influenced his environment not only by excitation, but also by disinhibition. He wanted to get at the mind-brain interface, and at the roots of clinical neurology — yet molecules and channels blossomed in his environment. There are not many of that breed of leaders among us. I knew him in his capacity as President of the European Neuroscience Association, as a 'co-speaker' at symposia, as a table neighbor at dinners inspired by neuroscience, and as a friend with whom I discussed culture and politics. In one such encounter, he surprised me by telling that he was against the whole concept of German unification. I wholeheartedly agreed. I knew him as one who confided, in response to my question why his cortex cerebri book was in German, that he could only really express himself in German. When I countered with something like: "well, ..." to express my doubt that this was the case, he insisted.

I was privileged to have known both Otto Creutzfeldt and Günter Baumgartner (an equally eminent neuroscientist, who was head of Neurology at the University Hospital in Zürich, and who died a few months before Otto, also from an incurable disease). As I witnessed in discussions with both together or with each individually, these men were bound by a deep and long-standing friendship. Interestingly, they shared several rather striking characteristics: they often expressed themselves in a mumbling and staccato manner, and sometimes almost uncomprehensibly so (just as Otto's handwriting was at times illegibly small). Thus, they were given to formulating concepts that were not quite perfected, but were always interesting and ultimately clear. It was as if they had a melody in their head, that was ready for internal humming but not for bringing outward — something many or at least some of you may recognize; if not, please believe me, it exists: I suffer from it to a considerable degree, and not only for melodies.

Now there will be place for our four invited speakers — turned writers — in this session/chapter on plasticity. If you did not know it, now is the time to learn that plasticity is what plasticists do.

Reference

Creutzfeldt OD (1981) Philosophische Probleme der Neurophysiologie. In: Roessner H (ed) Rückblick in die Zukunft. Severin & Siedler, München, pp 256–278

2.2 Computational Structure of Central Nervous System Axons Reflects Developmental Strategies

G.M. Innocenti
Institute of Anatomy, University Lausanne, Switzerland

Callosal axons interconnecting visual areas often terminate in multiple separate columns. The geometry of some of the arbors appears adequate to synchronously activate columns but seems to be unnecessarily redundant. It may result from the same developmental strategies which are responsible for the exuberance of callosal and intrahemispheric projections in development.

2.2.1 Introduction

A chalk-on-blackboard drawing from Cajal which until recently existed at the Instituto Cajal in Madrid (De Felipe and Jones 1992) still summarizes most of what has been published on the morphology of axons interconnecting the cerebral hemispheres. Recent neuroanatomical research (Hartenstein and Innocenti 1981; Hendry and Jones 1983; Fish et al. 1991) has added little. The immortality of this piece of work of Cajal's owes less to its solidity than to the inadequacy of tracing techniques thus far available for visualizing long axons in the central nervous system.

The studies reviewed here were performed with biocytin (Horikawa and Amstrong 1988; King et al. 1989). This tracer, when injected extracellularly, can be taken up by cell bodies and is very effectively anterogradely-transported in many, although possibly not all, axonal systems. Biocytin visualizes details of axonal morphology previously revealed only (but capriciously) by the Golgi method, by the intracellular injections of horseradish peroxidase (HRP) (but with limitation on transport distances) and, more recently, by phaseolus agglutin (but with limitations due to the slow transport rate and to other difficulties with this tracer in developmental studies). The results we obtained with biocytin raised the question of what determines the sometimes unexpectedly complex geometry of central nervous system axons.

2.2.2 Geometry of Callosal Axons Implements Topographical Mapping Rules Between the Two Hemispheres

The analysis of individual, biocytin-labeled axons requires reconstructions from serial sections. We used a computer-interactive microscope, originally designed for dendritic three-dimensional reconstruction and analysis by Glaser et al. (1983). The software was modified to cope with the complexity of the arbors, their extension, and the necessity of keeping track, as precisely as possible, of axon diameters.

Seventeen axons labeled by biocytin injections near the 17/18 border of adult cats were reconstructed (Houzel et al. 1992), seven of them from the callosal midline. Only rarely did the reconstructed axons implement point to point connections between the hemispheres. More often (13 of 17 cases) they ended in multiple (usually 2–5) "terminal columns", i.e., clusters of synaptic boutons occupying separate volumes of cortex. Terminal columns are 300–600 μm in diameter, are separated by 125–2700 μm wide free spaces, and can span up to 3500 μm tangentially across the cortex. These columns resemble those described by Jones et al. (1975) for somatosensory callosal projections in the monkey and by Gilbert and Wiesel (1979, 1983) and Martin and Whitteridge (1984) for local projections originating from layer III neurons of areas 17 and 18. The classical electrophysiological findings of Berlucchi and Rizzolatti (1968) strongly suggest that, as for local projections (Gilbert and Wiesel 1989), a terminal column corresponds to an orientation column or to part of it, with the same orientation as the column where the cell body is located.

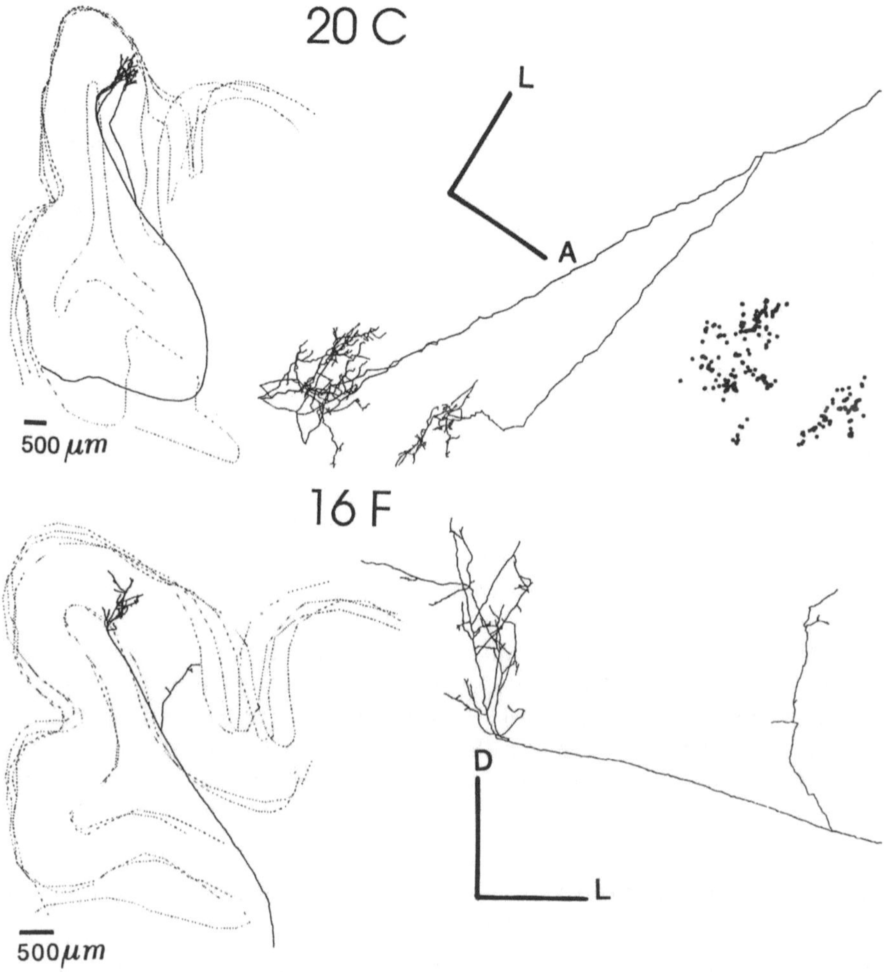

Fig. 2.2–1. Three-dimensionally reconstructed adult callosal axons filled anterogradely by a biocytin injection near the 17/18 border and terminating at homotopic location in the contralateral hemisphere. Axon 20C illustrates the parallel architecture, 16F the serial architecture (see text). Both axons are displayed in front views (*left*); enlarged top views (for 20C) and front views (for 16F) are shown on the *right*; terminal boutons (*dots*) are shown separately for 20C. *L*, Lateral; *A*, anterior; *D*, dorsal

Individual axons differ with respect to the complexity of their terminal arbor, their mediolateral and tangential spread, and the radial distribution of synapses. However, all of them have at least one focus of termination near the border between areas 17 and 18.

Terminal arbors express two types of architecture (Fig. 1). Parallel architecture is characterized by the formation in the white matter of primary or higher order branches of comparable length, which run close to each other over considerable distances and either terminate onto neighboring columns or converge onto the same column. Serial architecture is characterized by a tangentially running trunk, or main branch, from which collaterals depart that enter the cortex more or less radially. Although the arbor of some axons can be purely parallel (7 of 17) or purely serial (2 of 17), others have mixed parallel/serial types of architecture.

The parallel architecture is puzzling. Axons are costly in terms of production, transport, and maintenance of their components. Thus axonal geometry can be expected to be as parsimonious as possible. Indeed, the evolution of the nervous system seems to have favored shortening of axons, for example, with the formation of ganglionic masses in invertebrates (Cajal 1909) or the folding of cerebral cortex in mammals (Innocenti 1990). The parallel architecture seems to depart from this tendency by duplicating part of the axon. Perhaps this could be justified by considering that axons perform not only spatial but also temporal (Waxman 1978; Raymond and Lettvin 1978; Parnas 1979; Lüscher and Shiner 1990a,b), transformations of impulse trains generated at the cell body. The geometry of axonal arbors may therefore reflect specific computational operations in the temporal domain. The parallel architecture seems well suited for activating terminal columns synchronously, the serial architecture for the sequential activation of columns; indeed it resembles that of axons found in auditory brainstem structures of mammals and birds, where it establishes delay lines computing spatial position of the auditory stimuli (Carr and Konishi 1990).

2.2.3 Axonal Geometry Can Implement Temporal Rules of Neuronal Interaction

The presumable consequences of axonal geometry on the spatiotemporal activity patterns generated by callosal axons was studied with an ad hoc program (Maxsim) allowing us to simulate the propagation of an action-potential front in the arbor. At any given time after spike initiation the distance traveled by the spike front is directly proportional to the conduction velocity of the axon. In spite of several simplifications (see Fig. 2), the simulation produced conduction times from the callosal midline within the range found in electrophysiological experiments (Innocenti 1980; McCourt et al. 1990).

As expected, it was found that parallel architecture does favor the cotemporal activation of distinct columns. In the example shown in Fig. 2, columns a, b and c become simultaneously active within a 125 μs window

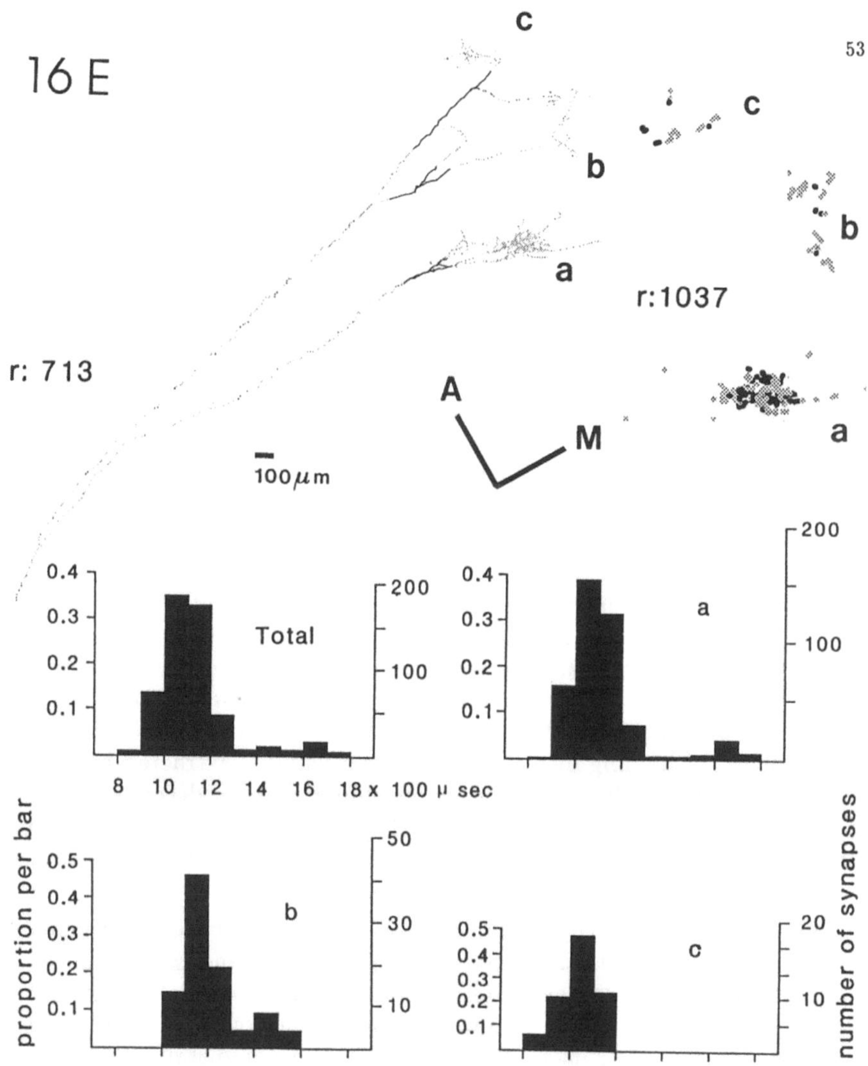

Fig. 2.2–2. Simulated invasion of a three-dimensionally reconstructed callosal axon (16E); parallel architecture) by an orthodromically propagating axon potential. *Upper panels*, top view of the axon (*left*) and enlarged view of its terminal boutons (*right*). The axonal segments and boutons activated within a 50 µs window are shown in *full black*; *r*, the time elapsed after initiation of the action potential; *a,b,c*, separate columns of termination. *Lower panels*, histograms of activation times for the whole arbor and for the individual terminal columns. The simulation is based on the following assumptions. Conduction velocity was considered to be related linearly to fiber diameter, i.e., $5.5\,\mathrm{m\,s^{-1}}\,\mu\mathrm{m}^{-1}$ (Waxman and Bennett 1972). Axons were considered to be myelinated and to have a g ratio (the ratio between axoplasm diameter and fiber diameter, inclusive of myelin) of about 0.7. In the corpus callosum of adult cats axons above 0.25 µm are almost all myelinated (Berbel and Innocenti 1988), and the diameter of our axon trunks (estimated with a resolution of 0.45 or 0.25 µm) ranged between 0.52–0.78 and 1.8–2.25 µm. The diameter of 0.25 µm is close to the limit of resolution of the optical microscope and was reached by preterminal branches (the branches which carry synaptic boutons) in our arbors. At around this diameter the conduction velocity is comparable in myelinated and unmyelinated axons (Waxman and Bennet 1972). No correction was introduced for the possible, small slowing down of spike conduction near branching points. No correction was introduced for tissue shrinkage (which was found to be in the order of 35%–40%) since, if identical shrinkage is assumed to occur for axon length and diameter, the correction does not affect the profiles of activation

during which about 53% of all boutons become active. This probably provides a lower estimate of the coactivation. In fact, the calcium currents generated by the spikes at the boutons and responsible for the transmitter release last several hundred microseconds (Llinas 1982). Therefore the activation of many boutons can overlap without being synchronous.

In addition to the parallel architecture, another structural arrangement appears to favor the coactivation of separate terminal columns. This is the differential diameter of axons feeding different columns, in axons with serial architecture. The simulations suggest that, for example, in some of these axons a decrease in axon diameter in the proximal column can counteract the delay introduced by the axonal segment connecting to more distal columns. Not all axons can activate columns in synchrony. In some axons, typically in some of those with serial architecture, delays of up to 2–3 ms were found between the earliest and the latest activated column.

In conclusion, one can consider all axons as delay lines generating differences in the timing of activation of different terminal columns (delta-t_a). The axons which coactivate columns specifically generate delta-t_a close to zero. The latter axons are interesting because it has recently been found that callosal connections, as other corticocortical connections, are involved in the synchronous activation of spatially separate cortical sites (Engel et al. 1991). It is believed that the synchronization could offer a solution to the problem of binding into a coherent form features extracted by different cortical columns. The structural arrangement that we describe represents an anatomical substrate which could allow a neuron with a given orientation selectivity in one hemisphere to synchronize the activity of neurons in several orientation columns of the other hemisphere.

Although the parallel architecture may be useful for coactivating columns is not economical in terms of axoplasm production and maintenance. We therefore wondered whether nature could not have constructed axons more sparingly. In the simulation shown in Fig. 3 the architecture of the same axon shown in Fig. 2 was manipulated by moving distally the origin of the first-order branches. This achieved an economy of about 8000 μm^3 of axoplasm, corresponding to 33% of the total axoplasm between the callosal midline and layer VI. The consequences for the temporal pattern of activity were negligible. Similar findings were obtained from the manipulation of other axons with parallel architecture.

2.2.4 Juvenile Callosal Axons Establish Early, Parallel Geometries

Searching for what may cause callosal axons to behave extravagantly, we looked at how the geometry of callosal arbors develops. Sites of area 17

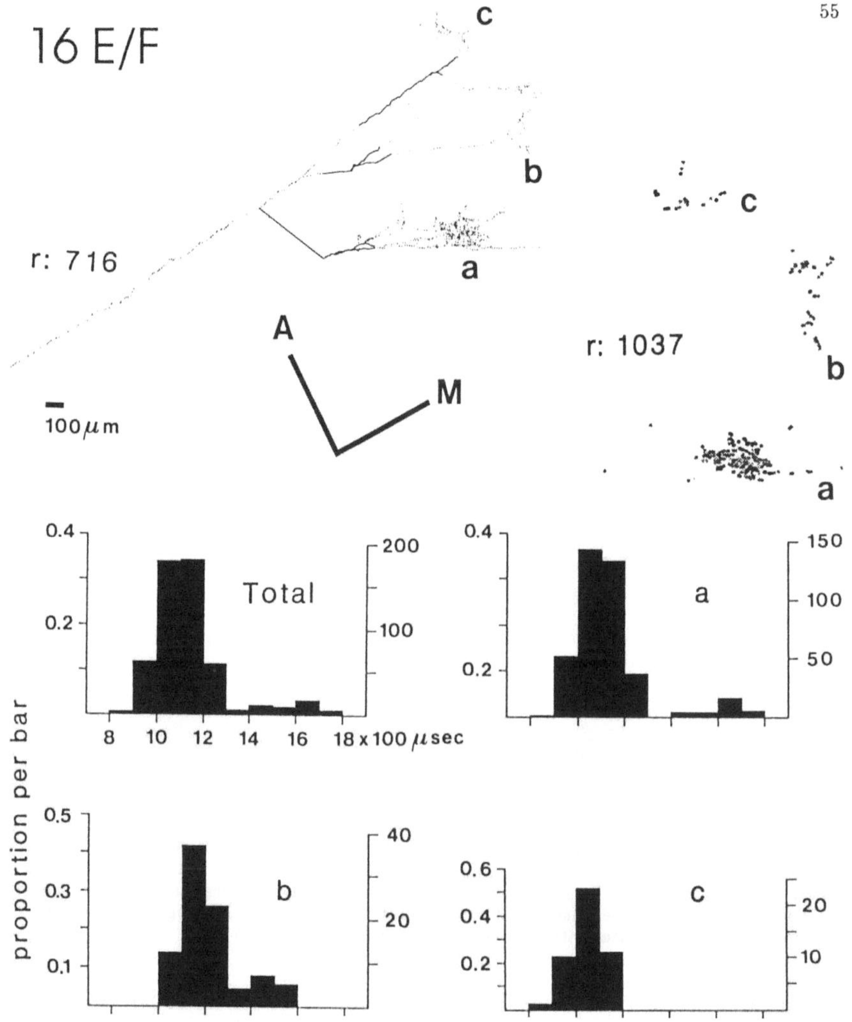

Fig. 2.2–3. Simulated invasion of the callosal axon 16E by an orthodromically propagating axon potential after altering its arbor in a way which decreases the amount of axoplasm subserving column *a* by about 8000 μm^3. Original diameter is preserved in the new branch connecting *a*. *Upper panels*, top view of the axon (*left*) and enlarged view of its terminal boutons (*right*). The axonal segments and boutons activated within a 50 μs window are shown in *full black*; *r* the time elapsed after initiation of the action potential; *a*,*b*,*c*, denote separate columns of termination. *Lower panels*, histograms of activation times for the whole arbor and for the individual terminal columns. Notice that the new structure does not alter original the temporal activation of the arbor shown in Fig. 2

which lose callosal projections in development and sites near the 17/18 border which maintain callosal connections were injected separately in kittens of different ages. Data analyzed thus far concern the first and second postnatal week.

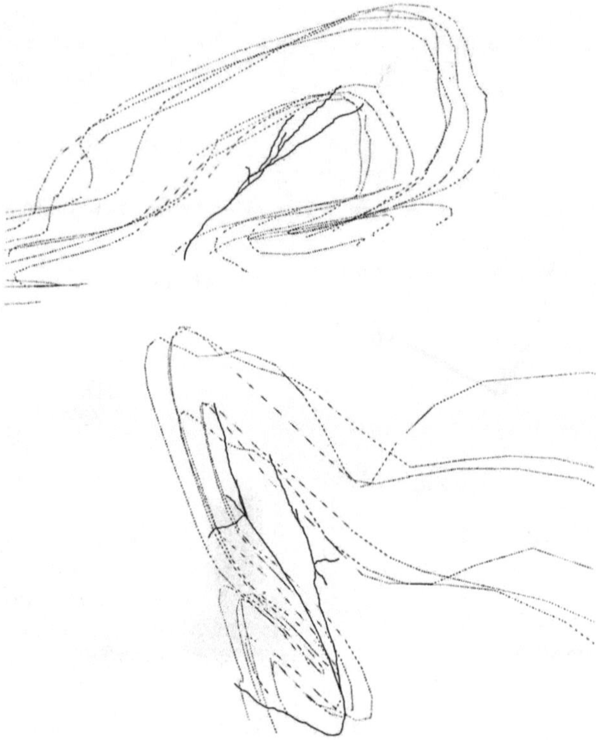

Fig. 2.2–4. Three-dimensionally reconstructed callosal axons filled anterogradely by a biocytin injection near the 17/18 border and terminating at homotopic location in the contralateral hemisphere. Kitten injected at *P1*, sacrificed at *P3*. Partially rotated, conical views of the reconstructions are shown. Notice the existence of parallel running branches which stop at the bottom of the gray matter or in the white matter. *Scale bar*, approximately 200 μm

As early as postnatal day 3 (Fig. 4) callosal axons establish bifurcations in the white matter, from which parallel branches originate, directed to the cortex. These branches are found in axons which are eliminated and in those which presumably are maintained. In both types of axons the bifurcations in the white matter are established before the axons have entered the cortex to any significant extent. Indeed, some of the branches enter layer VI, but most of them end in the white matter.

The branches in the white matter exceed in number those found in the adult, suggesting that the latter are the survivors of a process of selection. Furthermore, while most of the branches are directed toward the region of the 17/18 border, which receive callosal afferents in the adult, others are directed to the medial part of area 17 or to the lateral part of area 18, which do not. Thus in the development of callosal connections, two types of exuberance are found. One concerns the formation of transient transcallosal axons originating from medial/ventral parts of area 17 (Innocenti 1981) and the other the formation of transient branches in the white matter of the lateral gyrus, below areas 17 and 18.

What determines the formation of the branches is not known. Since the branches are established before the axons reach their final targets, it appears unlikely that they are determined by the same computational constraints which the morphology of adult axons appears to satisfy. Furthermore, not only callosal axons but also intracortical axons form early transient branches in the white matter of the lateral gyrus (Assal and Innocenti 1993). A target-searching strategy of the axons and/or signals coming from the white matter and possibly related to the presence of transient neuronal populations of the subplate might determine the bifurcations associated with axonal exuberance in development.

2.2.5 Conclusions

The structure of callosal axons implements two operations of a computational kind. One operation is the topographical mapping between neuronal populations of the two hemispheres. Different from what was previously believed, the mapping is only rarely point to point. Instead, each callosally projecting neuron can, through its axon, activate several, spatially distinct and widely distributed "points" of the contralateral hemisphere. The other operation appears to be a dispersion in the temporal domain of action potentials generated at the cell body. On first approximation each terminal bouton of an axonal arbor can be considered as an element in a matrix which multiplies the time of occurrence of an action potential (with the value of 1) by the conduction delays introduced by the geometry of the arbor. The geometry of many callosal axons appears to be suited to activate simultaneously spatially separate target neurons. The branching pattern of these axons is often non-economical in terms of the amount of axoplasm produced and is determined early in development. It appears to reflect the same developmental strategy and/or mechanisms which are responsible for the overproduction of axons and axonal branches (exuberance). Later in life, computational constraints might establish the criteria whereby certain juvenile axons and/or axonal branches are selected while others are eliminated.

Acknowledgments. D. Aggoun-Zouaoui, F. Assal, J.-C. Houzel, P. Lehmann, C. Milleret, J.-D. Nicoud, L. Tettoni have contributed to various parts of this work. The research was supported by the Swiss National Science Foundation grant no. 3129948.90, European Science Foundation Twinning program grant, and special research funds of the Lausanne University/Federal Institute of Technology.

References

Assal F, Innocenti GM (1993) Transient intra-areal axons in developing cat visual cortex. Cerebral Cortex 3: 290-303

Berbel P, Innocenti GM (1988) The development of the corpus callosum in cats: a light- and electron-microscopic study. J Comp Neurol 276: 132–156

Berlucchi G, Rizzolatti G (1968) Binocularly driven neurons in visual cortex of split-chiasm cats. Science 159: 308–310

Cajal RS (1909) Histologie du système nerveux de l'homme et des vertébrés, vol. 1. Maloine, Paris

Carr CE, Konishi M (1990) A circuit for detection of interaural time differences in the brain stem of the barn owl. J Neurosci 1030: 3227–3246

De Felipe J, Jones EG (1992) Santiago Ramón y Cajal and methods in neurohistology. Trends Neurosci 15: 237–246

Engel AK, König P, Kreiter AK, Singer W (1991) Interhemispheric synchronization of oscillatory neuronal responses in cat visual cortex. Science 252: 1177–1179

Fish SE, Rhoades RW, Bennett-Clarke CA, Figley B, Mooney RD (1991) Organization, development and enucleation-induced alterations in the visual callosal projection of the hamster: single axon tracing with Phaseolus vulgaris leucoagglutinin and Di-I. Eur J Neurosci 3: 1255–1270

Gilbert CD, Wiesel TN (1979) Morphology and intracortical projections of functionally charactersed neurones in the cat visual cortex. Nature 280: 120–125

Gilbert CD, Wiesel TN (1983) Clustered intrinsic connections in cat visual cortex. J Neurosci 3: 1116–1133

Gilbert CD, Wiesel TN (1989) Columnar specificity of intrinsic horizontal and corticocortical connections in cat visual cortex. J Neurosci 9: 2432–2442

Glaser EM, Tagamets M, McMullen NT, Van der Loos H (1983) The image-combining computer microscope — an interactive instrument for morphometry of the nervous system. J Neurosci Meth 8: 17–32

Hartenstein V, Innocenti GM (1981) The arborization of single callosal axons in the mouse cerebral cortex. Neurosci Lett 23: 19–24

Hendry SHC, Jones EG (1983) Thalamic inputs to identified commissural neurons in the monkey somatic sensory cortex. J Neurocytol 12: 299–316

Horikawa K, Armstrong WE (1988) A versatile means of intracellular labeling: injection of biocytin and its detection with avidin conjugates. J Neurosci Meth 25: 1–11

Houzel JC, Milleret C, Innocenti GM (1992) Morphology of callosal axons terminating in visual cortical areas 17 and 18 in the adult cat. Eur J Neurosci Suppl 5: 259

Innocenti, GM (1980) The primary visual pathway through the corpus callosum: morphological and functional aspects in the cat. Arch Ital Biol 118: 124–188

Innocenti GM (1981) Growth and reshaping of axons in the establishment of visual callosal connections. Science 212: 824–827

Innocenti GM (1990) Pathways between development and evolution. In: Finlay BL et al. (eds) The neocortex. Plenum Press, New York, pp 43–52

Jones E, Burton H, Porter R (1975) Commissural and cortico-cortical "columns" in the somatic sensory cortex of primates. Science 190: 572–574

King MA, Louis PM, Hunter BE, Walker DW (1989) Biocytin: a versatile anterograde neuroanatomical tract-tracing alternative. Brain Res 497: 361-367

Llinas RR (1982) Calcium in synaptic transmission. Scientific American 247/4: 38–47–65

Lüscher HR, Shiner JS (1990a) Computation of action potential propagation and presynaptic bouton activation in terminal arborizations of different geometries. Biophys J 58: 1377–1388

Lüscher HR, Shiner JS (1990b) Simulation of action potential propagation in complex terminal arborizations. Biophys J 58: 1381–1399

Martin KAC, Whitteridge D (1984) Form, function and intracortical projections of spiny neurones in the striate visual cortex of the cat. J Physiol 353: 463–504

McCourt ME, Thalluri J, Henry GH (1990) Properties of area 17/18 border neurons contributing to the visual transcallosal pathway in the cat. Vis Neurosci 5: 83–98

Parnas I (1979) Propagation in nonuniform neurites: form and function in axons. In: Schmitt FO, Worden FG (eds) The neurosciences. MIT Press, Cambridge, pp 499–512

Raymond SA, Lettvin JY (1978) Aftereffects of activity in peripheral axons as a clue to nervous coding. In: Waxman SG (ed) Physiology and pathobiology of axons. Raven, New York, pp 203–225

Waxman SG (1978) Variations in axonal morphology and their functional significance. In: Waxman SG (ed) Physiology and pathobiology of axons. Raven, New York, pp 169–190

Waxman SG, Bennett MVL (1972) Relative conduction velocities of small myelinated and non-myelinated fibres in the central nervous system. Nature 238: 217–219

2.3 Neurochemical Properties of Cytochrome Oxidase-Rich Puffs and Interpuffs in Primate Visual Cortex: Regulation by Visual Input

S.H.C. Hendry[1] and R.K. Carder[2]

[1]Zanvyl Krieger Mind/Brain Institute, Johns Hopkins University, Baltimore, Maryland
[2]Department of Psychology, UCLA, Los Angeles, California

Repeated patches of cytochrome oxidase (CO) staining, referred to as puffs or blobs (Horton and Hubel 1981; Humphrey and Hendrickson 1983; Wong-Riley and Carroll 1984), occupy layers II and III of monkey primary visual cortex (area 17). The puffs and surrounding interpuffs comprise distinct compartments, in which neurons respond to different visual stimulus features and display different patterns of connectivity. Color opponent responses to low spatial frequency stimuli of any orientation are properties of the majority of puff neurons, while responses to oriented stimuli at high spatial frequencies are characteristic of interpuff neurons (Livingstone and Hubel 1984; Tootell et al. 1988 a,b,c; Ts'o and Gilbert 1988). These differences in neuronal physiology correlate with differences in afferent connectivity, as the puffs are directly innervated by axons from the dorsal lateral geniculate nucleus (LGN; Fitzpatrick et al. 1983; Livingstone and Hubel 1982), while the interpuffs receive no geniculocortical terminations. From puffs and interpuffs, separate paths lead to distinct compartments in the second visual area (V2) and, directly or indirectly, to interconnected channels or streams in the temporal lobe (Livingstone and Hubel 1983, 1988).

When viewed in sections cut parallel to the pial surface, the puffs appear as rows (Figs. 1A, 2B), with a center-to-center spacing between rows equal to the width of an ocular dominance column (Horton and Hubel 1981;

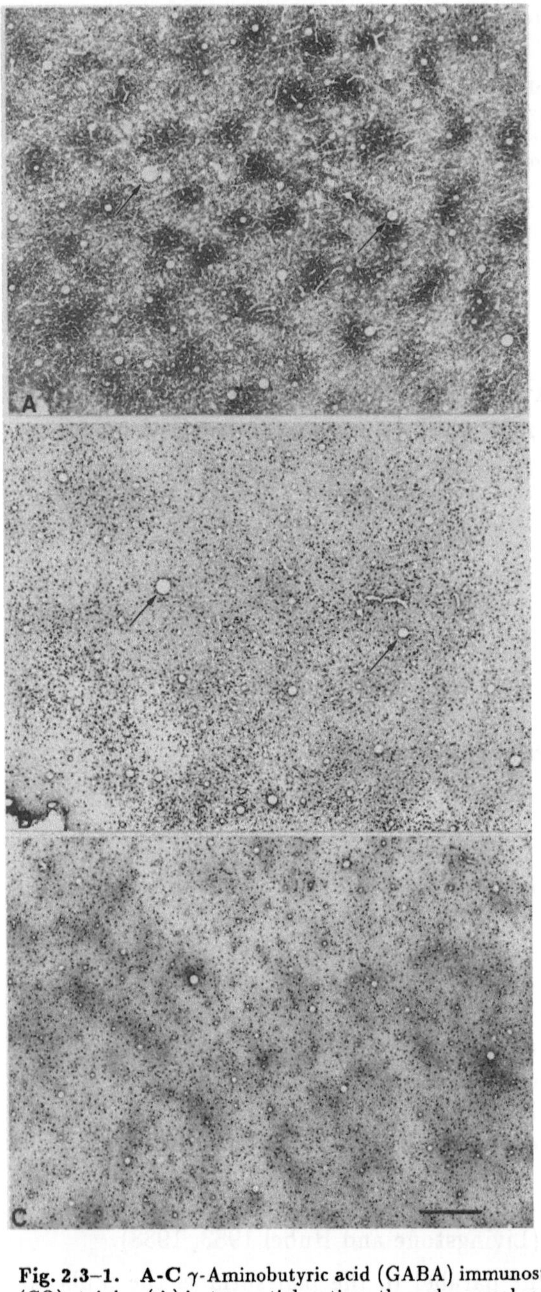

Fig. 2.3–1. **A-C** γ-Aminobutyric acid (GABA) immunostaining (**B**,**C**) and cytochrome oxidase (CO) staining (**A**) in tangential sections through normal monkey area 17. Comparison of adjacent sections stained for CO (**A**) and immunostained for GABA (**B**) show that immunoreactive somata are no denser in the puffs than in the surrounding interpuffs. *Arrows* indicate the same blood vessel profiles. However, when terminal immunostaining is apparent (**C**) patches are evident, which coincide with the positions of the CO puffs (not illustrated). *Scale bar*, 400 μm in **A** and **B**; 500 in **C**

Livingstone and Hubel 1984). Most puffs, in fact, occupy the centers of ocular dominance columns. From these data, the functionally distinct puffs and interpuffs can be seen as largely separate processing units for information falling upon each retina (Livingstone and Hubel 1987, 1988; Van Essen et al. 1992).

Studies of the neurotransmitters, receptors, and other molecules expressed by neurons in the CO-rich puffs and in the surrounding interpuffs have afforded two principal insights: (1) transmitter-specific populations and subpopulations preferentially occupy either puff or interpuff, suggesting that functional circuits in the two compartments are not equivalent and (2) neurochemical properties of puff and interpuff neurons are rapidly modified in the adult monkey by loss of input from one retina. These modifications are complex and specific for particular populations, suggesting they are part of a measured response by cortical neurons to compensate for the effects of monocular deprivation.

2.3.1 Normal Neurochemical Characteristics

With our current state of understanding, it is reasonable to divide cortical neurons into two, extremely broad transmitter-specific populations. The larger is composed of both pyramidal and spiny stellate neurons and appears to contain and use the excitatory amino acid glutamate or some derivative of glutamate as a neurotransmitter. The smaller population, comprising one out of five neurons in monkey area 17, contains the inhibitory amino acid GABA (Fitzpatrick et al. 1987; Hendry et al. 1987; Beaulieu et al. 1992). Neither the glutamate nor the GABA population is restricted to the puffs or interpuffs but, either as a whole or as subpopulations, each prefers one compartment over the other.

Inhibitory Circuits in Puffs and Interpuffs. GABA neurons are both the logical and historical start for studies of neurotransmitters in puffs and interpuffs. Immunostaining for GABA and its synthesizing enzyme, glutamic acid decarboxylase (GAD), in the supragranular layers of monkey area 17 reveals intensely stained periodic patches which correspond to the intensely stained CO puffs (Fig. 1C; Hendrickson 1985; Hendrickson et al. 1981). However, these patches are not regions of unusually high concentrations of GABA-immunoreactive somata; that is, GABA neurons are as common outside the puffs as in them (Fig. 1A,B; Fitzpatrick et al. 1987; Hendry et al. 1987; Beaulieu et al. 1992). The patches are, instead, collections of unusually large GABA terminals (Fitzpatrick et al. 1987). These are matched by aggregates of $GABA_A$ receptors, which are also more numerous, larger, or more immunoreactive within the puffs than in the surrounding interpuffs (Hendry et al. 1990).

Chemically distinct subpopulations of GABA neurons strongly prefer either puff or interpuff. Those GABA neurons which occupy the puffs are immunoreactive for the tachykinin neuropeptide family (Hendry et al. 1988a), for one calcium-binding protein, parvalbumin (Hendry et al. 1989; Van Broderode et al. 1990; Hendry and Carder 1990), and for a cell surface glycoprotein recognized by the monoclonal antibody, Cat-301 (Hendry et al. 1984, 1988b). In contrast, GABA neurons in interpuffs are immunoreactive for a second calcium-binding protein, calbindin (Fig. 2; Celio et al. 1986; Hendry et al. 1989; Van Brederode et al. 1990; Hendry and Carder 1993), and for a second neuropeptide, neuropeptide Y (Kuljis and Rakic 1989). These data suggest that either different classes of GABA neurons exist within puffs and interpuffs or that the same classes exhibit different neurochemical properties in the two compartments (see DeFelipe et al. 1990)

Excitatory Circuits in Puffs and Interpuffs. One strategy for examining the contribution of excitatory neurons to circuits in monkey area 17 is the localization of glutamate. Within layers II and III of monkey area 17, glutamate immunostaining is prominent within the puffs (Carder and Hendry 1993), as somata are more intensely immunostained and the neuropil is much denser in the puffs than in surrounding interpuffs. These results suggest that greater levels of excitatory neuronal elements exist in puffs, where they balance the prominent terminations of GABA neurons. Running radially through the puffs are bundles of apical dendrites, given off by pyramidal neurons in several layers, which are intensely immunostained for microtubule-associated protein 2 (Peters and Sethares 1991; Hendry and Bhandari 1992). As processes arising from excitatory cortical neurons (i.e., pyramidal neurons), these dendrites are in a position to sample from the afferent and intrinsic axons that terminate in the puffs.

2.3.2 Activity-Dependent Regulation in Puffs and Interpuffs

When vision in one eye is eliminated in adult monkeys, geniculocortical axons remain segregated into equal left- and right-eye dominance columns (Hubel et al. 1977; LeVay et al. 1980). However, the response to the intact eye is recorded in zones much wider than normal ocular dominance columns, while silent regions occupy narrow regions of cortex (LeVay et al. 1980). By these findings, the intact-eye columns appears to have undergone a functional expansion, despite the lack of geniculocortical axon sprouting. How is it possible to produce physiological changes in a hard-wired system? An unmasking of silent synapses or, more appropriately, a strengthening of previously ineffective synapses is usually offered as the best mechanism of using

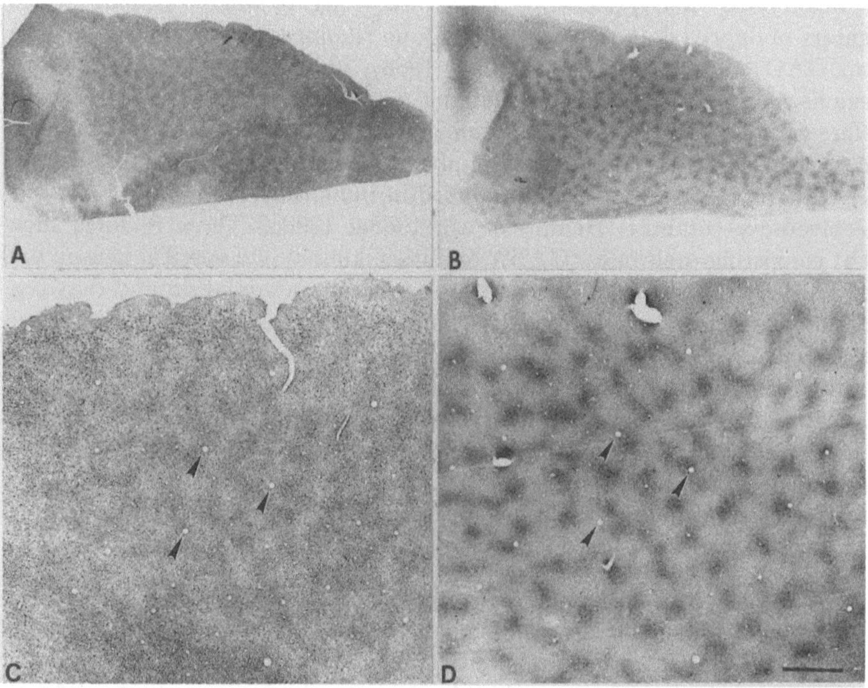

Fig. 2.3–2. **A-D** Calbindin immunostaining in normal monkey visual cortex. Comparison of calbindin immunostaining (**A,C**) and cytochrome oxidase (CO) staining (**B,D**) in adjacent sections shows intense immunostaining of somata and neuropil in the interpuffs. *Arrowheads* indicate the same blood vessel profiles in **C** and **D**. *Scale bar*, 1.8 mm in **A and B**; 500 μm in **C** and **D**

the available circuitry for novel functions; such unmasking or strengthening of synapses is certain to involve chemical changes in adult neurons.

Elimination of visual activity in one retina produces changes in the neurotransmitters, neurotransmitter enzymes and receptors, neuropeptides, second messenger molecules, cytoskeletal proteins, metabolic enzymes, and transcription factors in the adult monkey visual cortex (Benson et al. 1991; Carder and Hendry 1993; Chaudhuri and Cynader 1993; Graybiel and Ragsdale 1982; Hendrickson 1985; Hendry and Bhandari 1992; Hendry and Jones 1986, 1988; Hendry and Kennedy 1986; Hendry et al. 1988a, 1990; Horton 1984; Wong-Riley and Carroll 1984). Seen most dramatically in layer IVC, activity-dependent regulation of a broad range of molecules also occurs in the puffs and interpuffs of layers II and III. These changes are complex and involve both compartments and both sets of ocular dominance columns.

Reductions in Deprived-Eye Puffs. Neurons in puffs which lie at the centers of deprived-eye columns display no significant reductions in GABA and GAD (Hendry and Jones 1986, 1988), indicating that cells in these regions maintain something resembling normal GABA levels in the face of afferent inactivity. However, immunostaining for tachykinins, which coexist with GABA in puff neurons, and immunostaining for subunits of the $GABA_A$ receptor are markedly reduced in puffs that occupy the centers of deprived-eye columns (Hendry et al. 1988a, 1990). These findings show that coexisting molecules (GABA and tachykinins) in a cortical neuron can exhibit very different responses to a peripheral manipulation and that regulation of neurotransmitter (GABA) and receptor ($GABA_A$ subunits) need not occur in parallel. While GABA levels remain normal, the density of glutamate-immunoreactive somata and puncta and the intensity of those elements which persist are greatly reduced in deprived-eye puffs (Carder and Hendry 1993). Thus, the same manipulation elicits different reponses from neurons that contain these two amino acid neurotransmitters. As a result, an abnormal interaction between excitatory and inhibitory neurons might be expected in the deprived-eye puffs.

Reductions in Deprived-Eye Interpuffs. Although outside the regions innervated by geniculocortical axons, neurons of deprived-eye interpuffs display robust changes in immunoreactivity. In the deprived interpuffs, not only are tachykinin and $GABA_A$ receptor immunoreactivity reduced by elimination of retinal activity, but also GABA, GAD and calbindin immunoreactive levels are downregulated (Hendry and Jones 1986, 1988; Hendry et al. 1988a, 1990; Hendry and Carder 1992). These data demonstrate that the effects of monocular deprivation are transferred through intracortical circuits to neurons outside the regions of geniculocortical terminations. Moreover, the loss of visual drive appears to more keenly affect the GABA neurons of the interpuffs than those of the puffs (reduction in the former but not the latter), perhaps as a result of a competitive interaction in the interpuffs between elements previously driven by the deprived eye and those still driven by the normal eye. From such an interpretation, one would predict that puffs and interpuffs of the normal-eye columns also respond unequally to monocular deprivation (see below).

Changes in Normal-Eye Columns. Recent studies of neurotransmitter and protein immunoreactivity in normal-eye columns have shown that while the eye may be normal, the columns are not. That is, monocular deprivation leads to changes in immunocytochemically detectable levels of some proteins and peptides in the normal-eye column. The clearest result is the shift in calbindin immunoreactivity from the normal pattern of rich around the puffs and poor within the puffs to a decidedly abnormal pattern

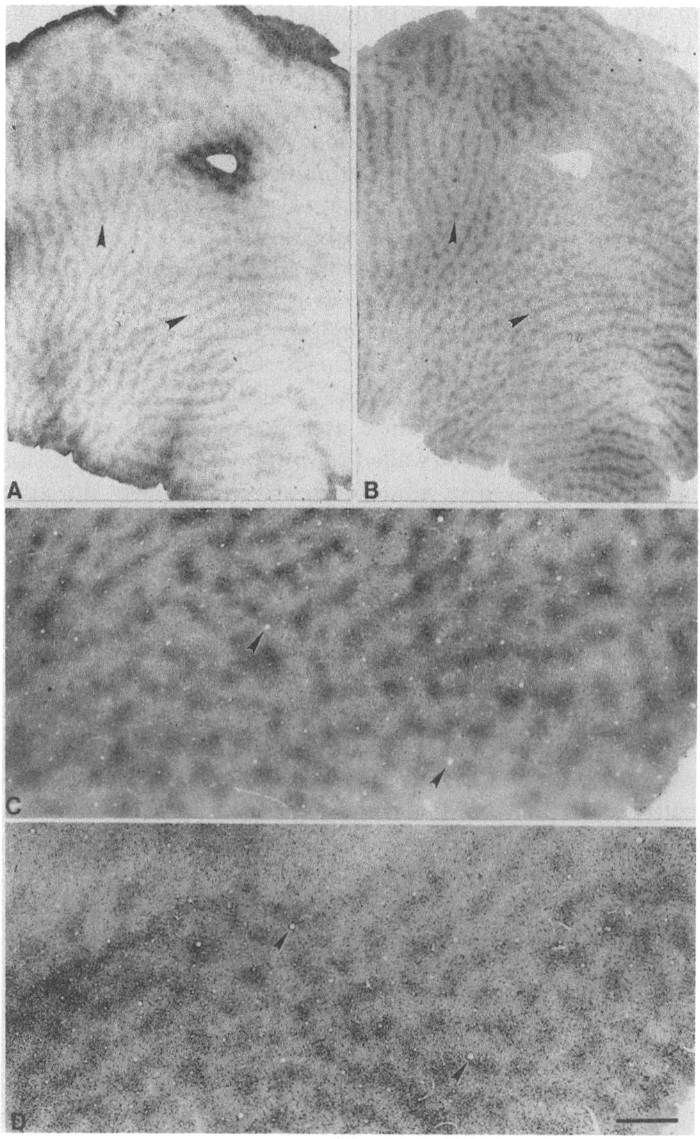

Fig. 2.3–3. **A-D** Calbindin immunostaining in adult monkeys monocularly deprived by the injection of tetrodotoxin into the vitreous body of one eye. Comparison of adjacent sections immunostained for calbindin (**A**,**D**) and stained for cytochrome oxidase (CO; **B**,**C**). Low magnification photomicrographs show intense immunostaining occupying the puffs in every other row, corresponding to the centers of normal-eye columns (*arrowheads* in **A** and **B**). High magnification photomicrographs (**C**,**D**) show that the puffs contain an abnormally high density of calbindin-positive somata. *Arrowheads* in **C** and **D** indicate the same blood vessel profiles. *Scale bar*, 2.5 mm in **A** and **B**; 600 μm in **C** and **D**

of very light around the puffs and very intense within the puffs (Fig. 3). The abnormally intense immunostaining of normal-eye puffs is the result of novel calbindin immunoreactivity within a subpopulation of puff neurons. However, those cells which express novel calbindin immunoreactivity are identical in size, laminar position, and neurochemical attributes to the calbindin cells normally found around the puffs. In the other compartment, the reduction in calbindin immunostaining in normal-eye interpuffs not only reinforces the finding that deprivation of one eye affects both sets of ocular dominance columns, it also more firmly establishes that what happens in one compartment need not occur in the other and can, in fact, be its opposite. Without a clear understanding of what calbindin does within a cortical neuron, any interpretation of these effects on the normal-eye column would be entirely speculative. At this point, the data provide evidence that in terms of their neurochemical attributes, neurons in adjacent ocular dominance columns are not walled off from one another, but that manipulations of one retina, even in adulthood, are felt by cells throughout the primate visual cortex.

2.3.3 Conclusions

The discovery of puffs in the monkey visual cortex helped cement a revolution in thought on the functional organization of the primate visual system. Detailed studies of the properties expressed by the neurons in and around the puffs indicate that puffs and interpuffs make up distinct compartments, with separate and unique inhibitory and excitatory circuits. Yet these circuits, or at least the molecules expressed by their individual elements, are capable of adapting to changes in retinal inputs. This adaptability of neurons is confined to neither puff nor interpuff, to neither deprived-eye column nor normal-eye column, but is shared by neurons at many separate levels in the analysis of binocular visual input.

References

Beaulieu C, Kisvárday Z, Somogyi P, Cynader M, Cowey A (1992) Quantitative distribution of GABA-immunopositive and -immunonegative neurons and synapses in the monkey striate cortex (area 17). Cerebral Cortex 2: 295–309

Benson DL, Isackson PJ, Gall CM, Jones EG (1991) Differential effects of monocular deprivation on glutamic acid decarboxylase and type II calcium-calmodulin-dependent protein kinase gene expression in the adult monkey visual cortex. J Neurosci 11: 31–47

Carder RK, Hendry SHC (1993) Neuronal characterization, compartmental distribution and activity-dependent regulation of glutamate immunoreactivity in adult monkey striate cortex. J Neurosci (in press)

Celio MR, Scharer L, Morrison JH, Norman AW, Bloom FE (1986) Calbindin immunoreactivity alternates with cytochrome oxidase-rich zones in some layers of the primate visual cortex. Nature 323: 715–717

Chaudhuri A, Cynader MS (1993) Activity-dependent expression of the transcription factor Zif 268 reveals ocular dominance columns in monkey visual cortex. Brain Res 605: 349–353
DeFelipe J, Hendry SHC, Hashikawa T, Molinari M, Jones EG (1990) A microcolumnar structure of monkey cerebral cortex revealed by immunocytochemical studies of double bouquet cell axons. Neurosci 37: 655–673
Fitzpatrick D, Itoh K, Diamond IT (1983) The laminar organization of the lateral geniculate body and the striate cortex in the squirrel monkey (*Saimiri sciureus*). J Neurosci 3: 673–702
Fitzpatrick D, Lund JS, Schmechel D, Towles AW (1987) Distribution of GABAergic neurons and axon terminals in the macaque striate cortex. J Comp Neurol 264: 73–91
Graybiel AM, Ragsdale CW Jr (1982) Pseudocholinesterase staining in the primary visual pathway of the macaque monkey. Nature 299 439–442
Hendrickson AE (1985) Dots, stripes and columns in monkey visual cortex. Trends Neurosci 8: 406–410
Hendrickson AE, Hunt SP, Wu J-Y (1981) Immunocytochemical localization of glutamic acid decarboxylase in monkey striate cortex. Nature 292: 605–607
Hendry SHC, Bhandari MA (1992) Neuronal organization and plasticity in adult monkey visual cortex: immunoreactivity for microtubule associated protein 2 (MAP 2). Vis Neurosci 9: 445–459
Hendry S, Carder RK (1992) Organization and plasticity of GABA neurons and receptors in monkey visual cortex. Prog Brain Res 90: 477–502
Hendry SHC, Carder RK (1993) Neurochemical compartmentation of monkey and human visual cortex: similarities and variations in calbindin immunoreactivity across species. Vis Neurosci 10: 1109–1120
Hendry SHC, Kennedy MB (1986) Immunoreactivity for a calmodulin-dependent protein kinase is selectively increased in macaque striate cortex after monocular deprivation. Proc Natl Acad Sci USA 83: 1536–1540
Hendry SHC, Jones EG (1986) Reduction in number of GABA immunostained neurons in deprived-eye dominance columns of monkey area 17. Nature 320: 750–753
Hendry SHC, Jones EG (1988) Activity-dependent regulation of GABA expression in the visual cortex of adult monkeys. Neuron 1: 701–712
Hendry SHC, Hockfield S, Jones EG, McKay R (1984) Monoclonal antibody that identifies subsets of neurons in the central visual system of monkey and cat. Nature 307: 267–269
Hendry SHC, Schwark HD, Jones EG, Yan J (1987) Numbers and proportions of GABA immunoreactive neurons in different areas of monkey cerebral cortex. J Neurosci 7: 1503–1520
Hendry SHC, Jones EG, Burstein N (1988a) Activity-dependent regulation of tachykinin-like immunoreactivity in neurons of the monkey primary visual cortex. J Neurosci 8: 1225–1238
Hendry SHC, Jones EG, Hockfield S, McKay RDG (1988b) Neuronal populations stained with the monoclonal antibody, Cat-301, in the mammalian cerebral cortex and thalamus. J Neurosci 8: 518–542
Hendry SHC, Jones EG, Emson PC, Lawson DEM, Heizmann CW, Streit, P (1989) Two classes of cortical GABA neurons defined by differential calcium binding protein immunoreactivities. Exp Brain Res 76: 467–472
Hendry SHC, Fuchs J, de Blas AL, Jones EG (1990) Distribution and plasticity of immunocytochemically localized $GABA_A$ receptors in adult monkey visual cortex. J Neurosci 10: 2438–2450
Horton JC (1984) Cytochrome oxidase patches: a new cytoarchitectonic feature of monkey visual cortex. Phil Trans R Soc Lond B 304: 199–253
Horton JC, Hubel DH (1981) Regular patchy distribution of cytochrome oxidase staining in primary visual cortex of macaque monkey. Nature 292: 762–764
Hubel DH, Wiesel TN, LeVay S (1977) Plasticity of ocular dominance columns in monkey striate cortex. Phil Trans R Soc Lond B 278: 377–409
Humphrey AL, Hendrickson AE (1983) Background and stimulus-induced patterns of high metabolic activity in the visual cortex (area 17) of the squirrel and macaque monkey. J Neurosci 3: 345–358
Kuljis RO, Rakic P (1989) Neuropeptide Y-containing neurons are situated predominantly outisde cytochrome oxidase puffs in macaque visual cortex. Vis Neurosci 2: 57–62
LeVay S, Wiesel TN, Hubel DH (1980) The development of ocular dominance columns in normal and visually deprived monkeys. J Comp Neurol 191: 1–51
Livingstone MS, Hubel DH (1982) Thalamic inputs to cytochrome oxidase-rich regions in monkey visual cortex. Proc Natl Acad Sci USA 79: 6098–6101
Livingstone MS, Hubel DH (1983) Specificity of cortico-cortical connections in monkey visual system. Nature 304: 531–534
Livingstone MS, Hubel DH (1984) Anatomy and physiology of a color system in the primate visual cortex. J Neurosci 4: 309–356

Livingstone MS, Hubel DH (1987) Psychophysical evidence for separate channels for the perception of form, color, movement and depth. J Neurosci 7: 3416–3468
Livingstone MS, Hubel DH (1988) Segregation of form, color, movement, and depth: anatomy, physiology, and perception. Nature 240: 740–749
Peters A, Sethares C (1991) The organization of pyramidal neurons in area 17 of monkey visual cortex. J Comp Neurol 306: 1–23
Tootell RBH, Hamilton SL, Switkes E (1988a) Functional anatomy of macaque striate cortex. IV. Contrast and magno-parvo streams. J Neurosci 8: 1594–1609
Tootell RBH, Silverman MS, Hamilton SL, De Valois RL (1988b) Functional anatomy of macaque striate cortex. V. Spatial frequency. J Neurosci 8: 1610–1624
Tootell RBH, Silverman MS, Hamilton SL, De Valois RL, Switkes E (1988c) Functional anatomy of macaque striate cortex. III. Color. J Neurosci 8: 1569–1593
Ts'o DY, Gilbert CD (1988) The organization of chromatic and spatial interactions in the primate striate cortex. J Neurosci 8: 1712–1727
Van Brederode JFM, Mulligan KA, Hendrickson AE (1990) Calcium-binding proteins as markers for subpopulations of GABAergic neurons in monkey striate cortex. J Comp Neurol 298: 1–22
Van Essen DC, Anderson CH, Felleman DJ (1992) Information processing in the primate visual system: an integrated systems perspective. Science 255: 419–423
Wong-Riley M, Carroll EW (1984) The effect of impulse blockade on cytochrome oxidase activity in the monkey visual system. Nature 307: 262–264

2.4 Studies on Regulation of Ocular Dominance Plasticity: Strategies and Findings

T. Kasamatsu
Smith-Kettlewell Eye Research Institute, San Francisco, California

2.4.1 Introduction: Activity Dependency

The study of developmentally regulated neural plasticity in mammalian visual cortex started 30 years ago when Wiesel and Hubel published a seminal paper on cortical effects of raising kittens with vision through one eye throughout early postnatal life (1963). They found that the majority of recorded cells in kitten visual cortex were exclusively responding to stimulation of the "experienced" eye, and that no response was obtained when the "deprived" eye was stimulated. The study of changes in ocular dominance or ocular dominance plasticity following monocular deprivation has quickly become the model of neurophysiological investigation of the age- and experience-dependent modification of neural connections in visual cortex. It was obvious from the beginning that binocular competition was a leading idea in understanding changes in ocular dominance, since little change was noted in the proportion of binocular cells when both eyes were lid-sutured (binocular deprivation, Wiesel and Hubel 1965).

2.4.2 Neurochemical Basis of Cortical Plasticity

If visual afferent is the sole input that activates visuocortical cells, activity dependency should also play a regulatory role, in addition to an executive one, in changes of ocular dominance. However, this is not quite so in the living brain: visuocortical cells are under the strong influence of many non-retinal inputs, constantly communicating with other cells, neuronal or glial, through various types of receptors, transporters, and ion channels. The necessity of non-retinal factors in the regulation of ocular dominance plasticity has been recognized for some time (e.g., Kasamatsu 1983). These are termed variously (e.g., central-core gating, Singer 1979; enabling factor, Cooper 1987), and here we call them categorically as neurochemical basis which operates primarily outside the retino-geniculo-occipital projection system.

In the brain the two elements, activity dependency as the specific and executive process and the neurochemical basis as its global regulatory mechanism, should be activated concurrently to bring about changes in ocular dominance following monocular deprivation. An example of this necessity was clearly shown in an early study in which a unique change in the ocular dominance distribution was found in noradrenaline (NA) infused visual cortex of visually normal kittens (Kuppermann and Kasamatsu 1984). The upshot of this study was that the same NA infusion induced no change if the kittens were kept in the dark during the NA infusion. (For further discussion on "activity dependence" and "cross-talk" between the two elements, see Kasamatsu 1991).

Among various neurochemical and cellular factors implicated until today as related to ocular dominance plasticity (e.g., γ-aminobutyric-acid type A ($GABA_A$) receptor, muscarinic receptors, astrocytes, *N*-methyl-D-aspartate (NMDA) receptors, nerve growth factor, and others), the NA-activated β-adrenoreceptor system has been studied most extensively. Therefore, I use this system as a prototype in reviewing below the strategies and findings in the study (Fig. 1).

2.4.2.1 Continuous Depletion of the Transmitter Pool

The study on the neurochemical basis of ocular dominance plasticity started with our first successful recording from visual cortex of kittens which had been injected daily with 6-hydroxydopamine (6-OHDA), a catecholamine (CA) related neurotoxin, into the lateral ventricle over a period of monocular lid suture for a few weeks (Kasamatsu and Pettigrew 1976). We found

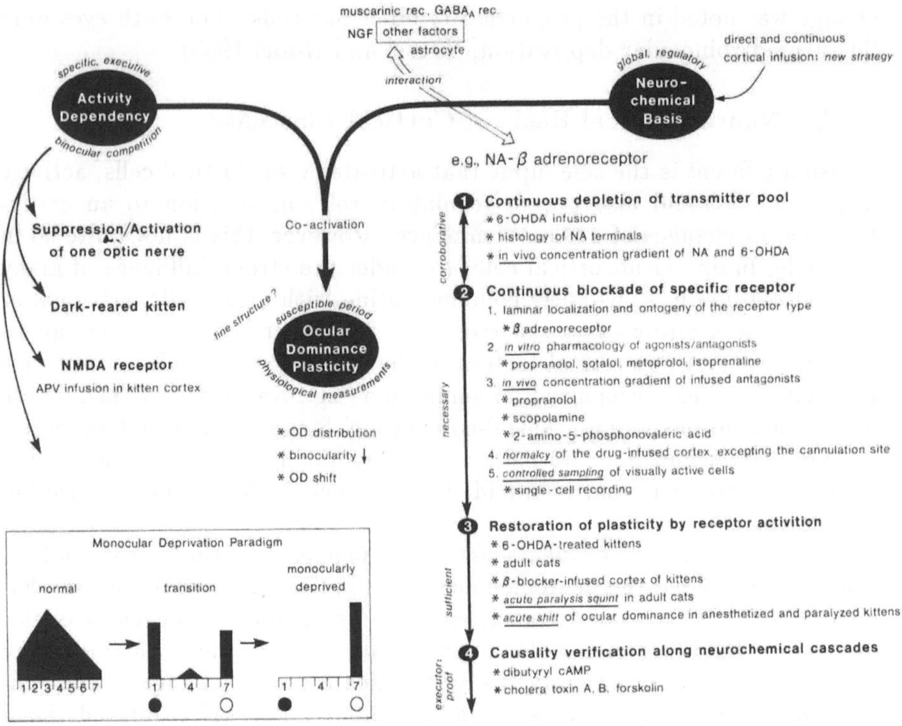

Fig. 2.4–1. Mechanisms underlying ocular dominance (OD) plasticity in visual cortex. We have proposed that the two processes, one determining ways in which binocular cells are stimulated by visual afferents, and the other regulating the cell's modifiability, should be concurrently set in action to change OD. *NA*, noradrenaline; *6-OHDA*, 6-hydroxydopamine; *cAMP*, adenosine 3',5'-cyclic monophosphate; *GABA*, gamma-aminobutyric-acid; *NGF*, nerve growth factor; *NMDA*, N-methyl-D-aspartate; *APV*, 2-amino-5-phosphonovaleric acid

that the expected shift in ocular dominance did not occur in the 6-OHDA injected kittens. Although this result was encouraging, a technical drawback was obvious here: we were not sure whether the disappearance of the cortical plasticity was due solely to the specific effect of 6-OHDA on CA terminals within visual cortex or to something else. Thus, we were driven to invent a new method of local and continuous infusion of 6-OHDA to limit the site of 6-OHDA action to visual cortex where the primary effect of monocular deprivation is thought to take place. The result of experiments with this new strategy was clearcut: in kitten visual cortex infused with 4 mM 6-OHDA in 0.4% ascorbate saline for a week, concurrently with monocular deprivation, the ocular dominance shift did not take place, and the proportion of binocular cells remained high as seen in the normal (Kasamatsu et al. 1979). The plasticity-reducing effect of 6-OHDA was indeed confined to a relatively small area within the infused hemisphere. This basic finding was replicated in other laboratories (Bear et al. 1983; Daw et al. 1983).

A few preconditions of the new strategy were also met experimentally in cat visual cortex. First, the presence of a dense network of CA terminals in kitten visual cortex was shown histologically by CA histofluorescence and electron microscopic methods (Itakura et al. 1981). Second, the extent of CA depletion or lesion of CA terminals was measured biochemically and histochemically in visual cortex infused continuously with 4 mM 6-OHDA for a week (Kasamatsu et al. 1981).

More importantly, the "critical concentration" of 6-OHDA at a recording site was estimated by determining an in vivo concentration gradient of 6-OHDA in the drug-infused hemisphere. In a cortical area where we found a large decrease in cortical plasticity (see Sect. 2.4.2.2 (5)), an estimated concentration of 6-OHDA over 7 days was as low as 6×10^{-6} M, indicating that 4 mM 6-OHDA supplied continuously at the infusion site was diluted at least ~ 700 times at a recording site to maintain an extracellular concentration close to the K_m (Michaelis-Menten constant) value of 6-OHDA (Kasamatsu et al. 1981).

The necessity of "continuous depletion" of the endogenous pool of CAs or nearly total lesion of CA terminals was well illuminated by the repeated failure of other investigators to reproduce the original finding when the extent of CA depletion was found insufficient, as evidenced by the significant amount of CAs biochemically recovered from the cortical tissue presumably affected by 6-OHDA (Bear and Daniels 1983; Bear et al. 1983; Daw et al. 1984, 1985a,b; Adrien et al. 1986; Trombley et al. 1986). The issue was especially critical for the proper interpretation of the findings in these experiments, in which 6-OHDA was initially injected either intraventricularly or systemically, and physiological recording was made some time after the end of the 6-OHDA treatment. To make the matter more complicated, in 6-OHDA infused kitten visual cortex, in which we first established the presence of specific degeneration of CA terminals near the infusion site, a few weeks later we found the reappearance of histochemically identified CA terminals and concurrent increases in endogenous NA and dopamine. The presence of regenerative CA terminals was undeniable, and the regeneration reached its near completion in ~ 24 weeks after the end of the initial 6-OHDA cortical infusion (Nakai 1987; Nakai et al. 1987).

2.4.2.2 Continuous Blockade of Specific Receptor

If a significant decrease in endogenous NA is responsible for the disappearance of cortical plasticity from the 6-OHDA infused visual cortex of kittens, essentially the same result should be obtained by continuous blockade of NA-related receptors in kitten visual cortex in which the CA terminals are left intact. This expectation was fulfilled when we obtained high binocularity in kitten visual cortex that had been infused with propranolol, a β-adrenoreceptor antagonist, during a week of monocular deprivation (Kasamatsu and Shirokawa 1985). The plasticity-suppressing effect of propranolol was concentration dependent ($\sim 6 \times 10^{-7}$ M for the half-maximum effect) and not due to its "local anesthetic" effects (Shirokawa and Kasamatsu 1986). The stereospecificity of propranolol in this effect was determined (Shirokawa and Kasamatsu 1987). The adequacy of our interpretation of the above findings was tested in the following five areas.

(1) *Afferents and Related Receptors: Laminar Distribution and Ontogeny.* The sensitivity of immature visuocortical cells to monocular deprivation varies depending on the cortical laminae (Shatz and Stryker 1978; Freeman and Olson 1982) and the presence of the high-susceptibility period in early postnatal life of kittens has been firmly established (Hubel and Wiesel 1970; Olson and Freeman 1980). Therefore, a candidate of the regulatory mechanism of ocular dominance plasticity must satisfy these conditions. In keeping with the above characteristics of the plasticity, the laminar distribution of both CA terminals, the majority of which are NA-containing ones, and β-adrenoreceptor binding sites is relatively high outside the granular layer and lowest in the granular layer (Itakura et al. 1981; Nakai et al. 1987). Though endogenous NA accumulates monotonously toward adulthood in cat visual cortex, the number of β-adrenoreceptor binding sites reaches the adult value around the 5th week and remains high throughout the 11th week, roughly corresponding to the ontogeny curve of the physiologically determined susceptibility (Jonsson and Kasamatsu 1983; see also Shaw et al. 1986, for a related finding.)

(2) *In Vitro Pharmacology of Agonists and Antagonists.* The study of physiological effects of cortically infused β-adrenoreceptor agonists/antagonists is very much facilitated by the accumulated knowledge on basic pharmacology of various kinds of agonists and antagonists and their pharmacokinetics in the rodent brain. In our physiological assay of the plasticity-suppressing effect of propranolol, for example, a part of its effects was reproduced by using sotalol, a β-blocker without "local anesthetic" effects (Shirokawa and Kasamatsu 1986) and metoprolol, an antagonist of β_1-adrenoreceptors (Shirokawa et al. 1989). On the other hand, tests of the plasticity-enhancing effects of the agonists are limited mainly to NA, and the similar effect of isoprenalin, for example, was confirmed only in a preliminary study (unpublished data).

(3) *In Vivo Concentration Gradient of Cortically Infused Antagonists.* As with 6-OHDA and NA, the in vivo concentration gradient of continuously infused propranolol was determined directly in cat visual cortex (Shirokawa and Kasamatsu 1986). Combined with the data obtained by the systematic sampling of individual cells (see Sect. 2.4.2.2 (5)), the critical concentration of propranolol at the recording site to obtain the half-maximum effect was found as low as $\sim 6 \times 10^{-7}$ M.

In addition to β-adrenoreceptors, the determination of the in vivo concentration was carried out for an antagonist of muscarinic cholinergic receptors and that of NMDA receptors. In the former the in vivo concentration gradient of continuously infused scopolamine was determined directly in cat visual cortex (Imamura and Kasamatsu 1989). The estimated concentration of scopolamine, 4 mM at the infusion site, for the half-maximum effect turned out to be at 4–13 x 10^{-5} M 2–4 mm from the infusion site, the value being 172 times higher than that of propranolol. An in vivo dilution factor of cortically infused 50 mM 2-amino-5-phosphonovaleric acid (APV, an NMDA receptor antagonist) over 4 days was calculated based on measurements of in vitro accumulation of APV by cortical slices carved from an area with a known distance from the APV infusion center and soaked up to 18 h (to reach the saturation) in a medium containing 150 μM APV (Bear et al. 1990). This measurement was totally inadequate for obtaining a dilution factor of 50 mM APV directly into the cortex over a period of 7 days since the maximal extracellular (i.e., bath) concentration of APV to which visuocortical cells were exposed was set at 150 μM, 333 times lower than 50 mM. Accordingly, a near-exponential decrease of the APV concentration close to the infusion site, the common behavior of various extracellularly injected chemical solutions, was not obtained in this study. Therefore, the critical concentration of APV for the significant decrease in cortical plasticity reported in their study might have been maximally 333 times higher than 20 μM, the value mentioned in the summary of the paper. The specificity of locally infused 50 mM APV has not been established yet.

(4) *Normalcy of the Drug-Infused Cortex.* Within a sphere of the drug-affected cortical mass, we should have independent evidence for the presence of normal cortical tissue outside the cannulation site and its immediate neighbor that may have been saved from the direct tissue damage of the cannulation yet affected by non-specific effects of a highly concentrated drug infused directly. The subject was discussed extensively in early studies on the specificity of directly injected 6-OHDA into the brain tissue (e.g., Hökfelt and Ungerstedt 1973). In our studies the matter has been always dealt with in two ways: (a) physiological assessment of receptive-field properties (such as orientation and direction selectivity, velocity preference, responsiveness to an optimally oriented moving light bar, ocular dominance, and others) of cells systematically recorded along a few recording tracks that had a known distance from the infusion center; (b) the reconstruction of recording tracks on histological sections stained for Nissl substances (and cytochrome oxidase, lately). A chance of encountering visually unresponsive cells and cells with abnormal receptive-field properties was higher than otherwise when a given track passed through an area that showed an abnormal appearance in Nissl histology.

(5) *Controlled Sampling of Visually Active Cells.* The currently employed assay method of ocular dominance plasticity is to compare visual responsiveness of single cells to stimulation of the monocularly deprived eye with that of the normal eye of the same animal. The premise of this assay is that receptive-field properties of visuocortical cells remain normal except for changes in ocular dominance since the animal receives proper visual stimulation through the normal, undeprived eye. On the other hand, binocular deprivation in young kittens is known to result in a measurable increase in the proportion of visually unresponsive cells and responsive cells with abnormal receptive-field properties, though binocularity remains relatively high (Wiesel and Hubel 1965). Therefore, the effect of "de facto binocular deprivation" is the commonest pitfall to be cleared before high

binocularity in visual cortex of the monocularly deprived preparation is taken as a genuine effect of any treatment received by such a cortex. In the present paradigm, this issue is very acute since the continuous infusion of a drug, through mechanical damage of cannulation or non-specific effects of the infused drug, may reduce the activity of visuocortical cells indiscriminatively, independently of which of the two eyes is visually deprived. One of preconditions necessary for remaining free from the effect of de facto binocular deprivation is to limit the sampling of visually active cells to a cortical area in which an estimated in vivo concentration of a locally infused drug is low and within the range for its specific action. This wish would impose on experimenters a rigid sampling practice to accumulate enough number of cells along a few recording tracks at most.

In addition, because the strength of ocular dominance plasticity differs in different cortical laminae, each recording track should go through the whole cortical depth collecting a comparable number of cells from the three sublaminae (supra-, infragranular, and granular layers) or at least making the number of recorded cells in the three subpopulations comparable between the experimental and control hemispheres.

Following the above thought, we have established the practice of accumulating 30 visually responding cells with an average intercellular spacing of $\sim 100\,\mu$m along a single track of ~ 3 mm or longer which is placed in the medial bank of the posterolateral gyrus obliquely through the three sublaminae, hopefully crossing the ocular dominance boundaries more than a few times (if the penetration is in the normal cortex). If the cell sampling is made by poking a small cortical area randomly, one may need to collect at least 300 cells, i.e., a sample size one order of magnitude larger than ours. This is impractical if not impossible to carry out. Some contradictory findings in the field, for example those about effects of APV infusion on ocular dominance plasticity (see Miller et al. 1989; Bear et al. 1990), may well be explained by the lack of the rigid sampling procedure in the latter.

2.4.3 Restoration of Cortical Plasticity by Receptor Activation

The two steps described above, continuous depletion of the transmitter pool and continuous blockade of related receptors, may be thought as reasonable tests of the "necessary" condition for the NA-activated β-adrenoreceptor system to regulate ocular dominance plasticity. Then, to restore the plasticity by selective activation of β-adrenoreceptors should comprise a "sufficient" test. The issue has been explored in the following five types of preparations.

(1) *6-OHDA Treated Kittens.* In an early study we showed that continuous cortical infusion with $\sim 50\,\mu$M NA restored ocular dominance plas-

ticity to kitten visual cortex which had been rendered aplastic due to the preceding injections of 6-OHDA into the lateral ventricle (Pettigrew and Kasamatsu 1978). The finding indeed indicated a specific role of NA within visual cortex among the two major biogenic CAs in the brain (i.e., NA and dopamine), since plastic changes in ocular dominance were restored only in a region of visual cortex affected by NA and not outside the region in the same kittens. The critical concentration of NA in the restoration of the cortical plasticity was calculated as 10^{-7} M based on a dilution factor of 1/500 at a recording site 2.5 mm from the center of the infusion (Kasamatsu et al. 1979).

(2) *Adult Cats.* In visual cortex of adult cats which received the continuous cortical infusion of NA, we found that brief monocular deprivation reduced the proportion of binocular cells significantly (Kasamatsu et al. 1979). This finding was interpreted as showing the initial stage of ocular dominance changes following monocular deprivation since a similar change in the ocular dominance distribution was customarily seen in visual cortex of freely behaving kittens after very brief monocular lid suture (1 day or less).

Essentially the same result was obtained when the normal adult cat was stimulated monocularly briefly (2 h a day for 6 consecutive days) combined with electrical stimulation of the locus coeruleus, a brainstem nucleus from which the central NA fibers originate (Kasamatsu et al. 1985).

(3) *Propranolol-Infused Visual Cortex of Kittens.* In kitten visual cortex infused with propranolol for a week, monocular lid suture in the second week usually resulted in little change in the ocular dominance distribution, indicating that the plasticity-suppressing effect of the β-blocker lasts for at least a week after the end of its infusion. In the similarly treated cortex, however, we found a clear sign of shift in ocular dominance if monocular deprivation was combined with the cortical infusion of NA for the second week (Shirokawa and Kasamatsu 1987).

(4) *"Acute Paralysis Squint" in Adult Cortex.* Previously we noted that the proportion of binocular cells was very low in cat visual cortex injected iontophoretically with concentrated NA throughout an overnight experiment under general anesthesia and paralysis (Heggelund and Kasamatsu 1981). The phenomenon was termed the "acute paralysis squint" effect since it resembled cortical effects of artificially created squint either surgically (Hubel and Wiesel 1965) or optically (Van Sluyters and Levitt 1980).

In a subsequent study (Heggelund et al. 1987), we extended this observation in the NA-infused visual cortex of adult cats. Under general anesthesia and paralysis, either of the two forms of the ocular dominance distribution, one with the normal pattern and the other dominated by group 1 and group 7 monocular cells, was readily obtained from the same cortex, depending on convergence or divergence of the visual axes of the two eyes controlled by a prism setting, which in turn kept paired receptive fields superimposed or widely separated. The effect reached a peak in about 20 h while the two eyes were stimulated with moving patterns before an assay of ocular dominance. These data indicate that the horizontal disparity up to several degrees, which is usually found in visual cortex of acutely anesthetized and paralyzed cats, is normally tolerated by binocular cells since the cortical susceptibility to the squint effect is maintained below a threshold. The threshold is, however, readily lowered through the activation of β-adrenoreceptors by NA infusion.

(5) *"Acute Shift" of Ocular Dominance in Anesthetized and Paralyzed Kittens.* Applying the same reasoning and strategies to kitten visual cortex, we have further shown that cortical NA infusion restores ocular dominance plasticity to the aplastic cortex of anesthetized and paralyzed kittens. The shift in ocular dominance toward the stimulated eye, an unmistakable signature of ocular dominance plasticity, with a marked reduction in binocularity (down to ~ 0.37) was attained following monocular exposure to TV shows for about 20 h (Imamura and Kasamatsu 1991a). Importantly, the plasticity-enhancing effect of (–)-NA was not shared by (+)-NA, a biologically inert stereoisomer of NA (Imamura and Kasamatsu 1988). In the same study, neither (–)-NA infusion into adult visual cortex combined with moving-pattern stimuli nor (–)-NA infusion into kitten visual cortex combined with monocular exposure to electronic noise on the same TV screen for about 20 h induced the shift in ocular dominance. Thus, to be consistent with our original thesis, we confirmed that both the elevated plasticity by activation of β-adrenoreceptors and strong binocular imbalance of cortical afferents are necessary to cause changes in ocular dominance.

Despite all the evidence presented above for the plasticity-enhancing effect of β-adrenoreceptor activation, any conclusion based on the comparison of pooled data between different conditions in a population study is prone to "sampling bias". Therefore, to push our theme one step further, we are currently examining how strongly and how rapidly one can change ocular dominance of single cells while physiological recording is made from them in NA-infused visual cortex of kittens ("on-line" modification). This matter has been discussed extensively elsewhere (Imamura and Kasamatsu 1991b) and a preliminary finding presented (Kasamatsu et al. 1993).

2.4.4 Verification of Causality Along Neurochemical Cascades

The causal link of the activation of β-adrenoreceptors with the elevation of ocular dominance plasticity beyond the usual threshold may be further strengthened. For example, one may directly stimulate adenylate cyclase, an enzyme crucial for the in vivo production of adenosine 3',5'-cyclic monophosphate (cAMP), by cortical infusion of aplastic visual cortex with either forskolin or cholera toxin. The former agent is known to work as a potent activator of adenylate cyclase in a G-protein independent manner and the latter in a G-protein dependent manner. We may also directly infuse the aplastic cortex with dibutyryl cAMP to increase an intraneuronal pool of cAMP. These pharmacological manipulations are combined with monocular lid suture, as done before, to examine how the infused drug affects the ocular dominance distribution. Thus, relying on the well-established biochemical cascades in the regulation of the protein kinase A system, we are now capable of probing neurochemical mechanisms underlying ocular dominance plasticity inside neuronal membrane. Correlational evidence has also been shown for the involvement of protein kinase C in ocular dominance plasticity (Sheu et al. 1990).

In kitten visual cortex infused with 6-OHDA for a week, we have shown that the subsequent infusion of such an aplastic cortex with 10^{-6} M dibutyryl cAMP clearly restored the cortical sensitivity to monocular deprivation, resulting in shift in ocular dominance (Kasamatsu 1980, 1983). The aplastic visual cortex of adult cats also served as a model in this matter. Again, in our preliminary study, we have obtained a significant decrease in binocularity in adult visual cortex infused with either forskolin or cholera toxin (unpublished observation).

2.4.5 Conclusions

Taking the NA-activated β-adrenoreceptor system as a model, I have reviewed how we proceeded to identify β-adrenoreceptors as a key element in the regulation of ocular dominance plasticity. The story, however, does not end here for the following two reasons. First, since more than a few second-messenger systems are biochemically identified, being functionally connected to various receptors and coupled among themselves, it is urgent to elucidate how different receptors interact in this matter. An example of such an effort is seen in our study on the role of muscarinic receptors, though secondary to that of β-adrenoreceptors, in the regulation of ocular dominance plasticity (Imamura and Kasamatsu 1989; Kasamatsu et al. 1991). Second, activation of second messengers, especially a cAMP-dependent one, inevitably led us to phosphorylation of phosphoproteins in the cell organelles. At the moment, we do not know much about likely target phosphoproteins because of the lack of a clear idea of biochemical/morphological processes which necessarily cause synaptic facilitation and synaptic elimination at the final stage of the execution of ocular dominance plasticity. A notable exception is, however, an intriguing suggestion of Aoki and Siekevitz (1985). They biochemically studied in kitten visual cortex the ontogenic change in the cAMP-stimulatable phosphorylation of a phosphoprotein, microtubule-associated protein 2 (MAP 2) and effects of dark rearing on it. Though the data points are few, their data indicate that the in vitro phosphorylation of MAP 2 was high in visual cortex of normal kittens during the postnatal susceptible period, and that the values were relatively reduced by dark rearing since birth. Molecular mechanisms regulating ocular dominance plasticity still stimulate our imagination.

Acknowledgments. Supported by USPHS grant EY 06733 and core grant EY 06883.

References

Adrien J, Blanc G, Buisseret P, Frégnac Y, Gary-Bobo E, Imbert M, Tassin J-P, Trotter Y (1986) Noradrenaline and functional plasticity in kitten visual cortex: a re-examination. J Physiol (Lond) 367: 73–98

Aoki C, Siekevitz P (1985) Ontogenetic changes in the cyclic adenosine 3',5'-monophosphate stimulatable phosphorylation of cat visual cortex proteins, particularly of microtubule-associated protein 2 (MAP 2): effects of normal and dark rearing and of the exposure to light. J Neurosci 5: 2465–2483

Bear MF, Daniels JD (1983) The plastic response to monocular deprivation persists in kitten visual cortex after chronic depletion of norepinephrine. J Neurosci 3: 407–416

Bear MF, Paradiso MA, Schwartz M, Nelson SB, Carnes KM, Daniels JD (1983) Two methods of catecholamine depletion in kitten visual cortex yield different effects on plasticity. Nature 302: 245–247

Bear MF, Kleinschmidt A, Gu Q, Singer W (1990) Disruption of experience-dependent synaptic modifications in striate cortex by infusion of an NMDA receptor antagonist. J Neurosci 10: 909–925

Cooper LN (1987) Cortical plasticity: theoretical analysis, experimental results. In: Rauschecker JP, Marler PR (eds) Imprinting and cortical plasticity, comparative aspects of sensitive periods. Wiley, New York, pp 177–192

Daw NW, Rader RK, Robertson TW, Ariel M (1983) Effects of 6-hydroxydopamine on visual deprivation in the kitten striate cortex. J Neurosci 3: 907–914

Daw NW, Robertson TW, Rader RK, Videen TO, Coscia CJ (1984) Substantial reduction of cortical noradrenaline by lesions of adrenergic pathway does not prevent effects of monocular deprivation. J Neurosci 4: 1354–1360

Daw NW, Videen TO, Parkinson D, Rader RK (1985a) DSP-4 (N-(2-chloroethyl)-N-ethyl-2-bromobenzylamine) depletes noradrenaline in kitten visual cortex without altering the effects of monocular deprivation. J Neurosci 5: 1925–1933

Daw NW, Videen TO, Rader RK, Robertson TW, Coscia CJ (1985b) Substantial reduction of noradrenaline in kitten visual cortex by intraventricular injections of 6-hydroxydopamine does not always prevent ocular dominance shifts after monocular deprivation. Exp Brain Res 59: 30–35

Freeman RD, Olson C (1982) Brief periods of monocular deprivation in kittens: effects of delay prior to physiological study. J Neurophysiol 47: 139–150

Heggelund P, Kasamatsu T (1981) Exogenous noradrenaline increases the neuronal platicity in cat visual cortex: localized, continuous microperfusion and iontophoresis. In: Fehér O, Joó F (eds) Cellular analogues of conditioning and neural plasticity, Adv Physiol Sci, vol 36. Pergamon Press, Oxford, pp 233–242

Heggelund P, Imamura K, Kasamatsu T (1987) Reduced binocularity in the noradrenaline infused striate cortex of acutely anesthetized and paralyzed, otherwise normal cats. Exp Brain Res 68: 593–605

Hökfelt T, Ungerstedt U (1973) Specificity of 6-hydroxydopamine induced degeneration of central monoamine neurons: an electron and fluorescence microscopic study with special reference to intracerebral injection on the nigro-striatal dopamine system. Brain Res 60: 269–297
Hubel DH, Wiesel TN (1965) Binocular interaction in striate cortex of kitten reared with artificial squint. J Neurophysiol 28: 1041–1059
Hubel DH, Wiesel TN (1970) The period of susceptibility to the physiological effects of unilateral eye closure in kittens. J Physiol (Lond) 206: 419–436
Imamura K, Kasamatsu T (1988) Acutely induced shift in ocular dominance during brief monocular exposure: effects of cortical noradrenaline infusion. Neurosci Lett 88: 57–62
Imamura K, Kasamatsu T (1989) Interaction of noradrenergic and cholinergic systems in regulation of ocular dominance plasticity. Neurosci Res 6: 519–536
Imamura K, Kasamatsu T (1991a) Ocular dominance plasticity restored by NA infusion to aplastic visual cortex of anesthetized and paralyzed kittens. Exp Brain Res 87: 309–318
Imamura K, Kasamatsu T (1991b) Ocular dominance plasticity: usefulness of anesthetized and paralyzed preparations. Japan J Physiol 41: 521–549
Itakura T, Kasamatsu T, Pettigrew JD (1981) Norepinephrine-containing terminals in kitten visual cortex: laminar distribution and ultrastructure. Neuroscience 6: 159–175
Jonsson G, Kasamatsu T (1983) Maturation of monoamine neurotransmitters and receptors in cat occipital cortex during postnatal critical period. Exp Brain Res 50: 449–458
Kasamatsu T (1980) A possible role for cyclic nucleotides in plasticity of visual cortex. Soc Neurosci Abstr 6: 494
Kasamatsu T (1983) Neuronal plasticity maintained by the central norepinephrine system in the cat visual cortex. In Sprague JM, Epstein AN (eds) Progress in psychobiology and physiological psychology, vol 10. Academic Press, New York, pp 1–112
Kasamatsu T (1991) Adrenergic regulation of visuocortical plasticity: a role of the locus coeruleus system. In: Barnes CD, Pompeiano O (eds) Progress in brain research, vol 88. Elsevier, Amsterdam, pp 599–616
Kasamatsu T, Pettigrew JD (1976) Depletion of brain catecholamines: failure of ocular dominance shift after monocular occlusion in kittens. Science 194: 206–209
Kasamatsu T, Shirokawa T (1985) Involvement of β adrenoreceptors in the shift of ocular dominance after monocular deprivation. Exp Brain Res 59: 507–514
Kasamatsu T, Pettigrew JD, Ary M (1979) Restoration of visual cortical plasticity by local microperfusion of norepinephrine. J Comp Neurol 185: 163–182
Kasamatsu T, Itakura T, Jonsson G (1981) Intracortical spread of exogenous catecholamines: effective concentration for modifying cortical plasticity. J Pharmacol Exp Ther 217: 841–850
Kasamatsu T, Watabe K, Heggelund P, Schöller E (1985) Plasticity in cat visual cortex restored by electrical stimulation of the locus coeruleus. Neurosci Res 2: 365–386
Kasamatsu T, Ohashi T, Imamura K (1991) Lithium reduces ocular dominance plasticity in kitten visual cortex. Brain Res 558: 157–162
Kasamatsu T, Niiyama K, Imamura K. Watanabe Y (1993) Rapid changes in ocular dominance of single cells in noradrenaline-infused visual cortex of kittens. Invest Ophthal Vis Sci 34: 792
Kuppermann BD, Kasamatsu T (1984) Enhanced binocular interaction in the visual cortex of normal kittens subjected to intracortical norepinephrine perfusion. Brain Res 302: 91–99
Miller KD, Chapman B, Stryker MP (1989) Visual responses in adult cat visual cortex depend on N-methyl-D-aspartate receptors. Proc Natl Acad Sci USA 86: 5183–5187
Nakai K (1987) Regenerative catecholamine-containing terminals in kitten visual cortex: an ultrastructural study. Neurosci Res 4: 475–485
Nakai K, Jonsson G, Kasamatsu T (1987) Norepinephrinergic reinnervation of cat occipital cortex following localized lesions with 6-hydroxydopamine. Neurosci Res 4: 433–453
Olson CR, Freeman RD (1980) Profile of the sensitive period for monocular deprivation in kittens. Exp Brain Res 39: 17–21
Pettigrew JD, Kasamatsu T (1978) Local perfusion of noradrenaline maintains visual cortical plasticity. Nature 271: 761–763
Shatz C, Stryker MP (1978) Ocular dominance in layer IV of the cat's visual cortex and the effects of monocular deprivation. J Physiol (Lond) 281: 267–283
Shaw C, Wilkinson W, Cynader M, Needler MC, Aoki C, Hall SE (1986) The laminar distributions and postnatal development of neurotransmitter and neuromodulator receptors in cat visual cortex. Brain Res Bull 16: 661–671
Sheu F-S, Kasamatsu T, Routtenberg A (1990) Protein kinase C activity and substrate (F 1/GAP-43) phosphorylation in developing cat visual cortex. Brain Res 524: 144–148

Shirokawa T, Kasamatsu T (1986) Concentration-dependent suppression by β-adrenergic antagonists of the shift in ocular dominance following monocular deprivation in kitten visual cortex. Neuroscience 18: 1035–1046
Shirokawa T, Kasamatsu T (1987) Reemergence of ocular dominance plasticity during recovery from the effects of propranolol infused in kitten visual cortex. Exp Brain Res 68: 466–476
Shirokawa T, Kasamatsu T, Kuppermann BD, Ramachandran VS (1989) Noradrenergic control of ocular dominance plasticity in the visual cortex of dark-reared cats. Dev Brain Res 47: 303–308
Singer W (1979) Central-core control of visual-cortex functions. In: Schmitt FO, Worden FG (eds) The neuroscience, fourth study program. MIT Press, Cambridge, MA, pp 1093–1109
Trombley P, Allen EE, Soyke J, Blaha CD, Lane RF, Gordon B (1986) Doses of 6-hydroxydopamine sufficient to deplete norepinephrine are not sufficient to decrease plasticity in the visual cortex. J Neurosci 6: 266–273
Van Sluyters RC, Levitt FB (1980) Experimental strabismus in the kitten. J. Neurophysiol 43: 686–699
Wiesel TN, Hubel DH (1963) Single-cell responses in striate cortex of kittens deprived of vision in one eye. J Neurophysiol 26: 1003–1017
Wiesel TN, Hubel DH (1965) Comparison of the effects of unilateral and bilateral eye closure on cortical unit responses in kittens. J Neurophysiol 28: 1029–1040

2.5 Homosynaptic Long-Term Depression and Its Relationship to Long-Term Potentiation in the Rat Neocortex In Vitro

A. Artola, S. Bröcher, T. Hensch, and W. Singer
Max Planck Institute for Brain Research, Frankfurt am Main, Germany

2.5.1 Introduction

Use-dependent, long-term changes of synaptic efficacy are thought to form a basis for learning and memory. The discovery of long-term potentiation (LTP; Bliss and Lømo 1973) provided a phenomenon that incorporates most of the features expected for a form of synaptic modification appropriate for learning and memory. In particular, LTP is input specific and, in the majority of cases, associative, i.e., modifications are restricted to the activated inputs and synapses potentiate only if they are active while the respective dendrite is sufficiently depolarized (for review see Bliss and Collingridge 1993).

Theoretical considerations suggested that these modifications should also include a decrease of synaptic strength. Long awaited, long-term depression (LTD) has only been demonstrated very recently. The first evidence came from studies in the cerebellum (Ito and Kano 1982; Ito et al. 1982; Sakurai 1987; Kano and Kato 1987). Later, LTD was also demonstrated in the hippocampus (Stanton and Sejnowski 1989; Chattarji et al. 1989). However, in both cases, induction of LTD required that the stimulation of the modifiable pathway was combined with activation of other inputs. In the cerebellum, the synaptic response of Purkinje cells (PC) to parallel fiber (PF) stimulation was only depressed when PFs were coactivated with climbing fibers (CF), which also converge on PC. In the hippocampus, tetanic stimulation of subicular afferents led to homosynaptic LTD, but only when the Schaffer collaterals were additionally activated in alternation. That a synaptic connection can undergo LTD following its own tetanic activation alone has only been demonstrated very recently, first in the neocortex (Artola et al. 1990a; Hirsch and Crépel 1990; see also Bindman et al. 1988) and then in the hippocampus (Dudek and Bear 1992; Mulkey and Malenka 1992; see also Dunwiddie and Lynch 1978), the striatum (Calabresi et al. 1992a) and the nucleus accumbens (Pennartz et al. 1993). This form of LTD is called "homosynaptic" since it only occurs at inputs whose activation contributed to the induction of the modification. It is thus phenomenologically different from another form of LTD, termed "heterosynaptic" LTD, which manifests itself at inactive inputs following tetanic stimulation of another exctitatory input to the same cell (for review see Artola and Singer 1993).

In most brain areas, the same synapses can undergo both homosynaptic LTD and LTP and similar tetanic stimuli can be applied to induce either modification. This raises the pivotal question: which parameters determine whether an excitatory synapse depresses or potentiates following its own activation? We have addressed this question in slices of the rat visual cortex.

2.5.2 Experimental Procedure

Slices (350 μm) of the rat visual cortex were prepared by standard techniques (Artola and Singer 1990) and were allowed to recover at room temperature. A single slice was then transferred to the recording chamber, where it was held completely submerged. The slice was maintained at 28°–30° and continuously perfused with a solution containing 124 mM NaCl, 5 mM KCl, 1.25 mM $NaHPO_4$, 2 mM $MgSO_4$, 2 mM $CaCl_2$, 26 mM $NaHCO_3$ and 10 mM *D*-glucose, saturated with 95% 0_2 and 5% CO_2. During exposure to high Ca^{2+}, only the extracellular concentration of Ca^{2+} ($[Ca^{2+}]_o$) was changed and raised to 4 mM. In some experiments, the $GABA_A$ receptor antagonist bicuculline (0.1–0.5 μM) was bath applied.

Intracellular recordings were obtained with 3M potassium acetate-filled electrodes (80–120 MΩ) from regular spiking cells in layer III (McCormick et

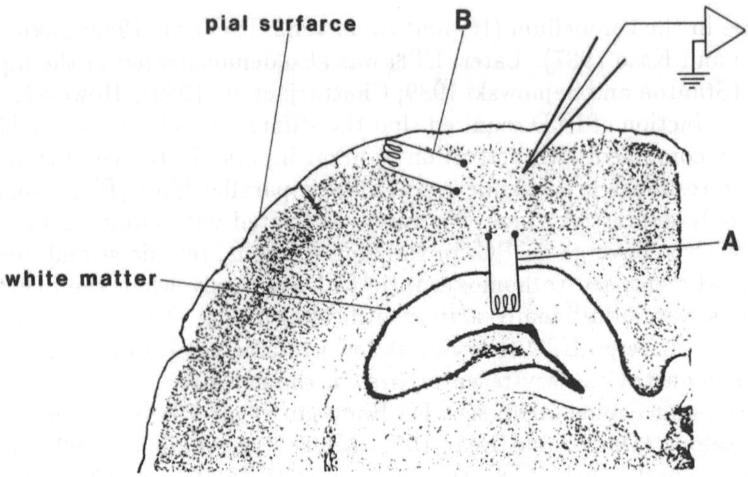

Fig. 2.5–1. Positions of recording and stimulating electrodes in the slice preparation

al. 1985). To examine the role of postsynaptic Ca^{2+} concentration ($[Ca^{2+}]_i$) on the induction of LTD, we injected neurons with high concentrations of one of the two Ca^{2+} chelators EGTA or BAPTA. Intracellular electrodes containing either 500 mM EGTA or 100 mM BAPTA and 3M potassium acetate were used. For injection, hyperpolarizing current pulses (1.5 nA, 0.5 s, 0.5 Hz) were applied for 10–20 min; an additional 30 min were allowed to elapse between injection and LTD induction to ensure diffusion of the chelators to the dendrites. The efficacy of the Ca^{2+} chelators was assessed by measuring the amplitude changes of the afterhyperpolarizing potentials (AHP) following current-induced depolarizing steps (500 ms).

Synaptic responses (corresponding to 50%–70% of maximum amplitude, 0.03 Hz) were elicited by electrical stimulation through bipolar tungsten electrodes that were placed in the underlying white matter (wm) and intracortically (ic) in layers I–II adjacent to the recorded neuron (Fig. 1). The distances between the recording electrode and the two stimulating electrodes were roughly the same. Tetanic stimulation was applied through the wm electrode and consisted of five 2-s-long stimulus trains (50 Hz) delivered at 10 s intervals. In most experiments, the stimulation intensity was raised during the tetanus to suprathreshold levels as indicated in the text and in the legend to the Fig. 2.

Synaptic responses were digitized online (digitizing rate of 0.2 ms) and averaged over five to ten successive trials. To assess post-tetanic modifications of the monosynaptic excitatory response (see Artola et al. 1990b), both the amplitude of the initial part of the postsynaptic potential (PSP)

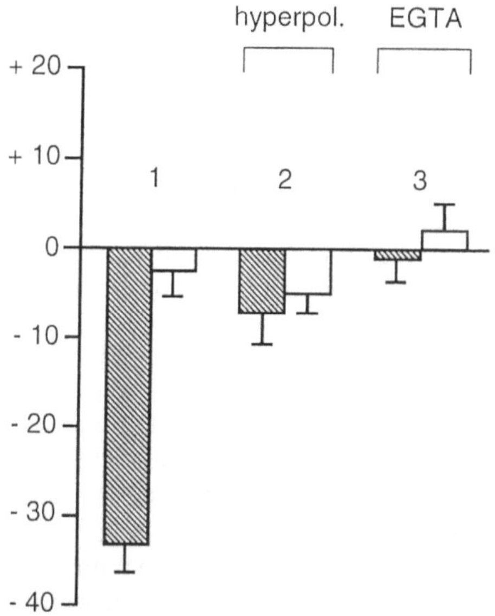

Fig. 2.5–2. Amplitude changes observed in responses to white matter (wm; *filled bars*) and intracortal stimulation (ic; *white bars*) 20–30 min after high intensity wm tetanus in control cells (group 1, $n=13$), in cells hyperpolarized by -40 mV below the resting membrane potential V_{mr} during tetanus (group 2, $n=6$), and in cells injected with *EGTA* (group 3, $n=8$). The amplitude change in the tetanized pathway of group 1 (long-term depression, LTD) differs significantly ($p<0.001$) from that in the non-tetanized pathway of the same group and from those in the tetanized pathways of groups 2 and 3. This indicates that homosynaptic LTD is input specific and is prevented by hyperpolarizing the postsynaptic cell during tetanus (group 2) or by injecting the Ca^{2+} chelator EGTA (group 3). The intensity required to induce LTD in normal Ringer solution depends on the tip separation of the bipolar stimulation electrode, i.e., on the ability to recruit the number of excitatory afferents required to reach the cooperativity threshold for the induction of LTD. In this set of experiments, the tip separation of the stimulating electrodes was 350 μm. The minimum stimulation intensity required for reliable LTD induction with this electrode corresponded to around three times the threshold intensity for spike generation. Stimulus intensity during tetanus was thus set to four times the threshold intensity. (From Artola et al. 1990 and Bröcher et al. 1992a)

and the initial slope were measured before (control) and 30 min after tetanus and expressed as percentages of control. Only the differences (+ for potentiation, – for depression) between the two responses are indicated in the text.

2.5.3 Induction of LTD and LTP in the Neocortex

Reduction of synaptic inhibition facilitates the induction of LTP in hippocampus (Wigström and Gustafsson 1983) and in neocortex (Artola and Singer 1987). Thus, in a first series of experiments, we investigated the long-term effect on synaptic strength of a constant wm, high-frequency stimulation in slices treated with various concentrations (0.1–0.3 μM) of the $GABA_A$ receptor antagonist bicuculline. Stimulus intensity during the conditioning tetanus was raised to twice the test intensity. The long-term effect of this high-frequency stimulation on the amplitude of the initial part of the synaptic response turned out to vary with the degree of blockade of the $GABA_A$ receptor-mediated synaptic inhibition. In cells bathed in normal Ringer solution tetanic stimulation failed to induce any long-term modification of the synaptic response ($-2.5 \pm 1.7\%$; $n = 12$; mean $\pm$ s.e.m). In cells in which bicuculline (0.1–0.2 μM) had only slightly reduced synaptic inhibition, the tetanus induced LTD ($-18.3 \pm 4.5\%$; $n = 7$) and in cells in which inhibition was strongly decreased by 0.3 μM bicuculline, it led to LTP ($+15.5 \pm 4.6\%$; $n = 20$; for details see Artola et al. 1990a; Artola and Singer 1990).

Disinhibition enhances the depolarizing response of the recorded neuron by two mechanisms: first, it reduces Cl^- conductances in the recorded neuron and, second, by increasing activity in polysynaptic pathways, it increases the excitatory input. Thus, with increasing concentrations of bicuculline, the same tetanic stimuli are likely to cause depolarizing responses of increasing amplitude. As bicuculline treatment should not alter the number of monosynaptically activated synapses, the occurrence of long-term modifications of the monosynaptic input appears to depend on the level of the postsynaptic depolarization. By analogy with the mechanisms for LTP induction (for review see Bliss and Collingridge 1993), this suggests that the mechanisms underlying LTD induction are also postsynaptic and involve a voltage-dependent threshold. This threshold, however, appears to be lower than that for LTP induction.

To test this hypothesis, a second series of experiments was performed in which the level of postsynaptic depolarization maintained during the tetanus was directly controlled by injecting current into the postsynaptic cell. LTD can be induced in normal medium provided that stimulus intensity during the tetanus is increased (see legend to Fig. 2). Hyperpolarizing the postsynaptic cell by -40 mV under the resting membrane potential (V_{mr}) throughout this high-intensity tetanus consistently prevented LTD (Fig. 2), whereas depolarizing it by $+20$ mV above V_{mr} led to the induction of LTP ($+27.2 \pm 4.2\%$; $n = 9$). Induction of both LTD and LTP was specific to the tetanized pathway. Responses evoked from ic stimulation were not modified (Fig. 2). This confirms that induction of homosynaptic LTD in the visual cortex requires a minimum postsynaptic depolarization and that homosynaptic LTD and LTP can be induced in the same pathway and with the same stimulation parameters if different levels of postsynaptic depolarization are maintained during tetanic stimulation.

2.5.4 Intracellular Injection of Ca^{2+} Chelators Blocks the Induction of Homosynaptic LTD

Recent evidence in cerebellum suggests that LTD induction depends on the presence of Ca^{2+} in the postsynaptic neuron (Sakurai 1990). As a direct test for this hypothesis, we examined the effect of the postsynaptic injection of Ca^{2+} chelators on the induction of homosynaptic LTD in the neocortex. Induction of LTD was prevented in cells in which either EGTA or BAPTA had led to a complete blockade of the early AHP (Fig. 2; for details see Bröcher et al. 1992).

These data indicate that effective buffering of free Ca^{2+} in the postsynaptic cell prevents LTD induction. Similar results have been obtained in the frontal cortex (Hirsch and Crépel 1992) and in the hippocampus (Mulkey and Malenka 1992), although two other studies (Kimura et al. 1990; Yoshimura et al. 1991) have reported that LTD can be successfully induced in cortical cells injected with EGTA or BAPTA. In these latter reports, however, the effectiveness of the injection procedure was assessed by measuring the level of spike adaptation during a depolarizing current pulse and the tetanus was applied before complete suppression of spike adaptation (see Fig. 2 in Kimura et al. 1990). This suggests the possibility that in these experiments buffering of intracellular Ca^{2+} may have been less complete than in our study. To test this hypothesis, tetanic stimulation was applied in two BAPTA-treated cells before early AHP were completely blocked. Despite the fact that BAPTA had already caused a near complete blockade of the early AHP, the tetanus induced a marked LTD that was indistinguishable from that obtained in nontreated cells (Bröcher et al. 1992).

2.5.5 Calcium-Induced LTD

These results confirm that in neocortex, a minimum $[Ca^{2+}]_i$ is required in the postsynaptic neuron for the induction of homosynaptic LTD. To test whether LTD induction actually depends on a surge of $[Ca^{2+}]_i$ in the postsynaptic neuron, we examined the influence of raising $[Ca^{2+}]_o$.

A marked (up to -35%) and lasting (>90 min after return to 2 mM $[Ca^{2+}]_o$) depression of synaptic responses can be produced in the absence of evoked synaptic activation by raising $[Ca^{2+}]_o$ transiently from 2 to 4 mM for 10 min. This depression is equally pronounced for the responses evoked by wm and ic stimulation. Ca^{2+}-induced LTD again depends on the membrane potential (V_m) of the postsynaptic neuron, because it is only obtained in cells whose V_{mr} is less negative than −79 mV and because it can be prevented by hyperpolarizing the postsynaptic neuron during bath application of high $[Ca^{2+}]_o$ (Artola et al. 1992). This indicates that the depression is caused by a postsynaptic Ca^{2+}-dependent process and is not due to direct effect of high $[Ca^{2+}]_o$ on presynaptic release. Therefore, this Ca^{2+}-induced LTD resembles tetanus-induced LTD. Further support for this comes from the fact that the two forms of LTD occlude each other (Artola et al. 1992).

Altogether, these results suggest that postsynaptic processes responsible for LTD are triggered by a rise of $[Ca^{2+}]_i$ and that the Ca^{2+} responsible for this rise enters, at least partly, through voltage-gated Ca^{2+} channels.

2.5.6 Discussion

Existence of Two Voltage-Dependent Thresholds: The ABS Rule. The requirements for LTD induction, postsynaptic depolarization and an increase of $[Ca^{2+}]_i$ within the postsynaptic dendrite qualitatively resemble those required also for the induction of associative LTP (for review see Bliss and Collingridge 1993). However, there is an important quantitative difference: LTP induction requires a stronger postsynaptic depolarization than induction of LTD. These results have led to a synaptic modification rule, referred to as the ABS (from Artola, Bröcher and Singer 1990a) rule in the literature (see Hancock et al. 1991), with two voltage-dependent thresholds for the induction of LTD and LTP, respectively (Fig. 3). If the first threshold (Θ^-) is reached, the activated synapses depress, and if the second threshold (Θ^+) is reached, which requires stronger depolarization, the activated synapses potentiate. Accordingly, there are two ranges of membrane potential in which synapses undergo no long-term modifications: first, when V_m remains below the threshold for LTD (Θ^-) and second, when V_m is around the second threshold (Θ^+) which separates the two ranges of V_m leading to LTD and LTP, respectively.

Evidence is now available that the ABS rule applies also to use-dependent synaptic modifications in other brain areas such as the prefrontal cortex, the striatum and the hippocampus. In slices of both visual (Artola et al. 1990a) and prefrontal cortex (Hirsch and Crépel 1991), a tetanic stimulus that would normally elicit LTP induces LTD when the depolarizing response to the tetanus is reduced by blocking *N*-methyl-D-aspartate (NMDA) receptors. In the striatum, the same tetanus which produces LTP in Mg^{2+}-free medium induces LTD in the presence of 1.2 mM $[Mg^{2+}]_o$ and no modification when the postsynaptic cell is hyperpolarized during the conditioning tetanus in the presence of 1.2 mM $[Mg^{2+}]_o$ (Calabresi et al. 1992a,b). Low $[Mg^{2+}]_o$ favours the opening of NMDA receptor-gated channels (Nowak et al. 1984; Mayer et al. 1984) and enhances the depolarizing response to the tetanus. Thus, the induction of LTD and LTP in the striatum shows the same voltage dependence as in the visual cortex. In area CA1 of the hippocampus, low-frequency stimulation (around 1 Hz) induces LTD, while high frequency stimulation ($>$ 10 Hz) induces LTP (Dunwiddie and Lynch 1978; Dudek and Bear 1992). Likewise, the long-term effect of a tetanus of a given frequency can be modified by current injection. A 5 Hz tetanus alone has no long-term effect on synaptic transmission, probably because this fre-

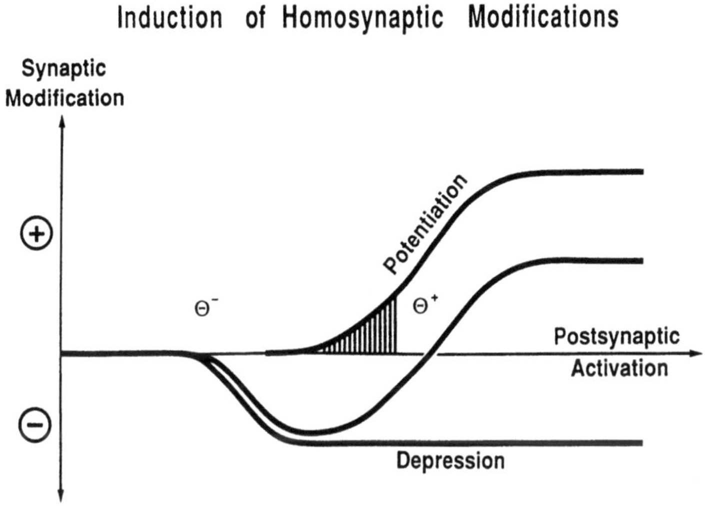

Fig. 2.5–3. Induction of homosynaptic modifications. The two different voltage-dependent thresholds (Θ^- and Θ^+) for the induction of homosynaptic long-term depression (LTD) and long-term potentiation (LTP). The *ordinate* indicates the direction of the synaptic gain change and the *abscissa* the membrane potential level of the postsynaptic neuron that is maintained during presynaptic activation. If the first threshold Θ^- is reached, a mechanism is activated which leads to a long-lasting depression of synaptic efficacy, and if the second threshold Θ^+ is reached, another process is triggered which leads to the potentiation of the synapse

quency is intermediate between the frequencies suitable for the induction of LTD and LTP (Dunwiddie and Lynch 1978; Dudek and Bear 1992), respectively, and thus produces a postsynaptic depolarization around the Θ^+ threshold. However, when paired with hyperpolarizing current injection, this same tetanus induces LTD, and when paired with depolarizing current injection, LTP (Stanton and Sejnowski 1989; Xie et al. 1992). Similarly, low-frequency tetani, which normally induce LTD (see above), are no longer effective if cells are hyperpolarized during stimulation (Mulkey and Malenka 1992).

The $[Ca^{2+}]_i$ Hypothesis. Several interpretations are possible in order to account for the observations that induction of both LTD and LTP require a postsynaptic depolarization and an increase of $[Ca^{2+}]_i$ and that LTP is induced at more depolarized levels of V_m than LTD. One proposal is that different levels of postsynaptic depolarization favour selectively the activation of different mechanisms which increase $[Ca^{2+}]_i$ and that the source of Ca^{2+} determines the direction of the synaptic change. Early evidence favoured this hypothesis, as it suggested that Ca^{2+} entering through NMDA receptor-gated channels mediates the induction of LTP (Collingridge et al. 1983; Artola and Singer 1987) while Ca^{2+} liberated from intracellular stores following activation of metabotropic glutamate (mGlu) receptors leads to

LTD. This view equated the LTP threshold Θ^+ with the activation threshold of NMDA receptor-gated conductances (Bear et al. 1987). Since a critical level of postsynaptic depolarization is needed to reduce the Mg^{2+} block of the NMDA receptor-gated channel (Nowak et al. 1984; Mayer et al. 1984), this interpretation accounted well for the fact that LTP induction required strong postsynaptic depolarization. However, evidence has since been obtained which is no longer compatible with this view. It has been demonstrated that the activation of NMDA receptor-gated channels, mGlu receptors and voltage-gated Ca^{2+} channels all contribute to the induction of both LTD and LTP (for review see Artola and Singer 1993).

Another suggestion is that strong depolarization leads to more pronounced increases of $[Ca^{2+}]_i$ than weak depolarization and that the induction of LTP requires a stronger increase of $[Ca^{2+}]_i$ than the induction of LTD. The two voltage-dependent thresholds for LTD and LTP would thus be related to corresponding changes of $[Ca^{2+}]_i$. This is essentially what Lisman (1979) postulated. According to his model, a moderate rise in $[Ca^{2+}]_i$ should lead to a predominant activation of phosphatases, while a stronger increase would favour activation of kinases. This proposal is compatible with the recent evidence that the efficacy of glutamate receptor-gated ion channels can be changed in opposite directions by dephosphorylation and phosphorylation (Wang et al. 1991; Greengard et al. 1991). The fact that a brief exposure to high $[Ca^{2+}]_o$ can induce LTD suggests that an appropriate rise in $[Ca^{2+}]_i$ is sufficient to trigger LTD. Additional evidence for this hypothesis comes from experiments in hippocampus. In CA1, depression of synaptic transmission is induced by strongly depolarizing the postsynaptic neuron, either with antidromic activation (Pockett and Lippold 1986; Pockett et al. 1990; Christofi et al. 1993) or current injection (Christofi et al. 1993). This depression depends on Ca^{2+} entry, because it is blocked if extracellular $[Ca^{2+}]$ is low (Christofi et al. 1993) and it is independent of the activation of Glu receptors, because it remains inducible if synaptic transmission is blocked during conditioning stimulation. Experiments in hippocampal slices also indicate that a rise in postsynaptic Ca^{2+} is necessary to induce LTP (Lynch et al. 1983) and that an increase in $[Ca^{2+}]_i$ alone may be sufficient to trigger the potentiation of synaptic transmission (Malenka et al. 1988). Finally, the following observations suggest that the induction of LTP needs a larger increase in $[Ca^{2+}]_i$ than the induction of LTD: first, in neocortex, incomplete buffering of intracellular Ca^{2+} by EGTA or BAPTA readily prevents the induction of LTP, but still allows for the induction of LTD (Bröcher et al. 1992; see also Kimura et al. 1990 and Yoshimura et al. 1991); second, in CA1 a 20 Hz tetanus normally induces LTP, but when $[Ca^{2+}]_o$ is lowered from 2.5 to 0.5 mM, it leads to LTD (Mulkey and Malenka 1992).

2.5.7 Conclusions

The results reviewed in this chapter suggest that: (a) a rise in $[Ca^{2+}]_i$ is necessary and sufficient to trigger both LTD and LTP, (b) the increase in $[Ca^{2+}]_i$ required to induce LTP has to be significantly larger than that necessary for LTD induction and (c) the amplitude of the $[Ca^{2+}]_i$ increase rather than the source of Ca^{2+} appears to be the relevant variable for the induction of synaptic modifications.

If all that is needed for LTD to occur is an appropriate surge of $[Ca^{2+}]_i$ in the respective postsynaptic profile, all conditions causing a sufficient increase in postsynaptic $[Ca^{2+}]_i$ should lead to LTD, irrespective of whether the synapses undergoing depression are active or inactive. Several observations indicate that this is indeed the case: there is evidence both from hippocampus and neocortex that tetanic stimulation of one excitatory input can cause depression of another, nonstimulated excitatory input to the same neuron (heterosynaptic LTD; for review see Artola and Singer 1993).

Finally, attributing a pivotal role to changes of $[Ca^{2+}]_i$ in triggering use-dependent synaptic modifications should not distract from the importance of the processes that preceed and follow the surge of $[Ca^{2+}]_i$. Indeed, under natural activation conditions, changes in synaptic gain can be expected to depend not only on the activity in the processing circuits proper, but also and to a critical extent on the many other factors that control the depolarization of dendritic compartments, the amplitude of $[Ca^{2+}]_i$ changes and the signalling cascades that are activated by the increase of intracellular Ca^{2+}.

References

Artola A, Singer W (1987) Long-term potentiation and NMDA receptors in rat visual cortex. Nature 330: 649–652

Artola A, Singer W (1990) The involvement of N-methyl-D-aspartate receptors in induction and maintenance of long-term potentiation in rat visual cortex. Eur J Neurosci 2: 254–269

Artola A, Singer W (1993) Long-term depression of excitatory synaptic transmission and its relationship to long-term potentiation. Trends Neurosci 11: 480–487

Artola A, Bröcher S, Singer W (1990a) Different voltage-dependent thresholds for inducing long-term depression and long-term potentiation in slices of rat visual cortex. Nature 347: 69–72

Artola A, Bröcher S, Singer W (1990b) An in vitro model of plasticity in neocortical slices. In: Squire L, Lindenlaub E (eds) The biology of memory. Schattauer, Stuttgart, pp 939–952

Artola A, Hensch T, Singer W (1992) A rise of $[Ca^{++}]_i$ in the postsynaptic cell is necessary and sufficient for the induction of long-term depression (LTD) in neocortex. Soc Neurosci Abstr 18: 1351

Bear MF, Cooper LN, Ebner FF (1987) A physiological basis for a theory of synapse modification. Science 237: 42–48

Bindman LJ, Murphy KPSJ, Pockett S (1988) Postsynaptic control of the induction of long-term changes in efficacy of transmission at neocortical synapses in slices of rat brain. J Neurophysiol 60: 1053–1065

Bliss TVP, Collingridge G (1993) A synaptic model of memory: long-term potentiation in hippocampus. Nature 361: 31–39

Bliss TVP, Lømo T (1973) Long-lasting potentiation of synaptic transmission in the dentate area of the anaesthetized rabbit following stimulation of the perforant path. J Physiol 232: 331–356

Bröcher S, Artola A, Singer W (1992) Intracellular injection of Ca^{++} chelators blocks induction of long-term depression in rat visual cortex. Proc Natl Acad Sci USA 89: 123–127

Calabresi P, Maj R, Pisani A, Mercuri NB, Bernardi G (1992a) Long-term synaptic depression in the striatum: physiological and pharmacological characterization. J Neurosci 12: 4224–4233

Calabresi P, Pisani A, Mercuri NB, Bernardi G (1992b) Long-term potentiation in the striatum is unmasked by removing the voltage-dependent magnesium block of NMDA recceptor channels. Eur J Neurosci 4: 929–935

Chattarji S, Stanton PK, Sejnowski TJ (1989) Commissural synapses, but not mossy fiber synapses, in hippocampal field CA3 exhibit associative long-term potentiation. Brain Res 495: 145–150

Christofi G, Nowicky AV, Bolsover SR, Bindman LJ (1993) The postsynaptic induction of non-associative long-term depression of excitatory synaptic transmission in rat hippocampal slices. J Neurophysiol 69: 219–229

Collingridge GL, Kehl SJ, McLennan H (1983) Excitatory amino acids in synaptic transmission in the Schaffer collateral-commissural pathway of the rat hippocampus. J Physiol 334: 33–46

Dudek SM, Bear MF (1992) Homosynaptic depression in area CA1 of hippocampus and effects of N-methyl-D-aspartate receptor blockade. Proc Natl Acad Sci USA, 89: 4363–4367

Dunwiddie T, Lynch G (1978) Long-term potentiation and depression of synaptic responses in the rat hippocampus: localization and frequency dependency. J Physiol 276: 353–367

Greeengard P, Jen J, Nairn AC, Stevens CF (1991) Enhancement of the glutamate response by cAMP-dependent protein kinase in hippocampal neurons. Science 253: 1135–1137

Hancock PJB, Smith LS, Phillips WA (1991) A biologically supported error-correcting learning rule. Neural Comp 3: 210–212

Hirsch JC, Crépel F (1990) Use-dependent changes in synaptic efficacy in rat prefrontal neurons in vitro. J Physiol 427: 31–49

Hirsch JC, Crépel F (1991) Blockade of NMDA receptors unmasks a long-term depression in synaptic efficacy in rat prefrontal neurons in vitro. Exp Brain Res 85: 621–624

Hirsch JC, Crépel F (1992) Postsynaptic calcium is necessary for the induction of LTP and LTD of monosynaptic EPSPs in prefrontal neurons. An in vitro study in the rat. Synapse 10: 173–175

Ito M, Kano M (1982) Long-lasting depression of parallel fiber-Purkinje cell transmission induced by conjunctive stimulation of parallel fibers and climbing fibers in the cerebellar cortex. Neurosci Lett 33: 253–258

Ito M, Sakurai M, Tongroach P (1982) Climbing fibre induced depression of both mossy fibre responsiveness and glutamate sensitivity of cerebellar Purkinje cells. J Physiol 324: 113–134

Kano M, Kato M (1987) Quisqualate receptors are specifically involved in cerebellar synaptic plasticity. Nature 325: 276–279

Kimura F, Tsumoto T, Nishigori A, Yoshimura Y (1990) Long-term depression but not potentiation is induced in Ca^{2+}-chelated visual cortex neurons. NeuroReport 1: 65–68

Lisman J (1989) A mechanism for the Hebb and the anti-Hebb processes underlying learning and memory. Proc Natl Acad Sci USA 86: 9574–9578

Lynch G, Larson J, Kelso S, Barrionuevo G, Schottler F (1983) Intracellular injections of EGTA block induction of hippocampal long-term potentiation. Nature 305: 719–721

Malenka RC, Kauer JA, Zucker RS, Nicoll RA (1988) Postsynaptic calcium is sufficient for potentiation of hippocampal synaptic transmission. Science 242: 81–84

Mayer ML, Westbrook GL, Guthrie PB (1984) Voltage-dependent block by Mg^{2+} of NMDA responses in spinal cord neurones. Nature 309: 261–263

McCormick DA, Connors BW, Lighthall JW, Prince DA (1985) Comparative electrophysiology of pyramidal and sparsely spiny stellate neurons of the neocortex. J Neurophysiol 54: 782–806

Mulkey RM, Malenka RC (1992) Mechanisms underlying induction of homosynaptic long-term depression in area CA1 of the hippocampus. Neuron 9: 967–975

Nowak L, Bregestovski P, Ascher P, Herbet A, Prochiantz A (1984) Magnesium gates glutamate-activated channels in mouse central neurons. Nature 307: 462–465

Pennartz CMA, Ameerun RF, Groenewegen HJ, Lopes da Silva FH (1993) Synaptic plasticity in an in vitro slice preparation of the rat nucleus accumbens. Eur J Neurosci 5: 107–117

Pockett S, Lippold OCJ (1986) Long-term potentiation and depression in hippocampal slices. Exp Neurol 91: 481–487

Pockett S, Brookes NH, Bindman LJ (1990) Long-term depression at synapses in slices of rat hippocampus can be induced by bursts of postsynaptic activity. Exp Brain Res 80: 196–200

Sakurai M (1987) Synaptic modification of parallel fibre-Purkinje cell transmission in in vitro guinea-pig cerebellar slices. J Physiol 394: 319–336

Sakurai M (1990) Calcium is an intracellular mediator of the climbing fiber in induction of cerebellar long-term depression. Proc Natl Acad Sci USA 87: 3383–3385
Stanton PK, Sejnowski TJ (1989) Associative long-term depression in the hippocampus induced by hebbian covariance. Nature 339: 215–218
Wang LY, Salter MW, MacDonald JF (1991) Regulation of kainate receptors by cAMP-dependent protein kinase and phosphatase. Science 253: 1132–1135
Wigström H, Gustafsson B (1983) Facilitated induction of hippocampal long-lasting potentiation during blockade of inhibition. Nature 301: 603–604
Xie X, Berger TJ, Barrionuevo G (1992) Isolated NMDA receptor-mediated synaptic responses express both LTP and LTD. J Neurophysiol 67: 1009–1013
Yoshimura Y, Tsumoto T, Nishigori A (1991) Input-specific induction of long-term depression in Ca^{2+}-chelated visual cortex neurons. NeuroReport 2: 393–396

Chapter 3

On the Input to the Visual Cortex

3.1 Introduction

V. BRAITENBERG
Max-Planck-Institut für biologische Kybernetik, Tübingen, Germany

Segregation is a word with a magic ring. Another one is correlation. Both have their separate followings, but they lead their adepts in different directions. There is the party of those who seem to think that an ever more subtle distinction of sensory signals conveyed to separate regions of the brain is all there is to know about the function of the brain. And they may not be far from the truth, since segregation of qualities is certainly the main part of the job if what you want is appropriate reactions.

The other party succumbs to the fascination of the old concept of form. Correlations in time and in space within a system uniformly fed by one sort of sensory input give away the presence and movement of objects in the outside world. Objects may be classified by their formal aspects in even more significants ways, and in a much greater variety, than by the quality of the signals they emit. This again is an important part of the truth, and much of the brain is undoubtedly devoted to the abstraction of form from the variety of sensory experience.

To say that segregation of different qualities in vision is the task of the retina, and abstraction of form that of the visual cortex is a truism and a falsehood at the same time. Physiologically speaking, it is obvious that the retina makes little sense without the visual cortex, and vice versa. And from the abstract standpoint of information handling, clearly the level of recognition defines the requirements for the alphabet of elementary symbols in which the input is coded, and is itself conditioned by that alphabet.

The two papers by Wässle et al. and by Lee provide an excellent opportunity for speculations on the relative merits of the segregationist and the correlationist approach. Together, as well as singly, they show how fruitful their combination can be.

3.2 Color Coding in the Primate Retina: Predictions and Constraints from Anatomy

H. WÄSSLE[1], U. GRÜNERT[1], P.R. MARTIN[2], AND B.B. BOYCOTT[3]
[1]Max-Planck-Institut für Hirnforschung, Frankfurt am Main, Germany
[2]Department of Physiology, University of Sydney, Australia
[3]Division of Anatomy and Cell Biology, UMDS (Guy's Campus), London, United Kingdom

Most mammals are dichromats and have two cone types, a short-wavelength sensitive (S-) cone and a middle-wavelength sensitive (M-) cone. Old World monkeys and humans have a third cone type, the long-wavelength sensitive (L-) cone, which is the result of a more recent mutation of the M-cone pigment (Mollon 1986; Mollon and Jordan 1988; Nathans 1989). The opsins of M- and L-cones differ by 15 out of 348 amino acids only (Nathans 1989), and there are no antibodies available which could discriminate the two types of cones. This makes the anatomical analysis of the M- and L-cone pathways of the primate retina more difficult. By contrast, specific antibodies are available for S-cones and S-cone bipolar cells, respectively; S-cones form about 10% of cones in primates (Wikler and Rakic 1990). Because the S-cone pathway has been studied in great detail recently (Kouyama and Marshak 1992), it not is dealt with here. It seems, like the rod pathway, to be a constant feature of all mammalian retinae.

A central theme of this chapter is to uncover whether the Old World monkey's acquisition of trichromacy has been achieved by changes in the circuitry of the mammalian retina, or whether, with minor modifications, these primates can use the general mammalian retinal wiring scheme. For a mutation producing a "new" cone type to be advantageous there must be neural pathways available for its unique access by the brain while the retina continues to function effectively. We have argued elsewhere (Wässle and Boycott 1991) that the primate eye and retina have evolved under selection pressure favoring high spatial resolution, culminating in an all cone fovea and in the well-known one-to-one connection between foveal ganglion cells and cones. We will show here that such a recent mutation in the cone pigment, superimposed onto this high acuity system — the midget system — is the basis for trichromacy in primates.

Most retinal ganglion cells, which project to the parvocellular layers of the monkey lateral geniculate nucleus, the so-called P-cells, have concentric receptive fields (RF); for example, type I cells (Wiesel and Hubel 1966) are excited by M-cone stimulation in their RF center and inhibited by L-cone stimulation in their RF surround or vice versa (Kaplan et al. 1990). Because this opponent organization of P-cell RFs is similar to the chromatic antagonism of goldfish or turtle retinal ganglion cells, the well-established color-selective wiring diagram of the fish or turtle retina has been considered to be a model of color-selectivity of the primate retina (Kolb 1991; Gouras 1992). In goldfish and turtle retinae there is good evidence that the chromatic antagonism may originate in the outer plexiform layer (OPL). Bipolar cells selecting several cones of one type provide the center response and horizontal cells selecting other cone types mediate the surround response (Stell and Lightfoot 1975; Scholes 1975; Ohtsuka and Kouyama 1986). Here we summarize some of our recent findings concerning connections of primate bipolar and horizontal cells and show that they are in clear contrast to the goldfish or turtle models (Boycott et al. 1987; Wässle et al. 1989; Boycott and Wässle 1991; Wässle and Boycott 1991; Martin and Grünert 1992).

3.2.1 Bipolar Cells

Polyak (1941) found bipolar cells in the primate retina, which synapse with only one cone pedicle; hence he called them "midget" in contrast to "diffuse" bipolar cells, which receive input from several cones. Later it was shown by electron microscopy that there are two types of midget bipolar cells: flat (putative OFF) and invaginating (putative ON) midgets (Kolb et al. 1969; Kolb 1970). P-cells of the primate *fovea*, the midget ganglion cells, receive direct input from only one midget bipolar cell (Kolb 1991; Kolb and De Korver 1991; Calkins et al. 1992). The center response of their RF therefore derives from a single cone, and this is commonly accepted as the color-selective wiring of the primate fovea (Wässle and Boycott 1991). "Diffuse" bipolar cells do not select specific cones (Boycott and Wässle 1991). Figure 1A shows the dendritic branches of a diffuse bipolar cell, drawn from a Golgi-stained retinal whole mount as viewed from the photoreceptor side. The fine dendritic branches terminate in aggregates, where they make contacts with individual cone pedicles. A midget bipolar cell was also stained in this field. It has a smaller cell body, and the dendrites terminate in a single aggregate. By changing the focal plane of the microscope, the cone mosaic above the two bipolar cells could also be seen and brought into register with the terminal aggregates (Fig. 1B,C). This diffuse bipolar cell obviously contacts all eight cones in its dendritic field, while the midget bipolar connects to one cone pedicle only. We observed six different morphological types of diffuse bipolar cells in macaque monkey retinae and analyzed the cone contacts of more than 50 cells; in no case did we find cone selectivity of their dendrites. Diffuse bipolar cells are present throughout the primate retina, and every cone pedicle is in contact with several of them. They probably represent the first stage of the achromatic luminance channel and provide input to ganglion cells of the M-pathway, such as parasol cells, as well as to other ganglion cell classes (Boycott and Wässle 1991; Lee et al. 1988).

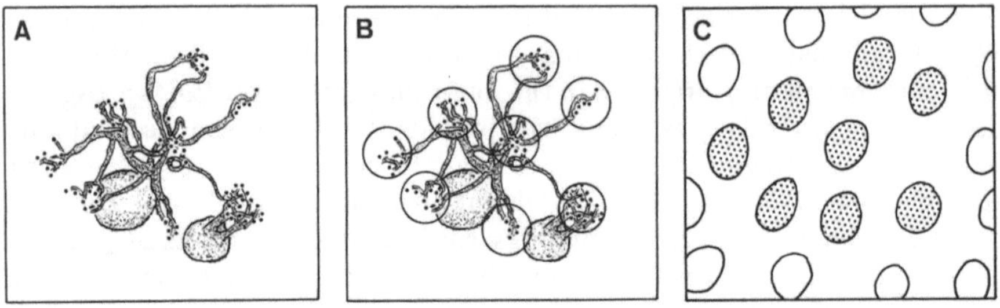

Fig. 3.2–1. **A** Drawing of two bipolar cells from a Golgi-stained whole mount of a monkey retina as viewed from the photoreceptor side. The cell body and the dendritic tree of the diffuse bipolar cell are in the center, the midget bipolar cell is in the lower right. **B** The terminal aggregates are surrounded by circles that represent cone pedicles. **C** Cones to which the two cells are connected are indicated by *hatching*. The diffuse bipolar cell contacts eight cones, the midget bipolar only one (width of the frame: 50 μm)

The situation has been less clear in *peripheral retina*. There are many reports from Golgi-stained retinae that midget bipolar cells are not restricted to the foveal region but are present all the way to the periphery of the primate retina (Polyak 1941; Rodieck 1988; Boycott and Hopkins 1991). Martin and Grünert (1992) demonstrated quantitatively that in the macaque monkey the density of cone bipolar cells is 2.5–4 times that of cones at all eccentricities. Hence, there could be two midget bipolar cells, an ON- and an OFF-cell for every cone all the way out into peripheral retina. We now have evidence from immunocytochemical labeling of midget bipolar cells that this is true (Wässle et al. 1994). Milam et al. (1993) stained primate retinae immunocytochemically with antibodies against recoverin and presented convincing evidence that OFF-midget bipolar cells were stained throughout the retina. We have used recoverin as a marker of OFF-midget bipolar cells and extended the study by Milam et al. (1993) by comparing midget bipolar density with that of cones.

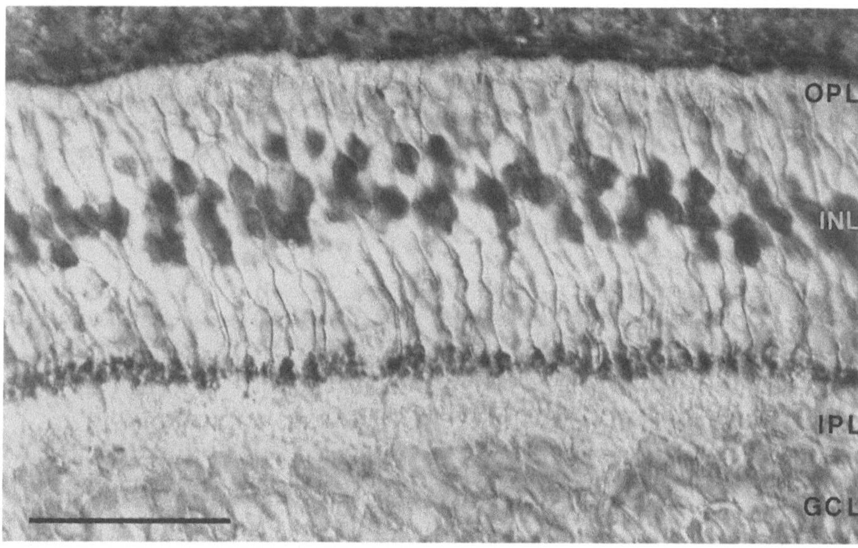

Fig. 3.2–2. Light micrograph of a vertical frozen section through a macaque monkey retina stained immunocytochemically with antibodies against recoverin. Putative OFF-midget bipolar cells are labeled. Recoverin is a calcium binding protein which was first isolated from photoreceptors (Lambrecht and Koch 1991), and this is the reason for the strong labeling at the top of this micrograph. *OPL*, outer plexiform layer; *INL*, inner nuclear layer; *IPL*, inner plexiform layer; *GCL*, ganglion cell layer. *Scale bar*, 50 μm

Figure 2 shows a vertical section through the central retina of a macaque monkey stained for recoverin immunoreactivity. The labeled bipolar cells have their cell bodies in the center of the inner nuclear layer (INL); their axon terminals are small and form a tightly packed band at the outer margin of the inner plexiform layer (IPL). From each cell body a single unbranched apical dendrite runs vertically up into the OPL, to approach a single cone pedicle. These are the morphological characteristics of OFF-midget bipolar cells. Their density and also the cone density were measured from horizontal sections, and the results are presented in Fig. 3. The cone density decreases from $8000/mm^2$ at 5 mm eccentricity to $3000/mm^2$ at 14 mm eccentricity (Fig. 3A). The OFF-midget bipolar density (Fig. 3B) closely

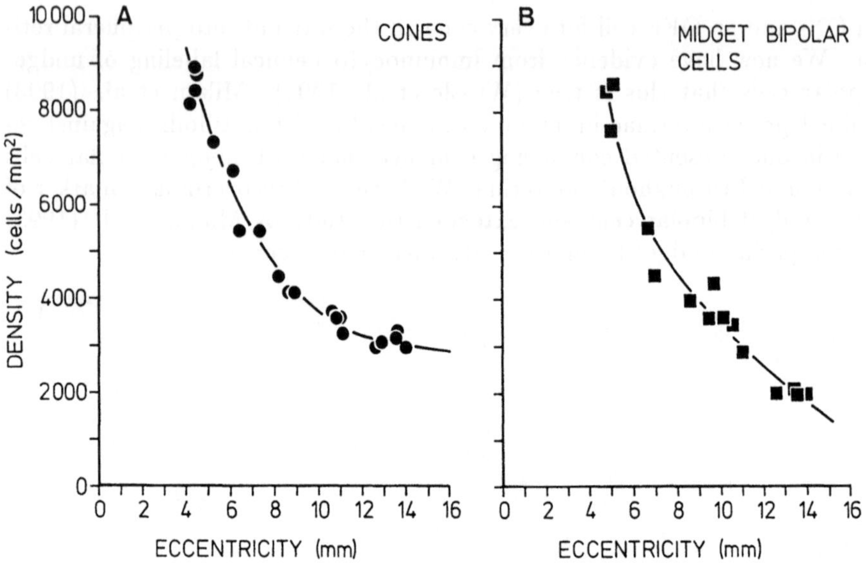

Fig. 3.2–3. Comparison of the cone density (**A**) and of the OFF-midget bipolar cell density (**B**). The cells were counted in horizontal sections immunostained with antibodies against recoverin. They were sampled along an intersect extending from the optic nerve head toward upper peripheral retina

follows the cone density out to an eccentricity of 9 mm. However, beyond that, and into *far peripheral* retina, OFF-midget bipolars comprise only 60% of the cone density. There are two possible explanations for this deviation. Midget bipolar cells in far peripheral retina make exclusive contacts with a single cone, comparable to their connectivity in the central retina, but only 60% of the cones are innervated. Alternatively there could be a convergence of several cones onto one midget bipolar cell. Polyak (1941) described polysynaptic forms of his midget bipolar cells which contact two or even three cone pedicles — these are, in all other respects, indistinguishable from "monosynaptic" midget bipolar cells (Boycott and Hopkins 1991; Hopkins and Boycott 1992). Figure 4A-C shows three midget bipolar cells drawn from *mid peripheral* retina of a Golgi-stained whole mount (eccentricity 6–8 mm); the monosynaptic type (Fig. 4A) made up 84% of all midget bipolars; 14% made contacts with 2 cones (Fig. 4B), and 2% with three cones (Fig. 4C). In *far peripheral* retina (eccentricity ~ 14 mm) 28 midget bipolar cells were encountered in the same Golgi-stained whole mount: 6 received input from two cones, 17 from three cones, and 5 from even four cones. Thus, with increasing eccentricity, there is a gradual increase in the proportion of midget bipolar cells contacting two or more cones and a decrease in midget bipolar cells contacting one cone. Fig. 4D shows two neighboring far peripheral midget bipolars; together they contact six adjacent cones. Peripheral midget bipolars never indicated cone selectivity; their dendritic branches always contacted neighboring cones. This does not necessarily exclude cone selectivity since M- and L-cones might have a clumped distribution (Mollon and Bowmaker 1992), but at the borders of such patches a mixed cone input occurs and the chromatic signal would be degraded.

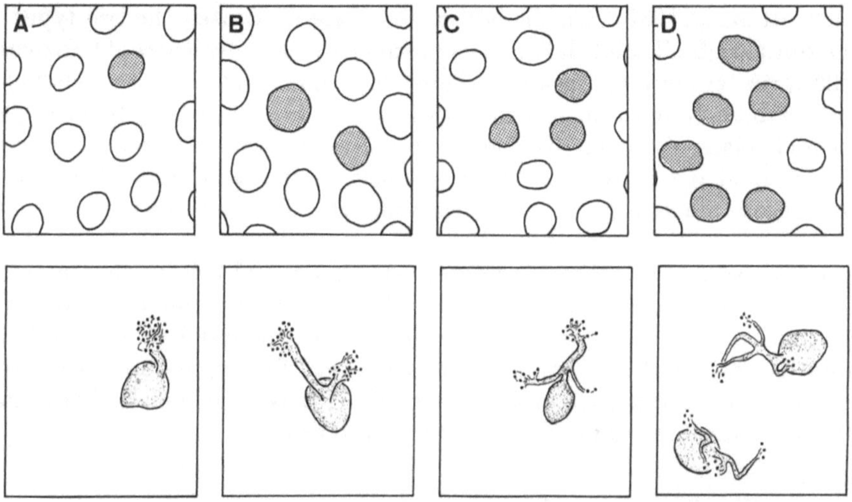

Fig. 3.2–4. Drawings of midget bipolar cells from two different areas of a Golgi-stained whole mount of a macaque monkey retina. *Top row*, cone mosaic; the cones contacted are indicated by hatching. *Bottom row*, cell bodies, dendrites, and terminal aggregates. **A** Midget bipolar cell which contacts a single cone pedicle. **B** "Double headed" midget bipolar cell. **C** "Triple headed" midget bipolar cell. **D** Two neighboring midget bipolars which each contact three cone pedicles (eccentricity: **A**–**C**, 6–8 mm; **D**, 14 mm; *frame size*: 40 × 35 μm)

The ON-midget bipolar cells can be labeled with antibodies against cholecystokinin (CCK; Wässle et al. 1994). They are also present throughout the retina, and every M- and L-cone pedicle is contacted by an ON- and an OFF-midget bipolar cell.

In summary, most ON- and OFF-midget bipolar cells of the *fovea* and *mid peripheral* retina exclusively contact single cone pedicles and occur at the same density as the cones. Thus the cone mosaic is transferred in complete detail into the IPL. In *far peripheral* retina there is a convergence onto individual midget bipolar cells of up to four cones. This might degrade the chromatic information.

3.2.2 Horizontal Cells

In the retina of the cat (Wässle et al. 1978), rabbit (Raviola and Dacheux 1990), and tree shrew (Müller and Peichl 1993) horizontal cells innervate *all* the cones in their dendritic fields, without any signs of cone selectivity. Each cone pedicle contacts between three and four horizontal cells of each type. Cone selectivity of primate horizontal cells, as suggested by Kolb (1991), would therefore be a major deviation from the general mammalian scheme (Boycott et al. 1987). On the left in Fig. 5 are illustrated the two types of horizontal cells (H_1 and H_2) of macaque monkey. The dendrites of horizontal cells form terminal aggregates, where they make contacts with individual cone pedicles (circled in the drawings in the middle of Fig. 5). By changing the focal plane of the microscope the cones can be brought into register with the terminal aggregates. The hatched cones on the right-hand side of Fig. 5 are those contacted by the two horizontal cells. It is obvious that all cones in their respective dendritic fields are contacted non-selectively by these two horizontal cells. We never observed a horizontal cell which contacted only 50% of the cones and would therefore be selective for L- or M-cones (Boycott et al. 1987; Wässle et al. 1989). Electrophysiological recordings from H_1 horizontal cells support the anatomical finding of a mixed M- and L-cone input (Dacheux and Raviola 1990). Primate H_1 and H_2 horizontal cells are therefore luminosity type horizontal cells.

3.2.3 The Receptive Field of P-Ganglion Cells and Their M- and L-Cone Inputs

In the *fovea* it has been shown by electron microscopy (Kolb 1991; Kolb and De Korver 1991; Calkins et al. 1992) that ON- or OFF-midget ganglion cells (P-cells) receive their major excitatory input from a single ON- or OFF-midget bipolar cell. Here the center response of the RF would be dominated by whichever (L- or M-) cone the midget bipolar cell happened to contact. Two models of the surround response can be considered. For simplicity we describe here a P-cell with M-cone input to the center. For the surround to be strictly color-selective it should receive input only from L-cones. Horizontal cells, because of their non-selective cone connections cannot provide this surround inhibition. Small-field amacrine cells which select L-cone midget bipolar cells in the IPL and inhibit the M-cone driven midget ganglion cell could do so (Reid and Shapley 1992). Many such small amacrines have been observed (Polyak 1941; Boycott and Dowling 1969; Kolb et al. 1992). If they follow the general mammalian rule they would use glycine as their transmitter (Vaney 1990; Wässle and Boycott 1991). If as Paulus and Kröger-Paulus (1983) and Lennie et al. (1992) have argued, the surround obtains mixed cone input, the horizontal cells

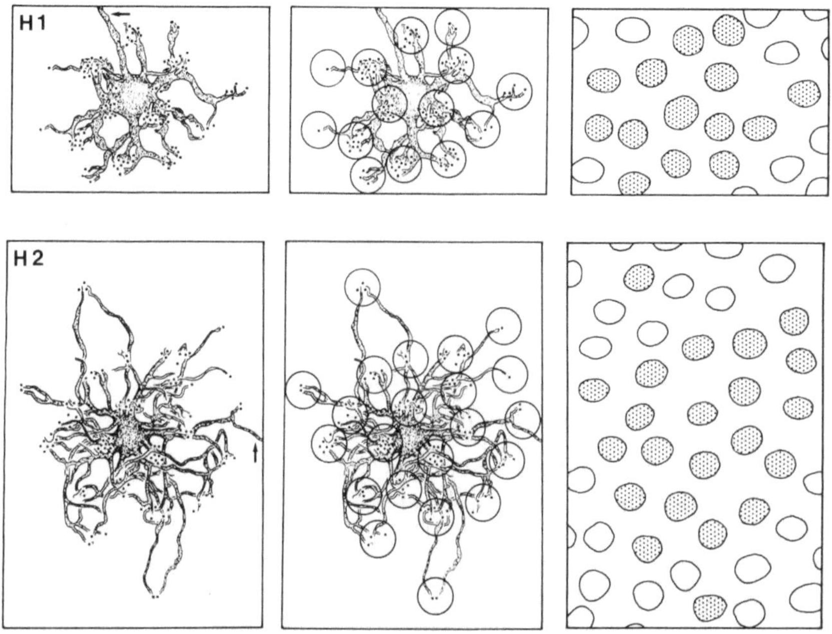

Fig. 3.2–5. Drawings of representative H_1 and H_2 horizontal cells from a Golgi-stained whole mount of a macaque monkey retina. Both cells are at 5.7 mm eccentricity; width of each frame, 60 µm. *Left*, dendritic morphology of the cells (*arrows* point to axons). *Middle*, terminal aggregates are surrounded by *circles* that represent cone pedicles. *Right*, cones to which the two cells are connected are indicated by *hatching*

could create this. The physiological results are still conflicting and do not exclude one or the other model; perhaps this distinction could be made pharmacologically. If the surround is the result of glycinergic inhibition by small-field amacrine cells, it should be blocked by strychnine, the glycine antagonist. However, if it is a contribution of the horizontal cells, which probably use GABA as their transmitter, surround responses should be blocked by specific GABA antagonists.

The situation becomes more complicated in *mid peripheral* retina, where there are more cones than there are ganglion cells (Wässle et al. 1990). Ganglion cells in mid peripheral retina have extended dendritic trees (Watanabe and Rodieck 1989); thus here there must be convergence of several midget bipolar cells to create the RF center of a P-cell. The dendrites of the P-cell therefore must *select* midget bipolar cells according to their M- or L-cone input. The nature of the L- and M-cone mosaic would be evident in the dendritic fields of cone selective P-cells. If L- and M-cones are clustered, then ganglion cells could have round dendritic fields, each ganglion cell receiving input from one such cluster. If L- and M-cones are distributed in a paracrystalline way, the neighboring cones would often be differing (L- or M-) types, and cone selective ganglion cells should be more interdigitated as they reach around each other to find their appropriate midget bipolar cell axon terminals.

In *far peripheral* retina the situation is further complicated by the above finding of cone convergence onto midget bipolar cells. If there is no cone selectivity of such "polysynaptic" midget bipolar cells in the OPL, P-cells would get a mixed cone input into their receptive field and would fail to produce a chromatic signal. The poor color vision in the human peripheral visual field suggests that this might be the case.

3.2.4 Evolution of the Midget System and of Trichromacy in the Primate Retina

We have shown that midget bipolar and midget ganglion cells are the major components for the transfer of (L- and M-) chromatic signals through the primate retina. One might argue they are a specialization of the primate retina not found in other mammalian retinae. It has also been suggested that the midget system might be a distinct feature of the primate *fovea.* We think this is not so for the following reasons. In all mammalian retinae, which have a specialized region of high density, such as the central area of the cat or the visual streak of the rabbit, comparable gradual anatomical changes from peripheral toward central retina occur: cell density increases, dendritic field sizes decrease, and the convergence decreases. The result of such changes is an increase of spatial resolution in central retina. The primate retina follows this general mammalian scheme. But it is special because it reached the anatomic limits of the reduction of dendritic fields and convergence; this is the one-to-one connection between ganglion cells and cones. Wässle and Boycott (1991) have suggested the midget system evolved to meet the requirements of high spatial resolution, and that it evolved before the mutation occurred which separated L- from M-cones. This late mutation, superimposed onto the one-to-one connection, established the chromatic system. The selective advantage of trichromatic vision could lead to proliferation of color pathways in cortical and perhaps even subcortical visual centers. However, the requirements of high acuity have meant that the action spectra of M- and L-cones had to remain close. The lack of a morphological distinctness between M- and L-cones, the common regular mosaic, and the retention of the full numbers of cones in dichromates, provide circumstantial support to this idea of their late evolution.

In *peripheral retina*, where dendritic fields of P-cells cover an area containing approximately 20 midget bipolar cell axons, one must postulate additional mechanisms to gain chromatically specific connections: ganglion cell dendrites must contact, for example, M-cone midget bipolars and avoid L-cone midget bipolar cells. It is unclear how such selectivity might evolve. Is it predetermined by genetic control such as the expression of different markers on chromatically different bipolar cells, or are the connections shaped after birth as a response to visual experience? The latter possibility, which involves synaptic plasticity, has the attraction that such mechanisms are also used for other processes and in other parts of the visual system such as the visual cortex. Hence, the mutation of the cone pigment may once more piggy-back on mechanisms which have evolved for other purposes.

References

Boycott BB, Dowling JE (1969) Organization of the primate retina: light microscopy. Philos Trans R Soc Lond B Biol Sci 255: 109–194

Boycott BB, Hopkins JM (1991) Cone bipolar cells and cone synapses in the primate retina. Vis Neurosci 6: 49–60

Boycott BB, Wässle H (1991) Morphological classification of bipolar cells of the primate retina. Eur J Neurosci 3: 1069–1088

Boycott BB, Hopkins JM, Sperling HG (1987) Cone connections of the horizontal cells of the rhesus monkey's retina. Proc R Soc Lond B 229: 345–379

Calkins DJ, Schein SJ, Tsukamoto Y, Masarachia P, Sterling P (1992) Parallel pathways to midget ganglion cells in macaque fovea (Abstract). Invest Ophthalmol Vis Sci Suppl 33: 2408

Dacheux RF, Raviola E (1990) Physiology of HI horizontal cells in the primate retina. Proc R Soc Lond B 239: 213–230

Gouras P (1992) Retinal circuitry and its relevance to diagnostic psychophysics and electrophysiology. Curr Opin Ophthalmol 3: 803–812

Hopkins JM, Boycott BB (1992) Synaptic contacts of a two-cone flat bipolar cell in a primate retina. Vis Neurosci 8: 379–384

Kaplan E, Lee BB, Shapley RM (1990) New view of primate retinal function. In: Osborne NN, Chader G (eds) Progress in Retinal Research, vol 9. Pergamon, Oxford, pp 273–336

Kolb H (1970) Organization of the outer plexiform layer of the primate retina: electron microscopy of Golgi-impregnated cells. Philos Trans R Soc Lond B Biol Sci 258: 261–283

Kolb H (1991) Anatomical pathways for colour vision in the human retina. Vis Neurosci 7: 61–74

Kolb H, DeKorver L (1991) Midget ganglion cells of the parafovea of the human retina: a study by electron microscopy and serial-section reconstructions. J Comp Neurol 303: 617–636

Kolb H, Boycott BB, Dowling JE (1969) A second type of midget bipolar cell in the primate retina. Philos Trans R Soc Lond B Biol Sci 255: 177–184

Kolb H, Linberg KA, Fisher SK (1992) Neurons of the human retina: a Golgi study. J Comp Neurol 318: 147–187

Kouyama N, Marshak DW (1992) Bipolar cells specific for blue cones in the macaque retina. J Neurosci 12: 1233–1252

Lambrecht HG, Koch KW (1991) A 26 kd calcium binding protein from bovine rod outer segments as modulator of photoreceptor guanylate cyclase. EMBO J 10: 793–798

Lee BB, Martin PR, Valberg A (1988) The physiological basis of heterochromatic flicker photometry demonstrated in the ganglion cells of the macaque retina. J Physiol 404: 323–347

Lennie P, Haake PW, Williams DR (1992) The design of chromatically opponent receptive fields. In: Landy MS, Movshon JA (eds) Computational models of visual processing. MIT Press, Cambridge, pp 71–82

Martin PR, Grünert U (1992) Spatial density and immunoreactivity of bipolar cells in the macaque monkey retina. J Comp Neurol 323: 269–287

Milam AH, Dacey DM, Dizhoor AM (1993) Recoverin immunoreactivity in mammalian cone bipolar cells. Vis Neurosci 10: 1–12

Mollon JD (1986) Understanding colour vision. Nature 321: 12–13

Mollon JD, Bowmaker JK (1992) The spatial arrangement of cones in the primate fovea. Nature 360: 677–679
Mollon JD, Jordan G (1988) Eine evolutionäre Interpretation des menschlichen Farbensehens. Die Farbe 35/36: 139–170
Müller B, Peichl L (1993) Horizontal cells in the cone-dominated tree shrew retina: morphology, photoreceptor contacts, and topographical distribution. J Neurosci 13: 3628–3646
Nathans J (1989) The genes for color vision. Sci Am 260: 42–49
Ohtsuka T, Kouyama N (1986) Electron microscopic study of synaptic contacts between photoreceptors and HRP-filled horizontal cells in the turtle retina. J Comp Neurol 250: 141–156
Paulus W, Kröger-Paulus A (1983) A new concept of retinal colour coding. Vision Res 23: 529–540
Polyak SL (1941) The retina. Chicago University Press, Chicago
Raviola E, Dacheux RF (1990) Axonless horizontal cells of the rabbit retina: synaptic connections and origin of the rod aftereffect. J Neurocytol 19: 731–736
Reid RC, Shapley RM (1992) Spatial structure of cone inputs to receptive fields in primate lateral geniculate nucleus. Nature 356: 716–718
Rodieck RW (1988) The primate retina. In: Steklis HD, Erwin J (eds) Comparative primate biology. Neuroscience vol 4, Liss, New York, pp 203–278
Scholes JH (1975) Colour receptors, and their synaptic connexions, in the retina of a cyprinid fish. Philos Trans R Soc Lond B Biol Sci 270: 61–118
Stell WK, Lightfoot DO (1975) Color-specific interconnections of cones and horizontal cells in the retina of the goldfish. J Comp Neurol 159: 473–502
Vaney DI (1990) The mosaic of amacrine cells in the mammalian retina. In: Osborne NN, Chader G (eds) Progress in retinal research, vol 9. Pergamon, Oxford, pp 49–100
Wässle H, Boycott BB (1991) Functional architecture of the mammalian retina. Physiol Rev 71: 447–480
Wässle H, Boycott BB, Peichl L (1978) Receptor contacts of horizontal cells in the retina of the domestic cat. Proc R Soc Lond B 203: 247–267
Wässle H, Boycott BB, Röhrenbeck J (1989) Horizontal cells in the monkey retina: cone connections and dendritic network. Eur J Neurosci 1: 421–435
Wässle H, Grünert U, Röhrenbeck J, Boycott BB (1990) Retinal ganglion cell density and cortical magnification factor in the primate. Vision Res 30: 1897–1911
Wässle H, Grünert U, Martin PR, Boycott BB (1994) Immunocytochemical characterization and spatial distribution of midget bipolar cells in the macaque monkey retina. Vision Res (in press)
Watanabe M, Rodieck RW (1989) Parasol and midget ganglion cells of the primate retina. J Comp Neurol 289: 434–454
Wiesel TN, Hubel DH (1966) Spatial and chromatic interactions in the lateral geniculate body of the rhesus monkey. J Neurophysiol 29: 1115–1156
Wikler KC, Rakic P (1990) Distribution of photoreceptor subtypes in the retina of diurnal and nocturnal primates. J Neurosci 10: 3390–3401

3.3 Ganglion Cells and Cortical Function

B.B. LEE
Department of Neurobiology, Max Planck Institute for biophysical Chemistry, Göttingen, Germany

3.3.1 Introduction

Recordings from primate ganglion cells offer an unique opportunity to combine results from different disciplines. Ganglion cell activity may be related to retinal anatomy and receptor activity but also may be related to processing at a cortical level and, eventually, to human performance, for it is now widely accepted that the visual system of the Old World primate provides an appropriate model for human vision. I argue here that a comparison of ganglion cell behavior and psychophysical performance can be used to constrain models of cortical function and to infer properties of cortical mechanisms.

It has been a traditional approach to cortical function to interpret cortical properties in terms of the neurons providing afferent input. For example, the Hubel and Wiesel (1962) model of simple and complex receptive fields was constructed by combining geniculate inputs. The validity of this model is still in dispute, demonstrating the difficulty of developing cortical receptive field models in the same way as for ganglion cells or geniculate neurons. This has several reasons. Firstly, techniques of linear systems analysis have proved very effective in describing responses of ganglion cells (e.g., Frishman et al. 1987). A large part of cell behavior can be captured by linear models, and those non-linearities which do occur are not so severe, so complex, or so numerous as to preclude relatively straightforward modification of the linear approach (Victor 1987). Such analysis has been less successful with cortical neurons. In the case of striate cortex, it has proved especially difficult to account for orientation specificity with a linear model. Secondly, peripheral sensory neurons can usually be divided into a limited number of relatively homogeneous classes, often conveniently segregated anatomically, as with the parvocellular (PC) and magnocellular (MC) pathways in the lateral geniculate nucleus of the primate. The number of cortical cell types is likely to be much larger and they are more difficult to separate anatomically. Lastly, peripherally it is usually clear which neurons provide output, and which are concerned with internal housekeeping. In the retina, the former are the ganglion cells and the latter the horizontal and amacrine cells. In the cortex, even if it can be determined which cells provide output, it is often unclear which of them carry an informationally important signal and which serve other purposes; for example, the output from area 17 back to the geniculate may serve as non-specific feedback control (Marrocco et al. 1982).

Neocortex has proved amenable to anatomical, pharmacological and other forms of manipulation; thus many aspects of its synaptology, circuitry, and cell biology have been revealed, but much of this knowledge has been difficult to interpret functionally. The most productive studies of cortical function have been achieved in a behavioral context, in which single unit activity in the cortex has been recorded during performance of a sensory or motor task, as reported by Goldberg in this volume. However, neuronal properties and models so defined are often heuristic in that they are difficult to interpret in terms of sensory input to the cortex or to the region studied.

A satisfactory description of cortical processes must include models of the transformations imposed on incoming signals. I show here three examples in which ganglion cell behavior can be used to suggest properties which might be sought at a cortical level. The first concerns a low-pass filter which must act on signals from the PC pathway. Secondly, responses of striate cortical cells to equal-luminance patterns can be interpreted in terms of responses of the MC pathway. Lastly, recent results on the physiological signals underlying hyperacuity tasks suggest cortical filters with specific, spatially constrained properties.

3.3.2 Low-Pass Filtering of the PC Pathway Signal

There is good evidence that detection of luminance and chromatic modulation is mediated by the MC and PC pathways, respectively (Lee et al. 1989b; Lee et al. 1990). However, psychophysical sensitivity to chromatic modulation falls off rapidly above 5 Hz (e.g., de Lange 1958; Fig. 1) while PC pathway cells respond to much higher frequencies. It is thus necessary to postulate low-pass filtering of the PC pathway signal at a central site. With red-green chromatic modulation it is not possible directly to determine the characteristic of this filter by comparing cell and psychophysical sensitivities since detection of red-green chromatic modulation may be contaminated at higher frequencies by intrusion of a luminance mechanism (Swanson et al. 1987), probably arising from residual responses in the MC pathway (Lee et al. 1989a). With chromatic modulation along a tritanopic confusion line, only PC pathway cells with short-wavelength (S-) cone input give a response. Comparison of physiological and psychophysical sensitivities can then reveal the filter characteristic.

This comparison is shown in Fig. 1 in the upper panel. The filled circles indicate mean responsivity of 13 cells with excitatory S-cone input as a function of temporal frequency. This is compared to psychophysical sensitivity to a similar stimulus. Care was taken in both sets of measurements to ensure that the tritanopic confusion line had been well determined. Cell responsivity rises to a peak about 10 Hz, whereas psychophysical sensitivity falls off rapidly above ca. 5 Hz. Other observers have yielded similar results. The ratios of psychophysical to physiological data give an estimate of the central attenuation of the S-cone signal, as plotted in the right-hand panel. We fitted the data with linear filters; the corner frequency of such a filter is that at which the signal is attenuated by 3 dB, the number of stages gives its steepness. Results such as those in Fig. 1 could be fitted with four or five filters with corner frequencies from 4–8 Hz, depending on the observer.

The central locus of this filter is unknown. Color-opponent neurons in layer 4 of primate area 17 still respond to chromatic modulation at high frequencies (Lennie et al. 1990). Following the temporal properties of chromatic neurons further into area 17 (or elsewhere) might permit localisation of this central filter. In summary, psychophysical flicker mechanisms have been well characterized, as have their possible substrates in the PC and MC pathways. These mechanisms have not yet been pursued into the cortex.

3.3.3 Responses to Equal-Luminance Borders

The minimally distinct border technique (MDB) is parent to many recent studies of performance with equal-luminance patterns. In this technique, an observer is required to adjust the relative intensities of two adjoining, differently colored fields until border distinctness is minimal. This is found to occur at equal luminance, and MDB shares with flicker photometry all the characteristics of a photometric method (Boynton and Kaiser 1968). At equal luminance, the border does not melt completely except with tritanopic pairs of lights. The residual distinctness is directly related to the |M–L| cone difference signal (Valberg and Tansley 1977).

There is good evidence that cells of the MC pathway provide the physiological substrate for MDB (Kaiser et al. 1990; Valberg et al. 1992). Its response to a moving border goes through a minimum at equal luminance. Furthermore, cells of this pathway give a residual response to equal-luminance borders which very closely matches psychophysical residual distinctness (Valberg et al. 1992). This is illustrated in Fig. 2. Equal-luminance borders were moved back and forth across an MC cell receptive field. With the Wh–607 nm and the Wh–498 nm border, there is a residual response to both directions of border movement. The residual response is absent to a pair of lights lying close to a tritanopic line, the Wh–570 nm border.

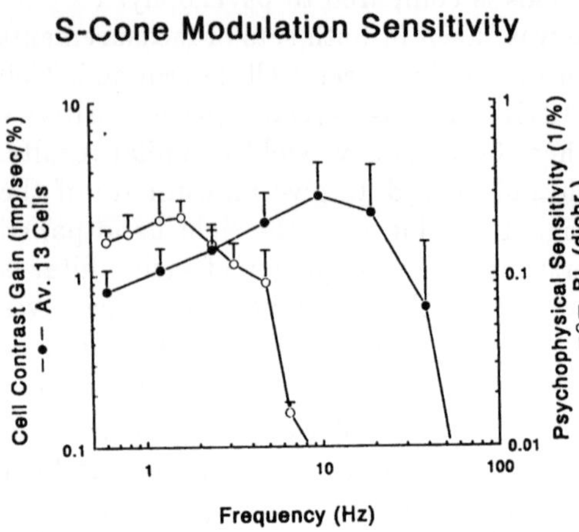

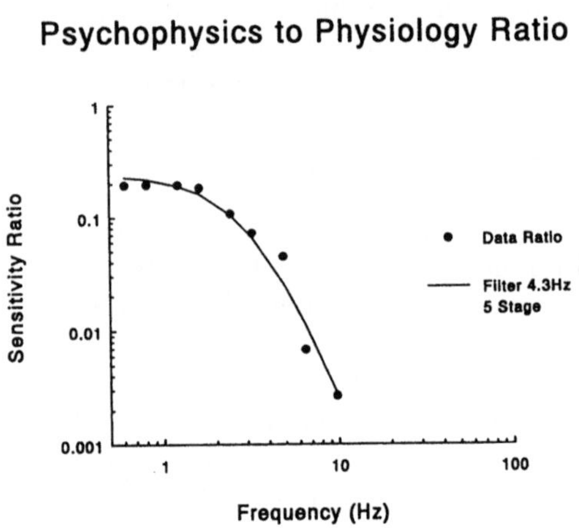

Fig. 3.3–1. *Top*, mean responsivity of 13 +S-(ML) cells plotted together with psychophysical sensitivities of a human observer. Modulation lay along a tritanopic confusion line to modulate only S-cones. Cell responsivity is the S-cone modulation required for a 20 impulse/s response. Human data are the mean of three thresholds using method of adjustment, again calibrated in terms of S-cone modulation. *Bottom*, points are ratios of the two data sets. The curve was fitted using a least-squares criterion and is the characteristic of a filter with parameters shown (Yeh, Lee, and Kremers, unpublished observations)

This property of the MC pathway is of importance when interpreting cortical responses. Responses to equal-luminance patterns have been reported from several cortical regions, including MT, where input from the MC pathway is thought to be dominant. In area 17 such responses were first reported by Gouras and Krüger (1979) and were attributed to PC pathway input. Figure 3 shows examples of such responses taken from Creutzfeldt et al. (1987). Equal-luminance borders, consisting of a square set in a differently colored surround, were moved across the receptive field of a cell from the awake behaving monkey. The cell can be seen to respond to both edges. The 550 nm/white edge is not far from a tritanopic confusion line

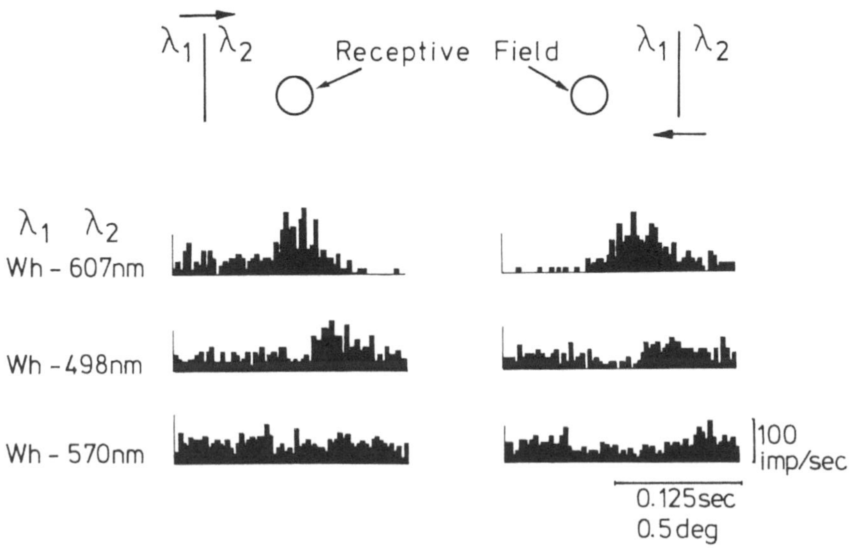

Fig. 3.3–2. Responses of MC pathway cell to equal-luminance borders. The border separated the colors shown. To the white-red and white-blue borders there is a response to both directions of border movement, but not to the white-yellow border, which lies close to a tritanopic confusion line. Average of 20 sweeps, 4 msec per bin (Lee, Valberg, and Kaiser, unpublished observations)

and gives a much smaller response than in the other two conditions. This illustrates how knowledge of the afferent signal may help in interpreting cortical behavior.

3.3.4 Ganglion Cell Responses and Hyperacuity

Human observers are able to make spatial judgements to a precision much finer than the cone mosaic. Recent results suggest that only cells of the MC-pathway deliver signals of sufficient spatial precision to provide a physiological substrate for hyperacuity tasks (Lee et al. 1993; Lee, Wehrhahn, Westheimer, Kremers, unpublished observations). In order to interpret ganglion cell signals in this way it is necessary to consider the central hyperacuity mechanism which acts upon them. The nature of the afferent signal can then be used to constrain central models and to suggest the kind of cortical spatial filter which might be involved.

A hyperacuity which is relatively simple to interpret in terms of cortical mechanism is the detection of small displacements. If an edge suddenly jumps across the retina, a movement of ca. 12 arc seconds can be detected in the fovea (Basler 1909), which is well within the hyperacuity range. Movement in one direction causes an increase in illumination of a strip of

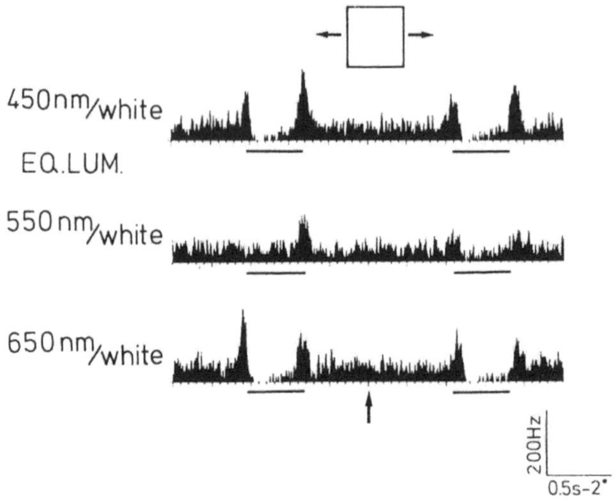

Fig. 3.3–3. Responses of cell from area 17 of the awake behaving macaque to equal-luminance borders. Differently colored squares were inserted into a white surround and moved back and forth across the receptive field while the animal was fixating. The red-white and blue-white contours evoke vigorous responses, but responses are less vigorous to the green-white stimulus, which is closer to a tritanopic confusion line. Note that this holds independent of movement direction. (From Creutzfeldt et al. 1987)

retina, and movement in the other direction a decrease, so that whether there is a signal from on- or from off-center cells gives an unambiguous cue as to movement direction. Measurement of responses of MC cells as a function of jump amplitude and contrast has shown that for detection of edge movement direction signals from only one or two cells are sufficient (Lee et al. 1993). How many cells are necessary, i.e., is it possible to estimate how many cells are actually used by a central detection mechanism?

Figure 4B shows psychophysical thresholds in the parafovea for detection of the direction of jump of a 40% contrast stimulus as a function of length. It can be seen that a decrease in length of 100 arc minutes causes only minor elevation in threshold until at lengths below about 20 minutes a more rapid rise occurs. It is possible to calculate the number of cells expected to signal the movement of edges of different lengths. An artificial matrix with similar statistics to those expected of on- or off-center MC cells in the parafovea is shown in Fig. 4A, with receptive field diameter of parafoveal MC cells also indicated. Density and distribution were derived from the literature (Perry et al. 1984; Silveira and Perry 1991; Wässle et al. 1981; Wässle et al. 1990). If an edge is placed on this matrix as indicated and displaced to left or right, it is possible to calculate how many cells will reach

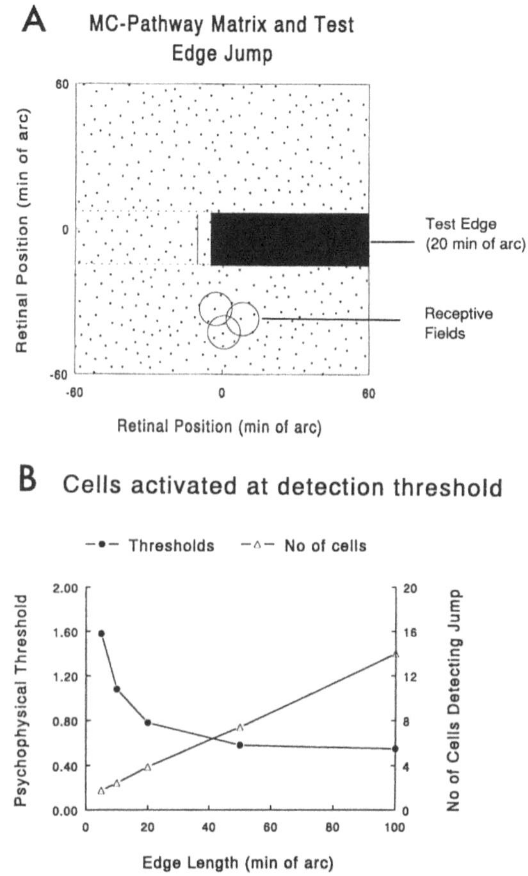

Fig. 3.3–4. A Ganglion cell matrix of density equivalent to that of on- or off-center MC-cells in the parafovea. Receptive field center diameter is indicated. An edge is considered as jumping left or right across this matrix. **B** Psychophysical thresholds for a 40% contrast edge plotted as a function of edge length and compared to the number of cells signalling the presence of a jump, calculated as described in the text. Use of a line rather than an edge has little effect on jump thresholds

threshold based on the physiological data described elsewhere (Lee et al. 1993). This number of cells is also plotted as a function of edge length in Fig. 4B. With short edges two or three cells are activated. With longer edges many more cells are activated, but this increase translates into only a minor threshold improvement. The data suggest that with short edges signals from few ganglion cells are necessary and sufficient for displacement detection, implying very efficient central use of the afferent signal. The lack of substantial threshold reduction with longer edges implies that central mechanisms are unable to integrate along the edge to an appreciable extent. This finding is often expressed in psychophysical models of hyperacuity by postulating that the spatial filters involved are orientation specific but are only two to three times longer than they are wide and thus cannot summate along extended line or edge segments (Wilson 1986); parafoveal

MC cells have diameters of ca. 12–14 arc minutes. This implies that cortical channels for hyperacuity should be embodied in orientation-specific neurons with small receptive fields not much longer than they are wide. Studies of single cells in monkey striate cortex have demonstrated some cells which meet these requirements (Parker and Hawken 1985). Such cells presumably receive limited convergence from a few MC cells.

3.3.5 Conclusions

I have argued that certain properties of cortical neurons can be inferred by comparing ganglion cell responses with psychophysical performance. Elsewhere, Lennie (1991) suggests a strongly model-based approach to elucidate cortical function. Both these suggestions are prompted by the difficulties encountered when trying to model cortical properties neurophysiologically. Of course, it is possible that such approaches are too simple to describe a cortical network in which, through distributed processing, cortical neurons tailor their properties to the behavioral task at hand. Although this implies a lack of cortical specificity reminiscent of early concepts of cortical function, modern neuroanatomical data suggest complex but highly specific cortical connectivity, and it is possible that a lack of neurophysiological specificity merely reflects a lack of sufficient sophistication in the stimuli presented. More close attention to the afferent input under conditions in which performance is precisely defined may help solve some of these difficulties.

Acknowledgements. I wish to thank the many colleagues who have contributed to the experiments described. In particular, I thank Gerald Westheimer for permission to include unpublished psychophysical data on the effect of length on displacement thresholds in the parafovea.

References

Basler A (1909) Über das Sehen von Bewegungen IV. Pflüg Arch Physiol 128: 427–430

Boynton RM, Kaiser PK (1968) Vision: the additivity law made to work for heterochromatic photometry with bipartite fields. Science 116: 366–368

Creutzfeldt OD, Weber H, Tanaka M, Lee BB (1987) Neuronal representation of spectral and spatial stimulus aspects in foveal and parafoveal area 17 of the awake monkey. Exp Brain Res 68: 541–564

de Lange H (1958) Research into the dynamic nature of the human fovea-cortex systems with intermittent and modulated light. Journal of the Optical Society of America 48: 779–789

Frishman LJ, Freeman AW, Troy JB, Schweitzer-Tong DE, Enroth-Cugell C (1987) Spatiotemporal frequency responses of cat retinal ganglion cells. J of Gen Physiol 89: 599–628

Gouras P, Krüger J (1979) Responses of cells in the foveal visual cortex of the monkey to pure color contrast. J Neurophysiol 42: 850–860

Hubel DH, Wiesel TN (1962) Receptive fields, binocular interaction, and functional architecture in the cat's visual cortex. J Physiol 160: 106–154

Kaiser PK, Lee BB, Martin PR, Valberg A (1990) The physiological basis of the minimally distinct border demonstrated in the ganglion cells of the macaque retina. J Physiol 422: 153–183
Lee BB, Martin PR, Valberg A (1989a) Nonlinear summation of M- and L-cone inputs to phasic retinal ganglion cells of the macaque. J Neurosci 9: 1433–1442
Lee BB, Martin PR, Valberg A (1989b) Sensitivity of macaque retinal ganglion cells to chromatic and luminance flicker. J Physiol, London 414: 223–243
Lee BB, Pokorny J, Smith VC, Martin PR, Valberg A (1990) Luminance and chromatic modulation sensitivity of macaque ganglion cells and human observers. J Opt Soc Am A 7: 2223–2236
Lee BB, Wehrhahn C, Westheimer G, Kremers J (1993) Macaque ganglion cell responses to stimuli that elicit hyperacuity in man: detection of small displacements. J Neurosci 13: 1001–1009
Lennie P (1991) The perceptual significance of cortical organization. In: Valberg A, Lee BB (eds) From pigments to perception. Plenum, New York, London, pp 249–260
Lennie P, Krauskopf J, Sclar G (1990) Chromatic mechanisms in striate cortex of macaque. J Neurosci 10: 649–669
Marrocco RT, McClurkin JW, Young RA (1982) Modulation of lateral geniculate cell responsiveness by visual activation of the corticogeniculate pathway. J Neurosci 2: 256–263
Parker A, Hawken M (1985) Capabilities of monkey cortical cells in spatial-resolution tasks. J Opt Soc Am A 2: 1101–1114
Perry VH, Oehler R, Cowey A (1984) Retinal ganglion cells that project to the dorsal lateral geniculate nucleus in the macaque monkey. Neuroscience 12: 1110–1123
Silveira LCL, Perry VH (1991) The topography of magnocellular projecting ganglion cells (M-ganglion cells) in the primate retina. Neuroscience 40: 217–237
Swanson WH, Ueno T, Smith VC, Pokorny J (1987) Temporal modulation sensitivity and pulse detection thresholds for chromatic and luminance perturbations. J Opt Soc Am A 4: 1992–2005
Valberg A, Tansley BW (1977) Tritanoptic purity-difference function to describe the properties of minimally distinct borders. J Opt Soc Am 67: 1330–1336
Valberg A, Lee BB, Kaiser PK, Kremers J (1992) Responses of macaque ganglion cells to movement of chromatic borders. J Physiol 458: 479–602
Victor JD (1987) The dynamics of the cat retinal X-cell centre. J Physiol 386: 219–246
Wässle H, Boycott BB, Illing R-B (1981) Morphology and mosaic of on- and off-beta-cells of the cat retina and some functional considerations. Proc Roy Soc B 212: 177–195
Wässle H, Grünert U, Röhrenbeck J, Boycott BB (1990) Cortical magnification factor and the ganglion cell density of the primate retina. Vision Res 30: 1897–1911
Wilson HR (1986) Responses of spatial mechanisms can explain hyperacuity. Vision Res 26: 453–469

Chapter 4

Microcircuitry I

4.1 Introduction: Searching for the Operational Principle of Cortex

J. Bullier
Cerveau et Vision, INSERM, Bron/Lyon, France

"From the anatomical and neurophysiological point of view, the generality or uniformity of function and structure of all cortical areas must be emphasized". This quotation from Otto Creutzfeldt's work (*Cortex Cerebri* 1993) summarizes one of the long-held views that has guided his investigations in different regions of the brain of various species.

Throughout the years, this opinion has held true and modern techniques of anatomy have further strengthened it. Thus, we know that throughout the cortex and across many species, 80% of the neurons are excitatory, use glutamate as a transmitter, and send axons to other parts of the brain as well as recurrent collaterals arborizing locally. The remaining neurons are inhibitory cells with local projections using GABA as a neurotransmitter (see Chap. 4.5).

Such a uniform anatomical *Bauplan* suggests that the principle of operation is also universal for all cortices. In contrast to our progress concerning the structural knowledge of cortical wiring, we must admit that in the last 30 years of neurophysiological research, we have not made major advances in the search for this operational principle. This is well illustrated by the fact that the questions posed in Chap. 4.3 are put in the same terms as in the early papers by Creutzfeldt and collaborators on the relative importance of excitation and inhibition for orientation selectivity in visual cortex neurons.

Most of the past work in anatomy and physiology of the cortex has demonstrated the presence of specific groups of cells and pathways. Words such as cortical columns, patchy organization, parallel pathways, anatomical segregation, and cell types well describe the successful areas of the last 20 years of cortical research. Teasing out separate functional subgroups is important in opening up the possibility of isolating circuits devoted to specific functions (see Chap. 4.5). Such discoveries tell us that the cortex usually does not put together apples and oranges, so to speak (although it sometimes does it!), but they do not provide us with the principle of operation of cortex.

The discovery of specific functional groups of cells in the cortex has led to the idea that the receptive field properties of cortical neurons could be explained by a sort of construction game. In this kind of model, the properties of cell B derive from those of cell A, because cell A drives cell B. A classical example of the construction game is the so-called hierarchical organization of simple and complex cells in area 17. This model has been refuted by almost every attempt to test it, but nonetheless remains solidly anchored to all the modern textbooks, for want of a better one.

To be able to play the construction game with cortical neurons, it is necessary to identify interconnected neurons. Physiological attempts to do so, whether by electrical stimulation or cross-correlation techniques, have been remarkably unsuccessful in the cortex. For interneuron distances of more than a few hundred microns, it is practically impossible to identify a pair of interconnected neurons. This does not mean that the activities of cortical neurons are independent, since cross-correlation studies demonstrate that the firing of many neurons within the same area or in different areas tend to be synchronized. Even when a pair of interconnected pyramidal neurons is identified by double intracellular recording, the amplitude of the excitatory postsynaptic potentials (EPSP) evoked in one cell by a spike in the other is of the order of a few hundred microvolts, suggesting that the synaptic effect of one neuron on another is very weak in the cortex. This weakness of neuronal interactions in the cortex contrasts with the situation in peripheral or thalamic regions, in which the construction game has been so successful in explaining the receptive field properties of different types of neurons.

The weak neuron-to-neuron interactions observed in the cortex are no real surprise to anatomists, who have recognized for some time the enormous convergence and divergence of corticocortical connections. Thus, one cortical pyramidal neuron projects to several thousand other pyramidal cells, and the contact between two cortical pyramidal neurons rarely exceeds a few synapses (see Chap. 4.6).

Because of the sparsity of cortical connections, it appears necessary to consider groups of interconnected neurons in order to understand the operational principles of cortex. Such groups of neurons, when firing together, have a chance to exert some action on other cortical neurons, whereas a single neuron has no chance. The study of interactions between groups of neurons has been facilitated by the recent revival of cortical intracellular recording, which had been largely left aside during the last 20 years because of the difficulties of in vivo intracellular recording. Much of the impetus for this revival has come from the development of the in vitro slice preparation, which enables the morphological, pharmacological and electrical characterization of various neuron types (see Chap. 4.2). The power of this approach is well demonstrated in Chap. 4.4, which offers a convincing explanation of the neuronal basis of spindle waves in the electroencephalogram (EEG), a question which was of great interest to Otto Creutzfeldt.

Chapter 4.3 demonstrates how successfully the fineness of the analysis provided by intracellular recording can be combined with the idea that, in cortex, groups of cells and not just single cells should be considered. They present an original solution to the nagging question of the importance of inhibition for orientation selectivity in visual cortex. Since the model proposed in the 1960s by Hubel and Wiesel, this question has been posed to every young investigator entering the field, in the same way that the old zen masters used the kô-ans, those impossible questions, to guide the uninitiated on the road to understanding the principles of life.

4.2 Dendritic Geometry of Pyramidal Neurons in the Visual Cortex of the Rat

A.U. LARKMAN, G. MAJOR, K.J. STRATFORD, AND J.J.B. JACK
University Laboratory of Physiology, Oxford University, United Kingdom

4.2.1 Introduction

For the past several years, a major interest of our laboratory has been the quantitative study of the soma-dendritic morphology of pyramidal neurons in the visual cortex of the albino rat. We have two broad objectives in mind for this type of work. The first is the hope that such information will be of use in understanding the connectivity of the cortex; for example, it may be helpful to have information about the number and extent of the dendrites of a particular type of neuron or to have estimates of the number of dendritic spines and their distribution within the depth of the cortex. Second, quantitative morphology is essential for developing electrical models of dendritic systems, which themselves can aid in the understanding of the integrative properties of neurons.

Our basic approach has been to inject individual neurons intracellularly with markers, originally using horseradish peroxidase, more recently biocytin. All our injections have been performed using slices of visual cortex maintained in vitro. This approach has, in our view, the overwhelming advantages of allowing individual neurons to be stained in isolation, allowing easy reconstruction across relatively thin tissue sections, and enabling us to make stable intracellular recordings from the labeled cells. The approach is not without its disadvantages, however, principally the loss of parts of the dendritic and axonal arbors during slice preparation and the risk of degenerative changes during maintenance in vitro. We suggest that simple precautions can minimize these problems (see Larkman and Mason 1990), although the unavoidable limitation remains that consideration of the axon arbors must be limited to very local collateral systems. To date, we have only investigated in detail pyramidal neurons in layers 2/3 and 5, and we will not deal here with axonal morphology at all.

4.2.2 Cell Classes

We considered that all of the pyramidal neurons in layer 2/3 could be lumped together in a single class, albeit showing considerable variation, while we split those in layer 5 into two classes. Obviously, other classifications and further subdivisions are possible, but we wished to keep the number of classes small. Additionally, the classification we have used correlates with some aspects of the intrinsic electrophysiology of these cells (Larkman and Mason 1990; Mason and Larkman 1990).

Morphologically, the layer 2/3 cells appeared to form a spectrum without obvious discontinuities. In general, the apical dendrite tended to branch to form the terminal arbor close to the upper boundary of layer 2/3 irrespective of the position of the soma. Thus, cells deep in the layer had longer apical trunks than more superficial cells; cells very close to the layer 1 boundary often had no trunks at all. Electrophysiologically, the cells were all similar in showing a non-bursting, adapting pattern of spike firing and minimal time-dependent rectification.

Within layer 5, however, the situation was rather different. Some cells had thick apical trunks that formed terminal arbors in layer 1, while others had slender, tapering trunks that ended, without an obvious arbor or tuft, at some level within layer 2/3. We felt that these represented separate cell classes which we termed 'thick' and 'slender' L5 cells respectively. Their intrinsic electrophysiological properties correlated with this classification: thick cells fired distinctive bursts of action potentials and, on average, had lower input resistances than slender cells, whose firing patterns were also more conventional and resembled layer 2/3 cells. However, in almost every respect that we examined, the slender L5 cells were more variable than the other classes, and it may be that further subdivision of this class would be useful. Examples of each of the three cell classes are shown in Fig. 1.

The splitting of layer 5 cells into these two classes has received further support from recent studies in our laboratory by Ekkehard Kasper and Joe Lübke, who showed by back-labeling with fluorescent microspheres that the thick, burst-firing cells essentially all project to the superior colliculus. Many thick, bursting cells also project to the pontine nuclei (Wang and McCormick 1993). None of the slender cells project to the colliculus, but many project via the callosum to the contralateral hemisphere. Thus, it appears that this classification, based initially on soma-dendritic morphology, aligns with those based on intrinsic electrophysiology and axonal projection targets. There has also been the suggestion that this sort of classification also correlates with differences in local axon arborization pattern (Chagnac-Amitai et al. 1990).

4.2.3 Dendritic Morphology

Despite the differences between the cell classes, there were many aspects of dendritic morphology common to all the cells studied.

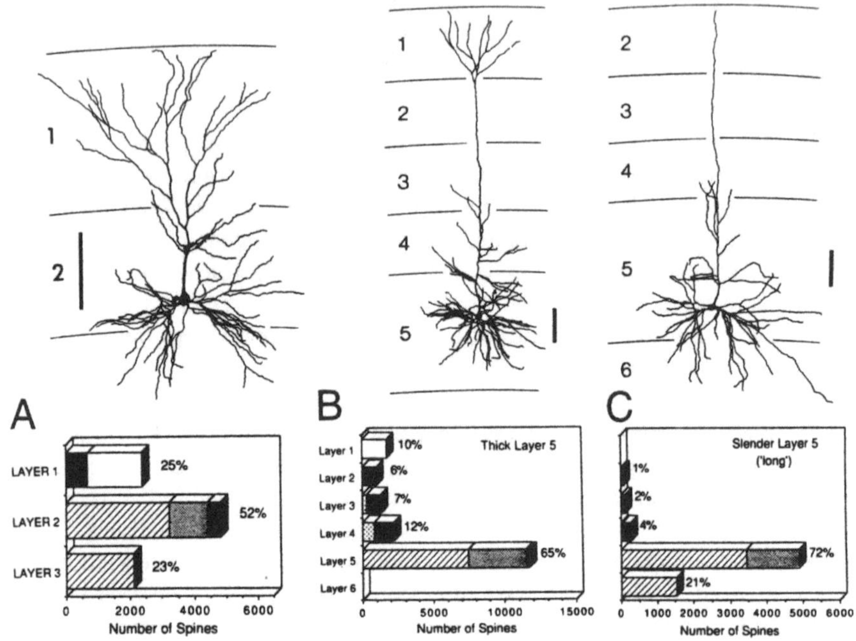

Fig. 4.2–1. Camera lucida reconstructions of examples of **A** layer 2/3, **B** thick layer 5 and **C** slender layer 5 cells. *Scale bars* 100 μm. Beneath each cell is a bar graph showing the absolute numbers and percentages of spines in each of the cortical layers. The contributions made by the different types of dendrite are shown in the bar graphs as follows: *hatched*, basals; *stippled*, obliques; *black*, apical trunk; *plain*, terminal arbor. (Modified from Larkman 1991c)

Basal Dendrites. The key feature of the basal dendrites of all the cell classes was that the majority of the dendritic branch points were located close to the soma. Thus, most of the thicker, intermediate segments were also short, while the thinner, densely spiny terminal segments were generally much longer and accounted for about 90% of the total dendritic length of the basal trees (Larkman 1991a). Most cells had between three and eight basal trees, and the diameters of the stem segments leaving the soma varied widely, even for a single cell, but it was striking that the thicker stem segments did not give rise to trees with thicker or longer terminal segments, but rather to trees with a greater number of terminal segments (Larkman 1991b). Most of the basal terminal segments on any individual cell ended at similar path lengths from the soma. In other words, the basal dendrites were all of similar length, which is not the impression given by Golgi studies of neocortical pyramidal cells (e.g., Davies and Katz 1983; Petit et al. 1988).

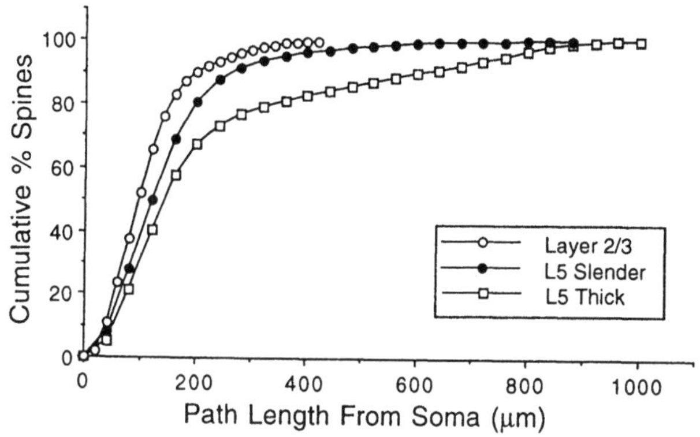

Fig. 4.2–2. The mean cumulative distribution of spines as a function of path length from the soma. Although both types of layer 5 cells can have small numbers of spines further than 700 μm from the soma, the majority are within 200 μm

Apical Dendrites. We divided the apical dendritic system into trunk, oblique and terminal arbor components. The great majority of apical oblique trees arose from the proximal part of the apical trunk, and the branching pattern of these proximal oblique trees was very similar to the basals. The most distal oblique trees tended to be less highly branched, with terminal segments that were shorter, thinner and less spiny — features they share with the terminal arbor segments. These properties led us to propose that the basal and the proximal majority of oblique dendrites together sample a roughly spherical volume of cortex centred about the soma.

4.2.4 Spine Distributions

The majority of excitatory synaptic inputs to cortical pyramidal cells are made via dendritic spines (Colonnier 1968; Diamond et al. 1970). Spines can therefore serve as markers, visible with the light microscope, of the locations of most excitatory inputs. We have attempted to estimate both the total numbers of spines on individual cells and their distribution with respect to dendrite type and cortical layer. The total numbers that we obtained were higher than previous Golgi-based estimates and ranged from a mean of 7950 ± 2700 for layer 2/3 cells, through 8650 ± 3100 for slender L5 cells to 14 950 ± 3350 for thick L5 cells (Larkman 1991c). Perhaps surprisingly, the basal and oblique dendrites together contributed over 65% of the total spines for all three cell classes (80% for layer 2/3 cells). In most cases, the majority of spines were located within the cortical layer containing the soma (Fig. 1); the only exceptions were when the soma was close to a laminar boundary.

We also examined the distribution of spines with respect to distance or more precisely, dendritic path length from the cell soma. We calculated the number of spines that lay within each of a number of concentric shells, each perhaps 10 or 20 μm in thickness, centred on the cell soma. The cumulative proportion of the total number of spines increased steeply and roughly linearly for the first 150–200 μm, depending on cell class, then showed a quite abrupt decrease in slope (Fig. 2). This inflection corresponds to the distance at which the basal and oblique dendrites terminate. For layer 2/3 and slender L5 cells, approximately 80% of the total spines occurred proximal to this point, and even for thick L5 cells the proportion was about 65%. This high proportion emphasizes the quantitative importance of the basal and proximal oblique dendrites.

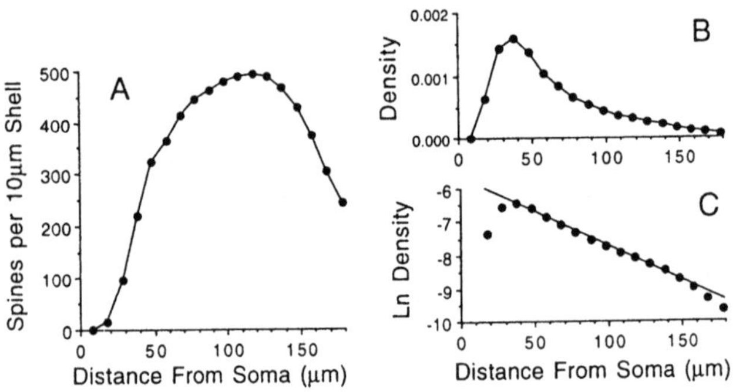

Fig. 4.2–3. **A** Graph showing the mean number of spines contained within each of a series of shells, 10 μm thick, centred on the soma for layer 3 cells. **B** Sampling density, defined as the number of spines per shell divided by the volume of that shell, as a function of distance from the soma. Units are μm^{-3}. **C** Semilogarithmic plot of **B**, showing how the sampling density declines approximately exponentially after an initial rapid rise. Units are $\ln(\mu m^{-3})$

We took this type of analysis a stage further in an attempt to estimate a measure of the intensity with which a given neuron sampled the volume of cortex around it. We restricted the analysis to the roughly spherical volume sampled by the basal and proximal apical trunk and oblique dendrites, and we assumed that the dendrites showed radial uniformity — that is they projected from the soma along radii without bias in favor of any particular direction. Again we calculated the number of spines occurring within shells, this time of 10 μm thickness (Fig. 3A). We then divided that number by the volume of the corresponding shell, to obtain an estimate of the number of spines per unit volume of cortex contributed by a given cell, as a function of distance from the soma. We found that this approximate measure of 'sampling density' increased rapidly to reach a maximum some 40 μm from the soma, then declined approximately exponentially until the end of the basal and oblique dendrites, 150–200 μm from the soma (Fig. 3B,C). The sampling density declined e-fold every 50 μm or so from the soma. These sorts of calculations may be useful in extending estimations of local cortical connectivity and connection probability and its dependence on distance, which can then be compared with estimates based on electrophysiology (see Mason et al. 1991; Nicoll and Blakemore 1993).

4.2.5 Electrical Geometry

Our second major interest is to develop computer models of the passive electrical geometry of the dendrites of pyramidal cells. It is clear that the membranes of pyramidal cell dendrites can contain a range of voltage-dependent ionic conductances and so may be capable of a range of non-passive behavior. Passive models are particularly relevant under the restricted and perhaps artificial conditions when the membrane potential is well below threshold and there is only minimal synaptic input. Nevertheless, these are the conditions that prevail during many experiments using in vitro slice preparations. These preparations are used for a sizeable proportion of cellular neuroscience investigations, and they have generated some major controversies. Some aspects of these controversies might be clarified if reliable passive electrical models of the neurons involved were available. In addition, it is important to establish firm foundations for single-cell models incorporating active properties. Thus, in our opinion, passive modelling remains a challenging area of study with extremely important, if circumscribed, implications.

The cable properties of neocortical pyramidal cells in slices were investigated in a careful study by Stafstrom et al. (1984). Unfortunately, these authors used only electrophysiological data, and their analysis hinged on the assumption that these cells could be represented as single equivalent cylinders. One of the requirements for such a representation is that all the dendrites must terminate at the same electrical distance from the soma. It is clear that pyramidal cell dendrites are usually of very different physical lengths, so we investigated whether this was also the case for their electrical lengths. The presence of dendritic spines will increase the relative electrical lengths of spiny segments. For many purposes, when considering the flow of current outward through the spine membrane, the resistance of the spine head membrane is so large relative to the axial resistance of the spine neck that the geometry of the spine can be ignored and the spine membrane area can be incorporated into the dendritic shaft. We use the following simple transformations:

$$F = \text{Area of segment including spines/area of segment ignoring spines}$$

$$l^* = lF^{2/3}$$

$$d^* = dF^{1/3}$$

where l^* and d^* are the transformed segment lengths and diameters. For many spiny segments, the value of F was close to 2, that is, the presence of dendritic spines approximately doubled the membrane area of the segment (Larkman et al. 1992).

The relative electrical lengths of dendritic segments will also depend upon their physical lengths and diameters, and, assuming that their membrane electrical properties are uniform, a simple transformation can be applied to each segment to take these into account:

$$l' = 2l/d^{1/2}$$

where l' is the relative electrical length, also known as the 'morphotonic length' (Larkman et al. 1992), l is the physical length and d the diameter of each segment.

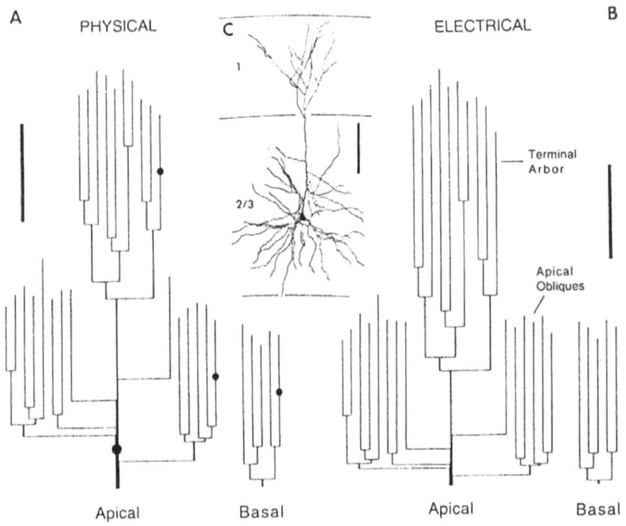

Fig. 4.2–4. **A** Dendrogram of the apical and one typical basal dendritic tree for a layer 2/3 cell (drawn in **C**). Dendritic lengths are shown to scale, and line thickness is an approximate indication of diameter. *Scale bars*, 100 μm. **B** Dendrograms showing relative electrical lengths for the same trees after the transformation of segment lengths according to their spine densities and diameters. The synaptic input sites used for the simulations shown in Fig. 6 are indicated by *filled circles*. *Scale bar*, 3.4 $\text{cm}^{1/2}$. (Modified from Larkman et al. 1992)

Since distal and terminal segments tend to be both thinner and more spiny than proximal segments, the effect of both these transformations is to make distal segments longer relative to proximal ones (Fig. 4). Thus, in electrical terms, the apical trunks are shorter relative to the basal and oblique dendrites than their physical lengths would suggest. However, the terminal arbor segments tend to become extremely long, and it is clear from Fig. 4 that the dendrites of conventionally shaped pyramidal cells are not all of the same electrical length. The apical trunk/terminal arbor system can be more than twice as long, in electrical terms, as the basals and proximal obliques. In general, therefore, these cells cannot be represented as single equivalent cylinders, and analyses based on that assumption will be in error. A further point is that the most proximal segments of the apical trunk, which are generally very thick and only sparsely spiny, are very short in relative electrical terms. Thus most of the oblique trees arise from the apical trunk electrically very close to the soma.

4.2.6 Electrical Models

While the simple morphological transformations outlined above can give useful insights into the relative electrical lengths of dendrites, some form of model that includes information about the electrical membrane properties of the cell is needed to predict its electrical responses. We have used both the conventional compartmental approach (Stratford et al. 1989) and also analytical mathematical solutions of the cable equations that can be applied to dendrites of arbitrary geometry (Major 1993; Major et al. 1993a,b). Compartmental models are highly flexible and can easily be adapted to include voltage-dependent conductances if required. On the other hand, the analytical solutions are very efficient computationally and give explicit information about the time constants and amplitudes of the exponential components generated by the model. For the applications to be described here, both approaches make use of the same information and yield identical results.

For each neuron, the model is constructed from a detailed representation of the soma-dendritic morphology, with spines incorporated as described above, together with the electrical parameters C_m (the specific membrane capacitance), R_m (the specific membrane resistivity) and R_i (the cytoplasmic resistivity). For all the cases considered here, these are taken to be the same for all the dendritic segments. There is now overwhelming evidence that during conventional sharp-electrode recording there is an additional conductance at the soma, perhaps in part due to an imperfect electrical seal between the electrode and the cell membrane (Iansek and Redman 1973; Durand and Carlen 1985; Blanton et al. 1989; Clements and Redman 1989; Coleman and Miller 1989; Edwards et al. 1989; Stratford et al. 1989; Pongracz et al. 1991; Spruston and Johnston 1992). We therefore include a 'somatic shunt conductance' (g_{shunt}) at the recording site. In earlier modelling work, it has often been assumed that conventional values could be used for C_m ($1.0\,\mu F\ cm^{-2}$) and R_i ($100\,\Omega\ cm$) and that R_m could be obtained from the cell membrane time constant (τ_m) using the relationship:

$$\tau_m = R_m \times C_m$$

However, the presence of a somatic shunt invalidates this relationship. Our approach has been to record the cell voltage response to a small, brief pulse of injected current and then to use an optimizing routine to vary all four electrical parameters of the model until it produces a response to an identical pulse that best matches that of the cell. This approach, pioneered by Clements and Redman (1989), avoids the difficulties associated with exponential 'peeling', but in some studies it has been assumed that the morphological measurements are exactly correct. This is very unlikely to be the case, and in our hands entirely plausible measurement errors, together with uncertainty over the exact value of C_m and the presence of noise in the recorded pulse response, could lead to widely different optimal parameter values, all of which gave indistinguishably good fits to the recorded voltage response (Stratford et al. 1989). To try and constrain this degree of non-uniqueness, we have tried several different approaches. During tight-seal, whole-cell recording, it is likely that g_{shunt} is zero and so can be omitted from the model, which enables the other parameters to be determined with greater precision. Working with Dr. P. Jonas and Prof. B. Sakmann in Heidelberg, we have recorded from, injected, and reconstructed hippocampal pyramidal neurons using whole-cell recording in slices. For these cells we were able to find a reasonably narrow range of values for C_m and R_i. Since C_m, and probably R_i, are unlikely to vary greatly between different pyramidal cell types, we have used these values to constrain the non-uniqueness shown by our original sample of neocortical cells. Additionally, we have used channel and neurotransmitter blockers to improve the linearity and reduce the noise in our sharp-electrode recordings. The consensus from these approaches is a range of parameter values that seem likely to be appropriate for these cells. These are:

$$C_m = 0.65 - 0.85\,\mu F cm^{-2}$$

$$R_i = 200 - 350\,\Omega cm$$

$$R_m = 30\,000 - 100\,000\,\Omega cm^2$$

We can now use these values to predict the likely passive electrical properties of neocortical pyramidal cell dendrites.

4.2.7 Implications of Electrical Geometry

Since pyramidal cell dendrites are of very different relative electrical lengths, there is little point in quoting single-number values for electrotonic lengths for these cells. It is of more value to derive length values for the basal/oblique and terminal arbor dendrites separately. Using values in the middle of the range quoted above ($C_m = 0.75\,\mu\text{Fcm}^{-2}$, $R_i = 275\,\Omega\text{cm}$ and $R_m = 65\,000\,\Omega\text{cm}^2$) for all the cell classes, the basal and oblique dendrites were in the region of 0.3 space constants (λ) long, while the terminal arbors were about 0.8 λ for layer 2/3 cells and 1.7 λ for thick layer 5 cells. Perhaps still more informative is to estimate the distribution of dendritic spines, serving as markers for excitatory synapses, with electrotonic distance from the soma. Cumulative spine distributions for the three cell classes are shown in Fig. 5. It is clear that although a minority of spines are quite remote, the majority are located very close to the soma in electrical terms. For all the cell classes, some 60% of spines are likely to be within 0.3 λ.

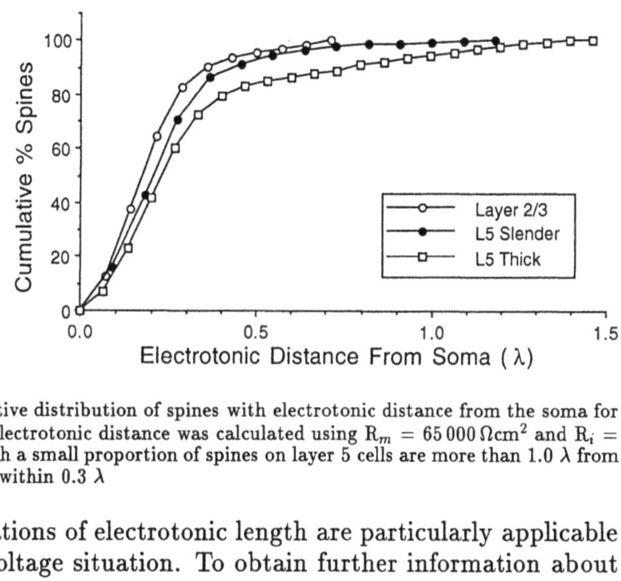

Fig. 4.2–5. Mean cumulative distribution of spines with electrotonic distance from the soma for the three cell classes. The electrotonic distance was calculated using $R_m = 65\,000\,\Omega\text{cm}^2$ and $R_i = 275\,\Omega\text{cm}$. Note that although a small proportion of spines on layer 5 cells are more than 1.0 λ from the soma, 60% or more are within 0.3 λ

The simple calculations of electrotonic length are particularly applicable to the steady state voltage situation. To obtain further information about the effects of cable filtering on individual synaptic inputs located at various points on the dendritic tree, it is necessary to specify a time course for the synaptic current. The faster the current, the greater will be the degree of smoothing by the dendritic cable. The picture emerging from the study of cortical pyramidal cells (Hestrin 1992) and other cell types, such as layer 4 stellate cells (Stern et al. 1992), cerebellar granule cells (Silver et al. 1992), mossy fiber inputs to CA3 pyramidal cells (Jonas et al. 1994) and patches from hippocampal cells (Colquhoun et al. 1992), is that synaptic currents mediated by AMPA-type glutamate receptors are likely to show extremely fast kinetics. Rise times of less than 1 ms and decay time constants of 1–4 ms seem likely.

We have explored the effects of cable filtering by placing fast synaptic inputs at various locations on the soma and dendrites of a model layer 2/3 cell. The cell and the synaptic locations we have used are shown in Fig. 4. Figure 6A shows the range of excitatory postsynaptic potentials (EPSP) shapes recorded at the soma in the absence of any somatic shunt, corresponding to the situation during whole-cell recording. In spite of the electrical compactness of the cell, the EPSP show a wide range of shapes. This differential filtering of inputs may be of considerable significance at the network level. The effect of introducing a 10 nS shunt at the soma, as might be induced during sharp-electrode recording, is shown for the same inputs in Fig. 6B. By comparing these waveforms with Fig. 6A, it can be seen that the main effect of the shunt is to increase the rate of decay of the synaptic potentials. This causes a large reduction in their time integrals or areas, but has a less dramatic effect on their peak amplitudes.

A further implication of the electrotonic structure of these cells and the likely rapid kinetics of fast glutamatergic transmission is the difficulty of achieving adequate voltage clamp of excitatory synaptic inputs. Many researchers are aware of the difficulty of clamping distal synaptic inputs, but it is becoming clear that even proximal synapses can be severely affected (see also Wheal et al. 1991). We have simulated the effects of perfect voltage clamp applied via a whole-cell patch pipette of zero series resistance at the soma. Even for proximal inputs under these ideal conditions, there is very little difference in the voltage transient occurring at the synaptic site whether the amplifier is in voltage clamp or current clamp mode (Fig. 6C). There is considerable voltage escape even for slower currents with time courses more like those mediated by *N*-methyl-D-aspartate (NMDA) receptors (Fig. 6C, inset). For fast inputs, the synaptic current recorded at the soma under voltage clamp is a severely attenuated and slowed version of the current at the synapse (Fig. 6D). If there is any uncompensated series resistance, the problem becomes even more marked (Fig. 6D, trace bs). Thus, even quite proximal inputs, recorded under good conditions, cannot be voltage clamped accurately and the assumption of adequate clamp will lead to large interpretational errors. For example, all the fast synaptic inputs shown in Fig. 6 were modelled as conductance changes with peak values of 0.4 nS. When injected into a basal dendrite and recorded with a somatic shunt, this gave a peak depolarization at the soma of some 130 μV, which is close to the mean quantal size we recorded for hippocampal pyramidal cells (Larkman et al. 1991). If the conductance of a single AMPA/kainate channel is about 5 pS (Stern et al. 1992), such a conductance change would be produced by the simultaneous opening of about 80 channels. However, even with perfect whole-cell voltage clamp, the peak synaptic current is attenuated by a factor of about 4 and might be interpreted as due to the

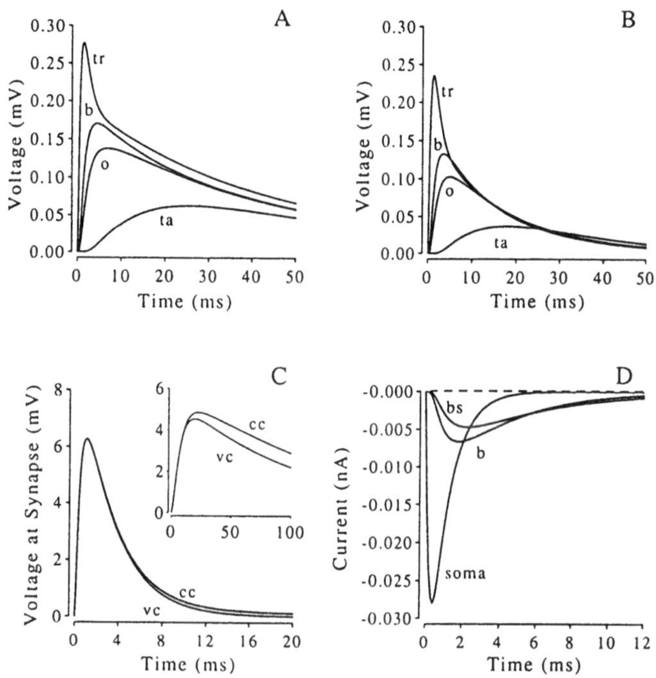

Fig. 4.2–6. A-D. Simulations of synaptic inputs onto a model of the layer 2/3 cell shown in Fig. 4, using $C_m = 0.75\,\mu F\ cm^{-2}$, $R_m = 65\,000\ \Omega cm^2$ and $R_i = 275\ \Omega cm$. Fast synaptic inputs were modelled using a double exponential function with time constants of 0.2 and 1 ms, with a peak conductance of 0.4 nS and a reversal potential of 70 mV positive to rest. **A** Excitatory postsynaptic potentials (EPSP) recorded at the soma following synaptic inputs to the proximal apical trunk (*tr*), basal (*b*), proximal oblique (*o*) and terminal arbor (*ta*) dendrites. Actual locations of input sites are shown in Fig. 4A. **B** As for A, but with a somatic shunt of 10 nS included in the model. **C** Voltage transients occurring at the synaptic input site (on a basal dendrite) during a fast synaptic input under current clamp (*cc*) and perfect voltage clamp applied at the soma (*vc*). Note that the voltage clamp is virtually ineffective in controlling the voltage at the synaptic site. The same is largely true even for slower, *N*-methyl-D-aspartate (NMDA)-type conductance changes (*inset*), here modelled as a double exponential function with time constants of 5 and 100 ms (Stern et al. 1992), a peak conductance of 0.1 nS and a reversal potential of 70 mV positive to rest. By contrast, a steady voltage displacement at the soma is only 7% less at the synaptic site, so the measured reversal potential for these synaptic inputs will be quite accurate. **D** Excitatory postsynaptic current (EPSC) recorded at the soma following fast synaptic conductance changes at the soma and on a basal dendrite (*b*) under perfect voltage clamp. Note that the current recorded for the basal input is severely slowed and shows considerable attenuation of the peak. If the voltage clamp is applied through an uncompensated series resistance of 10 MΩ, the attenuation and slowing of the basal input are even more severe (*bs*). *Horizontal dashed line* indicates zero current

simultaneous opening of perhaps 20 channels. With a 10 MΩ uncompensated series resistance, the peak current is only about 5 pA, which could lead to an estimate of only about 14 channels open at the peak. Estimates of the number of channels opening following the release of a quantum of transmitter have important implications for issues such as the likely level of quantal variance, which in turn may be relevant to controversies such as the quantal analysis of plastic changes like hippocampal long-term potentiation (LTP; Stevens 1993). Thus, the development of electrical models, based on detailed anatomical information, can be relevant to important areas of current research.

Acknowledgements. This work is supported by the Wellcome Trust and the Royal Society.

References

Blanton MG, Lo Turco JJ, Kriegstein AR (1989) Whole cell recording from neurons in slices of reptilian and mammalian cerebral cortex. J Neurosci Methods 30: 203–210

Chagnac-Amitai Y, Luhmann HJ, Prince DA (1990) Burst generating and regular spiking layer 5 pyramidal neurons of rat neocortex have different morphological features. J Comp Neurol 296: 598–613

Clements JD, Redman SJ (1989) Cable properties of cat spinal motoneurones measured by combining voltage clamp, current clamp and intracellular staining. J Physiol 409: 63–87

Coleman PA, Miller RF (1989) Measurement of passive membrane parameters with whole-cell recording from neurons of the intact amphibian retina. J Neurophysiol 61: 218–230

Colonnier M (1968) Synaptic patterns on different cell types in the different laminae of the cat visual cortex. An electron microscope study. Brain Res 9: 268–287

Colquhoun D, Jonas P, Sakmann B (1992) Action of brief pulses of glutamate on AMPA/kainate receptors in patches from different neurones of rat hippocampal slices. J Physiol 458: 261–287

Davies CA, Katz HB (1983) The comparative effects on early-life undernutrition and subsequent differential environments on the dendritic branching of pyramidal cells in rat visual cortex. J Comp Neurol 218: 345–350

Diamond J, Gray EG, Yasargil GM (1970) The function of the dendritic spine: an hypothesis. In: Anderson P, Jansen JKS (eds) Excitatory synaptic mechanisms. Universitetsforlaget, Oslo, pp 213–222

Durand D, Carlen P (1985) Electrotonic parameters of neurons following chronic ethanol consumption. J Neurophysiol 54: 807–817

Edwards FA, Konnerth A, Sakmann B, Takahashi T (1989) A thin slice preparation for patch clamp recordings from neurones of the mammalian central nervous system. Pflügers Archiv 414: 600–612

Hestrin S (1992) Activation and desensitization of glutamate-activated channels mediating fast excitatory synaptic currents in the visual cortex. Neuron 9: 991–999

Iansek R, Redman SJ (1973) An analysis of the cable properties of spinal motoneurones using a brief intracellular current pulse. J Physiol 234: 613–636

Jonas P, Major G, Sakmann B (1994) Quantal analysis of unitary EPSCs at the mossy fibre synapse on CA3 pyramidal cells of rat hippocampus. J Physiol (in press)

Larkman AU (1991a) Dendritic morphology of pyramidal neurones of the visual cortex of the rat: I. Branching patterns. J Comp Neurol 306: 307–319

Larkman AU (1991b) Dendritic morphology of pyramidal neurones of the visual cortex of the rat: II. Parameter correlation. J Comp Neurol 306: 320–331

Larkman AU (1991c) Dendritic morphology of pyramidal neurones of the visual cortex of the rat: III. Spine distributions. J Comp Neurol 306: 332–343

Larkman AU, Mason A (1990) Correlations between morphology and electrophysiology of pyramidal neurones in slices of rat visual cortex. I. Establishment of cell classes. J Neurosci 10: 1407–1414

Larkman A, Stratford K, Jack J (1991) Quantal analysis of excitatory synaptic action and depression in hippocampal slices. Nature 350: 344–347

Larkman AU, Major J, Stratford KJ, Jack JJB (1992) Dendritic morphology of pyramidal neurones of the visual cortex of the rat: IV. Electrical geometry. J Comp Neurol 323: 137–152

Major G (1993) Solutions for transients in arbitrarily branching cables. III. Voltage clamp problems. Biophys J 65: 469–491

Major G, Evans JD, Jack JJB (1993a) Solutions for transients in arbitrarily branching cables. I. Voltage recording with a somatic shunt. Biophys J 65: 423–449
Major G, Evans JD, Jack JJB (1993b) Solutions for transients in arbitrarily branching cables. II. Voltage clamp theory. Biophys J 65: 450–468
Mason A, Larkman AU (1990) Correlations between morphology and electrophysiology of pyramidal neurones in slices of rat visual cortex. II. Electrophysiology. J Neurosci 10: 1415–1428
Mason A, Nicoll A, Stratford K (1991) Synaptic transmission between individual pyramidal neurons of the rat visual cortex in vitro. J Neurosci 11: 72–84
Nicoll A, Blakemore C (1993) Patterns of local connectivity in the neocortex. Neural Computation 5: 665–680
Petit TL, LeBoutillier JC, Gregorio A, Libstug H (1988) The pattern of dendritic development in the cerebral cortex of the rat. Dev Brain Res 41: 209–219
Pongracz FS, Firestein S, Shepherd GM (1991) Electrotonic structure of olfactory sensory neurons analyzed by intracellular and whole cell patch techniques. J Neurophysiol 65: 747–758
Silver RA, Traynelis SF, Cull-Candy SG (1992) Rapid-time-course miniature and evoked excitatory currents at cerebellar synapses in situ. Nature 355: 163–166
Spruston N, Johnston D (1992) Perforated patch-clamp analysis of the passive membrane properties of three classes of hippocampal neurons. J Neurophysiol 67: 508–529
Stafstrom CE, Schwindt PC, Crill WE (1984) Cable properties of layer V neurons from cat sensorimotor cortex in vitro. J Neurophysiol 52: 278–289
Stern P, Edwards FA, Sakmann B (1992) Fast and slow components of unitary EPSPs on stellate cells elicited by focal stimulation in slices of rat visual cortex. J Physiol 449: 247–278
Stevens CF (1993) Quantal release of transmitter and long-term potentiation. Neuron 10 (Suppl): 55–63
Stratford K, Mason A, Larkman A, Major G, Jack J (1989) The modelling of pyramidal neurones in the visual cortex. In: Durbin R, Miall C, Mitchison G (eds) The computing neuron. Addison-Wesley, Wokingham, pp 296–321
Wang Z, McCormick DA (1993) Control of firing mode of corticotectal and corticopontine layer V burst-generating neurons by norepinephrine, acetylcholine, and 1S,3R–ACPD. J Neurosci 13: 2199–2216
Wheal HV, Simpson L, Phelps S, Stockley E (1991) Excitatory amino acid receptor subtypes and their roles in epileptiform synaptic potentials in the hippocampus. In: Wheal HV, Thomson AM (eds) Excitatory amino acids and synaptic transmission. Academic Press, London, pp 239–264

4.3 The Canonical Microcircuit: A Co-operative Neuronal Network for Neocortex

R.J. Douglas and K.A.C. Martin
Medical Research Council Anatomical Neuropharmacology Unit, Oxford, United Kingdom

Otto Creutzfeldt's work had particular relevance for our research in his emphasis on three important concepts and two important techniques. The techniques were the use of intracellular recording in the visual system and the use of computer-assisted methods for stimulus presentation, data capture and analysis. Although the aphorism *Vorsprung durch Technik* is well known on this side of the English Channel, it was only when we first began intracellular recording ourselves in 1979 that we came to appreciate the power of intracellular recording and automated data capture and the difficulties that Otto Creutzfeldt and his collaborators had to overcome to make their in vivo intracellular recordings. Their successes were considerable and their basic observations have stood the test of time.

On the conceptual side, Otto Creutzfeldt had three key ideas that were of direct relevance to our work. Firstly, the generality of the functional structure of the neocortex (Creutzfeldt 1977); secondly, that the neocortex was a co-operative neuronal network (Creutzfeldt et al. 1974); and thirdly, the significance of intracortical inhibition (Benevento et al. 1972). These three ideas are very much our current concerns in developing our models for the neocortex (Douglas et al. 1989), and it is evident that if these ideas form a central theme, then cortical circuitry, particularly visual cortical circuitry has to be viewed in a very different way to the way it has been looked at over the past 30 years. Visual physiologists have focused on the receptive field of single units as the prime source of explanations of cortical function and even structure. The best example of this is the simple cell, which surely must be the most studied, yet in some ways least understood cell in the history of the neocortex. The original model of the Hubel and Wiesel (1962) simple cell, devised on the basis of physiological observations, has remained a sufficient explanation of the basic connectivity, and it has served as a model of the basic principle of hierarchical feedforward processing through the visual areas.

However, the simple cell model and its progeny are essentially static and passive. Their properties are essentially dictated by their presynaptic elements. They do not predict what will happen when we move a stimulus across the receptive field and they predict that any recalibration of the system, induced by changes in contrast for example, will be carried out by the subcortical components of the visual circuits, not by the cortex itself. In such models, the cortical circuits do not play an active role. Furthermore, if we now wanted to build a simple cell we could not use these models as blueprints, because we would be short of some essential specifications. For instance, we would need to know how many neurons and synapses were involved and a whole host of biophysical details pertaining to input resistances, the magnitude of the synaptic conductances, and so on. Even if we had this information there might be the nagging doubt that we have not included any of the cortical circuitry that has been discovered since Hubel and Wiesel (1962) first proposed their simple cell model (see Martin 1988).

However, if we had a complete description of the circuitry underlying the simple cell, we might still be no further in the quest to understand the general principles of cortical organization and function. If we want to use visual cortex as a model for cortex, then clearly we do need a view of the connections and functions of the circuits that is not restricted to a particular receptive field type. It is here that we have to make the decision as to whether we continue to pursue the mapping of receptive fields, with all its accessibility and tradition, but its severe limitations in generality, or whether we try to arrive at a more basic and more general understanding of cortical circuits, in which receptive fields are not the central concern, but arise naturally from the deep structure of cortex. It is this alternative program that we have pursued and that led us to take seriously the notion that all of the neocortex has the same *Bauplan* and to develop our *canonical* microcircuit for neocortex (Douglas et al. 1989; Douglas and Martin 1991).

In his essay on "Generality of the functional structure of the neocortex," Otto Creutzfeldt (1977) reviewed the extensive historical evidence for the notion of a common *Bauplan* for all cortical areas. As early as the nineteenth century, neocortex was referred to as 'isocortex', and subsequent quantitative and qualitative research has supported this view. All cortical areas are topographically and reciprocally connected to the thalamus, and each area contains the same proportions of pyramidal (70%) and smooth GABAergic (20%) neurons; pyramidal neurons are the principal projection neurons in all areas; pyramidal neurons in the deep layer (layer 5 and 6) tend to project subcortically to thalamus and midbrain, whereas pyramids in the superficial layers (layers 2 and 3) connect to other cortical areas; layer 1 is sparsely populated, mainly by GABAergic neurons (see Creutzfeldt 1977; Douglas and Martin 1989). The thalamic afferents and the spiny neurons are demonstrably excitatory, whereas the GABAergic neurons are inhibitory in function. We became sufficiently convinced of the existence of a common *Bauplan* through our studies of the morphology and intracortical projections of physiologically identified neurons in the visual cortex of the sheep (Martin and Whitteridge, unpublished), cat (Martin and Whitteridge 1984; Martin 1988) and monkey (Freund et al. 1989; Anderson et al. 1993).

The intracortical synaptic connections and neurochemistry of these physiologically identified neurons have been extensively studied by ourselves over the past decade (see Martin 1988; Somogyi 1989; Douglas and Martin 1990). These studies have shown that the major targets of all neuronal types in the visual cortex are spiny neurons. The thalamic afferents form about 90% of their synapses with spiny neurons. The pyramidal cells form synapses mainly with other pyramidal cells. The GABAergic neurons (e.g., basket cells, chandelier cells) all form most of their synapses, or all of their synapses in the case of the chandelier cell, with spiny cells (pyramids and spiny stellates). The pattern of connectivity is highly convergent and divergent. For example, we found that single thalamic afferents frequently make only one synapse with any target neuron and maximally only eight synapses (Freund et al. 1985). Since each layer 4 neuron receives probably several hundred thalamic synapses, it is evident that many thalamic afferents converge on a cortical neuron. Conversely, each thalamic afferent must contact thousands of neurons. The co-operative neuronal network is almost tangible in the pattern of connections seen in the light and electron microscopes. The response of each neuron seems inevitably to be due to the co-operative action of hundreds of others. Similarly each neuron provides only a tiny fraction of the response of thousands of others. This pattern of connections forms the structural basis of the canonical circuit (Fig. 1).

The techniques for establishing the morphology, synaptic connections and neurochemistry have been developed to a fine degree over the past decade. Physiologically, however, the traditional methods of extracellular single unit recording are eminently suitable for receptive field mapping, but receptive fields are intrinsically bound to particular cortical areas. Thus another more general method needed to be applied for the analysis of all cortical areas. We chose to examine the response to an electrical pulse stimulus applied to the thalamocortical afferents (Douglas et al. 1989; Douglas and Martin 1991). Although electrical stimulation is unnatural, because it induces a highly synchronous and abnormally patterned activation of the thalamocortical afferents, its value in the analysis of circuitry is well proven (Eccles 1964). In the context of our studies, it offered a number of advantages. The stimulus was easily controlled and highly reproducible. Direct activation of the cortical afferents reduced the contribution of retinal and lateral geniculate processing to the cortical response and so was applicable to any cortical area. This was particularly important because the natural stimulus requirements are not yet known for most cortical areas. Pulse stimulation is also a popular engineering test signal because it offers many mathematical advantages in the analysis of systems (see Jack et al. 1975).

Our procedure was to record intracellularly from cortical neurons using horseradish peroxidase (HRP)-filled pipettes and record the response to an electrical pulse applied to the thalamic afferents. The form of the intracellular response was very similar to that found in motor cortex by Phillips (1959) and in visual cortex by Otto Creutzfedt and his collaborators (Watanabe et al. 1966), who showed that the response was strongly dominated by the inhibitory synapses. We discovered in addition that the form of the intracellular response to the pulse was strongly correlated with the laminar location of the pyramidal cells we recovered (Fig. 2). Pyramidal cells in the deep layer (which project principally to subcortical structures) respond to a pulse with a short latency excitatory postsynaptic potential (EPSP) followed by a rapid onset inhibitory postsynaptic potential (IPSP), which lasted for 100–200 ms. Compared to the deep layer pyramids, superficial layer pyramids (which project to other cortical areas) had a 2–3 times longer EPSP and a much more slowly evolving IPSP. This difference

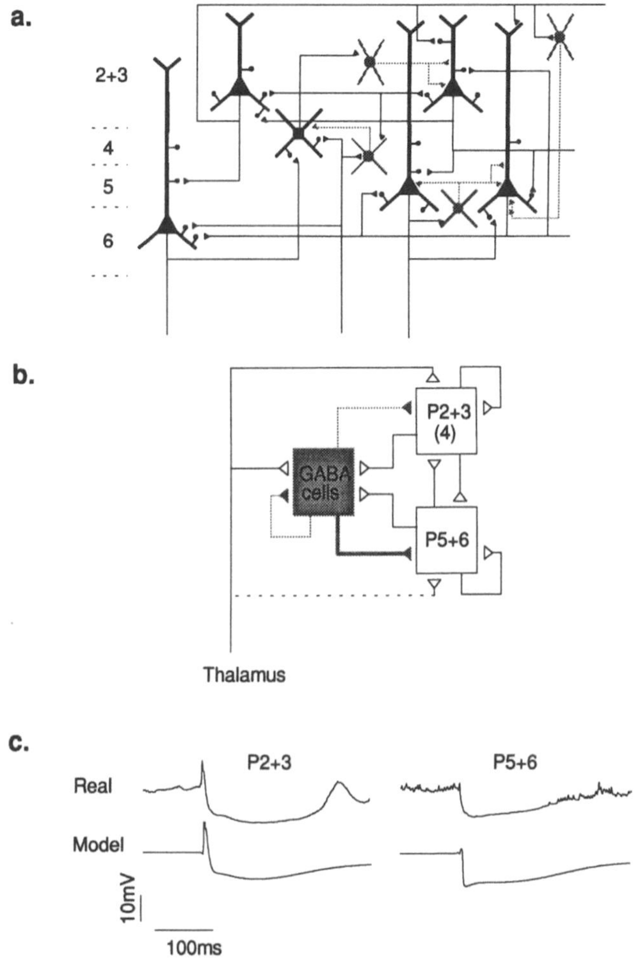

Fig. 4.3–1. The canonical microcircuit that successfully predicts the intracellular responses of cortical neurons to stimulation of thalamic afferents (Douglas et al. 1989). **a** Summary of defined connections between identified neurons that comprise the local microcircuit. Spiny (excitatory) neurons in *black*, smooth (GABAergic) inhibitory neurons in *grey*. **b** Three populations of neurons interact with one another: one population is inhibitory (GABA cells, solid synapses) and two are excitatory (open synapses), representing superficial (P2+3) and deep (P5+6) layer pyramidal neurons. Long-range pyramidal projections are not incorporated in the microcircuit because their functions are not modelled. The layer 4 spiny stellate cells are incorporated with the superficial group of pyramids. Each population receives excitatory input from the thalamus. The inhibitory inputs activate both $GABA_A$ and $GABA_B$ receptors on pyramidal and GABA neurons. **c** The pulse response of real and model neurons. Intracellular traces of real neurons are averaged. Responses of real layer 3 pyramidal neuron shows early depolarization and action potential (cropped) followed by sustained hyperpolarization that reaches maximum amplitude at a latency of about 100 ms. (In this and some other neurons, a rebound depolarization occurred during relaxation from hyperpolarization). In real layer 6 pyramidal neurons, the hyperpolarization evolved rapidly and reached a maximum at about 25 ms after the pulse. Responses of model neurons were similar in form to real neurons

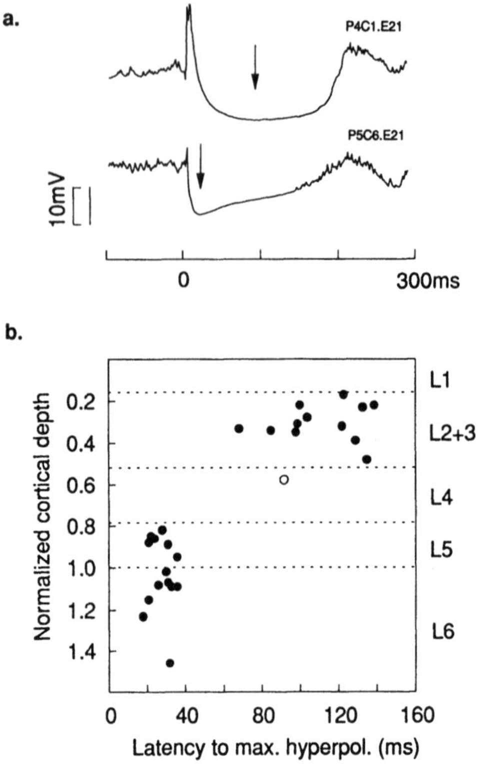

Fig. 4.3–2. Response of cortical neurons to electrical stimulation of the optic radiation. **a** Averaged intracellular response of typical superficial (*top trace*, layer 2/3) and deep (*bottom trace*, layer 5/6) neurons. Position of maximum hyperpolarization indicated by *arrows*. **b** Relationship between laminar position of pyramidal neurons and latency to maximum hyperpolarization for 26 pyramidal neurons (*filled circles*) and one spiny stellate cell (*open circle*)

in the pulse response is part of a consistent pattern of difference between superficial and deep layers that we have attempted to encapsulate in the canonical microcircuit (Fig. 1).

The pulse response appeared simple, but when we attempted to model it, we soon discovered many missing pieces. One example was the relative contributions of the two GABA receptor types, both of which were known to exist on cortical neurons, but whose contribution to the shape of the pulse response in vivo was unknown. To uncover these components, we returned to the intracellular recording pipette, but now attached an additional drug-filled multibarrel pipette to *spritz* GABA agonists and antagonists on the outside of the neuron. By this means we dissected the components of the inhibitory response and showed that the $GABA_A$ receptors were responsible for producing an early IPSP of short durations, whereas the $GABA_B$ receptors provided a more slowly evolving and long-lasting component (Douglas et al. 1989; Douglas and Martin 1991).

Otto Creutzfeldt explained his intracellular results by supposing that there was little excitatory convergence within the cortex and that most of the intracortical connections were inhibitory (Creutzfeldt et al. 1975). The pulse response, with its brief excitation followed by a long-lasting inhibition, seemed to be the temporal equivalent of the spatial distribution of inhibition that he and his colleagues had seen, with the spread of inhibition extending well beyond that of excitation. Unfortunately, our extensive anatomical studies showed that his view of extensive and powerful inhibitory circuits could not be supported. Instead, viewed from the microanatomy of the cortical circuits, it seemed that the excitatory circuits must always dominate the inhibitory circuits, both in magnitude and extent (Martin 1988). Nevertheless, our basic observations on the intracellular responses to visual and pulse stimulation were similar to those made by Otto Creutzfeldt and his colleagues. How, then, can structure and function be reconciled?

An important clue for the resolution of structure and function came from the pharmacological studies of the pulse response. In order to reveal the action of the excitatory pathways, we blocked the $GABA_A$ receptors, for it is the rapid onset of the $GABA_A$ response that terminates the EPSP generated by the pulse stimulation. When the $GABA_A$ receptors were blocked with N-methyl-bicuculline, the monosynaptic EPSP generated by the thalamic afferents remained small and unchanged, but the polysynaptic excitatory response generated by the intracortical circuits grew in magnitude and duration (Fig. 3; Douglas et al. 1989; Douglas and Martin 1991). This showed the latent power of the intracortical excitatory synapses that are normally kept in check by the rapidly acting $GABA_A$ component. When only $GABA_B$ receptors were active, inhibition still occurred, but it took far longer to suppress the excitatory response (Fig. 3). This provided a crucial insight to the possible mode of operation of the cortical circuits during natural stimulation. The lack of response of cortical neurons during nonoptimal stimulation was thought to be due to inhibition of powerful thalamic afferent excitation (e.g., Creutzfeldt et al. 1974, 1975). Our results suggested an alternative: that the lack of response was due to the intracortical excitatory circuits not providing sufficient amplification of a relatively weak thalamic excitation.

The compatibility of this view of the functional operation with our structural view was demonstrated through formal computer modelling (Douglas and Martin 1991). We attempted to provide as realistic a representation as we could of the components of the circuit and the synaptic connections between them. We simulated with remarkable fidelity the pulse response, including the striking difference between the superficial and deep layer pyramids and the effects on the shape of the pulse response of the GABA agonists and antagonists. The performance of the model showed how crucial the balance was between the excitatory and inhibitory populations, which are roped together like climbers on a rock face. If the activity of one goes up

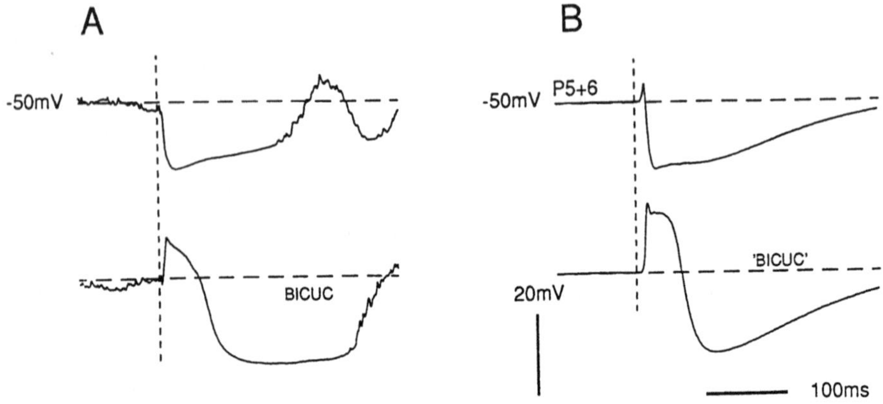

Fig. 4.3–3. **A,B** Comparison of the effect of blocking $GABA_A$ receptors with N-methyl-bicuculline between real neuron (**A**) and model neuron in the canonical microcircuit (**B**). *Top traces* show normal response of deep layer neuron to pulse stimulation of thalamic afferents (time applied indicated by *vertical dashed line*). *Bottom traces* show response during partial block of $GABA_A$ receptors, simulated in *right trace*

so does the other; if the activity in one group goes down, so does the other. Because of their intimate relationship, pulse activation of the thalamic afferents invariably sets in motion a sequence of excitation and inhibition in every neuron. The model also predicted that the excitation due to intracortical spiny neurons would greatly exceed that of the thalamic afferents. The conclusion is that even in the thalamorecipient layers the thalamic input does not provide the major excitation to any neuron. Instead, the cortical excitatory paths provide a means of amplifying the thalamic signal.

In this model, control of excitation is a much more synergistic process, which is in total contrast to all other models of cortical inhibition, where inhibition and excitation are seen as oppositional. High activity of the inhibitory population is associated with low activity in the excitatory population and vice versa. In Otto Creutzfeldt's view, the optimal response of simple cells was due to the excitatory action of a few geniculate afferents. Thus, to generate direction and orientation selectivity, the strong excitation provided by these thalamic afferents has to be quenched by equally strong inhibition. The problem is that our extensive experiments failed to detect the strong inhibition required. We did not find strong hyperpolarizations during non-optimal stimuli (Douglas et al. 1988, 1991), nor did we see powerful shunting inhibition (Berman et al. 1991; Douglas et al. 1988, 1991). The activity of synapses of the thalamic afferents is not 'vetoed' by inhibition at the level of the dendritic spines (Dehay et al. 1991). Our biophysical experiments on cortical slices in vitro showed that GABAergic inhibition, whether induced by GABA ionophoresis or synaptic activation, could not suppress strong excitation (Berman et al. 1992). Our computer simulations showed that even chandelier cell inhibition acting at the axon initial segment, the site of action potential generation, could not suppress moderate excitation (Douglas and Martin 1990). The problem is illustrated simply in Fig. 4, which shows that visually evoked inhibition in a cortical neuron can hardly offset even a small amount of excitatory current (0.2 nA) injected through the recording pipette (see Berman et al. 1991). Thus, the cortical circuits do not seem to be designed for suppressing the powerful excitation that most models suppose arrives through the thalamic afferents.

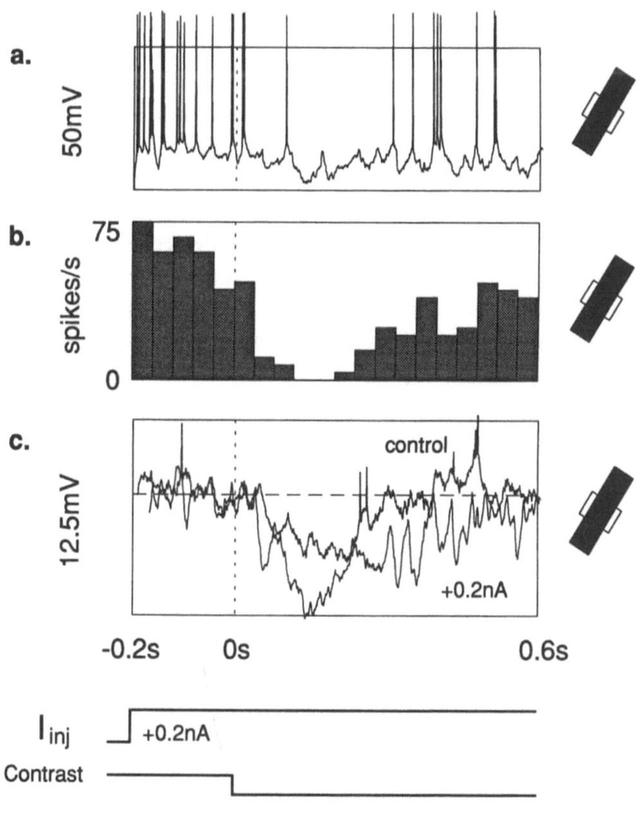

Fig. 4.3–4. Intracellular response of layer 5 pyramidal cells with an S1 receptive field to flashed bar at optimal orientation. Excitatory current (0.2 nA) was injected to raise the firing rate and then the bar was flashed (onset indicated by *vertical dashed line*). **a** Intracellular response (single sweep) showing visually induced inhibition of current induced discharge. **b** Histogram showing spike rate averaged over seven trials. Flash initially supressed firing (latency 40 ms), but the effect diminished from 250 ms onward. **c** Averaged membrane potentials (seven trials) associated with flash-induced inhibition of firing with (*thick line*) and without (*thin line*) injected current

The solution lies in the recurrent excitatory pathways of the local cortical circuits that are able to amplify the small thalamic signal (Douglas and Martin 1991). These circuits ensure that inhibition need only ever act against a small to moderate amount of excitation, never against the full-blown excitation that is seen during the optimal response to a natural stimulus. Apart from explaining the puzzle of the inordinate effectiveness of what are small magnitudes of inhibition (Douglas et al. 1988), this mechanism has the important effect of ensuring that the individual neurons in the cortical microcircuits are not subject to powerful inhibition. Even in a quiescent (inhibited) state they remain poised near threshold; their membrane potential is never strongly hyperpolarized. Under normal (non-pathological) conditions, the cortical circuits do not exhibit the profound swings in activity, suggesting that the cortical network could be regarded as an isopotential sheet that is "dented" into hills and valleys by the cortical activity initiated either by the thalamus or other cortical areas. The network as a whole attempts to maintain a level homeostatic plain. This implies that properties such as contrast gain control, signal-to-noise enhancement and construction of a whole from incomplete parts (completion) arise naturally out of the wiring of the cortical circuits.

Although our view of the connectivity and operation of the cortical circuits differs in detail from that of Otto Creutzfeldt, the two concepts he was grappling with, i.e., co-operativity and inhibition, find a solution in his third, the concept of a common cortical *Bauplan*. The canonical microcircuit binds together these three in a coherent whole. He may not have agreed with our solutions, but surely he would have approved of the general direction.

References

Anderson JC, Martin KAC, Whitteridge D (1993) Form, function and intracortical projection of neurons in the striate cortex of the monkey, Macacus nemestrinus. Cerebral Cortex 3: 412–420

Benevento LA, Creutzfeldt OD, Kuhnt U (1972) Significance of intracortical inhibition. Nature 283: 124–126

Berman NJ, Douglas RJ, Martin KAC, Whitteridge D (1991) Mechanisms of inhibition in cat visual cortex. J Physiol 440: 697–722

Berman NJ, Douglas RJ, Martin KAC (1992) GABA-mediated inhibition in the neural networks of visual cortex. In: Mize RR, Marc RE, Sillito AM (eds) Progress Brain Res 90: 443–476

Creutzfeldt OD (1977) Generality of the functional structure of the neocortex. Naturwissenschaften 64: 507–517

Creutzfeldt O, Kuhnt U, Benevento LA (1974) An intracellular analysis of visual cortical neurones to moving stimuli: responses in a co-operative neuronal network. Exp Brain Res 21: 251–274

Creutzfeldt O, Innocenti GM, Brooks D (1975) Neurophysiological experiments on the afferent and intrinsic connections in the visual cortex (area 17). In: Santini M (ed) Golgi Centennial Symposium, Proceeedings. Raven, New York, pp 319–337

Dehay C, Douglas RJ, Martin KAC, Nelson JC (1991) Excitation by geniculocortical synapses is not 'vetoed' at the level of dendritic spines in cat visual cortex. J Physiol 440: 723–734

Douglas RJ, Martin KAC (1990) Control of neuronal output by inhibition at the axon initial segment. Neural Computation 2: 283–292
Douglas RJ, Martin KAC (1991) A functional microcircuit for cat visual cortex. J Physiol 440: 735–769
Douglas RJ, Martin KAC, Whitteridge D (1988) Selective responses of cells do not depend on shunting inhibition. Nature 332: 642–644
Douglas RJ, Martin KAC, Whitteridge D (1989) A canonical microcircuit for neocortex. Neural Computation 1: 480–488
Douglas RJ, Martin KAC, Whitteridge D (1991) An intracellular analysis of the visual responses of neurones in cat visual cortex. J Physiol 440: 659–696
Eccles JC (1964) The physiology of synapses. Springer, Berlin Heidelberg New York
Freund TF, Martin KAC, Somogyi P, Whitteridge D (1985) Innervation of cat visual areas 17 and 18 by physiologically identified X- and Y-type thalamic afferents. II. Identification of postsynaptic targets by GABA immunocytochemistry and Golgi impregnation. J Comp Neurol 242: 275–291
Freund TF, Martin KAC, Soltesz I, Somogyi P, Whitteridge D (1989) Arborisation pattern and postsynaptic targets of physiologically identified thalamocortical afferents in striate cortex of the monkey. J Comp Neurol 28: 315–336
Hubel DH, Wiesel TN (1962) Receptive fields, binocular interaction and functional architecture in the cat's visual cortex. J Physiol (Lond) 165: 559–568
Jack JJB, Noble D, Tsien, RW (1975) Electric current flow in excitable cells. Oxford University Press, Oxford
Martin KAC (1988) From single cells to simple circuits in the cerebral cortex. The Wellcome Prize Lecture. Q J Exp Physiol 73: 637–702
Martin KAC, Whitteridge D (1984) Form function and intracortical projections of spiny neurones in striate visual cortex of the cat. J Physiol 353: 463–504
Phillips CG (1959) Actions of antidromic pyramidal volleys on single Betz cells in the cat. Q J Exp Physiol 44: 1–25
Somogyi P (1989) Synaptic organization of GABAergic neurons and $GABA_A$ receptors in the lateral geniculate nucleus and visual cortex. In: Lam DK-T, Gilbert CD (eds) Neural mechanisms of visual perception. Portfolio, The Woodlands Texas, pp 35–62
Watanabe SM, Konishi M, Creutzfeldt O (1966) Postsynaptic potentials in the cat's visual cortex following electrical stimulation of afferent pathways. Exp Brain Res 1: 272–283

4.4 Cellular Mechanisms of Thalamocortical Oscillations

D.A. McCormick, T. Bal, and M. von Krosigk
Section of Neurobiology, Yale University, School of Medicine, New Haven, Connecticut

Thalamic and cortical networks generate varying patterns of activity depending upon behavioral state, such as sleep, waking, attention, and sensory processing (reviewed in McCormick 1992; Steriade and Deschênes 1984). The cellular mechanisms of generation of these oscillations have been largely unknown, with the exception of spindle waves (Andersen and Andersson 1968; Steriade and Deschênes 1984). Understanding the cellular basis of forebrain oscillations is more than an academic exercise: they are intimately involved in the processes of sleep, they can be perverted to generate epileptic seizures, and they may even be involved in sensory-motor processing (Gloor and Fariello 1988; Gray et al. 1989; Steriade and Deschênes 1984).

Perhaps the best known oscillations in thalamocortical networks are those associated with slow wave sleep. During slow wave sleep, rhythmic activity appears in the electroencephalogram (EEG) in two prominent frequency ranges: delta waves which are 0.5–4 Hz slow waves appearing throughout the cortex, and which become prominent during deeper stages of sleep, and spindle waves, characterized by 7–12 Hz oscillations that wax and wane over a 2–3 s period and that reappear with a marked regularity of once every 5–20 s (reviewed in McCormick 1992; Steriade and Deschênes 1984). Higher frequency (30–70 Hz) oscillations also appear in both the EEG and local field potentials in the cerebral cortex in waking and attentive animals and can be driven by receptive field stimulation in sensory areas (e.g., Gray et al. 1989).

While delta wave oscillations have traditionally been viewed as cortical in origin, spindle waves are known to be generated in the thalamus (see Steriade and Deschênes 1984). Removal of the cerebral cortex leaves intact the generation of spindle wave oscillations in the thalamus, while large lesions of the thalamus abolish the occurrence of spindle waves in the cerebral cortex (see Andersen and Andersson 1968). Intracellular recordings from thalamic relay neurons in vivo reveal that spindle waves are generated by rebound burst discharges occurring in response to the arrival of barrages of inhibitory postsynaptic potentials (IPSPs). Further investigations have revealed that these IPSPs are likely generated by burst discharges in the nucleus reticularis of the thalamus (nRt; Mulle et al. 1986). Attempts to surgically isolate the nRt from the rest of the thalamus have been reported to abolish spindle wave generation in thalamic relay nuclei, while leaving spindle-like oscillations intact in the nRt, suggesting a "pacemaker" role of this nucleus in the generation of these characteristic oscillations (Steriade et al. 1985, 1987).

Recently we have described an in vitro slice preparation which spontaneously generates spindle oscillations in vitro (von Krosigk et al. 1993). Sagittal slices of the ferret dorsal lateral geniculate nucleus (LGNd) generate spontaneous spindle waves of 6–9 Hz, lasting approximately 1–3 s and appearing about once every 5–30 s, as found in vivo (Fig. 1A). Intracellular recordings from thalamic relay cells in the A-lamina of the ferret LGNd while simultaneously recording extracellularly from nearby relay cells revealed that spindle oscillations are associated with barrages of IPSPs, presumably mediated in large part by $GABA_A$ receptors (γ-aminobutyric acid;

Fig. 1B). On occasion (about one out of two to five) these IPSPs removed enough inactivation of the low-threshold Ca^{2+} current to result in a rebound low-threshold Ca^{2+} spike (Fig. 1C), resulting in a burst of action potentials (Fig. 1D).

Intracellular recordings from the GABAergic neurons of the perigeniculate nucleus (PGN; which is similar in anatomy and function to the nRt) reveal barrages of excitatory postsynaptic potentials (EPSP) arriving in marked synchrony with each burst of unit activity in the dorsal lateral geniculate nucleus (LGNd; Fig. 1H). These EPSPs presumably arise from the burst discharges from relay cells since: (a) cortical afferents are inactivated by the slice procedure; (b) the EPSPs are in synchrony with bursts of action potentials in relay cells; (c) the EPSPs are blocked by CNQX (6-cyano-7-nitroquinoxaline-2,3-dione); and (d) activation of relay cells with glutamate results in barrages of EPSPs identical to those occurring during spindle waves and which reverse at 0 mV, as expected for AMPA/kainate receptor mediated EPSPs (α-amino-3-hydroxy-5-methyl-4-isoxazole propionic acid; von Krosigk et al. 1993).

These intracellular recordings suggest that spindle waves are generated in local regions of the thalamus as an interaction between the relay excitatory cells and PGN GABAergic neurons. Indeed, knife cuts placed between the PGN and the relay lamina of the LGNd result in abolition of spindle wave generation in vitro (von Krosigk et al. 1993). We would like to propose the following scenario for the generation of spindle waves in the thalamus. First, a critical number of neighboring relay or PGN (nRt) cells generate a synchronized burst of activity, either spontaneously or in response to activation of cortical or retinal afferents. The activation of PGN cells results in inhibition of a number of relay cells, a few of which generate a rebound Ca^{2+} spike and associated burst of action potentials (Fig. 1C). These bursts of action potentials then excite even more PGN neurons during the next cycle, and thus more cells are recruited into this oscillation. Through the anatomical divergence of connections between relay cells and PGN neurons, more and more cells are recruited into this oscillation as the spindle wave grows. However, at the same time, PGN cells are undergoing a progressive hyperpolarization, perhaps generated by a Ca^{2+}-activated K^+ current (Fig. 1F,G). At the initial stages of the spindle wave this hyperpolarization facilitates burst firing in PGN cells, although as the spindle wave progresses, the hyperpolarization begins to reduce the number of PGN cells involved in the oscillation owing to the membrane potential in these cells being subthreshold for the activation of the low-threshold Ca^{2+} spike by the barrages of EPSPs. In this manner, the spindle wave dampens or "winds down". This hypothetical framework forms the basis for ongoing experiments into the cellular mechanisms of spindle wave generation in our laboratory.

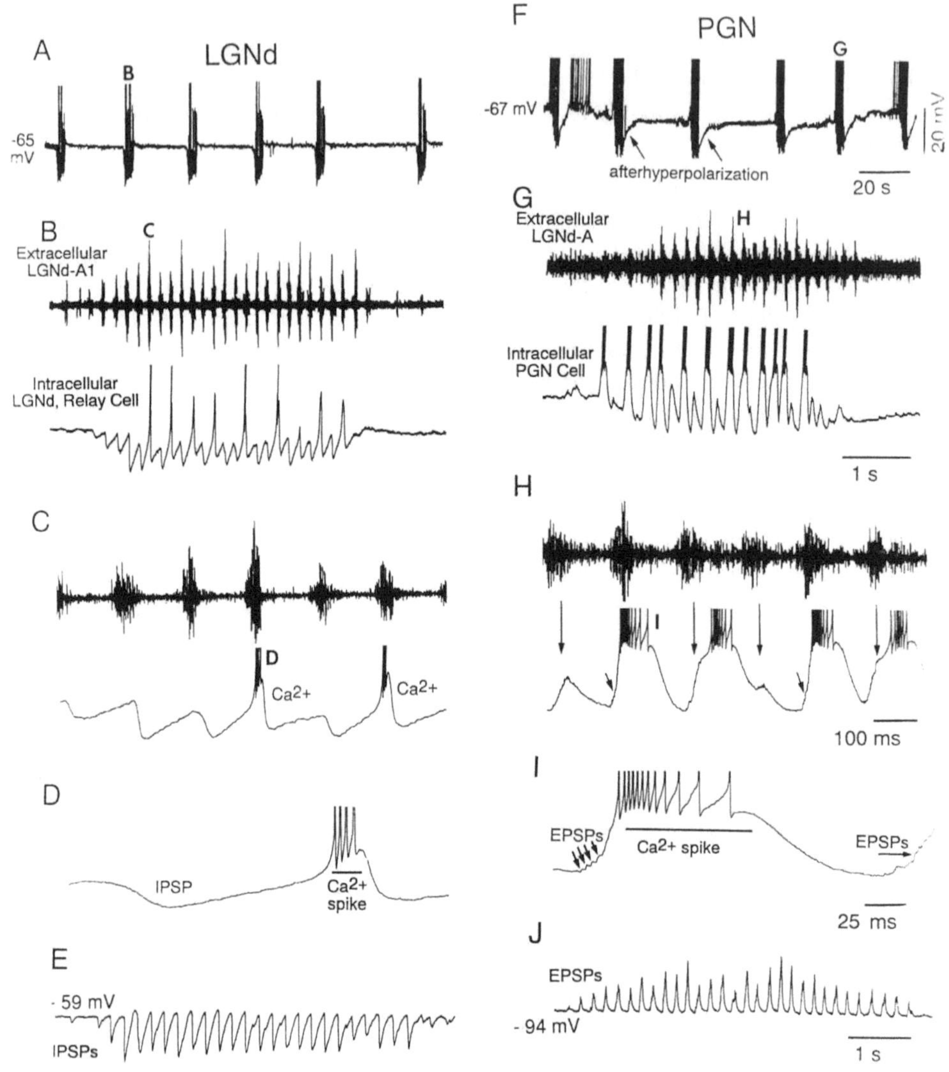

Fig. 4.4–1. Spindle oscillations in ferret geniculate slices. **A** Intracellular recording from an LGNd relay cell in lamina A reveals the recurrence of spindle waves approximately once every 20 s. **B-D** Detail of one spindle wave and parts of the spindle wave recorded both intracellularly in a relay neuron and locally as extracellular multiple unit activity in the LGNd. **E** Depolarization to –59 mV with the intracellular injection of current completely inactivates the low threshold Ca^{2+} current and reveals the barrage of IPSPs associated with a spindle wave. **F** Intracellular recording from a morphologically identified perigeniculate (PGN) GABAergic cell during generation of spindle waves. **G** Expansion of one spindle wave in the PGN cell and the simultaneously recorded multiple unit activity in the relay lamina (lamina A). **H** Close examination of the simultaneous recordings reveals that each burst of activity in the relay lamina is associated with a barrage of EPSPs in the PGN cells, which then activate, on occasion, a low-threshold Ca^{2+} spike and a high frequency burst discharge. **I** In some cases groups of three to five EPSPs arrive at the same frequency at which relay cells generate action potentials during a burst (compare **D** and **I**). **J** Hyperpolarization to –94 mV prevents the activation of low-threshold Ca^{2+} spikes and reveals the underlying barrages of EPSPs. Spindle oscillations are successfully recorded in at least one slice in 78 out of 85 experiments. (From von Krosigk et al. 1993)

Since spindle waves are a network phenomenon, they are influenced by probably every feature of the thalamic synaptic circuit. How the various features of thalamic physiology affect the generation of spindle waves is currently under investigation. One important feature is the intrinsic properties of the neurons involved in the generation of spindle waves. Through intracellular recordings in vitro, we have found that nRt/PGN neurons have a propensity intrinsically to generate rhythmic bursts of action potentials through an interaction of the low threshold Ca^{2+} current and a Ca^{2+}-activated K^+ current (Bal and McCormick 1993). This intrinsic oscillation can occur, depending upon membrane potential, at any frequency between 0.5 and 12 Hz, suggesting that these cells have the intrinsic ability to participate on a cycle-by-cycle basis in the generation of spindle waves (Bal and McCormick 1993).

Perversion of Spindle Waves into Absence Seizures. Absence seizures are characterized by a striking 3 cycle-per-second oscillation (referred to as spike-and-wave discharge) in the EEG while the patient stares aimlessly and perhaps blinks. Animal models of these seizures reveal that they are generated in the thalamus and cerebral cortex as an interaction between the GABAergic neurons of the nRt, the relay cells, and cortical circuitry (Gloor and Fariello 1988). Of particular relevance here are two findings: (a) application of large doses of penicillin, a weak $GABA_A$-antagonist, to cats (intramuscularly) slowly converts spindle oscillations into absence seizures; and (b) administration of $GABA_B$ antagonists to rats with absence-like seizure activity results in abolition of this synchronized discharge (see Gloor and Fariello 1988; Hosford et al. 1992).

In ferret LGNd slices we found that application of $GABA_A$-receptor antagonists result in a pronounced enhancement of the $GABA_B$-mediated IPSP and a subsequent enhanced rebound burst discharge in relay cells. This enhanced rebound Ca^{2+} spike not only results in a markedly enhanced number of spikes generated during each burst discharge but also results in the generation of a rebound burst discharge in a majority of relay cells with each cycle of the oscillation, thereby resulting in a highly synchronized, slow (2–4 Hz) oscillation which resembles that of absence seizures (von Krosigk et al. 1993). Interestingly, block of $GABA_B$ receptors results in block of these absence-like seizures in thalamic slices, while block of $GABA_B$ receptors in normal slices does not abolish normal spindle waves, suggesting that these receptors may be useful in the treatment of absence seizures (Hosford et al. 1992).

Delta-Frequency Oscillations in Thalamic and Cortical Cells. Extracellular and intracellular recordings from thalamic relay cells and some cortical pyramidal cells in cats and primate during slow wave sleep reveal that

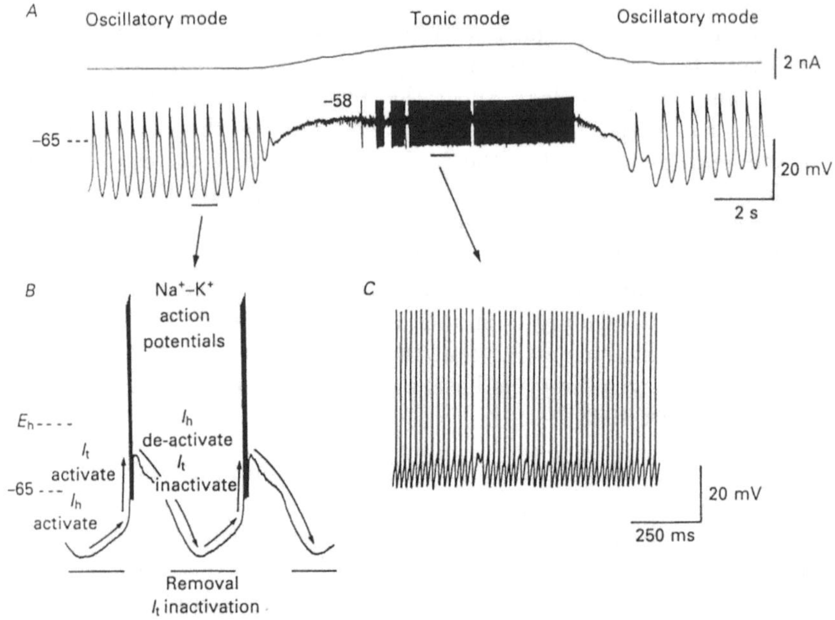

Fig. 4.4–2. Two different firing modes of thalamic relay neurons and the proposed ionic substrate of rhythmic burst firing. **A** This cat LGNd neuron generate rhythmic burst firing at a rate of about 2 Hz. Depolarization of the cell to −58 mV with the intracellular injection of current (*top trace*) halts the rhythmic activity and switches the neuron to the tonic, or single-spike, mode of action potential generation. Removal of the depolarization reinstates the oscillatory activity. **B** Expanded trace of oscillatory activity and the proposed currents which largely mediate it. Activation of the low threshold calcium current (I_t) depolarizes the membrane toward threshold for a burst of Na^+- and K^+-dependent fast action potentials. The depolarization deactivates the portion of I_h that was active immediately before the Ca^{2+} spike. Repolarization of the membrane due to I_t inactivation is followed by a hyperpolarizing overshoot, due to the reduced depolarizing effect of I_h. The hyperpolarization in turn deinactivates I_t and activates I_h, which depolarizes the membrane toward threshold for another Ca^{2+} spike. **C** Expanded trace of single spike activity. (From McCormick and Pape 1990)

many neurons generate repetitive bursts of action potentials in relation to the generation of the EEG (Steriade and Deschênes 1984; Hirsch et al. 1983). Intracellular recordings, both in vivo and in vitro, in thalamic relay cells have demonstrated rhythmic burst discharge in these cells that is generated intrinsically in each cell through an interaction between the low-threshold Ca^{2+} current and a hyperpolarization-activated cation current (Fig. 2, McCormick and Pape 1990). The activation of a low-threshold Ca^{2+} spike results in deactivation of some portion of a hyperpolarization activated cation current (known as h-current), allowing the cell to hyperpolarize substantially after the Ca^{2+} spike inactivates. This hyperpolarization, however, once again activates the h-current, which again depolarizes the cell and generates another low-threshold Ca^{2+} spike, thereby starting the cycle again. This cycle will continue as long as the cells are left in a hyperpolarized state. Depolarization of thalamic relay cells, either through the intracellular injection of current or through the actions of various neurotransmitters, results in an abolition of rhythmic burst discharge through inactivation of the low-threshold Ca^{2+} current and a switch to the tonic mode of action potential generation (Fig. 2).

Stimulation of afferents to the nucleus reticularis of the thalamus can result in synchronized slow oscillations in thalamic relay cells (Steriade et al. 1991). Furthermore, rhythmic burst firing in these cells can occur in relation to the generation of slow waves of the EEG (Steriade et al. 1993a,b,c). However, large lesions of the thalamus leave a neocortex which still generates delta wave oscillations (Steriade et al. 1993b). Together, these results indicate that thalamic neurons may be involved in the generation of the delta waves of the EEG, although the degree of their involvement and their role is not yet clear.

4.4.1 Conclusion

Synchronized rhythmic oscillations are a common feature of thalamic neurons and neuronal circuits. Through an interaction of intrinsic properties and axonal connections the thalamus contributes to the generation of spindle waves, absence seizures, and perhaps delta waves. The awake and attentive state is associated with an abolition of these oscillations due in part to depolarization of thalamic relay and nRt neurons. The abolition of slow oscillations signifies the preparation of thalamocortical circuits for behavior and cognition.

References

Andersen P, Andersson SA (1968) Physiological basis of the alpha rhythm. Appleton-Century-Crofts, New York

Bal T, McCormick DA (1993) Mechanisms of oscillatory activity in guinea pig nucleus reticularis thalami in vitro: a mammalian pacemaker. J Physiol 468: 669–691

Gloor P, Fariello RG (1988) Generalized epilepsy: some of its cellular mechanisms differ from those of focal epilepsy. Trends Neurosci 11: 63–68

Gray CM, Konig P, Engel AK, Singer W (1989) Oscillatory responses in cat visual cortex exhibit inter-columnar synchronization which reflects global stimulus properties. Nature 338: 334–337

Hirsch JC, Fourment A, Marc ME (1983) Sleep-related variations of membrane potential in the lateral geniculate body relay neurons of the cat. Brain Res 259: 308–312

Hosford DA, Clark S, Cao Z, Wilson WA (1992) The role of $GABA_B$ receptor activation in absence seizures of lethargic (lh/lh) mice. Science 257: 398–401

McCormick DA (1992) Neurotransmitter actions in the thalamus and cerebral cortex and their role in neuromodulation of thalamocortical activity. Progress in Neurobiology 39: 337–388

McCormick DA, Pape H-C (1990) Properties of a hyperpolarization-activated cation current and its role in rhythmic oscillation in thalamic relay neurons. J Physiol (Lond) 431: 291–318

Mulle C, Madariaga A, Deschênes M (1986) Morphology and electrophysiological properties of reticularis thalami neurons in cat: in vivo study of a thalamic pacemaker. J Neurosci 6: 2134–2145

Steriade M, Deschênes M (1984) The thalamus as a neuronal oscillator. Brain Res Rev 8: 1–63

Steriade M, Deschênes M, Domich L, Mulle C (1985) Abolition of spindle oscillations in thalamic neurons disconnected from nucleus reticularis thalami. J Neurophysiol 54: 1473–1497

Steriade M, Domich L, Oakson G (1987) The deafferented reticular thalamic nucleus generates spindle rhythmicity. J Neurophysiol 57: 260–273

Steriade M, Curró Dossi R, Nunez A (1991) Network modulation of a slow intrinsic oscillation of cat thalamocortical neurons implicated in sleep delta waves: cortical potentiation and brainstem suppression. J Neurosci 11: 3200–3217

Steriade M, Nunez A, Amzica A (1993a) A novel slow oscillation (<1 Hz) of neocortical neurons in vivo: depolarizing and hyperpolarizing components. J Neurosci 13: 3252–3265

Steriade M, Nunez A, Amzica A (1993b) Intracellular analysis of relations between the slow (<1 Hz) neocortical oscillation and other sleep rhythms of the electroencephalogram. J Neurosci 13: 3266–3283

Steriade M, Contreras D, Curro Dossi R, Nunez A (1993c) The slow (<1 Hz) oscillation in reticular thalamic and thalamocortical neurons. Scenario of sleep rhythms generation in interacting thalamic and neocortical neurons. J Neurosci 13: 3284–3299

von Krosigk M, Bal T, McCormick DA (1993) Cellular mechanisms of a synchronized oscillation in the thalamus. Science 261: 361–364

4.5 The Order in Synaptic Circuits of the Cerebral Cortex: Lessons from the Hippocampus

P. Somogyi

Medical Research Council, Anatomical Neuropharmacology Unit,
Oxford University, United Kingdom

4.5.1 Introduction: What Is the Basic Cortical Circuit?

Neuronal diversity in the cerebral cortex has fascinated scientists for more than a hundred years, but a basic circuit accounting for interconnections of the existing neuronal types in the cortex is still missing. The structural and functional relationships among cortical cells and their subcortical inputs was a major interest of Otto Creutzfeldt and his associates, who pioneered many approaches for the elucidation of visual cortical circuitry in functional terms (Creutzfeldt 1993; Creutzfeldt and Ito 1968; Creutzfeldt et al. 1974, 1975). In a symposium in his honor it is most appropriate to provide a brief summary of recent insights into synaptic relationships of cortical neurons.

In relation to the main neurotransmitters of its resident neurons the cortex is a simple structure neurochemically consisting of two major cell types: principal cells (pyramidal cells, spiny stellate cells, hippocampal granule cells), which usually have densely spiny dendrites and release excitatory amino acids as transmitters, and non-principal, or local circuit cells, which generally have sparsely spiny or smooth dendrites (for review see McCormick 1992; Somogyi 1989). The latter population forms about 20% of cortical neurons (Gabbott and Somogyi 1986), and releases γ-aminobutyric acid (GABA) as an inhibitory transmitter. The thalamic input to the cortex and long-range associational connections between different cortical areas are mediated by excitatory amino acid mechanisms (for review see McCormick 1992; Tsumoto 1992). Evidence is strong for glutamate as an excitatory amino acid used in cortical connections, and it may be the only acidic amino acid released by thalamo-cortical and corticocortical terminals. Therefore, defining cortical circuitry involved in fast signal processing comes down to defining the origin, relationship and properties of GABAergic and glutamatergic synaptic junctions. Much of the complexity in cortex is due to the diversity of sources of both the glutamatergic and the GABAergic inputs. For example, stereological estimates show that an average cortical neuron receives about 5800 synapses in the cat striate cortex (Beaulieu et al. 1992). Of these, about 960 originate from local, GABAergic, inhibitory neurons, each one making between one and ten synaptic contacts on a single postsynaptic cell. Assuming an average of five, it follows that at least 200 inhibitory neurons, belonging to distinct functional classes, may converge onto a single postsynaptic cortical cell.

The numerical relationship between the terminals using these two classes of transmitters is remarkably stable across species and different cortical areas. About 17% of synaptic boutons are immunopositive for GABA (Beaulieu and Somogyi 1990; Beaulieu et al. 1992) and presumably release it as transmitter. Most of these terminals form morphologically identifiable Gray's type 2 synapses (Gray 1959). Almost all the remaining synaptic boutons form Gray's type 1 synaptic junctions, exhibiting thick postsynaptic densities. Several lines of evidence suggests that the overwhelming majority of these boutons store and release glutamate as neurotransmitter (Liu et al. 1989, Bramham et al. 1990). An additional important element to this somewhat simplified cortical circuit is provided by a sparse, but diverse array of subcortical, nonthalamic, monoaminergic inputs from the reticular formation which, due to lack of space, are not discussed here.

A basic cortical circuit is defined here as the sum of: (a) a subpopulation of principal cells which can be considered homogeneous in terms of their main output, (b) the main excitatory inputs that lead to the activation of these cells, and (c) the local circuit inhibitory neurons involved in the regulation of both the inputs and outputs of these principal cells. A major difficulty in accounting for the relationships of these three elements *in the neocortex* is that a similar basic circuit is superimposed several times in each column, from layer to layer. In addition, the connections between layers complicate the recognition of the general layout of the basic circuit. Another complexity is introduced by the rich interconnections of the principal cells whose local recurrent axon collaterals overlap in tissue space with the afferents activating the same cells, or may even provide the main excitatory drive to a population of cells. Therefore, in order to define the synaptic relationships in the basic cortical circuit we have investigated the simpler archicortical area of the hippocampal formation.

The hippocampal formation has two great advantages for the study of basic cortical circuits (for review see Amaral and Witter 1989; Andersen 1975; Witter 1989). (a) The cell bodies of principal cells are arranged into a single layer, resulting in the tangential alignment of functionally equivalent segments of the dendritic fields of individual cells into distinct, well-recognizable laminae. (b) The recurrent local axon collaterals of the principal cells are separated into distinct laminae in the CA1 area and the dentate gyrus. The dentate gyrus in nonprimate species is the only cortical area in which the recurrent collaterals of the principal cells are almost completely separated from the major excitatory inputs to the principal cells (Ramón y Cajal 1893). Although this makes the dentate gyrus unusual in that its principal cells are not normally directly interconnected, the laminar segregation of extrinsic and recurrent inputs offers a unique opportunity to study the relationship of principal cells and local inhibitory cells, both anatomically and physiologically (for review see Scharfman 1992; Mody et al. 1992). The following section focuses on two recent studies of the dentate gyrus (Han et al. 1993; Halasy and Somogyi 1993b) and their implications for other hippocampal areas and the neocortex.

4.5.2 Synaptic Connections in the Dentate Gyrus

Overview. The dentate gyrus is considered the gate to the hippocampal formation through which input from the neocortex arriving via the entorhinal afferents activates the trisynaptic circuit from the dentate area through the CA3 area to the CA1 region (Fig. 1; for review see Andersen 1975; Buzsaki 1989). The principal cells are the granule cells, arranged densely in a cell body layer. Their spiny dendrites ramify in the molecular layer and their axons run in the opposite direction to the CA3 area, where they provide the mossy fiber input to the proximal dendrites of pyramidal cells and local circuit neurons. The granule cell layer is followed by the so-called hilus, containing another set of principal cells, the mossy cells (Amaral 1978;

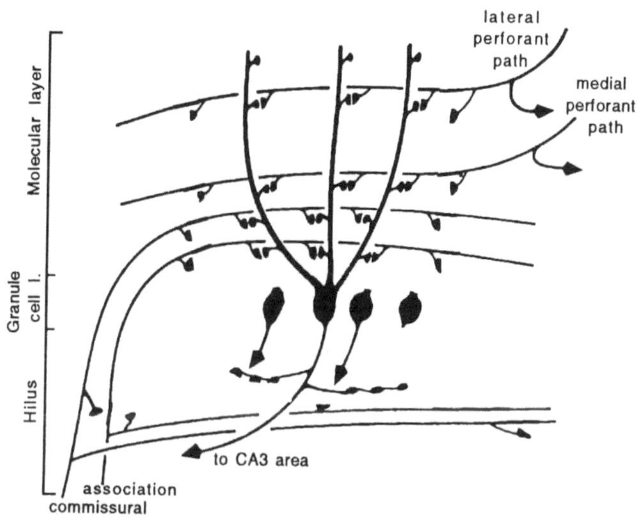

Fig. 4.5–1. Schematic view of the main connections of granule cells in the rat dentate gyrus, demonstrating the lamination of excitatory amino acid inputs to the dendritic field and the spatial segregation of local axon collaterals. Entorhinal afferents innervate the outer two thirds of principal (granule) cell dendrites in the molecular layer, whereas the granule cell local axons mainly innervate the hilus, then run to the CA3 area where they provide the mossy fiber input to proximal dendrites. The hilus contains the excitatory mossy cells (not shown) and a variety of other cell types (see Figs. 2,3). The inner one third of the molecular layer is innervated by ipsilateral association fibers, provided by hilar mossy cells and CA3c pyramidal cells and by commissural fibers from the contralateral hippocampal formation. The association and commissural pathways overlap (not shown)

Frotscher et al. 1991), the local recurrent axon collaterals of granule cells and a variety of cell types whose identity and synaptic relationships are not fully determined (Amaral, 1978; Frotscher 1991; Gulyas et al. 1992), but they are probably all innervated by the granule cells. The entorhinal corticocortical afferents terminate in the outer two thirds of the molecular layer; the medial entorhinal area innervating the middle one-third, and the lateral entorhinal area innervates the outermost one-third of the molecular layer (Hjorth-Simonsen and Jeune 1972). The inner one third of the molecular layer is innervated by ipsilateral association fibers, provided by hilar mossy cells (Buckmaster et al. 1992) CA3c pyramidal cells (Li et al. 1993) and commissural fibers, probably originating from similar cells of the contralateral hilus as the ipsilateral association pathway (Hjorth-Simonsen and Laurberg 1977).

The Question. Is it possible to discern any general rules in the interrelationships between extrinsic and intrinsic excitatory amino acid pathways, the inputs and outputs of local circuit inhibitory neurons and the inputs and output of the principal granule cells? Local circuit neurons were identified on the basis of their electrophysiological characteristics in intracellular recordings of hippocampal slices in vitro (Han et al. 1993). The cells discussed below generally had short duration action potentials with a fast repolarization phase, and each action potential was followed by a fast, deep afterhyperpolarization. Apart from one cell type, and in contrast to granule cells, the inhibitory neurons showed little spike frequency accommodation. The synaptic inputs of the cells were studied by electrical stimulation of afferent pathways. The cells were identified by intracellular injection of biocytin followed by visualization using avidin-horseradish peroxidase (HRP) conjugates. The nature of transmitter in the terminals of the intracellularly recorded and morphologically identified cells was determined by immunoreacting their labeled synaptic boutons for GABA.

Definition of Cells According to Output. We have found that in the dentate gyrus the principal granule cells receive inputs from at least five distinct types of GABAergic neurons, terminating in four out of five cases, in mutually exclusive domains on the surface of granule cells (Figs. 2,3). They are: (a) hilar cell with cell body and dendrites restricted to the hilus and having ascending axons to the outer two thirds of the molecular layer, where they terminate in conjunction with the perforant pathway (*HIPP* cell); (b) molecular layer cell with both axon and dendrites restricted to the outer two thirds of the molecular layer in association with the perforant pathway (*MOPP* cell); (c) hilar cells with axon restricted to the inner one third of the molecular layer terminating in conjunction with the commissural and association pathways (*HICAP* cell) (d) ***basket cells***, terminating on the somata and proximal dendrites of granule cells; (e) ***axo-axonic cells***, terminating exclusively on the axon initial segments of the granule cells. Basket (Ribak et al. 1978) and axo-axonic cells (Soriano et al. 1990; Soriano and Frotscher 1989) are known to be GABAergic, MOPP and HICAP cells were found to store GABA in their terminals (Halasy and Somogyi 1993b) and more recently HIPP cells were also shown to be immunoreactive for GABA (K. Halasy, E. Buhl, P. Somogyi, unpublished observation). This is in agreement with physiological evidence that some hilar cells generate inhibitory postsynaptic potentials (IPSP) in granule cells (Scharfman et al. 1990; for review see Misgeld et al. 1992; Scharfman 1992).

Inputs. The analysis of the dendritic distribution of these cell types is most revealing with regard to the spatial selection of inputs (Han et al. 1993). The dendrites of the HIPP cell are restricted to the hilus, hence it has ample access to the recurrent axon collaterals of granule cells but cannot receive significant direct input from entorhinal afferents. Thus, HIPP cells are feedback regulators of the granule cell/perforant path synapse (Figs. 2,3).

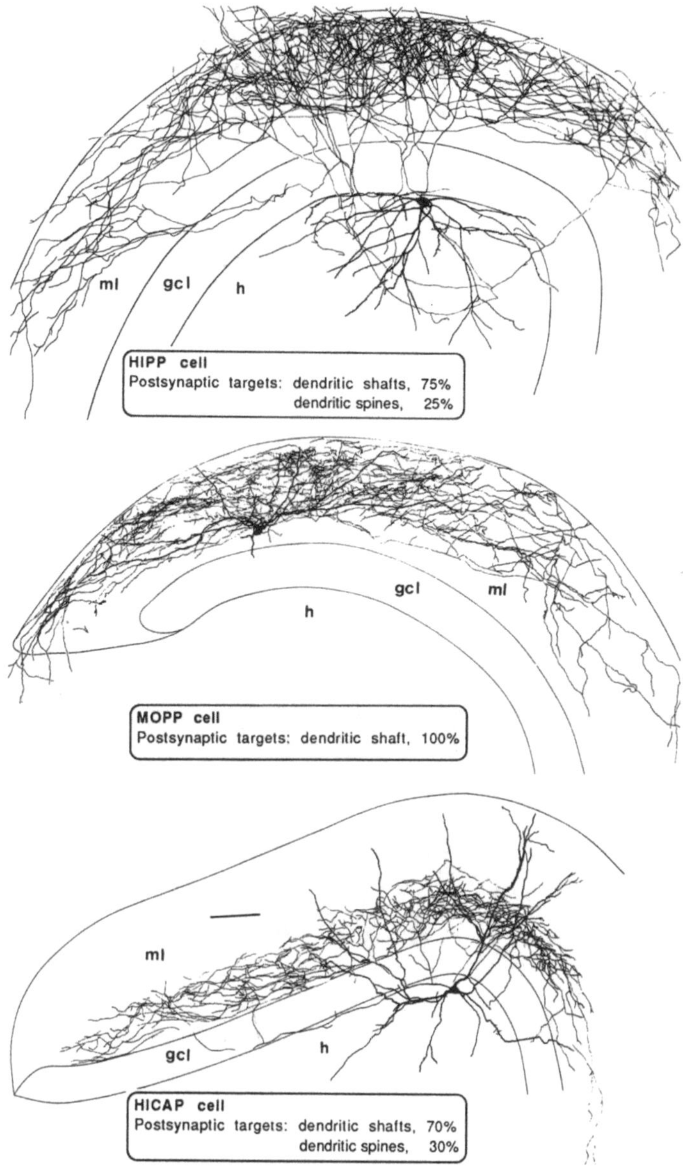

Fig. 4.5–2. Reconstructions of three types of local circuit GABAergic neurons terminating in the dendritic field of granule cells and demonstrating selective axonal and dendritic patterns. Cells were recorded in vitro and visualized with biocytin. Synaptic targets were identified by electron microscopy. HIPP and MOPP cells innervate the outer two thirds of the molecular layer in conjunction with the perforant path. HICAP cells innervate the inner one third of the molecular layer in conjunction with the commissural/association pathway. The dendrites of HIPP cells are restricted to the hilus, whereas those of MOPP cells are confined to the outer molecular layer. HICAP cells have access to all dentate excitatory amino acid afferents. *Scale* for all cells, 100 μm. (Data based on Han et al. 1993; Halasy and Somogyi 1993b)

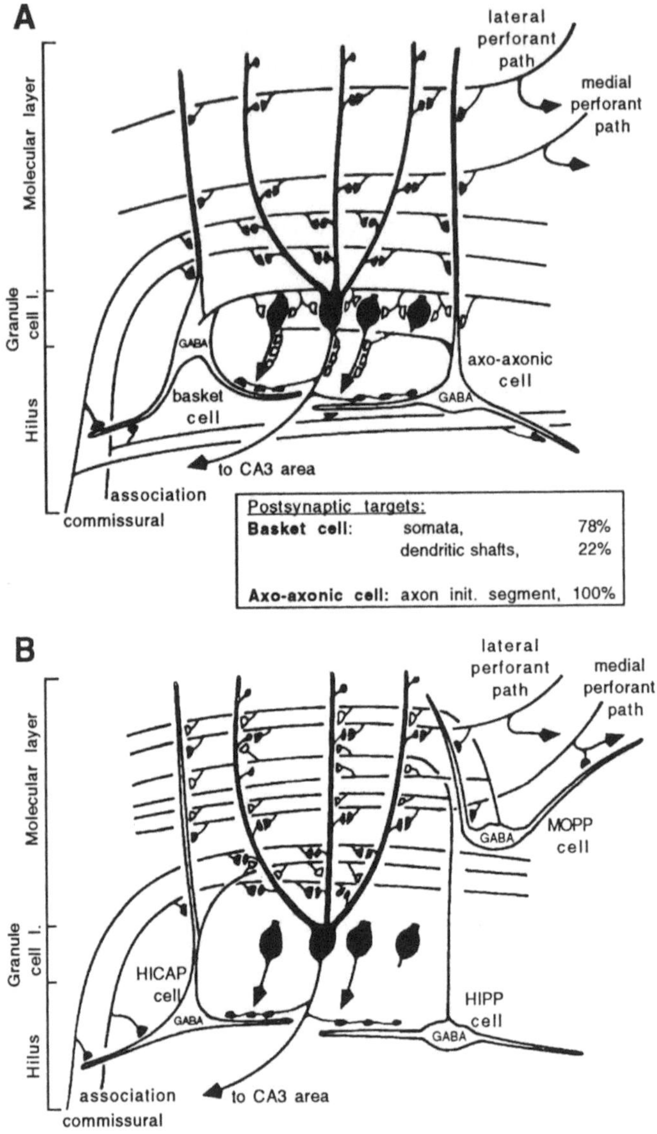

Fig. 4.5–3. Relationships between inputs and subdivisions of the multiple GABAergic innervation of dentate granule cells. **A** There are two levels of output control by the basket and axo-axonic cells, both of them probably receiving the full complement of dentate afferents as well as recurrent granule cell input. **B** Three types of GABAergic neurons (open cells and terminals), innervating the dendritic region, associate their terminals with particular excitatory amino acid pathways (see also Fig. 2). The distribution of dendrites suggests feedforward activation of MOPP cells, feedback (recurrent) activation of HIPP cells and both types of inputs to HICAP cells; the latter could mediate inhibitory cross-influence by the perforant path on the association/commissural input to granule cells

In contrast, MOPP cells have access to entorhinal input that also activates the granule cells but cannot receive significant recurrent input from the granule cells on which they terminate (Fig. 2). Thus, MOPP cells may serve feedforward regulation of the granule cell/perforant path synapse (Fig. 3). Basket, axo-axonic and HICAP cells distribute dendrites in all layers of the dentate gyrus, thus they are in a position to receive input from all the extrinsic afferent pathways (see e.g., Buzsaki 1984; Buzsaki and Eidelberg 1981; Zipp et al. 1989), and from the recurrent collaterals of granule cells (Fig. 3). The relative strength of these inputs is not yet known. The most striking of the latter three cell types is the HICAP cell, which clearly receives perforant path input on its dendrites in the outer two thirds of the molecular layer but restricts its efferent synapses to the inner one-third of the dendritic region of granule cells, where other excitatory pathways terminate. In theory, the perforant path input could provide an inhibitory cross-influence on the commissural/association input to granule cells in a feedforward manner. Much more information is needed, however, on the relative timing and the strength of activation of the different cell types before real functional schemes can be drawn up.

Outputs. The basket and axo-axonic cell types establish their synapses at sites thought to be the most effective place for inhibition to control the final output of granule cells. So called "basket cells" have been frequently described, but without identification of their synaptic output they were not differentiated from axo-axonic or other neuronal types that do not contribute to somatic synapses. It has been suggested that only those cells that provide a substantial proportion of their synaptic terminals to the cell bodies of other neurons should be called basket cells (Halasy and Somogyi 1993b). Axo-axonic cells were identified more recently in the dentate gyrus (Soriano and Frotscher 1989). From our sample it is now apparent that the same individual axo-axonic cells innervate both granule cells and one of their excitatory input cells, the hilar mossy cells.

Although the dual GABAergic innervation of proximal parts of the cortical principal cells by basket and axo-axonic cells is a universal feature of all cortical areas (see Somogyi 1989), the reason for the segregation of inputs is not known. Physiologically identified inhibitory cells of the hippocampal pyramidal layer were found to evoke only fast IPSPs in pyramidal cells in vitro (Miles 1990; Miles and Wong 1984). Recently, Buhl et al. (1993) recorded simultaneously from both a dentate axo-axonic cell and one of its postsynaptic granule cells, followed by anatomical identification of the synaptic contacts that exerted the postsynaptic action on the granule cell through the release sites on the axon initial segment. Action potentials in the axo-axonic cell produced very fast, brief IPSPs in the granule cell with a reversal potential of –78 mV. Identified basket cells evoked similar postsynaptic effect in the CA1 region (Buhl et al. 1993), and we suggested that the effects of both cell types were mediated through $GABA_A$ receptors. It is possible that the need arose for two different GABAergic output control cells in cortex because the two cells play roles in different aspects of the response of postsynaptic principal cells during their synaptic activation. Axo-axonic cells are ideally placed for synchronization of the output of a large population of principal cells, whereas basket cells may control the degree of depolarization spreading to the initial segment through the soma. In other words, basket cells may rescale excitatory postsynaptic potentials (EPSP) arriving from the dendrites at the somatic level, and therefore indirectly control the threshold of spike generation in the initial segment.

The GABAergic HICAP, MOPP and HIPP cells terminate in precise spatial conjunction with particular excitatory amino acid inputs to the same cell, thus they are in a position to interact selectively with those specific inputs. This idea is reinforced by the observed pairing of a glutamatergic and a GABAergic synapse on a single dendritic spine, although it is probably better to consider the effect of these inputs in a statistical manner and estimate the overall effect of synapses on a dendritic region rather than on a single spine. Measurements of the number of GABAergic synapses on different parts of granule cells show that in numerical terms most of the GABAergic transmission takes place at the dendrites. There are at least three times as many GABAergic synapses on dendrites than on the somata and axon initial segments of granule cells (Halasy and Somogyi 1993a). As in the neocortex (Beaulieu and Somogyi 1990; Beaulieu et al. 1992) the overall proportion of GABA immunopositive synapses is also 17% in the dentate gyrus (Halasy and Somogyi 1993a).

4.5.3 How General Is the Selective GABAergic Innervation of Dendritic Regions?

As elsewhere in the cortex, in the CA1 area the stratum pyramidale is innervated by the somatic basket cells with a small overflow to the adjacent dendritic layers (Ramón y Cajal 1893). Preliminary studies show that in addition there are GABAergic neurons whose axonal output is biased for the dendritic laminae. A local circuit neuron, the bistratified cell, provides its synaptic boutons, in two parallel bands, more to the strata radiatum and oriens than to the stratum pyramidale (Halasy et al. 1993). An even more specialised cell type, the butterfly cell almost completely avoids terminating in the stratum pyramidale and innervates the full depth of the strata radiatum and oriens in precise conjunction with the glutamatergic Schaffer collateral/commissural input (Halasy et al. 1993). Cell bodies of all the above three cell types occur in or near the pyramidal cell layer. Further

GABAergic cells reside in the strata lacunosum moleculare (Lacaille and Schwartzkroin 1988a,b), radiatum and oriens (Lacaille and Williams 1990), but there is only limited information on their axonal termination, although probably they innervate mostly the dendritic regions (see e.g., Kawaguchi and Hama 1987, 1988).

In the neocortex, in addition to basket and axo-axonic cells, there are local circuit neurons that make synapses mainly with dendrites and dendritic spines (for review see Somogyi 1989; and P. Somogyi, Z. Kisvárday, D. Beroukas, unpublished results). It is not yet known whether their terminals are co-distributed in conjunction with particular excitatory inputs to the same postsynaptic cells. In general, because the dendritic regions of many different cell types overlap, very little is known about the precise distribution of particular afferent synapses on the surface of neocortical neurons. Nevertheless, the limited information that is available, particularly from the laminar selectivity of afferents, indicates that stratified dendritic innervation by both glutamatergic and GABAergic afferents exists in many cortical areas.

4.5.4 What Are the Possible Functional Consequences of the Conjoint Excitatory and Inhibitory Dendritic Inputs?

Any hypothesis must consider the effect of the inhibitory cell in the context of the effect on the excitatory input located on the same dendritic segments.

Selective Subcortical State Control. Innervation of nonprincipal cells in neocortex and hippocampus by nonthalamic subcortical afferents (Freund 1992; Mulligan and Tork 1987) provides an opportunity for different subcortical afferents to control particular excitatory inputs selectively via the GABAergic cells that terminate in conjunction with them.

Presynaptic Inhibition of Glutamate Release. Glutamate release is reduced by the activation of presynaptic $GABA_B$ receptors (Thompson and Gähwiler 1992). The costratification of GABAergic axons and glutamatergic inputs provides an opportunity for GABAergic influence on a particular pathway and not on the others. However, the GABAergic terminals in the dendritic region are neither in preferential membrane apposition nor do they form axo-axonal synapses with glutamatergic terminals. Instead the anatomically defined synaptic junctions are formed with postsynaptic dendrites and spines, so GABA can reach glutamatergic terminals only by diffusion from the presumed release site at the synaptic junction. Obviously, the probability of GABA diffusion to nearby terminals is greater, and the costratification may thus ensure preferential influence on some, but not on other input pathways.

Regulation of NMDA Receptor Activation. The activation of *N*-methyl-D-aspartarte (NMDA) receptor mediated excitatory postsynaptic effect is voltage dependent, increasing with depolarization, and GABAergic hyperpolarizing and shunting influences decrease the NMDA receptor mediated effect (Davies et al. 1991; Staley and Mody, 1992). The selective pairing of GABAergic and glutamatergic inputs could lead to the differential adjustment of the probability of NMDA receptor activation and/or effect at the synapses of different pathways that converge on to the same cell. This could result in a differential, long term modification of synaptic strengths of only some inputs to a cell, as has been demonstrated for long term potentiation (for review see e.g., Bliss and Collingridge 1993).

Increasing the Dynamic Range of EPSP. Postsynaptic GABAergic influences reduce both the amplitude and the duration of glutamate-mediated EPSPs. The *pathway specific* GABA-mediated events, with appropriate gain and timing (Staley and Mody, 1992) could rescale the EPSPs arriving at the same dendritic segment and thereby extend their dynamic range. This mechanism would work by reducing EPSPs, which are suprathreshold in the absence of GABAergic influence, into the subthreshold domain and render them available for spatial and temporal summation. This may be a general role for those GABAergic terminals which are selectively co-aligned with glutamatergic terminals in the dendritic tree in many parts of the central nervous system. The arrangement found in the dentate gyrus and elsewhere is highly appropriate for the separate adjustment of the gain of different corticocortical excitatory inputs. The downward rescaling would also dampen spurious variability in EPSP strength and make inputs more reliable. The mechanisms discussed above are not mutually exclusive. The recognition of the order in synaptic circuits opens them up for physiological exploration.

4.5.5 Conclusions

What are the lessons from the hippocampus that are likely to be applicable to cortical circuits in general? Studies of the hippocampal circuitry lead to the following predictions: (a) Populations of principal output cells of the cortex receive segregated excitatory glutamatergic input on their dendrites. (b) Excitatory inputs to a cortical area always activate both excitatory and inhibitory cells in parallel. (c) Some inhibitory cells are connected primarily in a feedforward manner, others primarily in a feedback manner, but some types receive both inputs. (d) Inhibitory cells terminating on the somatic region (basket cells), or on the axon initial segment (axo-axonic cells) of principal cells receive the same excitatory inputs as the principal cells and in addition a recurrent input from the cells that they innervate. (e) Some inhibitory cells terminate only on the dendritic tree of principal cells in conjunction with subsets of excitatory inputs, and their inhibitory effect is likely to be pathway specific. Certain elements of these predictions remain to be proven in the neocortex, but the strategy applied in the hippocampus could provide useful guidelines for their investigation.

References

Amaral DG (1978) A Golgi study of cell types in the hilar region of the hippocampus in the rat. J Comp Neurol 182: 851–914

Amaral DG, Witter MP (1989) The three-dimensional organization of the hippocampal formation: a review of anatomical data. Neuroscience 31: 571–591

Andersen P (1975) Organization of hippocampal neurons and their interconnections. In: Isaacson RI, Pribram KH (eds) The hippocampus. Plenum, New York, pp 155–175

Beaulieu C, Somogyi P (1990) Targets and quantitative distribution of GABAergic synapses in the visual cortex of the cat. Eur J Neurosci 2: 296–303

Beaulieu C, Kisvárday Z, Somogyi P, Cynader M, Cowey A (1992) Quantitative distribution of GABA-immunopositive and -immunonegative neurons and synapses in the monkey striate cortex (area 17). Cerebral Cortex 2: 295–309

Bliss TVP, Collingridge GL (1993) A synaptic model of memory: long-term potentiation in the hippocampus. Nature 361: 31–39

Bramham CR, Torp R, Zhang N, Storm-Mathisen J, Ottersen OP (1990) Distribution of glutamate-like immunoreactivity in excitatory hippocampal pathways: a semiquantitative electron microscopic study in rats. Neuroscience 39: 405–417

Buckmaster PS, Strowbridge BW, Kunkel DD, Schmiege DL, Schwartzkroin PA (1992) Mossy cell axonal projections to the dentate gyrus molecular layer in the rat hippocampal slice. Hippocampus 2: 349–362

Buhl EH, Halasy K, Somogyi P (1993) Hippocampal unitary IPSPs: identified sources and number of release sites. Proc 16th Eur Neurosci Ass Meeting 225: Abstr 873

Buzsaki G (1984) Feed-forward inhibition in the hippocampal formation. Prog Neurobiol 22: 131–153

Buzsaki G (1989) Two-stage model of memory trace formation: a role for "noisy" brain states. Neuroscience 31: 551–570

Buzsaki G, Eidelberg E (1981) Commissural projection to the dentate gyrus of the rat: evidence for feed-forward inhibition. Brain Res 230: 346–350

Creutzfeldt OD (1993) Cortex cerebri. Performance, structural and functional organization of the cortex. (English translation) Mary Creutzfeldt, Göttingen

Creutzfeldt O, Ito M (1968) Functional synaptic organization of primary visual cortex neurones in the cat. Exp Brain Res 6: 324–352

Creutzfeldt OD, Kuhnt U, Benevento LA (1974) An intracellular analysis of visual cortical neurones to moving stimuli: responses in a co-operative neuronal network. Exp Brain Res 21: 251–274

Creutzfeldt O, Innocenti GM, Brooks D (1975) Neurophysiological experiments on afferent and intrinsic connections in the visual cortex (area 17). In: Santini M (ed) Golgi Centennial Symposium Proceedings. Raven, New York, pp 319–337

Davies CH, Starkey SJ, Pozza MF, Collingridge GL (1991) $GABA_B$ autoreceptors regulate the induction of LTP. Nature 349: 609–611

Freund TF (1992) GABAergic septal serotonergic median raphe afferents preferentially innervate inhibitory interneurons in the hippocampus and dentate gyrus. In: Ribak CE, Gall CM, Mody I (eds) The dentate gyrus and its role in seizures. Elsevier, Amsterdam, pp 79–91 (Epilepsy Res, Suppl 7)

Frotscher M (1991) Target cell specificity of synaptic connections in the hippocampus. Hippocampus 1: 123–130

Frotscher M, Seress L, Schwerdtfeger WK, Buhl E (1991) The mossy cells of the fascia dentata: a comparative study of their fine structure and synaptic connections in rodents and primates. J Comp Neurol 312: 145–163

Gabbott PLA, Somogyi P (1986) Quantitative distribution of GABA-immunoreactive neurons in the visual cortex (area 17) of the cat. Exp Brain Res 61: 323–331

Gray EG (1959) Axo-somatic and axo-dendritic synapses of the cerebral cortex: an electron microscope study. J Anat 93: 420–433
Gulyas AI, Miettinen R, Jacobowitz DM, Freund TF (1992) Calretinin is present in non-pyramidal cells of the rat hippocampus I. A new type of neuron specifically associated with the mossy fibre system. Neuroscience 48: 1–27
Halasy K, Somogyi P (1993a) Distribution of GABAergic synapses and their targets in the dentate gyrus of rat: a quantitative immunoelectron microscopic analysis. J Hirnforschung 34: 299–308
Halasy K, Somogyi P (1993b) Subdivisions in the multiple GABAergic innervation of granule cells in the dentate gyrus of the rat hippocampus. Eur J Neurosci 5: 411–429
Halasy K, Buhl EH, Somogyi P (1993) Target specificity of physiologically identified inhibitory GABAergic interneurons in the rat hippocampus. Soc Neurosci Abstr 19: 351
Han Z-S, Buhl EH, Lorinczi Z, Somogyi P (1993) A high degree of spatial selectivity in the axonal and dendritic domains of physiologically identified local-circuit neurons in the dentate gyrus of the rat hippocampus. Eur J Neurosci 5: 395–410
Hjorth-Simonsen A, Jeune B (1972) Origin and termination of the hippocampal perforant path in the rat studied by silver impregnation. J Comp Neurol 144: 215–232
Hjorth-Simonsen A, Laurberg S (1977) Commissural connections of the dentate area in the rat. J Comp Neurol 174: 591–606
Kawaguchi Y, Hama K (1987) Two subtypes of non-pyramidal cells in rat hippocampal formation identified by intracellular recording and HRP injection. Brain Res 411: 190–195
Kawaguchi Y, Hama K (1988) Physiological heterogeneity of nonpyramidal cells in rat hippocampal CA1 region. Exp Brain Res 72: 494–502
Lacaille J-C, Schwartzkroin PA (1988a) Stratum lacunosum-moleculare interneurons of hippocampal CA1 region. I. Intracellular response characteristics, synaptic responses, and morphology. J Neurosci 8: 1400–1410
Lacaille J-C, Schwartzkroin PA (1988b) Stratum lacunosum-moleculare interneurons of hippocampal CA1 region. II. Intrasomatic and intradendritic recordings of local circuit synaptic interactions. J Neurosci 8: 1411–1424
Lacaille J-C, Williams S (1990) Membrane properties of interneurons in stratum oriens-alveus of the CA1 region of rat hippocampus in vitro. Neuroscience 36: 349–359
Li X-G, Somogyi P, Ylinen A, Buzsaki G (1993) The hippocampal CA3 network: an in vivo intracellular labeling study. J Comp Neurol (in press)
Liu C-J, Grandes P, Matute C, Cuenod M, Streit P (1989) Glutamate-like immunoreactivity revealed in rat olfactory bulb, hippocampus and cerebellum by monoclonal antibody and sensitive staining method. Histochemistry 90: 427–445
McCormick DA (1992) Neurotransmitter actions in the thalamus and cerebral cortex and their role in neuromodulation of thalamocortical activity. Prog Neurobiol 39: 337–388
Miles R (1990) Variation in strength of inhibitory synapses in the CA3 region of guinea-pig hippocampus in vitro. J Physiol 431: 659–676
Miles R, Wong RKS (1984) Unitary inhibitory synaptic potentials in the guinea-pig hippocampus in vitro. J Physiol 356: 97–113
Misgeld U, Bijak M, Brunner H (1992) Granule cell inhibition and the activity of hilar neurons. In: Ribak CE, Gall CM, Mody I (eds) The dentate gyrus and its role in seizures. Elsevier, Amsterdam, pp 113–118 (Epilepsy Res, Suppl 7)
Mody I, Kohr G, Otis TS, Staley KJ (1992) The electrophysiology of dentate gyrus granule cells in whole-cell recordings. In: Ribak CE, Gall CM, Mody I (eds) The dentate gyrus and its role in seizures. Elsevier, Amsterdam, pp 159–168 (Epilepsy Res, Suppl 7)
Mulligan KA, Tork I (1987) Serotonergic axons form basket-like terminals in cerebral cortex. Neurosci Lett 81: 7–12
Ramón y Cajal S (1893) Estructura del asta de ammon y fascia dentata. Anal Soc espan Historia natural 22: 53–114
Ribak CE, Vaughn JE, Saito K (1978) Immunocytochemical localization of glutamic acid decarboxylase in neuronal somata following colchicine inhibition of axonal transport. Brain Res 140: 315–332
Scharfman HE (1992) Differentiation of rat dentate neurons by morphology and electrophysiology in hippocampal slices: granule cells, spiny hilar cells and aspiny 'fast-spiking' cells. In: Ribak CE, Gall CM, Mody I (eds) The dentate gyrus and its role in seizures. Elsevier, Amsterdam, pp 93–109 (Epilepsy Res, Suppl 7)
Scharfman HE, Kunkel DD, Schwartzkroin PA (1990) Synaptic connections of dentate granule cells and hilar neurons: results of paired intracellular recordings and intracellular horseradish peroxidase injections. Neuroscience 37: 693–707

Somogyi P (1989) Synaptic organisation of GABAergic neurons and $GABA_A$ receptors in the lateral geniculate nucleus and visual cortex. In: Lam DK-T, Gilbert, CD (eds) Neural mechanisms of visual perception. Portfolio, The Woodlands Texas, pp 35–62
Soriano E, Frotscher M (1989) A GABAergic axo-axonic cell in the fascia dentata controls the main excitatory hippocampal pathway. Brain Res 503: 170–174
Soriano E, Nitsch R, Frotscher M (1990) Axo-axonic chandelier cells in the rat fascia dentata: Golgi-electron microscopy and immunocytochemical studies. J Comp Neurol 293: 1–25
Staley KJ, Mody I (1992) Shunting of excitatory input to dentate gyrus granule cells by a depolarizing $GABA_A$ receptor-mediated postsynaptic conductance. J Neurophysiol 68: 197–212
Thompson SM, Gähwiler BH (1992) Comparison of the actions of baclofen at pre- and postsynaptic receptors in the rat hippocampus in vitro. J Physiol 451: 329–345
Tsumoto T (1992) Long-term potentiation and long-term depression in the neocortex. Prog Neurobiol 39: 209–228
Witter MP (1989) Connectivity of the rat hippocampus. In: Chan-Palay V, Kohler C (eds) The hippocampus — new vistas. Alan R Liss, New York, pp 53–69
Zipp F, Nitsch R, Soriano E, Frotscher M (1989) Entorhinal fibers form synaptic contacts on parvalbumin-immunoreactive neurons in the rat fascia dentata. Brain Res 495: 161–166

4.6 Constraints to a Random Plan of Cortical Connectivity

A. Schüz and V. Braitenberg
Max Planck Institute for biological Cybernetics, Tübingen, Germany

4.6.1 Introduction

The cerebral cortex distinguishes itself from other parts of the brain by some unique structural features, which we took as a guideline for the selection of a functional model (Braitenberg and Schüz 1991). Here, we want to take up the thread of a previous paper (Schüz 1992) and focus on a controversial issue, that of the degree of order which prevails in the cortex. The issue itself is reflected in the amount of detail which must be described in order to capture the relevant features of cortical anatomy.

As a preliminary exercise, we list a number of expressions that come up in this controversy. A real understanding of some of these would touch on issues of mathematics and physics, which are beyond the scope of this paper, but they serve to illustrate two opposite philosophies: (a) wiring, circuits, dedicated line, specificity, selectivity, modularity, precision, determinism, (b) random, stochastic, statistical, diffuse, distributed, global, disorderly, cooperative.

Some of these concepts could be defined more rigorously in quantitative terms (e.g., "disorder" in terms of entropy) and some admit of various interpretations. Thus, "specificity" of the connections as well as "selectivity" may mean different things: the plan of neural wiring could specify for each neuron the type and precise location of the neurons to which it is connected (e.g., in the visual system of the fly; Braitenberg 1967) or just the location, with a free choice of the neural type (e.g., thalamic afferents to the visual cortex; Peters 1979), just the type of target neuron, leaving open its location (e.g., mossy fiber-granular cell system in the cerebellum; Ramón y Cajal 1911; Bloedel and Courville 1981), or the location, but with some leeway within a certain region of imprecision (e.g., callosal connections in the cortex; Innocenti 1981). In other cases, the terms "selectivity" or "specificity" could refer to the site of termination on the target neurons (e.g., in case of the chandelier cells; Somogyi 1977) or simply to a statistical preference for some populations of target neurons without excluding the others (e.g., the thalamic afferents to the somatosensory cortex; White 1989). Also, the concept of a "dedicated line" is difficult to define if the scope of a mechanism is not known. Worse still are the difficulties in defining the shades of meaning of some of the words under (b). Some of them properly belong to the realm of statistics, while others are hardly better than colloquial. Since, depending on one's attitude, terms under (a) and under (b) may have positive or negative value, we shall try to avoid this terminology altogether and to define some objective measures instead.

4.6.2 An Elementary Description of the Cerebral Cortex

Although our experience is mainly with rodent brains, a glance at other mammalian species shows that the following facts generally hold true and lead to a view of the cortex which has gradually evolved through the works of Sholl (1956), Cragg (1967), Braitenberg (1974, 1977, 1978a), Szentágothai (1978), Peters and Jones (1984), Martin (1988), and others.

1. In the cerebral cortex, both short- and long-range connections abound. This is very different from the situation in the cerebellar cortex, where all connections are short range. Quite likely there is a continuum of ranges of corticocortical fibers, the distinction of an intracortical system from one that traverses the white matter being perhaps rather artificial. The important point, however, is that, as far as is known, no other part of the brain projects so profusely onto itself.

2. Most of the synapses in the cortex are between elements of the same kind, the pyramidal cells. We shall call these connections "homotypical". They produce all of the long-range and most of the short-range

connections. This is again, if not unique, very characteristic of the cortex. Elsewhere, the dominant type of synapse is "heterotypical", connecting elements of one kind to elements of another kind. Heterotypical connections predominate for instance in the retina and in the cerebellar cortex.

3. In the system of pyramidal cells, divergence and convergence are nearly maximal. The number of synaptic partners is not much smaller than the number of synapses of an individual neuron, which is about 8000 in the mouse. This implies that the influence of individual pyramidal cells on each other is usually weak.

4. Most of the synapses between the elements in the cortex are excitatory.

5. The large majority of the synapses in the pyramidal cell system reside on dendritic spines, and these also constitute the majority of synapses in the cortex.

These five points taken together characterize connectivity in the cerebral cortex. Although knowledge of neuroanatomy is far from complete, it seems that in no other part of the brain are they all valid together. Points 3 and 5 are of particular relevance for the interpretation of cortical structure which we have proposed on various occasions. Both rely on circumstantial evidence and alternative interpretations are possible. The relevance of point 5 lies in the assumption that the synapses on spines are modifiable in strength. Although ultimate proof is lacking this assumption is widely accepted and we do not intend to go into the arguments here. We want, instead, to elaborate on point 3, which is related to the issue addressed in Sect. 4.6.1.

4.6.3 Maximum Divergence and Convergence

The reasons why we adopted the hypothesis of maximum divergence and convergence are first of all geometrical in nature. It is clear that the contrary idea of minimum divergence and convergence, namely each neuron giving all of its synapses to one particular other neuron, is absurd. This would imply identity of axonal and dendritic trees in the style of the climbing fibers on cerebellar Purkinje cells.

This can be excluded by the sheer difference in the size of dendritic and axonal trees. The range within which an axonal tree of a pyramidal cell ramifies far exceeds that of a dendritic ramification, as the more and more sophisticated methods of intracellular injections have made abundantly clear. Such an axon cannot help but distribute its contacts onto a considerable number of other neurons.

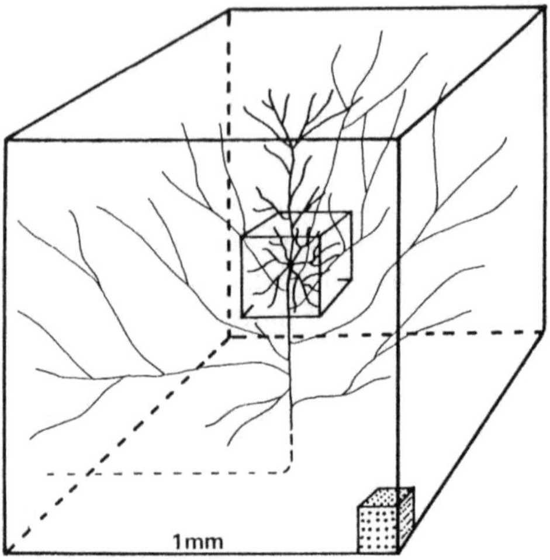

per 1 mm³	per neuron
9×10^4 neurons	-
0.4 km of dendrite	4 mm of dendrite
3-4 km of axon	10-20 mm of local axon
7×10^8 synapses	8.000 postsynaptic sites
	4.000 presynaptic sites on local axon

Fig. 4.6–1. The complex neighborhood relationships of an individual pyramidal cell. The size of the cube is chosen such as to just enclose all of the local axonal ramifications of this neuron. The *smaller cube* in the center is just big enough to enclose all the basal dendrites of this neuron. Below *on the right*, some measures for the average pyramidal cell in the mouse cortex are listed. *On the left*, the corresponding total numbers in the volume within which such a neuron ramifies are given. The *small cube* in the *lower right corner* illustrates the density of neurons. The numbers are taken from Schüz and Palm (1989) and Braitenberg and Schüz (1991)

This number could still be kept relatively small if axons and dendrites tended to take parallel courses or if synapses on axons clustered in such a way as to strongly favor particular dendrites which they happen to pass. However, this is not what is suggested by anatomy. When looking at the feltwork of fine axons and dendrites in light and electron microscopical sections, the prevailing impression is that of processes running in all directions, with no obvious alignment between axons and dendrites. If the verticality of the apical dendrites and of some vertical axonal segments provides an opportunity for strong contacts (Szenthágothai 1975), this does not seem to be the prevalent type of synaptic arrangement (Cragg 1967).

Furthermore, unmyelinated segments of axon collaterals of pyramidal cells carry synapses without any apparent order. Clusters of synapses that can occasionally be seen do not seem to go beyond the expected fluctuations of a random distribution (Hellwig et al. 1994).

All this speaks in favor of dendrites of pyramidal cells making their way through the thicket of axon collaterals (which appear earlier in development) without paying much attention to the orientation of these collaterals.

Figure 1 shows a cube of side 1 mm, a volume just big enough to include all the (local) ramifications of an average-size pyramidal cell in the mouse cortex. Although the neuron is richly ramified and has a total length of axon which is 10–20 times the side length of this cube, it only contributes a 1/100 000th of the axons ramifying there. Within the same cube, the processes of nearly 500 000 other neurons, some with cell bodies within the cube and some at various distances from it, are interwoven with each other at the highest possible packing density.

The smaller cube in the center of Fig. 1 focuses on the postsynaptic aspect. It is just big enough to completely contain the basal dendrites of one pyramidal cell. Although there are 4000–5000 postsynaptic sites on it, these are only 1/1000th of all the postsynaptic sites contained in this cube.

It can be shown that a single collateral passing through such a basal dendritic tree has a very low probability of hitting this tree at all and an even much lower one of hitting it more than once. Even if this dendritic tree were contained within a rich ramification of a single axon, the probability of it being contacted more than three times remains small ($p = 0.145$; Braitenberg 1978b). Thus, one of the points of cortical connectivity seems to be dispersion and mixing of signals.

The concept which fits the above description of the cortex is that of an associative memory (Braitenberg 1977). Models of this sort rely on a large number of similar elements connected into a network, with the strength of the connections increasing between elements that tend to be active together, as proposed by Hebb (1949). In this way, correlations between events in the outside world are represented as dynamic relations between the elements. The larger the number of elements, the more information the network will be able to deal with. The more complete the pattern of connections the more certainly the network can detect and incorporate every possible correlation. The system of pyramidal cells with their abundant and presumably modifiable interconnections seems an ideal substrate for learning of this kind.

4.6.4 Constraints

However, it is clear that for any number of elements approaching that of neurons in realistic nerve nets, complete connectivity of each element with every other one is impossible.

Consider the cerebral cortex of man with its 10^{10} neurons there were a synapse for each pair of them, assuming $1\,\mu m^3$ as the volume in which one synapse is found on average (a value pretty well established for a variety of species), the 10^{20} synapses necessary for complete connectivity would occupy a volume of $100\,m^3$, almost 100 000 times the volume of the human brain. In reality, the total number of synapses in the human cortex is of the order of 10^{14}. Apparently, only a small fraction of the complete set of connections is anatomically realized in the brain. The question is, then, according to what principle is the selection made.

If absolutely nothing is known about the statistical correlation of events which are likely to be stored in an associative memory, it is probably a good strategy to choose the connections entirely by chance. However, the brain lives in a real world, about which plenty of information is stored genetically. Quite probably, there is a preselection of connections which reflects the structure of the environment and predisposes the associative memory to learn patterns which are likely to occur, without, however, specifying them in detail.

This provides an interpretation of some brain structures which are already present at birth. The patterns of cortical interconnections via the white matter and the varying morphology of neuronal elements in different cortical areas and layers are part of the congenital equipment.

These preordained constraints of an associative memory should somehow be related to the redundancy of the messages coming from the environment. Since we have only intuitive notions about these, any attempt at relating brain structure to the reality of the outside world must necessarily be very sketchy. We shall consider a few examples below.

Cortical Areas and Columns. If nothing were known about the structure of the cortical grey matter, the pattern of the long-fiber connections in the white matter alone would force one to postulate distinct cortical areas. In fact, sensory input is not diffusely spread over the cortex, but segregated in narrow regions according to sensory modality. Equally impressive are the corticocortical connections which are not diffuse but organized in a pattern of bundles connecting specific areas to each other. In the associative memory, many of the correlations to be discovered refer to events within one cortical modality and will therefore be handled by local intracortical connections. Correlations of a different kind may be discovered between sensory modalities or between different areas handling the same sensory modality in different contexts (e.g., visual information in the context of language or in the context of object recognition). Clearly, this two-stage organization of the associative memory reflects the structuring of the information it deals with and is more efficient than a global random organization.

As an example, consider an object in the right upper corner of the visual field. If it moves, it is likely to be in a nearby position in the next moment. Therefore, neighborhood correlations must be important in the visual system. On the other hand, the position of the object in visual space has nothing to do with the pitch of the sound it emits and therefore no geometric correlations between the location of signals in visual and acoustic space are to be expected. When organization within one area is periodic in a so-called columnar fashion, it is likely that homologous positions within different columns (representing, for instance, a certain orientation in visual space) will show strong functional correlations and indeed such a column-to-column order of connections has been described (Livingstone and Hubel 1984a). Also, between different areas, specific column-to-column projection patterns may be apparent (Livingstone and Hubel 1984b).

Dendritic and Axonal Fields. Clearly, in the cortical associative memory, correlations between events can only be discovered and stored if the dendritic tree of one overlaps the axonal tree of another. Once it is known that intra-areal coordinates represent space or frequency or any other sensory order, the constraints imposed by this geometry acquire a very concrete meaning. The size of dendritic and axonal arborizations may have something to do with the regions within which visual input is integrated, with the maximum disparity relevant for stereoscopic vision, or with the critical bands in acoustic perception, etc. The comparison of these psychophysical quantities with neuroanatomy would certainly be worthwhile, but has not yet been exploited much.

Variation of Neuronal Shapes. Ramón y Cajal's cortex studies (1911) are still the richest source of cytoarchitectonic variations as they appear in the Golgi picture. A glance at his illustrations reveals, for instance, a striking variation of both the length and the density of dendritic processes within the basal dendritic tree of pyramidal cells. The extremes of this variation obviously predispose the network to detect correlations of different sorts: local events in a smaller set of afferent fibers in the case of the very concentrated dendritic trees and more diffuse events in a larger set of fibers in the case of the looser dendritic trees.

The variation of the shape of the axonal trees is perhaps even more striking. Among other things, their distribution in the cortex defines the functional relations between cortical layers.

Besides genetic prestructuring, there is evidence for an early shaping of the cortical network which is dependent on activity (Innocenti 1986; Payne et al. 1988). The phenomenon of pruning of axonal trees has been described for the thalamocortical terminations (LeVay et al. 1978) and it is possible that this is also the mechanism behind the clustering of collaterals of pyramidal cells which is sometimes observed (DeFelipe et al. 1986; Martin 1984). Such early learning or "imprinting" would certainly represent an effective means of optimizing an associative network. However, as mentioned above, even in such cases the connections are probably specified only with respect to region rather than to individual neurons. This is supported by a study by Kisvárday et al. (1986) and McGuire et al. (1991), who found in such clusters only one (and in one case two) synapses on each of the postsynaptic dendrites investigated.

Cyto- and Myeloarchitectonics. The variations of dendritic and axonal morphology are undoubtedly the main source of the subtle variations described in the literature on cyto- and myeloarchitectonics. It is reasonable to assume that where dendritic processes are long and numerous, the spacing of the neurons in the Nissl picture should be loose. Also, there is a general correlation of cell body size with the size of dendritic and axonal ramifications of the same neuron. Thus, undoubtedly the variety in the Golgi picture must underlie the variety in the Nissl picture. The myelin picture, which is mainly described in terms of the relative prominence of the two so-called stripes of Baillarger, may be taken as reflecting the thickness, length, and degree of myelination of intracortical axon collaterals of pyramidal cells. The relation of the Nissl and the myelin picture of one and the same area is not immediately obvious. Hellwig (1993) showed that relatively simple assumptions make it possible to predict one given the other. He assumed the dependence of the amount of myelin on cell size already mentioned and, in addition, a tendency of the horizontal collaterals to leave the axon a certain distance below the cell body. Quantifying these assumptions in an appropriate way, the density and average size of cell bodies in various layers in the Nissl picture was translated by a computer program into the expected density of myelinated fibers at different depths of the cortex. The result for most areas was remarkably similar to the actual myelin picture. Since Hellwig neglected nonpyramidal cells in his analysis, we may conclude that most of the areal variation of the "architectonic" kind is due to local variations in the morphology of pyramidal cells. Thus, what the areal variations show are locally different statistical constraints on the randomness of the basic network of the cortex. What this means in terms of function remains obscure. It must be said, however, that we have no real understanding of the importance of intracortical myelination, which is so abundant in some species (including man and cat) and almost absent in others (e.g., rodents).

References

Bloedel JR, Courville J (1981) Cerebellar afferent systems. In: Brookhart JM, Mountcastle VB (eds) Handbook of physiology, sect 1, The nervous system. Vol II. Motor control. American Physiol Soc, Bethesda, Maryland

Braitenberg V (1967) Patterns of projection in the visual system of the fly. I. Retina-lamina projections. Exp Brain Res 3: 271–298

Braitenberg V (1974) Thoughts on the cerebral cortex. J Theor Biol 46: 421–447

Braitenberg V (1977) On the texture of brains. Springer, Berlin Heidelberg New York

Braitenberg V (1978a) Cortical architectonics: general and areal. In: Architectonics of the cerebral cortex. Raven, New York, pp 443–465
Braitenberg V (1978b) Cell assemblies in the cerebral cortex. In: Heim R, Palm G (eds) Lecture notes in biomathematics (21). Theoretical approaches to complex systems. Springer, Berlin Heidelberg New York
Braitenberg V, Schüz A (1991) Anatomy of the cerebral cortex. Statistics and geometry. Springer, Berlin Heidelberg New York
Cragg BG (1967) The density of synapses and neurons in the motor and visual areas of the cerebral cortex. J Anat 101, 4: 639–654
DeFelipe J, Conley M, Jones EG (1986) Multiple, long-range focal collateralization of axons arizing from single neurons in monkey sensory-motor cortex. J Neurosci 6: 3749–3766
Hebb DO (1949) Organization of behavior. A neuropsychological theory, 2nd ed (1961). Wiley, New York
Hellwig B (1993) How the myelin picture of the human cerebral cortex can be computed from cytoarchitectural data. A bridge between von Economo and Vogt. J Hirnforsch 34: 387–402
Hellwig B, Schüz A, Aertsen A (1994) Synapses on axon collaterals of pyramidal cells are spaced at random intervals. A Golgi study in the mouse cerebral cortex. Biol Cybern (in press)
Innocenti G (1981) Growth and reshaping of axons in the establishment of callosal connections. Science 212: 824–827
Innocenti G (1986) General organization of callosal connections in the cerebral cortex. In: Peters A, Jones EG (eds) Cerebral cortex, vol 5. Sensory-motor areas and aspects of cortical connectivity. Plenum, New York
Kisvárday ZF, Martin KAC, Freund TF, Maglóczky Z, Whitteridge D, Somogyi P (1986) Synaptic targets of HRP-filled layer III pyramidal cells in the cat striate cortex. Exp Brain Res 64: 541–552
LeVay S, Stryker MP, Shatz CJ (1978) Ocular dominance columns and their development in layer IV of the cat's visual cortex: a quantitative study. J Comp Neurol 179: 223–244
Livingstone MS, Hubel DH (1984a) Specificity of intrinsic connections in primate visual cortex. J Neurosci 4: 2830–2835
Livingstone MS, Hubel DH (1984b) Anatomy and physiology of a color system in the primate visual cortex. J Neurosci 4: 309–356
Martin K (1984) Neuronal circuits in cat striate cortex. In: Peters A, Jones EG (eds) Cerebral cortex, vol 2. Functional properties of cortical cells. Plenum New York
Martin K (1988) From single cells to simple circuits in the cerebral cortex. Q J Exp Physiol 73: 637–702
McGuire BA, Gilbert CD, Rivlin PK, Wiesel TN (1991) Targets of horizontal connections in macaque primary visual cortex. J Comp Neurol 305: 370–392
Payne B, Pearson H, Cornwell P (1988) Development of visual and auditory cortical connections in the cat. In: Peters A, Jones EG (eds) Cerebral cortex, vol 7. Development and maturation of cerebral cortex. Plenum, New York
Peters A (1979) Thalamic input to the cerebral cortex. Trends Neurosci 2: 183–185
Peters A, Jones EG (1984) Classification of cortical neurons. In: Peters A, Jones EG (eds) Cerebral cortex, vol 1. Cellular components of the cerebral cortex. Plenum, New York
Ramón y Cajal S (1911) Histologie du système nerveux de l'homme et des vertébrés (translated by L. Azoulay). Maloine, Paris
Schüz A (1992) Randomness and constraints in the cortical neuropil. In: Aertsen A, Braitenberg V (eds) Information processing in the cortex. Experiments and theory. Springer, Berlin Heidelberg New York, pp 3–21
Schüz A, Palm G (1989) Density of neurons and synapses in the cerebral cortex of the mouse. J Comp Neurol 286: 442–455
Sholl DA (1956) The organization of the cerebral cortex. Methuen, London, Wiley, New York
Somogyi P (1977) A specific axo-axonic interneuron in the visual cortex of the rat. Brain Res 136: 345–350
Szenthágothai J (1975) The "module-concept" in cerebral cortex architecture. Brain Res 95: 475–496
Szenthágothai J (1978) Specificity versus (quasi-) randomness in cortical connectivity. In: Architectonics of the cerebral cortex. Raven, New York, pp 77–97
White EL (1989) Cortical circuits. Synaptic organization of the cerebral cortex, structure, function and theory. Birkhäuser, Boston

Chapter 5

Microcircuitry II

5.1 Introduction: Finding the Operational Principle of the Cortex

J. Bolz
Cerveau et Vision, INSERM, Bron/Lyon, France

Some readers might find it surprising to see in this book two chapters with almost identical headings. I think, however, this is as it should be: it emphasizes the important contributions which Otto Creutzfeldt made to this field. Otto Creutzfeldt pioneered the studies on the biophysical properties and the mechanisms of synaptic integration of cortical neurons, and he was the first to examine in a rigorous fashion the role of excitatory and inhibitory interactions in generating the characteristic response properties of neurons in the visual cortex. For Otto Creutzfeldt this was not "l'art pour l'art" since he was always concerned about how detailed knowledge of the processing mechanism of cortical cells and circuits might lead to an understanding of the general operating principles of the cortex (see his book *Cortex Cerebri*, Creutzfeldt 1983, 1993). He certainly was interested in the forest but he did not ignore the trees.

In the introduction to Chap. 2, my colleague Jean Bullier points out that cross-correlation experiments in the cortex very rarely identify pairs of interconnected neurons (see in this context also Chap. 5.5 by Tadaharu Tsumoto and Yoshio Hata). He therefore concludes that rather than searching for interactions between individual neurons one should search for interactions between groups of neurons. At first it might seem very difficult, with the techniques currently available, to monitor and manipulate the activity of populations of functionally related neurons. The cortex might be the right place, however, so start such an endeavor. The reason for this optimism has to do with the functional architecture of the cortex. It is now well-established that a characteristic organizing principle of the cortex is the clustering of cells with similar response properties. It is therefore possible, for instance, to focally inactivate pharmacologically a group of neurons with similar functional properties, e.g., within an individual cortical layer, and examine the effect of inactivation on the properties of the recipient cell in another cortical layer. Using this approach Torsten Wiesel, Charles Gilbert and myself were able, at least in some cases, to relate functional properties of cortical neurons to particular interlaminar cortical connections (see Bolz et al. 1989 for a review). A similar approach has been used by Ulf Eysel and his collaborators to study the role of lateral inhibition in the cortex (see Chap. 5.6).

The inactivation technique, in combination with single unit recordings, allows one to study the interactions between a group of neurons and an individual cell. This approach is not suitable, however, for examing the interactions between several populations of neurons. The contribution by Birgit Albowitz and Ulrich Kuhnt demonstrates how voltage sensitive dyes and arrays of photodetectors can be used to monitor optically the activity patterns of many cortical neurons simultaneously.

It should be obvious from the two chapters on cortical microcircuitry that we are just beginning to understand this issue. These days some neurobiologists believe that we have basically understood how the brain analyses, e.g., motion and color. They then go on to suggest that we have to create a new science to study the global functions of the brain. I do not share this view. The discovery of new and unexpected properties of cortical neurons (see Chap. 5.2 by Barry Connors and collaborators) and a deeper insight into the nature by which cortical neurons integrate their synaptic inputs (see Chap. 5.4 by David Ferster and collaborators) will change our view of how the cortex operates. We shall perceive the forest differently when we have learned more about the trees.

References

Bolz J, Gilbert DG, Wiesel TN (1989) Pharmacological analysis of cortical circuitry. Trends Neurosci 12: 292–296

Creutzfeldt OD (1983) Cortex cerebri. Leistung, strukturelle und funktionelle Organisation der Hirnrinde. Springer, Berlin Heidelberg New York

Creutzfeldt OD (1993) Cortex cerebri. Performance, structural and functional organization of the cortex (English translation). Mary Creutzfeldt, Göttingen

5.2 Layer I and the Excitable Apical Dendrite: Substrates for Intracortical Communication

B.W. CONNORS[1], L.J. CAULLER[2], H.G. KIM[1], AND I. BÜLTHOFF[1]

[1]Department of Neuroscience, Brown University, Providence, Rhode Island

[2]Cognition and Neuroscience Program, University of Texas at Dallas, Richardson, Texas

5.2.1 Introduction

The neocortex has intersecting vertical and horizontal components. Pyramidal cells, with their extended apical dendrites, are the most obvious vertical element. A large variety of axon types contribute to the horizontal organization of neocortex. At the confluence of these two systems, synapses of laminated axons converge on elongated pyramidal cells, allowing single neurons to sample activity from all layers of the cortex. The result is rich potential for the recombination and transformation of inputs, presumably a central factor in the evolutionary success of the neocortex.

Here we discuss selected features of the interface between horizontal and vertical organizations in neocortex. The interface itself is the most superficial layer of the neocortex, layer I. Within it is a horizontally directed system of axons that mediate projections from other neocortical areas. These axons synapse upon the most distal dendritic terminations of pyramidal neurons whose somata lie in layers II through V. The elongated layer V pyramidal neurons are faced with a complex task: to collect synaptic inputs from layers VI through I and transform the innumerable spatial and temporal input patterns into a useful output pattern at the axon initial segment. Recent studies have begun to clarify the mechanisms of layer I apical dendrite interactions.

Our experiments were performed on the primary somatosensory cortex of the rat. Whole-animal and in vitro slice methods were used to examine the anatomy of intracortical axonal systems in layer I, the structure of the large layer V pyramidal cells on which some of these axons terminate, and the physiological properties of the layer I synapses and the apical dendrites and somata of layer V cells. Detailed results are described elsewhere (Amitai et al. 1993; Cauller and Connors 1992, 1994, and submitted; Kim and Connors 1994).

5.2.2 Structure and Function of Horizontal Layer I Connections

Neocortical layer I is sometimes called the "molecular layer" because of its conspicuous lack of neuronal somata and its dense neuropil of axons and dendrites. In most areas of neocortex it is the thinnest layer, 100–150 μm thick, but it contains perhaps the most diverse set of afferent systems (Vogt 1991). Among its afferents are extensive sets of horizontally directed axons (Jones and Powell 1970; Fleischauer and Laube 1977). To illuminate the organization of horizontal axons in layer I we cut thin slices from paraformaldehyde-fixed neocortex, tangential to the pial surface, and placed small lumps of the lipophilic fluorescent dye 1,1'-dioctadecyl-3,3,3',3'-tetramethylindocarbocyanine perchlorate (Di-I) into layer I. The dye was allowed to transport for as long as 6 weeks.

Di-I-filled axons extended for distances up to 4 mm within layer I, although they more typically ended within 2 mm. A most striking feature of these fibers was the lack of orientation bias; labeled axons extended radially from the injection site in all directions with equal density. Axons often extended fine branches that coursed in all directions, while still remaining within the thin plane of layer I or superficial layer II. Varicosities were usually apparent, with spacings of 10–20 μm. These are presumably en passant synapses (Vaughan and Peters 1973).

The origins of long horizontal fibers within layer I have not been extensively studied. One prominent source may be the neocortex itself. Axonal connections between neocortical areas are usually reciprocal, and in many sensory systems of the cortex it appears that the connections from higher order areas to primary areas (the "backward" projections) terminate heavily in layers I and II. In a unique study, Rockland and Virga (1989) reconstructed single labeled axons and showed that backward projections from secondary to primary visual cortex in primates frequently run in layer I for distances of 1–4 mm.

We have traced some of the intracortical projections terminating in layer I of the rat SI cortex (Cauller and Connors, submitted). Brief administrations of horseradish peroxidase (HRP) to the pia of SI yielded retrogradely labeled neurons, mostly in layer V, of parietal cortex down to the rhinal sulcus, including areas SII and lateral parietal areas. Labeled neurons were also found in layers III–V of the agranular frontal (primary motor) cortex. Injections of fluorescent dextrans were made into either motor cortex or SII and adjacent lateral areas. Anterogradely labeled axons could be followed through layers V, VI, and the white matter, into SI and ultimately up to layers II and I. Within layer I the fluorescently labeled fibers coursed up to 1 mm horizontally and had branching and varicosity patterns similar to those of the Di-I-labeled axons described above. The results demonstrate that SI of the rat receives reciprocal connections from all major ipsilateral cortical areas to which it projects, and that many of these backward connections extend horizontally within layer I.

The physiology and pharmacology of layer I synapses are relatively obscure. Their size and ultrastructure are not remarkable among neocortical synapses (Jones and Powell 1970; Vaughan and Peters 1973), however their long distance from the somata of layer V cells makes their effectiveness questionable. To test the strength of the input, in vitro slices were bisected so that only a bridge of layer I connected two otherwise intact regions of cortex (Fig. 1, top). This allowed the horizontal layer I pathway to be stimulated selectively. Serial extracellular field potentials were recorded, and current-source density analysis confirmed that only layer I and upper layer II excitatory synapses were activated in this preparation (Cauller and Connors 1992).

Intracellular recordings using sharp microelectrodes were obtained from somata during specific layer I activation, and neurons were stained with biocytin for identification and reconstruction (Cauller and Connors 1994). The essential result was that pyramidal cells in layers II, III, and V (Fig. 1, bottom) could be strongly and rapidly excited as long as they had substantial dendritic branches within layer I. Layer I evoked EPSPs had relatively fast rise times, they were usually strong enough to evoke action potentials even from the relatively negative resting potentials of neurons in vitro, and they were long lasting.

When recording from somata of layer V cells, there were several characteristics that distinguished layer I evoked excitatory postsynaptic potentials (EPSPs) from those activated from pathways in deeper layers. First, the amplitudes of short-latency layer I EPSPs were virtually insensitive to

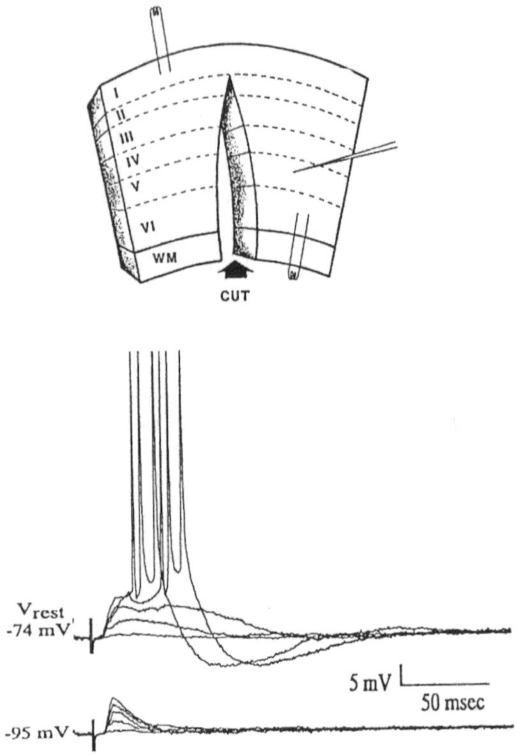

Fig. 5.2–1. Schematic representation of neocortical slice with a vertical cut to isolate the horizontal layer I pathway (*top*). Stimuli could be applied to layer I on the left side of the cut while recording was done on the opposite side of the cut. *Bottom panel*, intracellular recordings from the soma of a layer V intrinsically bursting neuron, while stimulating the layer I pathway. Superimposed recordings are shown as stimulus amplitude is increased stepwise, at resting potential and during hyperpolarization. (Cauller and Connors 1994)

alterations of somatic voltage. Second, obvious inhibitory postsynaptic potentials (IPSPs) were not generated by layer I stimulation, in contrast to their prevalence when stimulating deeper layers (see Connors et al. 1988). Instead of a late inhibitory phase, layer I EPSPs sometimes had multiple components that could last 100–200 ms. These long-latency EPSPs were abolished by hyperpolarizing the somatic membrane. Pharmacological analysis indicated that layer I excitation was mediated primarily by non-NMDA *N*-methyl-D-aspartate) glutamate receptors, although the late EPSP components were very sensitive to NMDA receptor antagonists. When strong inhibition was activated by deep layer stimulation, the EPSPs from layer I were very effectively abolished.

We conclude that the horizontal pathways of layer I can generate fast, effective excitation of pyramidal cells even when somata are more than 1 mm away in layer V. Layer I synapses mediate conventional, glutaminergic transmission. The prevalence of GABAergic neurons in layer I (Winer and Larue 1989; Martin et al. 1989) suggests that inhibitory interactions are also common there, but that they are too distant from somatic sites to be directly observed.

5.2.3 Passive Cable Models of Apical Dendrites

The apical dendrite of a layer V cell is relatively long, tapering, and branching, and an electrophysiologist's intuition might predict that synaptic inputs to its distal tuft would be greatly attenuated at the soma. Short of actually doing experiments, the most rational approach to quantifying dendritic function is cable analysis (Rall 1977). Arbitrarily shaped dendrites can be studied using compartmental methods, made practical by fast, cheap computers (Segev et al. 1989). Despite great advances in computational methods, dendrite modelling is still onerous territory because of uncertainties about the realistic values of key parameters.

The analysis of dendrites is greatly simplified by assuming that their constituent membrane is electrically passive. We have studied models of layer V pyramidal cells with passive apical dendrites (Cauller and Connors 1992; Cauller, Bülthoff and Connors, unpublished). Neurons were stained with biocytin, reconstructed using a computer-aided microscope system (Neurolucida), and modelled using the program NEURON (Hines 1989). Values for specific membrane resistance, cytoplasmic resistivity, and tissue shrinkage were varied across wide ranges, spine contributions were accounted for with a simple transformation, and specific membrane capacitance was assumed to be $1\,\mu F/cm^2$. In agreement with similar studies of layer V cells in visual cortex (Stratford et al. 1989; Larkman et al. 1992), our results implied that the distal tufts of passive apical dendrites should greatly attenuate even very strong synaptic input.

The strong attenuation of signals from apical tufts to soma, assuming passive membrane, is well illustrated by responses to frequency-modulated sinusoidal inputs. Figure 2A shows the transfer properties of pyramidal cell dendrites as a function of signal frequency. Roll-off is steep even at physiologically moderate frequencies, and especially so from distal apical branches. Attenuation becomes progressively less sensitive to variations in membrane resistivity as frequency increases. The most important conclusion in this context is that the passive dendrite model cannot account for the speed and effectiveness of layer I synapses measured experimentally.

5.2.4 Active Membrane Properties of Apical Dendrites

The passive model of the apical dendrite and the robust experimental effect of layer I excitation are incompatible. Resolution is provided by direct measurements from apical dendrites, which repudiate the passive model. Although early somatic measurements sometimes led to suggestions that apical dendrites of neocortex might have nonpassive properties (Purpura 1967; Deschenes 1981), direct recordings from dendrites themselves proved

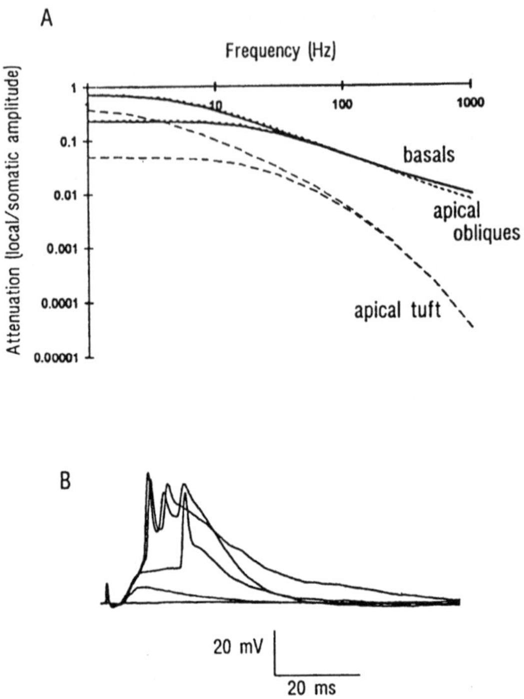

Fig. 5.2–2. A Results from modelling study of a layer V pyramidal cell, assuming passive dendritic membrane. Sinusoidal currents were applied to distal segments of the apical tufts, apical oblique branches, and the basal dendrites. Peak voltages measured at the soma were normalized to the local dendritic voltages. In each case specific membrane resistivity was set to either 10 or 100 KΩcm^2. Note that basal and oblique dendrites are very similar, but that apical tufts attenuate much more strongly. (Cauller, Bülthoff and Connors, unpublished)
B Whole-cell recording from the apical trunk of a layer V pyramidal cell, within layer IV, while stimulating the layer I pathway. As stimulus intensity was increased, fast and slow spike components were recruited. (Kim and Connors 1994)

technically difficult to achieve and verify. Sharp intracellular microelectrodes and dye injections of neocortical dendrites have now been performed both in vivo (Pockberger 1991) and in vitro (Amitai et al. 1993). The limited results demonstrated that many apical dendrites of pyramidal neurons from layers III and V are highly excitable, although the form and degree of excitability are very variable from cell to cell.

Routine recordings from apical dendrites have been achieved with whole-cell patch methods in slices of neocortex (Kim and Connors 1994). Recordings were made from the main apical trunk within layer IV, at distances of 100–500 μm from the soma in layer V, and confirmed with biocytin injections. Apical dendrites displayed a range of excitable properties whose complexity correlated with neuronal structure. The simplest responses were relatively small-amplitude, short-lasting spikes that could fire repetitively. These were recorded from cells with relatively thin apical trunks and few (mean of 5.5) oblique dendritic branches. Other dendrites generated a mixture of fast tetrodotoxin (TTX) sensitive spikes and slower, high-threshold spikes and plateaus (Fig. 3). The more complex spiking patterns were generated by thicker dendrites with more (mean of 12.0) oblique branches. This correlation between spiking pattern and morphology is similar to that

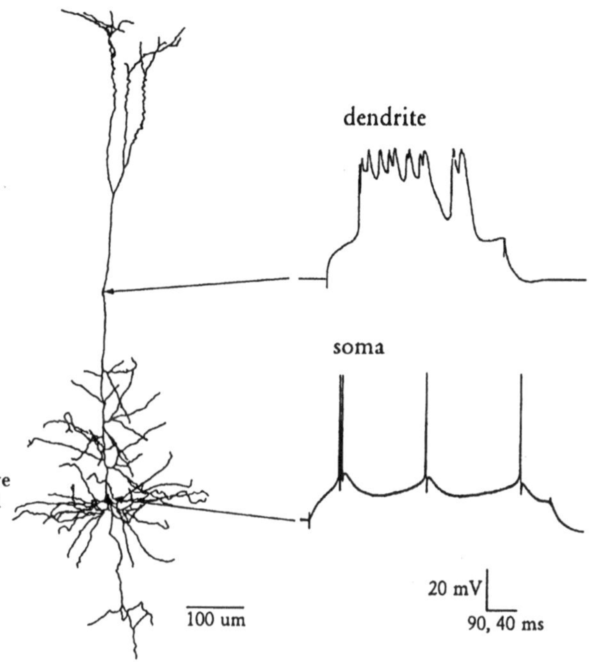

Fig. 5.2–3. Comparison of active responses in the soma and apical dendritic trunk of a layer V pyramidal cell. Composite of recordings from two cells; reconstructed neuron shows approximate positions of recordings. (Kim and Connors, unpublished)

observed from somatic recordings in layer V cells (Connors and Gutnick 1990).

Fast, low-threshold dendritic spikes proved to be TTX sensitive, while high-threshold slow spikes and plateaus were insensitive to TTX but suppressed by cobalt and cadmium, but not nickel. All tested dendrites seemed to have some component of high-threshold calcium currents; however, layer V cells with relatively complex apical dendrites apparently have the most. Dendrites also seem to have significant densities of voltage-dependent sodium channels (Huguenard et al. 1989), but their locations and densities cannot be definitively inferred. The time course of dendritically recorded sodium spikes implies that they are generated at some distant site; this site may be the soma/initial segment or distal dendritic branches (Amitai et al. 1993).

Layer I evoked EPSPs recorded from apical trunks display prominent fast and slow spiking components, at relatively low threshold (Fig. 2B). It seems likely that voltage-dependent dendritic currents contribute significantly to the synaptic currents generated within distal apical dendrites. Thus nonlinear postsynaptic currents can greatly amplify the effects of distal synaptic events, allowing them to propagate much farther and faster than passive membrane could possibly allow.

5.2.5 Conclusions and Speculations

The functions of backward projections within neocortex are still obscure (Bullier et al. 1988; Mignard and Malpeli 1991). However, our results imply that long horizontal axons within neocortical layer I subserve an important component of the intracortical (backward) pathway. The system is mediated primarily by glutaminergic synapses, and it can strongly excite pyramidal cells from layers II through V. The effectiveness of layer I inputs at the level of the soma depends strongly upon voltage-dependent sodium and calcium channels within the apical dendrite. Besides boosting the efficacy of distal synapses, the active channels within apical dendrites may offer a substrate for selective regulation of inputs to pyramidal cells. Synaptic inhibition placed proximally could serve specifically to disengage the apical dendrite from the soma, without affecting basal dendritic inputs. Alternatively, modulation of apical dendritic potassium channels could vary the electrotonic properties of apical dendrites, and thus alter the gain of the layer I influence.

Acknowledgements. These studies were supported by the National Institutes of Health and the Office of Naval Research.

References

Amitai Y, Friedman A, Connors BW, Gutnick, MJ (1993) Regenerative activity in the apical dendrites of pyramidal cells in neocortex. Cerebral Cortex 3: 26–38

Bullier J, McCourt ME, Henry GH (1988) Physiological studies on the feedback connection to the striate cortex from cortical areas 18 and 19 of the cat. Exp Brain Res 70: 90–98

Cauller LJ, Connors BW (1992) Functions of very distal dendrites: experimental and computational studies of layer I inputs to layer V pyramidal neurons in neocortex. In: McKenna T, Davis J, Zornetzer SF (eds) Single neuron computation. Academic Press, pp 199–230

Cauller LJ, Connors BW (1994) Synaptic physiology of horizontal afferents to layer I of primary somatosensory cortex in rats. J Neurosci (in press)

Connors BW, Gutnick MJ (1990) Intrinsic firing patterns of diverse neocortical neurons. Trends Neurosciences 13: 99–104

Connors BW, Malenka RC, Silva LR (1988) Two inhibitory postsynaptic potentials, and $GABA_A$ and $GABA_B$ receptor-mediated responses in neocortex of rat and cat. J Physiol (Lond) 406: 443–468

Deschenes M (1981) Dendritic spikes induced in fast pyramidal tract neurons by thalamic stimulation. Exp Brain Res 43: 304–308

Fleischhauer K, Laube A (1977) A pattern formed by preferential orientation of tangential fibres in layer I of the rabbit's cerebral cortex. Anat Embryol 151: 233–240

Hines M (1989) A program for simulation of nerve equations with branching geometries. Inter J Biomed Comp 24: 55–68

Huguenard JR, Hamill OP, Prince DA (1989) Sodium channels in dendrites of rat cortical pyramidal neurons. Proc Nat Acad Sci USA 86: 2473–2477

Jones EG, Powell TPS (1970) Electron microscopy of the somatic sensory cortex of the cat. II. The fine structure of layer I-II. Philos Trans R Soc Lond B 257: 13–21

Kim HG, Connors BW 91994) Apical dendrites of the neocortex: correlation between sodium- and calcium-dependent spiking and pyramidal cell morphology. J Neurosci (in press)

Larkman AU, Major G, Stratford KJ, Jack JJB (1992) Dendritic morphology of pyramidal neurones of the visual cortex of rat. IV.: Electrical geometry. J Comp Neurol 323: 137–152

Martin KAC, Friedlander MJ, Alones V (1989) Physiological, morphological, and cytochemical characteristics of a layer I neuron in cat striate cortex. J Comp Neurol 282: 404–414
Mignard M, Malpeli JG (1991) Paths of information flow through the visual cortex. Science 251: 1249–1251
Pockberger H (1991) Electrophysiological and morphological properties of rat motor cortex neurons *in vivo*. Brain Res 539: 181–190
Purpura DP (1967) Comparative physiology of dendrites. In: Quarton et al. (eds) The neurosciences. Rockefeller University Press, New York, pp 372–393
Rall W (1977) Core conductor theory and cable properties of neurons. In: Kandel ER (ed) Handbook of physiology, 1. The nervous system, vol 1. American Physiological Society, Bethesda, pp 39–98
Rockland KS, Virga A (1989) Terminal arbors of individual "feedback" axons projecting from V2 to V1 in the Macaque monkey: a study using immunohistochemistry of anterogradely transported Phaseolus vulgaris-leucoagglutinin. J Comp Neurol 285: 54–72
Segev I, Fleshman JW, Burke W (1989) Compartmental models of complex neurons. In: Koch C, Segev I (eds) Methods in neuronal modelling. MIT Press, Cambridge, Massachusetts, pp 63–96
Stratford K, Mason A, Larkman A, Major G, Jack J (1989) The modelling of pyramidal neurones in visual cortex. In: Durbin R, Miall C, Mitchison G (eds) The computing neuron. Addison-Wesley, Wokingham, UK, pp 296–321
Vaughan DW, Peters A (1973) A three dimensional study of layer I of the rat parietal cortex. J Comp Neurol 149: 355–370
Vogt BA (1991) The role of layer I in cortical function. In: Peters A (ed) Cerebral cortex, vol 9. Plenum, New York, pp 49–80
Winer JA, Larue DT (1989) Populations of GABAergic neurons and axons in layer I of rat auditory cortex. Neuroscience 33: 499–515

5.3 Spatiotemporal Activity Patterns in the Neocortical Slice: Fast Optical Recordings with Voltage-Sensitive Dyes

B. Albowitz and U. Kuhnt
Department of Neurobiology, Max Planck Institute for biophysical Chemistry, Göttingen, Germany

Information processing in the neocortex is mediated by afferent innervation and complex intrinsic circuitry. The morphology of this circuitry and the physiological properties of its elements and its synaptic connections are known to some detail (summarized in Creutzfeldt 1983). However, the impact of a known neuronal pathway for information processing is not clear. In order to determine the effectiveness of different structures, it is necessary to record simultaneously from many neuronal elements and to compare spatiotemporal activity patterns under different experimental conditions. We used voltage sensitive dyes and a fast optical recording technique to register spatiotemporal activity patterns in the neocortical slice of the guinea pig. The fluorescence signals that are recorded with this method correlate linearly with summated changes of membrane potentials (Ross et al. 1977; Albowitz et al. 1990).

Two problems were of particular interest to us. Firstly, we wanted to compare activity patterns evoked by stimulation of the white matter (activating vertically oriented afferent and efferent fibers) and of the neocortical surface (activating predominantly horizontal fibers of layer I). Horizontal fibers of layer I are abundant and make many synaptic contacts on apical dendrites (Larkman 1991). The impact of this strong synaptic input on neocortical activity is little known. Secondly, we wanted to determine how these different afferent inputs influence the horizontal spread of evoked activity, and which pathways are used for the propagation of signals in the horizontal direction. In this context it was important to determine how far the spread of activity in the horizontal direction is caused by the distribution of afferent fibers, on the one hand, and by intracortical, horizontally oriented connections on the other hand.

For a detailed description of the methods used, see Albowitz and Kuhnt (1993a). Briefly, coronal slices (350 μm) were prepared from guinea pig somatosensory or visual neocortex. Slices were stained with the fluorescent voltage-sensitive dye RH795 (0.1 mg/ml). Recordings were made in a superfusion chamber perfused with a medium containing (in mM): NaCl (124), KCl (5), NaH_2PO_4 (1.25), $MgSO_4$ (2), $CaCl_2$ (2), $NaHCO_3$ (26), and glucose (10). The medium was gassed with O_2/CO_2 and kept at a pH of 7.4. A recording electrode for intracellular or extracellular field potentials was placed in layer III. Stimulation electrodes (tungsten in glass, 100–500 kΩ, tip diameter below 30 μm) were inserted at the border between grey and white matter (WM) and in layer I (L1). The sites were stimulated separately at a rate of 0.125 Hz each. The pulse width was 30 μs. Single or double pulses separated by 50 ms were used. An inverted microscope projected voltage-dependent fluorescence changes on to a 10×10 photodiode array. Signals from each photodiode were separately current-to-voltage converted and amplified. They were then multiplexed, A/D converted, and processed by a computer. The temporal resolution of the system was 0.4 ms if all 100 channels were used. According to the objective of the microscope used ($\times$16 or 25), the spatial resolution of one photodiode was 94 or 60 μm. Thus, activity was detected simultaneously from a total area of $(940\,\mu m)^2$ or $(600\,\mu m)^2$. Larger areas of the slice could be monitored by moving the photodiode array with respect to the slice. Dye bleaching, staining, and illumination irregularities were corrected off-line. Light intensities were measured as relative fluorescence changes (dF/F), where F is the fluorescence light intensity of

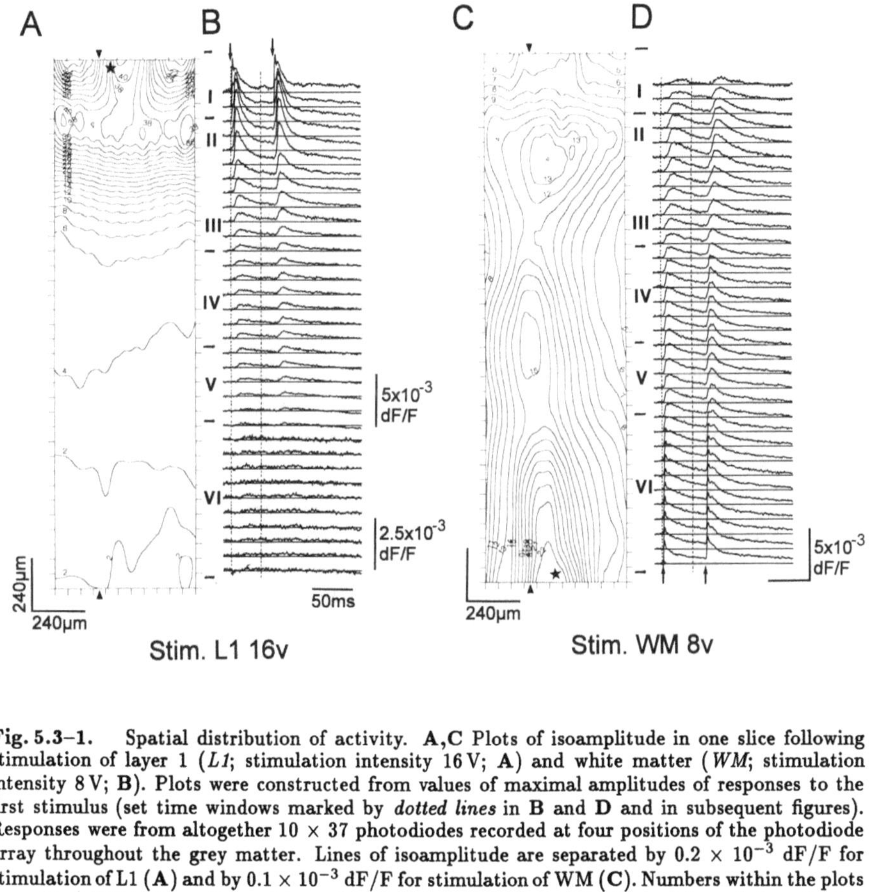

Fig. 5.3–1. Spatial distribution of activity. **A,C** Plots of isoamplitude in one slice following stimulation of layer 1 (*L1*; stimulation intensity 16 V; **A**) and white matter (*WM*; stimulation intensity 8 V; **B**). Plots were constructed from values of maximal amplitudes of responses to the first stimulus (set time windows marked by *dotted lines* in **B** and **D** and in subsequent figures). Responses were from altogether 10 × 37 photodiodes recorded at four positions of the photodiode array throughout the grey matter. Lines of isoamplitude are separated by 0.2×10^{-3} dF/F for stimulation of L1 (**A**) and by 0.1×10^{-3} dF/F for stimulation of WM (**C**). Numbers within the plots are x 10^{-4} dF/F. *Stars*, the positions of stimulation electrodes; *ticks*, the positions of photodiodes. Cortical layers are marked at the right edges of the plots. **B,D** Original records from a vertical row of photodiodes (marked with *triangles* in **A** and **C**) to a double pulse stimulus in L1 (**B**) and WM (**D**). Time points of stimulation are marked by *arrows*. Light decrease (corresponding to depolarisation) is plotted upwards (as in all following figures). Note that records from layer VI following stimulation of LI are plotted with twice the gain of the other records. (From Albowitz and Kuhnt 1993a)

the stained slice during illumination without evoked neuronal activity, and dF is the fluorescence change during evoked neuronal activity. A decrease of fluorescence (plotted upwards in all records shown) corresponds to membrane depolarisation, an increase of fluorescence corresponds to membrane hyperpolarisation.

5.3.1 Vertical Distribution of Activity

Figure 1 shows the distribution of activity dependent fluorescence changes following stimulation of L1 (Fig. 1A,B) and WM (Fig. 1C,D) in a vertical strip of the neocortical slice. The plots of isoamplitude (Fig. 1A,C) were constructed from values of maximal amplitudes from each photodiode record in response to the first stimulus of a double pulse. Stimulation of L1 evoked strong but localised activity in upper cortical layers. Stimulation of WM evoked activity throughout the depth of the cortex with maxima in layer II/III, at the border of layers IV and V, and close to the stimulation site in layer VI. Records from one row of photodiodes (marked by a triangle in Fig. 1A,C) in response to the double pulse stimulus are shown in Fig. 1B,D. Close to the stimulation sites, evoked fluorescence changes consisted of two components: an early fast rising part of short duration and a longer lasting second component. It was suspected that these two components reflect presynaptic and postsynaptic activity, respectively. The mixture of pre- and postsynaptic elements in responses is a common problem inherent to neocortical slice preparations because stimulation of L1 or WM activates not only afferent but also efferent fibers and possibly dendritic elements. In order to separate nonsynaptic components from postsynaptic activity, recordings in standard perfusion medium were compared with recordings in calcium-free medium, in which synaptic transmission is blocked (Fig. 2).

Traces of nonsynaptic activity in calcium-free medium (Fig. 2A,E) were subtracted from those of "mixed" responses in standard medium (also shown in Fig. 2A,E), resulting in traces of response components of exclusively postsynaptic origin (Fig. 2B,F). Records in calcium-free medium consisted predominantly of early, fast rising activity. Consequently, the fast rising parts were lacking from postsynaptic response components and thus confirmed the presynaptic nature of fast activity. Plots of maximal amplitudes from one vertical row of photodiode records from mixed and calcium-free responses (Fig. 2C,G) and from postsynaptic response components (Fig. 2D,H) show that maximal postsynaptic activity was elicited in upper layer II/III for both stimulation sites. Large parts of the evoked activity in lower cortical layers following stimulation of WM were of nonsynaptic origin (presynaptic and antidromic). The location of maximal postsynaptic activity did not coincide with the site of termination of thalamocortical afferents in layer IV (Garey 1976). Possibly, the di- and/or polysynaptic pathways from layer IV spiny stellate cells to dendrites and cell bodies in upper layer III are quantitatively more important. Alternatively, the strong postsynaptic response in upper layer II/III following stimulation of WM might be mediated by afferents other than those from the thalamus (e.g., corticocortical) which terminate in all cortical layers (Garey 1976). With

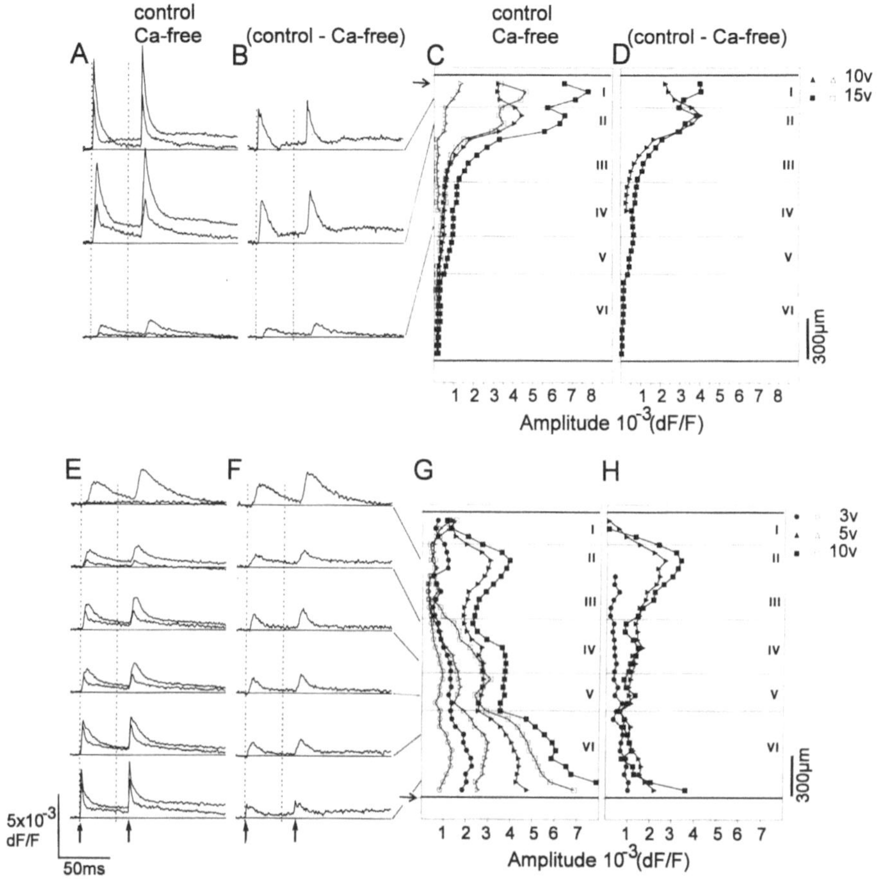

Fig. 5.3–2. Responses in standard and calcium-free medium. **A,B,E,F** Selected records from a row of photodiodes representing the vertical stimulation axis following double pulse stimulation of L1 with 15 V (**A,B**) and of WM with 10 V (**E, F**). **A,E** The responses in standard and calcium-free medium are superimposed. **B,F** Difference of the responses in standard and calcium-free medium. *Arrows*, time points of stimulation. **C,D,G,H** Maximal amplitudes from the response to the first stimulus from a vertical row of photodiodes representing the vertical stimulation axis following-double pulse stimulation of L1 with two stimulation intensities (**C,D**) and WM with three stimulation intensities (**G,H**). **C,G** Maximal amplitudes from responses in standard (mixed responses, *closed symbols*) and calcium-free (nonsynaptic responses, *open symbols*) medium. **D,H** Difference in values for maximal amplitudes of responses in standard and calcium-free medium (postsynaptic responses). *Arrows*, positions of stimulation electrodes. (From Albowitz and Kuhnt 1993a)

stimulation of L1, most of the evoked activity was postsynaptic. Presumably, these postsynaptic responses reflect the activation of apical dendrites of pyramidal cells by horizontal fibers of layer I. According to the present measurements, this is a very strong synaptic input; however, its effectiveness is locally strongly restricted. Responses declined toward lower cortical layers, and usually no activity persisted in layer VI. This is probably due to the remote site of layer I horizontal fiber synapses, EPSPs are likely to decline strongly before reaching the cell soma. The coincidence of maximal postsynaptic activity in upper layer II/III following stimulation of both L1 and WM might have some interesting consequences. Horizontal fibers of layer I consist largely of "association" fibers from other cortical areas mostly of the ipsilateral hemisphere (e.g., Rockland and Vigra 1989) and of noradrenergic fibers from mesencephalic areas (Morrison et al. 1981, Itakura et al. 1981). If activated in isolation, their effect is probably low due to the remote site of their synapses. If, however, layer I horizontal fibers are activated coincidentally with thalamocortical afferent fibers, the synaptic response to both inputs might be facilitated. Indeed, long-term changes of synaptic efficacy have been reported following the coactivation of afferent pathways from the thalamus and neighbouring cortical sites (Iriki et al. 1991). Further, noradrenergic agonists do facilitate long-term potentiation (Bröcher et al. 1992). In this context it might also be of importance, that long-term potentiation in the adult cat was located predominantly in layer III (Komatsu et al. 1988).

5.3.2 Horizontal Distribution of Activity

Figure 3 illustrates the spread of activity in the horizontal direction. When comparing the records from a row of photodiodes representing layer III (Fig. 3C), spread of activity elicited from WM was wider (Fig. 3B) than spread of activity elicited from L1 (Fig. 3A). The plots of maximal amplitudes (Fig. 3D,E) revealed different regions for the distribution of activity. Close to the stimulation site, amplitudes were high ("central region"). About 500 μm medial and lateral to the maximum amplitudes declined strongly, but not down to the baseline level of activity. Instead, activity of low amplitude persisted for more than 2 mm from the stimulation site ("peripheral regions"). On occasion, secondary though smaller maxima of activity were observed in these peripheral regions. These secondary maxima were located approximately 1 mm from the primary maximum. The width of peripheral regions was wider for activity following stimulation of WM. In contrast, the width of central regions (measured as half-amplitude width from plots of amplitude against distance) was wider for activity elicited by stimulation of L1. Central regions were widest in layer III, followed by layers V and IV.

The differences between central and peripheral regions in their rates of decline and their extent with different stimulation sites suggested that they

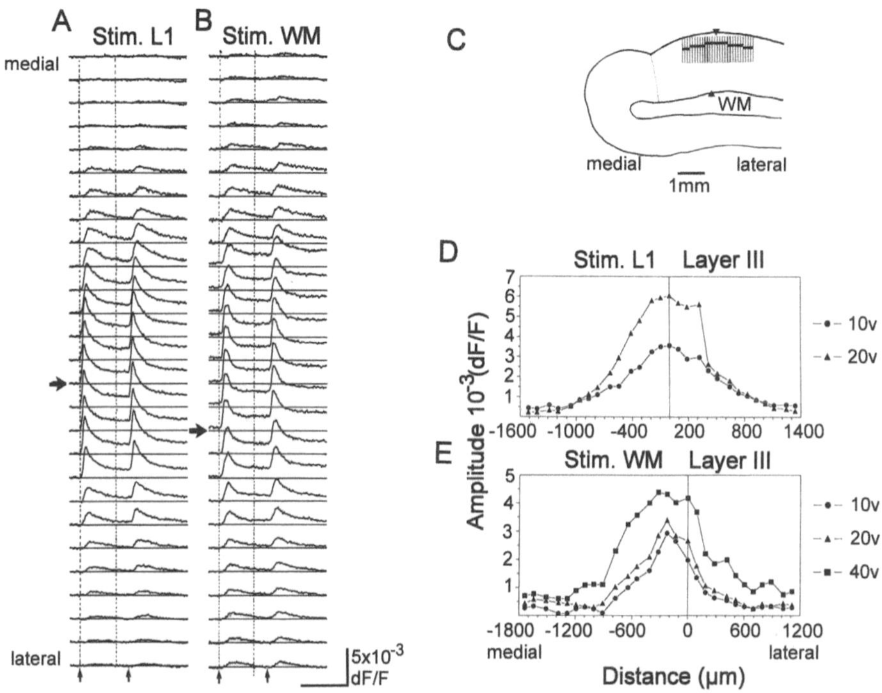

Fig. 5.3–3. Mediolateral distribution of activity. **A, B** Records from a row of photodiodes representing upper layer III (site of maximal amplitudes in the vertical direction). The respective photodiodes are marked *black* in **C**, which shows a schematic drawing of the slice with positions of the photodiode matrix and stimulation electrodes. Responses to double pulse stimulation of L1 with 20 V (**A**) and of WM with 40 V (**B**). Position of stimulation electrodes is marked by *large arrows* (see also *triangles* in **C**). Time points of stimulation are marked by *small arrows*. **D, E** Maximal amplitudes of the response to the first stimulus (set time windows indicated by *stippled lines* in **A** and **B**) in L1 at two different stimulation intensities (**D**) and in WM at three different stimulation intensities (**E**). Records are from the horizontal row of photodiodes marked *black* in **C** and plotted in **A** and **B**. The zero-line corresponds to the positions of the location of the vertical stimulation axis (see also *triangles* in **C**). (From Albowitz and Kuhnt 1993b)

are mediated by different pathways. Theoretically, the extent of horizontal spread is determined by both the distribution of afferents and by the distance and intensity of intracortical connections. We investigated the role of these two pathways for the spread of activity in the horizontal direction by comparing the extent of nonsynaptic activity (i.e., the maximal extent of afferent fibers) in calcium-free perfusion medium to that of mixed activity in standard medium. Nonsynaptic activity was recorded up to about 500 μm from the stimulation site. Thus, the maximal extent of afferent fibers correlated with the extent of the central region. Onset latencies of responses in standard medium were almost constant within this region and increased in peripheral regions. In conclusion, activity within the central region is probably mediated by monosynaptic afferent innervation. The central region was slightly wider with stimulation of L1, possibly due to the wider extent of layer I horizontal fibers than afferents from the white matter (Morrison et al. 1981). Single afferent fibers from the white matter have terminal fields of up to 500 μm (Gilbert and Wiesel 1979). Since stimulation of the white matter activates many such fibers with only partly overlapping terminal fields, the total width of stimulated afferents might reach about 1 mm, the width of the central region.

Beyond the extent of nonsynaptic activity, i.e., in peripheral regions, responses were likely mediated by intracortical horizontal connections. Both horizontally oriented interneuronal connections (Martin and Whitteridge 1984, Somogyi et al. 1983) as well as short- and long-distance axon collaterals of pyramidal neurons (Gilbert and Wiesel 1983) could potentially mediate spread toward peripheral regions. The large extent of peripheral regions, however, requires that long-distance horizontal pathways contribute significantly to horizontal spread of activity. The existence of long-distance horizontally oriented axon collaterals is well documented (Gilbert et al. 1990). Frequently these fibers interconnect patches of neurons separated by about 1 mm (e.g., Kisvárday and Eysel 1992). It has been suspected that these pathways serve a modulatory function only, because they are comparatively sparse (Creutzfeldt et al. 1977). The present findings, however, indicate that a substantial amount of information is transmitted via these fibers. Of particular interest are the secondary peaks that were observed in some preparations. These secondary peaks might reflect the "patchy" innervation of a group of neurons by axon collaterals from pyramidal neurons in the primary activated site. Long-distance patchy horizontal connections are known to exist in upper as well as lower cortical layers (Boyd and Matsubara 1991). It is not clear whether horizontal spread of activity occurs primarily in either upper or in lower layers and is then locally transmitted in the vertical direction. Alternatively, activity might spread more or less independently along upper and lower layers. Measurements of peak latencies revealed that responses were always earliest in upper cortical layers at peripheral sites. This indicates that horizontal spread occurs primarily along layer III, and that activity is then transmitted locally throughout the depth of the cortex. The present investigations have further clarified the quantitative impact of different afferent and of some intracortical pathways in the neocortical slice. Further work, using pharmacological tools and lesion experiments will have to define the exact pathways used for spread of activity.

References

Albowitz B, Kuhnt U (1993a) Evoked changes of membrane potential in guinea pig sensory neocortical slices: an analysis with voltage sensitive dyes and a fast optical recording method. Exp Brain Res 93: 213-225

Albowitz B, Kuhnt U (1993b) The contribution of intracortical connections to horizontal spread of activity in the neocortex as revealed by voltage-sensitive dyes and a fast optical recording method. J Europ Neurosci 5: 1349-1359

Albowitz B, Kuhnt U, Ehrenreich L (1990) Optical recording of epileptiform voltage changes in the neocortical slice. Exp Brain Res 81: 241-256

Boyd J, Matsubara J (1991) Intrinsic connections in cat visual cortex: a combined anterograde and retrograde tracing study. Brain Res 560: 207-215

Bröcher S, Artola A, Singer W (1992) Agonists of cholinergic and noradrenergic receptors facilitate synergistically the induction of long-term potentiation in slices of rat visual cortex. Brain Res 573: 27-36

Creutzfeldt OD (1983) Cortex cerebri. Leistung, strukturelle und funktionelle Organisation der Hirnrinde. Springer, Berlin Heidelberg New York

Creutzfeldt OD, Garey LJ, Kuroda R, Wolff JR (1977) The distribuition of degenerating axons after small lesions in the intact and isolated visual cortex of the cat. Exp Brain Res 27: 419-440

Garey LJ (1976) Synaptic organization of afferent fibres and intrinsic circuits in the neocortex. In: Creutzfeldt O (ed) Electrical activity from the neuron to the EEG and EMG. A. Morphological basis of EEG-mechanisms. Handbook of electroencephalography and clinical neurophysiology, vol 2. Elsevier, Amsterdam, pp 57-85

Gilbert CD, Wiesel TN (1979) Morphology and intracortical projections of functionally characterised neurons in the cat visual cortex. Nature 280: 120-125

Gilbert CD, Wiesel TN (1983) Clustered intrinsic connections in cat visual cortex. J Neurosci 3: 1116-1133

Gilbert CD, Hirsch JA, Wiesel TN (1990) Lateral interactions in the visual cortex. Cold Spring Harbor Symposia on Quantitative Biology 55: 663-677

Iriki A, Pavlides C, Keller A, Asanuma H (1991) Long-term potentiation of thalamic input to the motor cortex induced by coactivation of thalamocorticcal and corticocortical afferents. J Neurophysiol 65: 1435-1441

Itakura T, Kasamatsu T, Pettigrew JD (1981) Norepinephrine containing terminals in kitten visual cortex: laminar distribution and ultrastructure. Neurosci 6: 159-175

Kisvárday ZF, Eysel UT (1992) Cellular organization of reciprocal patchy networks in layer III of cat visual cortex (area 17). Neurosci 46: 275-286

Komatsu Y, Fujii K, Maeda J, Sakaguchi H, Toyama K (1988) Long-term-potentiation of synaptic transmission in kitten visual cortex. J Neurophysiol 59: 124-141

Larkman AU (1991) Dendritic morphology of pyramidal neurones of the visual cortex in the rat. III. Spine Distribution. J Comp Neurol 306: 332-343

Martin KAC, Whitteridge D (1984) Form, function and intracortical projections of spiny neurons in the striate visual cortex of the cat. J Physiol 353: 463-504

Morrison JH, Molliver ME, Grzanna R, Coyle, JT (1981) The intra-cortical trajectory of the coeruleo-cortical projection in the rat, a tangentially organized cortical afferent. Neurosci 6: 139-158

Rockland KS, Vigra A (1989) Terminal arbors of individual "feedback" axons projecting from area V2 to V1 in the Macaque monkey: a study using immunohistochemistry of anterogradely transported phaseolus vulgaris-leucoagglutinin. J comp Neurol 285: 54-72

Ross WN, Salzberg BM, Cohen LB, Grinvald A, Davila HV, Waggoner AS, Wang CH (1977) Changes in absorption, fluorescence, dichroism, and birefringence in stained giant axons: optical measurement of membrane potential. J Membr Biol 33: 141-183

Somogyi P, Kisvárday ZF, Martin KAC, Whitteridge D (1983) Synaptic connections of morphologically identified and physiologically characterized large basket cells in the striate cortex of the cat. Neurosci 10: 261-294

5.4 The Linearity of Synaptic Mechanisms in Simple Cells of Area 17 of Cat Visual Cortex

D. FERSTER, H.S. WHEAT, AND B. JAGADEESH
Department of Neurobiology and Physiology, Northwestern University, Evanston, Illinois

To understand the origin of receptive field properties in neurons of the visual cortex, it is first necessary to understand the mechanisms by which neurons integrate their synaptic inputs. In their first model of orientation selectivity, for example, Hubel and Wiesel used simple addition to combine the excitatory inputs to simple cells from different neurons in the lateral geniculate nucleus (LGN). Nonlinear synaptic mechanisms such as shunting inhibition or excitatory facilitation, however, support far more complex computations than can simple linear mechanisms. Several more recent models of cortical function have relied on these nonlinear mechanisms to generate orientation and direction selectivity (Torre and Poggio 1978; Koch and Poggio 1985; Wehmeier and Koch 1988). To determine which of these strategies are actually used by cortical neurons to construct their receptive fields, we have used intracellular recording techniques to study the integration of synaptic potentials.

5.4.1 Inhibitory Interactions

Postsynaptic inhibition can operate by at least two distinct mechanisms, one linear and one nonlinear. Hyperpolarizing inhibitory postsynaptic potentials (IPSPs) interact linearly with excitatory postsynaptic potentials (EPSPs); the inhibitory and excitatory synaptic currents sum to produce a net change in membrane potential. Shunting inhibitory synapses interact nonlinearly with EPSPs; the shunt-induced increase in membrane conductance inhibits the neuron by reducing the amplitude of EPSPs by a constant multiplicative factor (Fatt and Katz 1953; Blomfield 1974). This nonlinear property of shunting inhibition has been the basis for models of synaptic interaction in which shunting inhibition acts as an AND-NOT gate for excitatory inputs (Torre and Poggio 1978; Koch et al. 1983). The AND-NOT gate has in turn been applied to models for the generation of orientation and direction selectivity in cortical neurons (Koch and Poggio 1985). Using intracellular recordings, we have searched for shunting inhibition by examining the effect of visually evoked inhibition on the size of electrically evoked EPSPs (Ferster and Jagadeesh 1991).

Our probe for shunting inhibition was a test EPSP generated in simple cells of layer 4 by electrical stimulation of the LGN. If any visual stimulus, presented simultaneously with the delivery of the electrical stimulus, were to evoke shunting inhibition within the receptive field, then the amplitude of the test EPSP should be systematically decreased by the visual stimulus. The visual stimuli consisted of moving bars of light projected onto a tangent screen, the image of which was focused onto the retina. The bar moved back and forth across the receptive field at a frequency slightly different from that of the electrical stimulus. This asynchronous presentation of electrical and visual stimuli meant that each successive electrical stimulus was delivered with the visual stimulus in a different position within the receptive field.

Before applying our electrical test stimulus, we first applied visual stimuli alone in order to determine the structure of the receptive field. The intracellular records of Fig. 1 illustrate a simple cell whose receptive field was divided into four subregions, two ON regions and two OFF regions. A bar of the preferred orientation flashed in either ON region (Fig. 1A, 2 and 4) evoked depolarization at light on and hyperpolarization at light off. Flashing the bar in either OFF region (Fig. 1A, 1 and 3) evoked depolarization at light off and hyperpolarization at light on. When the bar was swept across the receptive field (Fig. 1B, top), a series of depolarizations and hyperpolarizations occurred as the bar encountered each of the subfields in turn. Sweeping the bar across the receptive field in the nonpreferred orientation evoked only a small depolarization, which is barely visible in the individual traces (Fig. 1B, bottom).

Once the structure of the receptive field was determined, the electrical stimulus was delivered to the LGN to evoke test EPSPs while the visual stimulus was swept repeatedly through the receptive field. Sample responses to the test EPSP are shown in Fig. 2A. Both the early and late EPSP components of the response (labeled 1 and 2) are mediated by monosynaptic connections (see figure legend). While the shape of the test EPSP varied little from trial to trial, the amplitudes of both EPSP components varied systematically with the position and orientation of the visual stimulus as it moved through the receptive field. This can be seen in the four graphs of Fig. 2B in which the amplitude of the test EPSP is plotted against the bar position at the time each electrical stimulus was delivered. Four separate graphs are shown for four different directions of bar motion.

In the nonpreferred orientation the bar evoked no consistent change in the size of the test EPSP, indicating that there is no significant shunting inhibition evoked by this stimulus (Fig. 2B, bottom). The bar of the preferred orientation did evoke large decreases in the size of the test EPSP (Fig. 2B, top), but this decrease was almost certainly not caused by shunting inhibition. Since the regions of the receptive field in which the bar

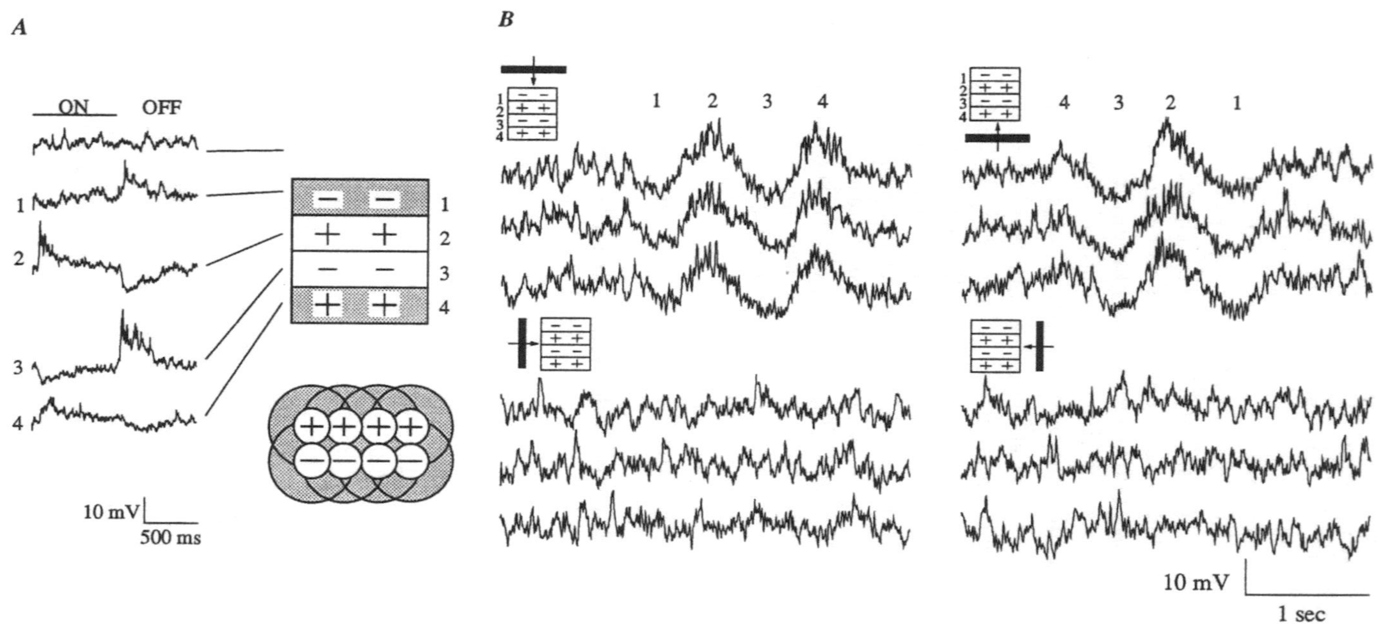

Fig. 5.4–1. **A** Schematic diagram of the receptive field of a simple cell in area 17 of cat visual cortex. To the left of the diagram are shown intracellular responses of the cell to a stationary bright bar flashed in each of the four subregions (*bottom four traces*), together with the response to the bar flashed outside of the receptive field (*top trace*). The horizontal bar at the top of the traces indicates the time during which the stimulus bar was on. Each trace represents the average of the response to four separate presentations of the stimulus. **B** The response of the cell to the bar being swept across the receptive field in four different directions, indicated by the *inset* above each set of traces. The responses to three consecutive stimulus presentations are shown for each direction of motion. Resting potential was −60 mV

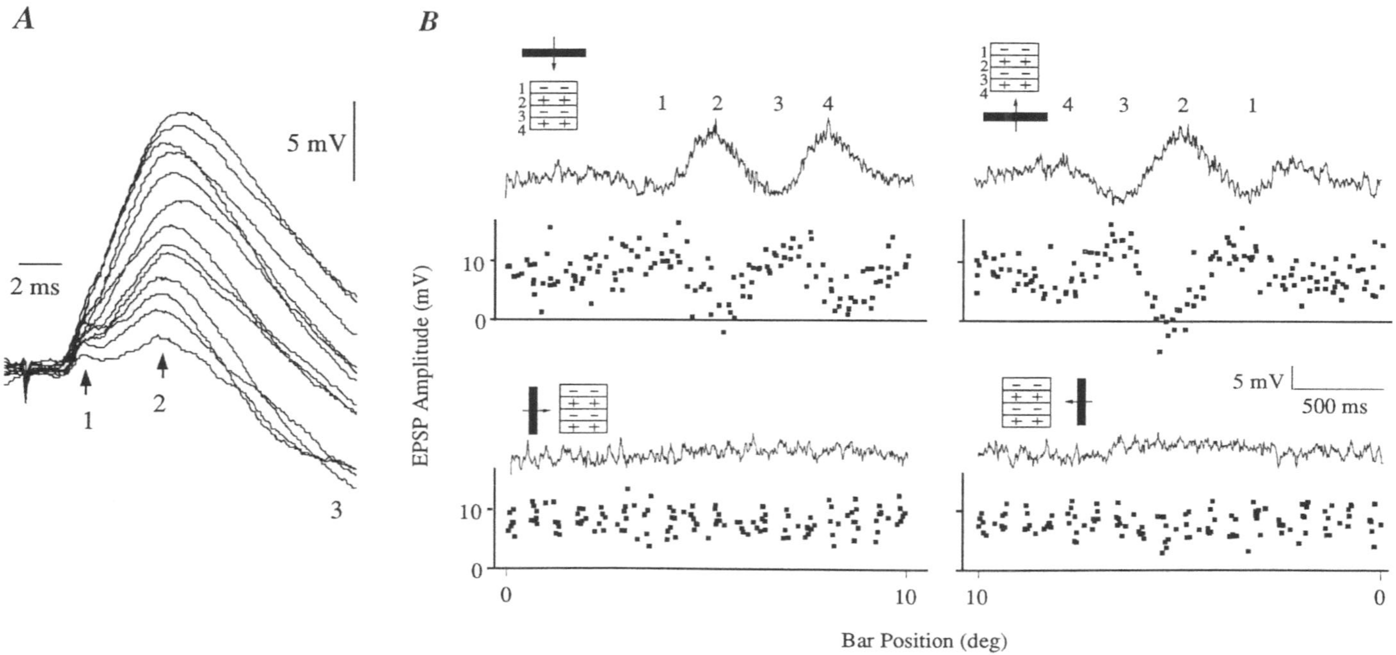

Fig. 5.4–2. **A** Test EPSPs generated in the simple cell of Fig. 1 by electrical stimulation of the LGN. The three components of the response labeled *1–3* correspond to a monosynaptic EPSP from the LGN (*1*), a monosynaptic EPSP following antidromic activation of corticogeniculate neurons and their intracortical collaterals (*2*; Ferster and Lindstrom 1985), and the beginning of a disynaptic IPSP, which lasted for 200 ms and reached an amplitude of 15 mV (*3*). Records were obtained while the visual stimulus was moving back and forth across the receptive field. The baselines of each trace have been matched, although during recording the test EPSPs were superimposed on the slow, visually evoked fluctuations in membrane potential. **B** Four graphs of test EPSP amplitude (measured at point *2* in *A*) against the position of the visual stimulus. The four different graphs correspond to four different directions of stimulus motion. Above each plot is shown the membrane potential changes evoked by the visual stimulus alone. The graph and the continuous curve are plotted in register: for a given bar position, the instantaneous membrane potential in the absence of electrical stimulation is located directly above the corresponding test EPSP amplitude. During the recording of all traces in the figure the cell was held hyperpolarized with –120 pA current injected through the recording electrode. The amplitudes of both components of the test EPSP are similarly affected by bar position. Graphs of test EPSP amplitude versus bar position based on measurements of EPSP amplitudes at point *1* in *A* are nearly identical to those shown

evoked these decreases were the same ones in which the bar evoked maximal excitation, the decreases probably reflect the decrease in driving force on excitatory postsynaptic currents (EPSCs) caused by the visually evoked depolarization. That this is the case is shown in Fig. 3, where the amplitude of the test EPSP is plotted against the absolute membrane potential recorded just prior to the delivery of the electrical stimulus (base membrane potential). The linear relationship between the test EPSP amplitude and the base membrane potential is exactly what would be expected if the fluctuations in the test EPSP amplitude were caused by a drop in EPSC driving force associated with depolarization.

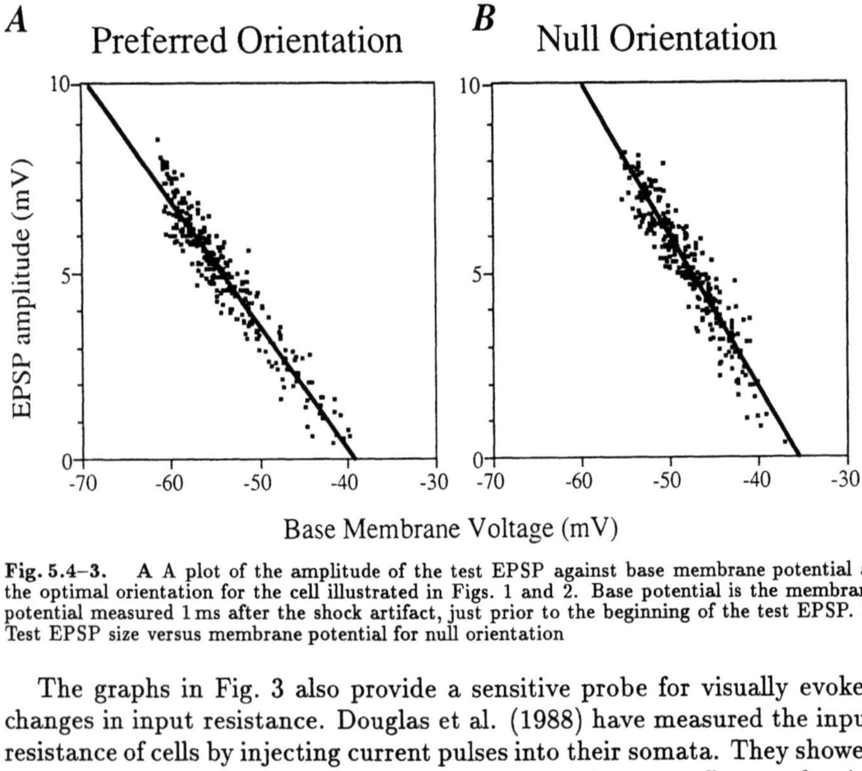

Fig. 5.4–3. **A** A plot of the amplitude of the test EPSP against base membrane potential at the optimal orientation for the cell illustrated in Figs. 1 and 2. Base potential is the membrane potential measured 1 ms after the shock artifact, just prior to the beginning of the test EPSP. **B** Test EPSP size versus membrane potential for null orientation

The graphs in Fig. 3 also provide a sensitive probe for visually evoked changes in input resistance. Douglas et al. (1988) have measured the input resistance of cells by injecting current pulses into their somata. They showed that visual stimuli of the nonpreferred orientation have no effect on the size of the depolarizations occurring in response to a current pulse of constant size. Since the current pulses are delivered to the soma, they indicate that shunting IPSPs do not occur in the soma or in regions of the dendritic tree closely electrically coupled to the soma. In contrast, the test EPSP is "delivered" throughout the dendritic tree, wherever terminals of geniculate relay cells and corticogeniculate neurons make contact (McGuire et al. 1984; Freund et al. 1985; Einstein et al. 1987). Shunting inhibition generated by the visual stimulus at any point within the dendritic tree would therefore have been detectable as a decrease in slope of the relationship between test EPSP amplitude and base membrane potential. That the slope of this relationship is the same for the preferred and nonpreferred orientation of the visual stimulus (Fig. 3) indicates that no orientation-specific shunting was evoked, and that no decrease in input resistance has occurred.

We have obtained similar results to those of Fig. 2 from a total of seven cortical neurons, five simple cells (including the one illustrated in Figs. 1–3), and one complex cell, each with resting potentials below −50 mV. In each a decrease in test EPSP amplitude of over 50% was observed during visual stimulation, but only during depolarizing phases of the visual response.

The two types of inhibitory mechanisms, hyperpolarizing and shunting, differ in the spatial extent of their effects as well as in the linearity of their interactions with EPSPs. Hyperpolarizing IPSPs act directly on the membrane potential by virtue of the current that enters the neuron through their ion channels and travels to the soma. As a result, hyperpolarizing IPSPs counteract the effects of all EPSPs reaching the soma. In contrast, shunting inhibitory synapses by themselves admit no current into the cell since the reversal potential is near rest. It is through the decrease in input resistance that shunting synapses act, and this decrease affects only those EPSPs generated nearby in the dendritic tree (Koch et al. 1983). Since we have found no shunting effects in our experiments, however, it appears that the IPSPs evoked in cortical neurons by moving bars of light interact linearly with EPSPs generated throughout the dendritic tree.

5.4.2 Excitatory Interactions

The test EPSP experiment may rule out the presence of shunting inhibition in the responses to visual stimulation in simple cells. But it does not address the fundamental question of whether EPSPs combine linearly with one another, or whether there is instead substantial facilitation among different excitatory inputs. To examine this issue we have examined the generation of direction selectivity in simple cells to see whether it depends on linear or nonlinear synaptic mechanisms.

By definition, if a cell sums its inputs linearly, its response to two simultaneously presented stimuli should equal the sum of the responses to the two stimuli when they are presented individually. The stimuli that we have used are drifting and flashing sinusoidal gratings. Drifting gratings can be expressed as a linear combination of stationary gratings. Thus if a cell is linear, its response to drifting gratings should be equal to a linear combination of its response to stationary gratings. Linear models of direction selectivity have been tested extensively with extracellular recording techniques (Ganz and Felder 1984; Emerson et al. 1987; Reid et al. 1987; Albrecht and Geisler 1991; Reid et al. 1991; Tolhurst and Dean 1991; McLean et al. 1993). However, while the linear model correctly predicts the preferred direction of motion, it consistently underestimates the amplitude of the direction selectivity, suggesting that nonlinear mechanisms contribute to the generation of direction selectivity (Reid et al. 1987; Reid et al. 1991; Tolhurst and Dean 1991). Tests of linearity based on extracellular recording are limited, however, by the highly nonlinear relationship between membrane potential and spike frequency. Whenever the membrane potential falls below threshold and spike frequency falls to zero, extracellular recordings provide no information about membrane potential, which is the variable that must be measured for a proper test of the linearity of synaptic summation. To obtain a complete description of the membrane potential change requires intracellular recording.

Simple cells respond to drifting gratings with a periodic modulation of their membrane potential at the temporal frequency of the visual stimulus, as shown in Fig. 4A. For the simple cell illustrated, the membrane potential fluctuated both above and below the resting potential (arrowheads). The depolarizing phases are larger in amplitude than the hyperpolarizations. The top trace shows the response to the preferred direction of motion, and the bottom trace the response to the nonpreferred direction. The peak-to-peak amplitude of the two responses are 11.2 and 4.8 mV, giving this cell a direction index of 0.40. The direction index (DI) is defined as (R_{pref}-R_{null}/(R_{pref}+R_{null}), where R_{pref} is the amplitude of the response to the preferred direction, and R_{null} is the amplitude of the response to the nonpreferred direction.

The linearity of the spatial summation underlying direction selectivity is tested by comparing the responses to drifting gratings with the responses to contrast-modulated stationary gratings (Fig. 4B). The response to contrast-modulated stationary gratings like the response to drifting gratings was a periodic modulation of membrane potential at the stimulus temporal frequency. The amplitude of the response to the stationary grating depended strongly on its spatial phase. When the individual grating bars were centered on the cell's subregions, the maximum response was evoked (phase = 90°). Smaller responses were evoked when the grating bars straddled the adjacent subregions (phase = 0°).

For directional cells, the latency of the peak of the response was also different at each spatial phase. In the top trace of Fig. 4B, for example, the positive phase of the response consists of a rounded peak lasting about one-half of a cycle (250 ms). In the fifth trace, however, the positive phase consists of a shorter, earlier peak followed by a gradual decline in membrane potential. Similar shifts in the latency of the responses to gratings at

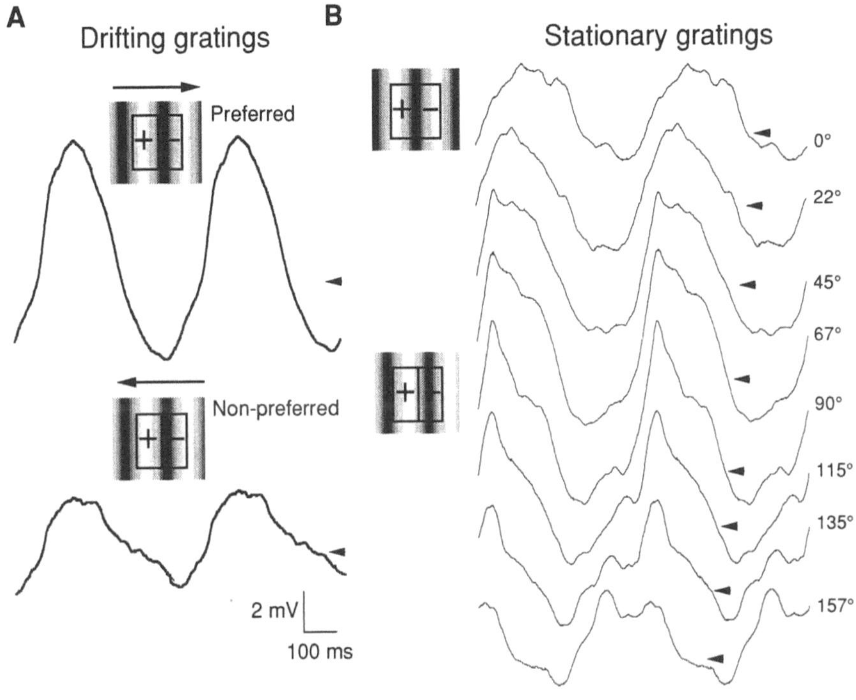

Fig. 5.4–4. **A** Fluctuations in membrane potential evoked by an optimally oriented sine wave grating, drifting in two directions at a temporal frequency of 2 Hz. The traces are each averages of response to 56 cycles of the grating presented in eight-cycle trials, with the first cycle eliminated. The average response to one cycle of the grating is repeated in the figure for clarity. The resting membrane potential, −60 mV, is marked with an *arrow*. Action potentials were removed from individual traces before averaging using the median filter. **B** Fluctuations in membrane potential evoked by a 2 Hz contrast-modulated stationary sine wave grating, presented at eight different spatial phases. Traces are the average of responses to 35 cycles of the grating. *Arrows*, peaks of the two components making up the depolarizing phase of the responses. Sine wave gratings were displayed on an oscilloscope (Tektronix 608, P31 phosphor) using a Picasso stimulus generator (Innesfree, Cambridge, Massachusetts). The peak contrast was 64% and the mean luminance 20 cd/m^2

different spatial phases have been observed extracellularly and have been proposed as the mechanism by which direction selectivity is produced in simple cells (Reid et al. 1987; Ferster 1988; Bekkers and Stevens 1990; Albrecht and Geisler 1991; Reid et al. 1991; Tolhurst and Dean 1991).

As stated above, since drifting gratings can be expressed as a linear combination of stationary gratings, if the cell's responses to the stimuli are linear, the responses to the drifting gratings must be a scaled sum of the responses to the stationary gratings. The linear relationship between the stationary and drifting gratings can be expressed as follows:

$$L(x) = C \cdot \sin\left(\omega_{sf} x \pm \omega_{tf} t \pm \frac{\pi}{2}\right) = \frac{1}{4}\sum_{n=0}^{7}\left(C \cdot \sin\left(\omega_{sf} x + n\frac{\pi}{8}\right) \cdot \sin\left(\omega_{tf} t \pm n\frac{\pi}{8}\right)\right)$$

where $L(x)$ is the modulation of the screen luminance about the mean, measured in the direction perpendicular to the grating bars, with contrast C, spatial frequency ω_{sf}, and temporal frequency ω_{tf}. Therefore a linear cell's responses to the drifting and stationary gratings shows the following analogous relationship:

$$R\left(C \cdot \sin\left(\omega_{sf} x \pm \omega_{tf} t \pm \frac{\pi}{2}\right)\right) = \frac{1}{4}\sum_{n=0}^{7} R\left(C \cdot \sin\left(\omega_{sf} x + n\frac{\pi}{8}\right) \cdot \sin\left(\omega_{tf} t \pm n\frac{\pi}{8}\right)\right),$$

where $R(x)$ is the response to stimulus x. These equations show that the responses to the stationary gratings must be shifted in time with respect to each other before they are summed, with the two directions of the temporal shift corresponding to the two directions of motion of the drifting gratings. For rightward motion, each trace has been shifted later in time with respect to the trace above it by 31.25 ms, or 1/16th of a temporal cycle (Fig. 5A, left and right columns). For leftward motion, each trace is shifted earlier in time.

Note that the leftward shift of the traces (Fig. 5A, right column) increases the difference in latency among the depolarizing phases of the different responses. The rightward shift (Fig. 5A, left column) brings the depolarizing phases more into line with one another. This difference between the right and left columns accurately predicts the direction selectivity of the cell (Fig. 5B). When the two columns of traces in Fig. 5A are summed to yield the linear prediction of the cell's response to drifting gratings, the resulting sum for the right column is much smaller than the sum for the left column (Fig. 5B, thin traces). These predicted responses were very close in shape and amplitude to those of the measured responses to drifting gratings (heavy traces). Note that the preferred direction was correctly predicted by the model. Furthermore, the direction index derived from the responses to the stationary gratings (0.47) agreed closely to the measured direction index (0.40).

Linear summation also correctly predicts the responses of weakly directional simple cells to drifting gratings (not shown). Unlike in directional cells, the latency of the peak of the depolarizing phase of the responses of nondirectional cells did not change with the spatial phase of the stationary grating. As a result, when the responses to stationary gratings were shifted in time and then summed to synthesize the response to drifting gratings,

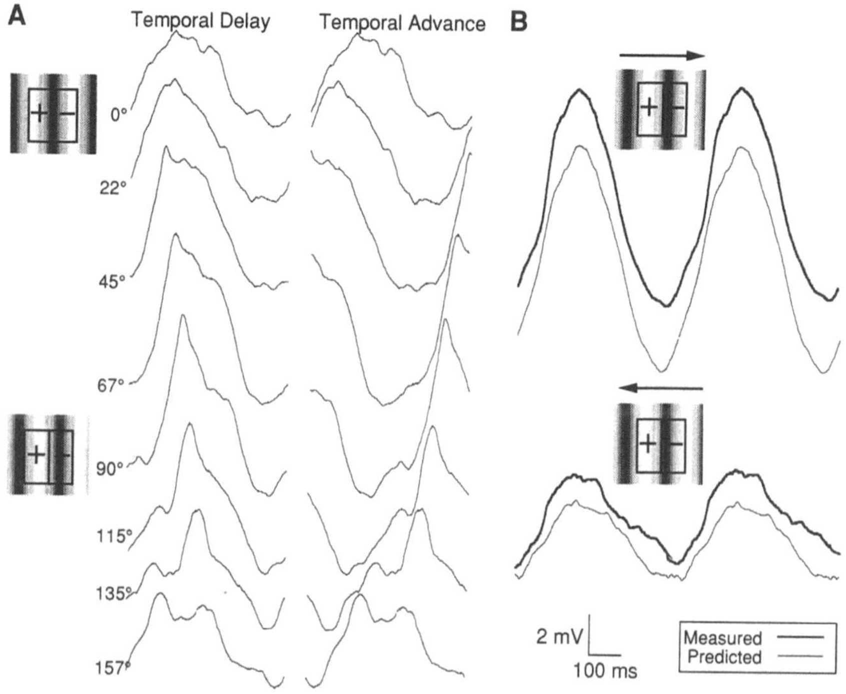

Fig. 5.4–5. **A** The traces of Fig. 4B repeated with each trace displaced in time from the previous by 1/16th of a cycle (31.25 ms). *Left*, the traces have been delayed with respect to the previous trace; *right*, the traces have been advanced. One cycle of each response is shown. **B** *Thick traces*, copies of the responses to drifting gratings from Fig. 4A. *Thin traces*, scaled sums of the traces in the left (*top*) and right (*bottom*) columns in **A**

the direction of the temporal shift had little effect on the size of the sum: the predictions for the two directions of motion are quite similar in size and shape to each other, yielding a directional index near zero.

The linear model accurately predicted the directionality for the 11 cells in our sample. In Fig. 6, the linear prediction of the direction index derived from the response to stationary gratings (predicted direction index) is plotted against the direction index of the response to drifting gratings (measured direction index). Nearly all of the points lie above the X axis, showing that the model correctly predicted the preferred direction in almost all cases. In addition, the points lie closely along a line of slope one (solid diagonal line), indicating that the model correctly predicted the directionality of most cells. The success of the model in predicting directionality of changes in membrane potential contrasts with extracellular results. Applying the model to extracellular recordings results in an underestimation of

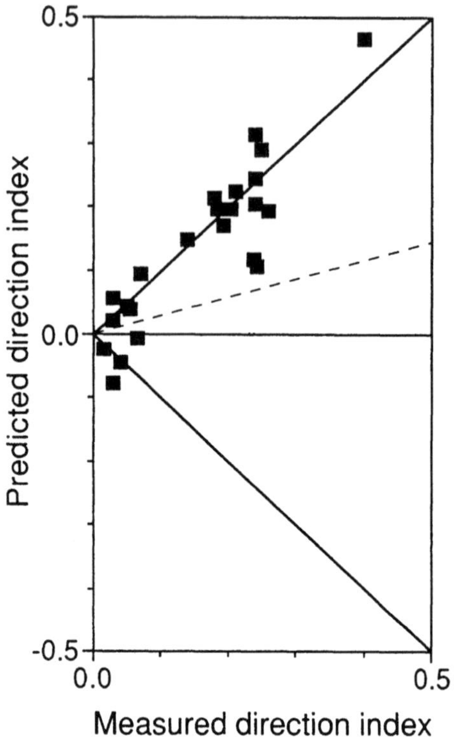

Fig. 5.4–6. A plot of the model's predictions of directionality against the measured directionality of 11 cells. The plotted direction index is defined as DI = (Pref − Null)/(Pref + Null), where Pref and Null are the peak-to-peak amplitude of the responses to the preferred and null directions of motion. More than 11 points are plotted since several neurons were tested at more than one temporal frequency, spatial frequency, or contrast. *Dashed line*, the model's prediction for extracellular recordings (from Reid et al. 1991). (The slope of the line is the median ratio of the predicted and measured direction indices for 37 cells tested at 155 different spatial and temporal parameters)

the amplitude of directionality, on average by a factor of 3 (dashed line in Fig. 6A, from Reid et al. 1991, Fig. 8E).

One surprising aspect of the result is that the maximum directionality of synaptic potentials (0.4) was so much weaker than that observed for action potentials recorded intracellularly (0.8) or extracellularly (1.0; Reid et al. 1987; Albrecht and Geisler 1991; Reid et al. 1991; Tolhurst and Dean 1991). This difference could be explained by a simple application of the spike threshold (Albrecht and Geisler 1991; McLean et al. 1993). Even if the changes in membrane potential have only moderate directionality, if the response to the nonpreferred direction of motion were below threshold, the direction index measured extracellularly would be 1. In this way, directionality in simple cells would arise initially from linear summation of postsynaptic potentials of different latency evoked from different parts of the receptive field. The directionality would then be enhanced by the visually nonspecific nonlinearity of the spike generating mechanism.

5.4.3 Conclusions

Intracellular experiments now show that direction selectivity and orientation selectivity (Ferster 1987; Douglas et al. 1988; Ferster and Jagadeesh 1992) may arise from linear synaptic mechanisms. Thus, the picture of a simple cell that emerges from direct examination of synaptic potentials is a simple one: EPSPs and IPSPs add linearly according to simple circuit principles to create intricate receptive field properties.

Acknowledgements. This work was supported by NEI grant R01 EY04726 and by a Development Award from the McKnight Endowment Fund for Neuroscience (to D.F.).

References

Albrecht DG, Geisler WS (1991) Motion selectivity and the contrast-response function of simple cells in the visual cortex. Vis Neurosci 7: 531–546

Bekkers JM, Stevens CF (1990) Two different ways evolution makes neurons larger. Prog Brain Res 83: 37–45

Blomfield S (1974) Arithmetical operations performed by nerve cells. Brain Res 69: 115–124

Douglas RJ, Martin KAC, Whitteridge D (1988) Selective responses of visual cortical cells do not depend on shunting inhibition. Nature 332: 642–644

Einstein G, Davis TL, Sterling P (1987) Pattern of lateral geniculate synapses on neuron somata in layer IV of the cat striate cortex. J Comp Neurol 260: 76–86

Emerson RC, Citron MC, Vaughn WJ, Klein SA (1987) Nonlinear directionally selective subunits in complex cells of cat striate cortex. J Neurophysiol 58: 33–53

Fatt P, Katz B (1953) The effect of inhibitory nerve impulses on a crustacean muscle fiber. J Physiol (Lond) 121: 374–389

Ferster D (1987) The origin of orientation selective EPSPs in simple cells of cat visual cortex. J Neurosci 7: 1780–1791

Ferster D (1988) Spatially opponent excitation and inhibition in simple cells of the cat visual cortex. J Neurosci 8: 1172–1180

Ferster D, Jagadeesh B (1991) Nonlinearity of spatial summation in simple cells of areas 17 and 18 of cat visual cortex. J Neurophysiol 66: 1667–1679

Ferster D, Jagadeesh B (1992) EPSP-IPSP interactions in cat visual cortex: an in vivo whole-cell patch study. J Neurosci 12: 1262–1274

Ferster D, Lindstrom S (1985) Synaptic excitation of neurones in area 17 of the cat by intracortical axon collaterals of cortico-geniculate cells. J Physiol (Lond) 367: 233–252

Freund TF, Martin KAC, Somogyi P, Whitteridge D (1985) Innervation of cat visual areas 17 and 18 by physiologically identified X- and Y-type thalamic afferents. II. Identification of postsynaptic targets by GABA immunocytochemistry and Golgi impregnation. J Comp Neurol 242: 275–291

Ganz L, Felder R (1984) Mechanism of direction selectivity in simple neurones of the cat's visual cortex analyzed with stationary flashed sequences. J Neurophysiol 51: 294–324

Koch C, Poggio T (1985) The synaptic veto mechanism: does it underlie direction and orientation selectivity in the visual cortex? In: Rose D, Dobson VG (eds) Models of the visual cortex. Wiley, New York, 408–419

Koch C, Poggio T, Torre V (1983) Nonlinear interactions in a dendritic tree: localization, timing, and role in information processing. Proc Nat Acad Sci USA 80: 2799–2802

McGuire BA, Hornung JP, Gilbert CD, Wiesel TN (1984) Patterns of synaptic input to layer 4 of the cat striate cortex. J Neurosci 4: 3021–3033

McLean J, Raab S, Palmer A (1993) Contribution of linear mechanisms to the specification of local motion by simple cells in areas 17 and 18 of the cat. Vis Neurosci (in press)

Reid RC, Soodak RE, Shapley RM (1987) Linear mechanisms of direction selectivity in simple cells of cat striate cortex. Proc Nat Acad Sci USA 84: 8740–8744

Reid RC, Soodak RE, Shappley RM (1991) Directional selectivity and spatiotemporal structure of receptive fields of simple cells in cat striate cortex. J Neurophysiol 66: 505–529
Tolhurst DJ, Dean AF (1991) Evaluation of a linear model of directional selectivity in simple cells of the cat's striate cortex. Vis Neurosci 6: 421–428
Torre V, Poggio T (1978) A synaptic mechanism possibly underlying direction selectivity to motion. Proc R Soc Lond B 202: 409–416
Wehmeier U, Koch C (1988) A computer simulation of cortical orientation selectivity in the cat visual system. Soc Neurosci Abstr 14: 898

5.5 Widespread Horizontal Interactions in Kitten Visual Cortex and Their Shrinkage with Development

T. Tsumoto and Y. Hata
Department of Neurophysiology, Biomedical Research Center, Osaka University, Medical School, Suita, Japan

5.5.1 Introduction

Neurons in the cat visual cortex are known to respond selectively to elongated edge of particular orientation and to be organized to make up orientation columns (Hubel and Wiesel 1962, see Gilbert 1983 for review). Furthermore, these columns are suggested to be organized in a regular fashion so as to form the "hypercolumn" which includes a complete set of orientation columns for a restricted region in the visual field together with ocular dominance columns for both eyes (Hubel and Wiesel 1974). In a theoretical study horizontal interactions between adjacent neurons were suggested to play a role in forming selective response properties, such as orientation selectivity, of visual cortical neurons (von der Malsburg 1973). In fact, physiological studies demonstrated that such interactions operate mainly between neurons with matched orientation preferences (Toyama et al. 1981; Ts'o et al. 1986; Hata et al. 1991).

In kittens, however, previous studies reported that a large proportion of visual cortical neurons are not orientation selective or even visually unresponsive just after opening of the eyes and that the proportion of orientation selective neurons increases gradually to adult level during first four weeks of postnatal development (Hubel and Wiesel 1963; Blakemore and

Van Sluyters 1975; Tsumoto and Suda 1982; Albus and Wolf 1984; see Frégnac and Imbert 1984 for review). Considering that selective responses of visual cortical neurons are formed on the basis of specific pattern of neural interactions, it is reasonable to expect that a developmental change in interactions may be a basis for the acquisition of selective responsiveness of visual cortical neurons. In the present study we therefore explored development of horizontal interactions between neurons, using cross-correlation analysis of spike trains of pairs of neurons recorded simultaneously with two independent electrodes. We focused on local interactions within 1 mm, which corresponds to the extent of a hypercolumn of the cat visual cortex (Albus 1975; Schoppmann and Stryker 1981; Löwel et al. 1987).

Parts of the present results have been presented previously (Hata et al. 1988, 1991, 1993).

5.5.2 Methods

Sixty-one kittens at postnatal ages from the 8th day to 9th week and 32 adult cats were used in this series of experiments. Experimental procedures for preparing and maintaining animals, extracellular single-unit recordings, cross-correlation analysis, and histology have been described previously (Hata et al. 1991, 1993). The animals were initially anesthetized with ketamine HCl (20–30 mg/kg, i.m.), and then with 1%–2% halothane in a mixture of 70% N_2O and 30% O_2 during surgical operations. During recordings they were anesthetized with 70% N_2O and 30% O_2. If the heart rate was changed suddenly by noxious stimuli, 0.1%–0.5% halothane was added to the gas to maintain an adequate depth of anesthesia. Spike discharges were recorded extracellularly from a pair of cells in the primary visual cortex using two glass micropipettes filled with 0.5 M sodium acetate with 2% pontamine sky blue (Fig. 1). For a pharmacological activation of cortical cells, double-barrel micropipettes attached to a recording pipette were used for an iontophoretic application of sodium L-glutamate (50 mM, pH 8.0). Retaining currents of 10–20 nA were used routinely to prevent the drug from leaking.

Trains of spikes obtained from two cells were fed into a computer, which provided on-line computation and display of cross-correlograms. To collect a sufficient number of spikes from each cell in a limited time of recordings, cells were activated by visual stimuli. Cells that showed only weak visual responses were activated by an iontophoretic administration of glutamate in addition to visual stimulation. In pharmacological activation of cells care was taken to minimize the amount of glutamate so as to enhance the visual response but not to activate cells with rates of discharge greater than 10 Hz except when visual stimuli also activated them. The criteria for signifi-

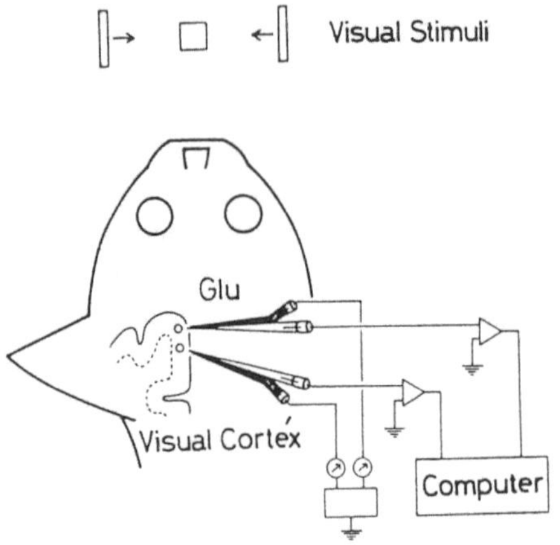

Fig. 5.5–1. Schematic depiction of experimental procedures. Recording electrodes were usually attached side by side by micropipettes containing sodium L-glutamate (Glu) for activation of neurons which had low or almost no activity

cance of features in the cross-correlograms have been described previously (Tsumoto et al. 1978; Hata et al. 1993). Features in the original cross-correlograms were interpreted to be due to neural interactions only when these features were not seen in the corresponding shuffled cross-correlogram.

5.5.3 Results

From the primary visual cortex of kittens at the 2nd to 9th postnatal week and adult cats, we recorded activities of 843 pairs of cells. Recording sites of 799 cell pairs were identified histologically, and 750 of these cell pairs were found to be separated horizontally by less than 1 mm. This report deals with these 750 pairs of cells.

Types of Significant Correlograms. Of the 750 cell pairs 173 showed correlated firings, indicating neural interactions in the cross-correlogram, and the rest had no significant correlation. The significant correlations were classified into six types according to the following features observed in the cross-correlograms; (a) a peak at the zero time bin (central peak), (b) a peak displaced from the zero time bin but its shoulder straddling the zero time (partially displaced peak), (c) double peaks consisting of central and displaced peaks (double peaks), (d) a trough displaced from the zero time bin (displaced trough), (e) a peak completely displaced from the zero time bin (displaced peak), and (f) double peaks consisting of two displaced peaks to both sides of the zero time bin (two opposed peaks). The proportions of cell pairs with each type of correlation are given in Table 1.

Figure 2 shows an example of cell pair showing a central peak in the cross-correlogram. This pair of cells was recorded from a kitten at the

	Postnatal weeks			
	2nd–3rd	4th–6th	7th–9th	Adult
Central peak	7.4	10.6	11.4	16.5
Partially displaced peak	0.8	6.8	3.6	3.1
Double peaks	3.3	0.6	1.4	2.7
Displaced trough	0.0	1.9	2.9	3.1
Displaced peak	3.3	2.5	0.7	0.0
Two opposed peaks	2.5	1.2	0.0	0.0
Total number of pairs	122	161	140	327

Table 5.5–1. Number of cell pairs showing each type of correlated firings (percentage of the total number of pairs in each age group)

8th postnatal week. Recording sites of both cells were located in layer V and separated horizontally by about 140 μm in the cortex. The cross-correlogram of this pair showed a peak at the zero time bin with width of 10 ms, indicating that the two cells tended to fire at the same time. The shuffled cross-correlogram was quite flat, indicating that the visual stimuli did not significantly contribute to the peak in the cross-correlogram (not shown). The type of correlated firing shown in Fig. 2A is explained most likely as a product of common inputs to a pair of cells (Moore et al. 1970). Such a center-peaked correlogram was found in 96 of the 750 pairs of cells (Table 1). In 83 of the 96 cell pairs, the peak width was shorter than 76 ms while in the other 13 pairs it was much wider (80–420 ms). Such a wide peak could be due to slow depolarizations of membrane potential which might be caused by non-specific modulatory inputs to cortical cells (see Hata et al. 1991). Therefore, these 13 pairs were discarded from further analysis, and the 83 pairs were judged as having common inputs.

The second type of significant cross-correlograms had a peak that shifted to the left or right from the zero time but whose shoulder straddled the zero. This partially displaced peak is most reasonably interpreted as the product of combination of excitatory linkage and common inputs (Hata et al. 1991). Twenty-seven pairs of cells showed this type of correlated firings (Table 1). The third type of cross-correlograms is characterized by a peak displaced from the zero time together with a central peak ("Double peaks" in Table 1). The peak displaced from the zero could be interpreted as the product of excitatory synaptic connection between the two cells (Moore et al. 1970). Thus, this type of correlograms seen in 16 pairs was also interpreted as the product of a combination of excitatory linkage and common inputs. In 43 cross-correlograms of these two types, the peak latencies of the displaced peak were below 6 ms in the 34 pairs. Thirty-three of these 34 pairs had a peak of less than 40 ms width, while the other one had quite a wide peak

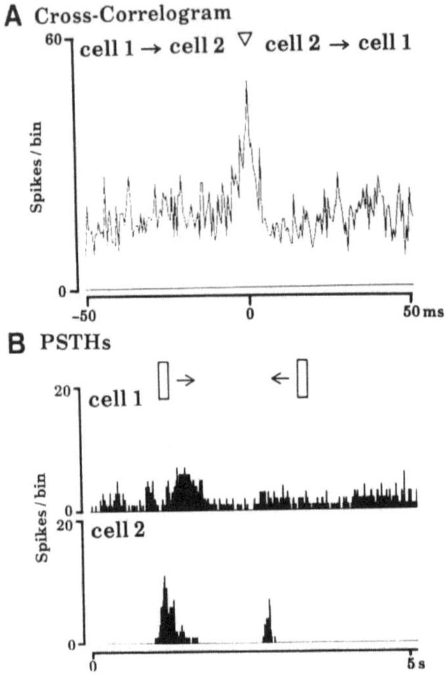

Fig. 5.5–2. A An example of a cross-correlogram showing central peak. **B** Peristimulus time histograms (*PSTHs*) obtained with stimulation by a light slit moving back and forth over their receptive fields. The orientation of the light slit is indicated at the top. Cross-correlograms in this and the following figure represent the number of spikes per bin that occurred in cell 1 before (*left*) and after (*right*) spikes of cell 2. Number of spikes: cell 1, 6915; cell 2, 1644. Bin width, 0.5 ms

(120 ms width). Therefore, those 33 pairs were judged to have an excitatory linkage and common inputs, but the other 10 pairs were not included in this group.

Seventeen pairs showed a displaced trough with an onset latency of 1–8 ms and a duration of 3–50 ms. Such a feature could be interpreted as a product of inhibitory linkage, based on the similarity of the shape and time course of the trough to those of inhibitory postsynaptic potentials. The peak completely displaced from the zero time bin was observed in correlograms from 9 pairs (Table 1). This type of peak could be interpreted as the product of direct excitatory synaptic connection between the two cells (Moore et al. 1970). The peak latencies were within 6 ms in seven pairs and were longer than 20 ms in the other two pairs. The former seven pairs had a peak width less than 10 ms and were judged to have an excitatory linkage between two cells. The latter two pairs were excluded from further analysis because the very long latencies might be due to polysynaptic interactions.

Figure 3 illustrates an example of cross-correlograms showing two opposed peaks obtained from a kitten at the 8th postnatal week. Both cells of the pair were located in layer II/III of the cortex and the horizontal distance between the two recording sites was unmeasurably small. In this case, preferred orientation of cells 1 and 2 was about the same (Fig. 3B).

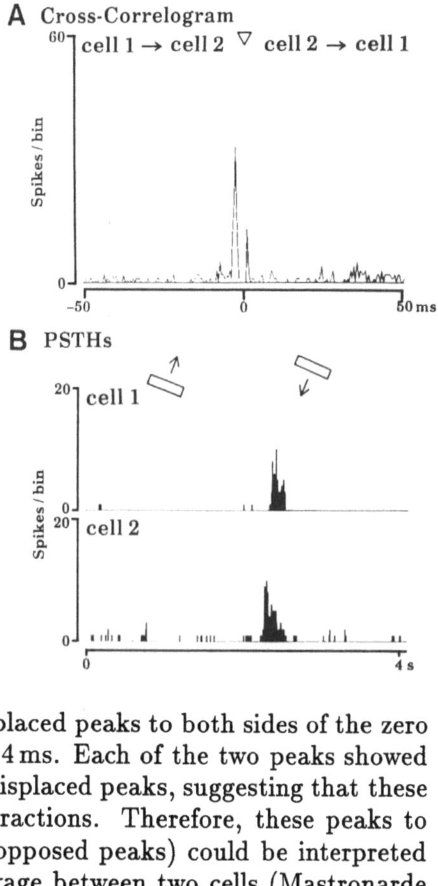

Fig. 5.5–3. A,B An example of correlated firings showing two opposed peaks. Conventions are the same as in Fig. 2. Number of spikes in **A**: cell 1, 197; cell 2, 514. Bin width, 0.5 ms

The cross-correlogram showed two displaced peaks to both sides of the zero time bin with peak latencies of 2 and 4 ms. Each of the two peaks showed similar width and latency to those of displaced peaks, suggesting that these two peaks may reflect excitatory interactions. Therefore, these peaks to both sides of the zero time bin (two opposed peaks) could be interpreted as a product of mutual excitatory linkage between two cells (Mastronarde 1983). This type of correlated firings were observed in five pairs in the present study (Table 1). The peak latencies of all ten peaks were within 8 ms.

Three Types of Functional Interactions. All the pairs with correlated firings were classified into three groups for further analysis according to the functional interpretation of cross-correlograms described above. In this classification, pairs with partially displaced peak or double peaks were counted as both of common inputs and excitatory linkage. The probability of encountering common inputs in the 2nd–3rd weeks (9.8%) was significantly lower ($p < 0.05$, χ^2-test) than that in the adult cats (17.7%). The value tended to increase with development and attained to 15.7% in the 7th–9th weeks. The same tendency was found in the probability of inhibitory link-

Horizontal distance	Postnatal weeks 2nd–3rd	4th–6th	7th–9th	Adult
-200 μm	5/23 (22%)	10/39 (26%)	7/27 (26%)	25/64 (39%)
-400 μm	7/38 (18%)	12/37 (32%)	8/40 (20%)	25/83 (30%)
-600 μm	4/24 (17%)	4/32 (13%)	9/36 (25%)	11/90 (12%)
-800 μm	0/12 (0%)	5/17 (29%)	1/18 (6%)	5/61 (8%)
-1000 μm	0/1 (0%)	0/11 (0%)	0/5 (0%)	2/29 (7%)

Table 5.5–2. Proportion of cell pairs showing significant correlation in each horizontal distance and age group. Numbers of cell pairs with significant correlation and of cell pairs sampled in each group are indicated as numerator and denominator, respectively.

age: it was not observed in the 2nd–3rd weeks, whereas in the 7th–9th weeks and in the adult it was seen at 2.9% and 3.1%, respectively, of the total number of pairs. On the other hand, the probability of excitatory linkage was lower in the 7th–9th weeks (4.3%) than in the younger group. The values in the 2nd–3rd weeks (9.0%) and the 4th–6th weeks (9.3%) were significantly higher than that in the adult cats ($p < 0.05$ and $p < 0.02$, respectively, χ^2-test).

Horizontal Spread of Correlated Firings in the Cortex. To explore the extent of horizontal neural interactions in the primary visual cortex, data were analyzed in terms of horizontal distance of the cell pairs. In the 2nd–3rd weeks, 19% of the pairs with horizontal distance less than 600 μm showed correlated firings but no correlated firing was found in pairs with horizontal distance over 600 μm (Table 2). This tendency was observed in cell pairs with common inputs and excitatory linkage. No inhibitory linkage was observed in this age group. In the 4th–6th weeks, correlated firings were found in cell pairs with horizontal distance up to 720 μm. Within this range there was no tendency of the probability to decrease with distance. For example, it was 29% between 600 and 800 μm. This is due mainly to the distribution of common inputs. In the 7th–9th weeks, on the other hand, correlated firings were observed mostly in cell pairs with horizontal distance less than 600 μm. Of the cell pairs with horizontal distance less than 600 μm 23% (24/103) had correlated firings while the proportion decreased to 4% (1/23) in the horizontal distance over 600 μm ($p < 0.05$, χ^2-test). In the adult cats, correlated firings were observed most frequently in cell pairs with horizontal distance less than 400 μm. Therefore it is possible to suggest that the horizontal extent of correlated firings might decrease gradually with development after the 7th week.

Relationship Between Correlated Firings and Orientation Preference. Many of the correlated firings were found in cell pairs with similar orientation preferences throughout all the age groups. However, the extent of tuning toward the difference less than 22° seemed rather broad in the 4th–6th weeks and became sharper in the 7th–9th weeks. In the 4th–6th weeks, for example, correlated firings were observed in a pair with almost orthogonal preferred orientations although they were never seen in the 7th–9th weeks and in the adult.

The tendency that the tuning of probability of correlated firings toward the same or similar orientation preference is broader in the 4th–6th weeks than in the 7th–9th weeks was clearly observed in pairs with common inputs. To test the sharpness of the tuning we compared the proportion of common inputs pairs in the groups of orientation difference of 0°–22° and 23°–45°. In the 7th–9th weeks, the proportion of common input in the group of orientation difference less than 22° (16/43) was significantly higher ($p < 0.05$, χ^2-test) than that within 23°–45° (3/25). In the 4th–6th weeks, however, we found no statistically significant difference in the proportion of common inputs between orientation difference groups of 0°–22° (15/55) and 23°–45° (4/22). Therefore it is possible to state that the proportion of common inputs was broadly tuned toward the same orientation preference in the 4th–6th weeks and that the tuning became sharper in the 7th–9th weeks.

Excitatory linkage showed the same tendency, although the limited number of pairs did not allow us to draw a conclusive statement. Inhibitory linkage was observed in pairs with orientation preferences up to 45° in the 4th–6th and 7th–9th weeks. In the 7th–9th weeks, three of the four pairs with inhibitory linkage had orientation differences between 23°–45° and only one pair was seen in the range less than 22°. Such a tendency that inhibitory interactions could be found more frequently between cell pairs with somewhat different preferred orientations than between cells with similar preferences had been described previously (Hata et al. 1988, 1991).

5.5.4 Discussion

Developmental Changes in Horizontal Spread of Correlated Firings. In the present study, correlated firings were observed mostly in pairs with horizontal separation less than 600 μm in the 7th–9th weeks and 400 μm in the adult as far as the local interaction within the extent of 1 mm is concerned. On the other hand, in the 4th–6th weeks the horizontal distance between correlated pair of cells was not restricted to within 600 μm. This difference in horizontal spread of interaction in the 4th–6th weeks and the later age may have been even underestimated because the area of the primary visual cortex is smaller in the 4th–6th weeks and increases during subsequent development. Assuming that the primary visual cortex increases in size with development in parallel with the whole brain, we could estimate that the values of distance measured in the kitten visual cortex may be about 1.6 (2nd–3rd weeks), 1.3 (4th–6th), and 1.1 (7th–9th) times longer in terms of retinal representation than those in adult cortex (Rose and Goodfellow 1973). Therefore, the measured value of horizontal extent of interactions in the 4th–6th weeks should be taken as representing wider extent of visual field representation than the corresponding value in the older age.

Anatomically, it is well known that cortical development involves an elimination of neural structures, such as synapses and dendritic spines beyond about 6 weeks of age (Cragg 1975; Lund et al. 1977; Winfield 1981). In addition, it has been reported that retraction of axon collaterals takes place in this developmental stage in both geniculocortical (LeVay et al. 1978) and intracortical connections (Meyer and Ferres-Torres 1984; Callaway and Katz 1990; Luhmann et al. 1990). These reports seem consistent with the present observations that the horizontal separation between correlated cells in the 4th–6th weeks is wider on average than that in the 7th–9th weeks and the adult. Therefore it is possible to conclude that the decrease in horizontal spread of correlated firings observed after the 7th week may reflect the relatively widespread connections in the visual cortex in the early stage and the subsequent regression of the relatively long connections during the late stage of development.

Relationship with Orientation Preference. Regarding orientation preference, we found that correlated firings are observed more frequently between neurons with similar orientation preferences, but we also observed that the tuning toward the same orientation preference was broader in the 4th–6th weeks than in the 7th–9th weeks. In the adult cats correlated firings were observed mainly between neurons with similar orientation preferences, and the degree of tuning toward the same orientation preference was sharper than that in the 7th–9th weeks. In other words, there are functional interactions between neurons with quite different orientation preferences before the 6th week of age, and thereafter such interactions decrease in frequency or disappear during development.

Patchy intracortical connections in supragranular layer of the visual cortex are thought to link columns of the same orientation preference (Gilbert and Wiesel 1983; Ts'o et al. 1986). These orientation-specific connections have been shown to emerge by retraction of exuberant connections and by proliferation of axonal branches in their target regions during development (Callaway and Katz 1990). These findings suggest that horizontal connections in the supragranular layer might be rather diffuse initially but later become specific by retracting connections which link neurons with unmatched orientation preferences. This possibility is quite consistent with the present results. Although the present study focused on more local interactions, it seems reasonable to assume similar mechanisms in formation of local connections. In fact, a retraction of exuberant axon collaterals has been reported to occur in some types of nonpyramidal cells in the range of horizontal distance observed in the present study (Meyer and Ferres-Torres 1984). Also, the present results show an increase in the probability of correlated firings between neurons with similar orientation preferences during development in addition to the decrease in those with quite different preferences. This suggests that interactions between neighboring neurons with similar orientation preferences may be strengthened functionally during development.

References

Albus K (1975) A quantitative study of the projection area of the central and paracentral visual field in area 17 of the cat. II. The spatial organization of the orientation domain. Exp Brain Res 24: 181–202

Albus K, Wolf W (1984) Early post-natal development of neuronal function in the kitten's visual cortex: a laminar analysis. J Physiol (Lond) 348: 153–185

Blakemore C, Van Sluyters RC (1975) Innate and environmental factors in the development of the kitten's visual cortex. J Physiol (Lond) 248: 663–716

Callaway EM, Katz LC (1990) Emergence and refinement of clustered horizontal connections in cat striate cortex. J Neurosci 10: 1134–1153

Cragg BG (1975) The development of synapses in the visual system of the cat. J Comp Neurol 160: 147–166

Frégnac Y, Imbert M (1984) Development of neuronal selectivity in primary visual cortex of cat. Physiol Rev 64: 325–434

Gilbert CD (1983) Microcircuitry of the visual cortex. Ann Rev Neurosci 6: 217–247

Gilbert CD, Wiesel TN (1983) Clustered intrinsic connections in cat visual cortex. J Neurosci 3: 1116–1133

Hata Y, Tsumoto T, Sato H, Hagihara K, Tamura H (1988) Inhibition contributes to orientation selectivity in visual cortex of cat. Nature 335: 815–817

Hata Y, Tsumoto T, Sato H, Tamura H (1991) Horizontal interactions between visual cortical neurones studied by cross-correlation analysis in the cat. J Physiol (Lond) 441: 593–614

Hata Y, Tsumoto T, Sato H, Hagihara K, Tamura H (1993) Development of local horizontal interactions in cat visual cortex studied by cross-correlation analysis. J Neurophysiol 69: 40–56

Hubel DH, Wiesel TN (1962) Receptive fields, binocular interaction and functional architecture in the cat's visual cortex. J Physiol (Lond) 160: 106–154

Hubel DH, Wiesel TN (1963) Receptive fields of cells in striate cortex of very young, visually inexperienced kittens. J Neurophysiol 26: 994–1002

Hubel DH, Wiesel TN (1974) Sequence regularity and geometry of orientation columns in the monkey striate cortex. J Comp Neurol 158: 267–294

LeVay S, Stryker MP, Shatz CJ (1978) Ocular dominance columns and their development in layer IV of the cat's visual cortex: a quantitative study. J Comp Neurol 179: 233–244

Löwel S, Freeman B, Singer W (1987) Topographic organization of the orientation column system in large flat-mounts of the cat visual cortex: a 2-deoxyglucose study. J Comp Neurol 255: 401–415

Luhmann HJ, Singer W, Martinez-Millan L (1990) Horizontal interactions in cat striate cortex. I. Anatomical substrate and postnatal development. Eur J Neurosci 2: 344–357

Lund JS, Boothe RG, Lund RD (1977) Development of neurons in the visual cortex (area 17) of the monkey (Macaca nemestrina): a Golgi study from fetal day 127 to postnatal maturity. J Comp Neurol 176: 149–188

Mastronarde D (1983) Interactions between ganglion cells in cat retina. J Neurophysiol 49: 350–365

Meyer G, Ferres-Torres R (1984) Postnatal maturation of nonpyramidal neurons in the visual cortex of the cat. J Comp Neurol 228: 226–244

Moore GP, Segundo JP, Perkel DH, Levitan H (1970) Statistical signs of synaptic interaction in neurons. Biophys J 10: 876–900
Schoppmann A, Stryker MP (1981) Physiological evidence that orientation columns are revealed in the cat's visual cortex by the 2-deoxyglucose method. Nature 293: 574–576
Toyama K, Kimura M, Tanaka K (1981) Organization of cat visual cortex as investigated by cross-correlation technique. J Neurophysiol 46: 202–214
Ts'o DY, Gilbert CD, Wiesel TN (1986) Relationships between horizontal interactions and functional architecture in cat striate cortex as revealed by cross-correlation analysis. J Neurosci 6: 1160–1170
Tsumoto T, Suda K (1982) Laminar differences in development of afferent innervation to striate cortex neurons in kittens. Exp Brain Res 45: 433–446, 1982
Tsumoto T, Creutzfeldt OD, Legendy CR (1978) Functional organization of the corticofugal system from visual cortex to lateral geniculate nucleus in the cat (with an appendix on geniculo-cortical mono-synaptic connections). Exp Brain Res 32: 345–364
von der Malsburg C (1973) Self-organization of orientation sensitive cells in the striate cortex. Kybernetik 14: 85–100
Winfield DA (1981) The postnatal development of synapses in the visual cortex of the cat and the effects of eyelid closure. Brain Res 206: 166–171

5.6 Large Basket Cells and Lateral Inhibition in Cat Visual Cortex

U.T. Eysel, Z.F. Kisvárday, F. Wörgötter, and J.M. Crook
Department of Neurophysiology, Faculty of Medicine, Ruhr University, Bochum, Germany

5.6.1 Introduction

The majority of cells in the visual cortex (80%) are excitatory, and their connectivity is governed by a vertical, columnar, and horizontal long-ranging patchy systems. The lateral excitatory network has been extensively described in terms of structure and possible function (Gilbert and Wiesel 1979, 1983, 1989; Martin and Whitteridge 1984; Kisvárday and Eysel 1992). However, the remaining 20% of the visual cortical cells are GABAergic (Gabbot and Somogyi 1986). These cells show characteristic structural specializations (Somogyi 1986) and include cells with long-ranging horizontal connections; the possible functional contributions of these cells are less well understood. Otto Creutzfeldt was among those who first observed inhibitory postsynaptic potentials in visually stimulated striate cortical cells and discussed physiological functions in terms of directional tuning and orientation specificity (Benevento et al. 1972; Creutzfeldt et al. 1974), a notion that was later supported by several groups (Morrone et al. 1982; Sillito 1975, 1977, 1979, 1984; Eysel et al. 1987, 1988, 1990; Crook and Eysel 1992). We have now looked in detail into the structure, connectivity and possible functional role of large GABAergic basket cells that support a horizontally long-ranging axonal system (for review see Kisvárday 1992). Here we present a comprehensive view of the lateral extent, connectivity, boutonal distribution, and functional specificity of the horizontally directed axons of large basket cells and suggest possible physiological functions on the basis of inactivation studies in the cat visual cortex (for review see Eysel, 1992).

5.6.2 Methods

Our results are based on physiological and anatomical work performed in the anesthetized cat visual cortex. In one set of experiments, labeling of ensembles of large basket cells was achieved by microiontophoretical injection of biocytin at sites of known orientation and direction preference. In another series of experiments, mapping and physiological testing of single cells were performed with glass micropipettes and remote γ-aminobutyric acid (GABA) inactivation of the visual cortex was achieved with multibarrel pipettes suited for multi-unit recordings and GABA microiontophoresis. All methods are described in full detail elsewhere (Eysel et al. 1988, 1990; Crook et al. 1991; Kisvárday and Eysel 1992; Crook and Eysel, 1992; Kisvárday et al. 1993).

5.6.3 Anatomy

The typical large basket cell in layer III of the cat visual cortex is characterized by a large soma size (20–30 μm), dendritic field of about 700 μm diameter, and axons horizontally extending over 1–2 mm. The terminals of a single basket cell contact the perisomatic region of hundreds of pyramidal and dozens of nonpyramidal cells including other basket cells.

Statistics. The axons and dendrites of large basket cells in layer II of area 17 were reconstructed, and boutons were plotted following biocytin labeling. The cell shown in Fig. 1 displays the typical horizontally long ranging axonal system, and the boutonal distribution discloses a reduced number of boutons in the immediate surrounding of the soma (Fig. 2A). This mapping of the boutonal distribution allowed a statistical analysis in relation to the map of the average orientation specificity on the cortical surface to estimate the possible inhibitory impact of large basket cells on cells with similar and dissimilar orientation preferences in relation to their own.

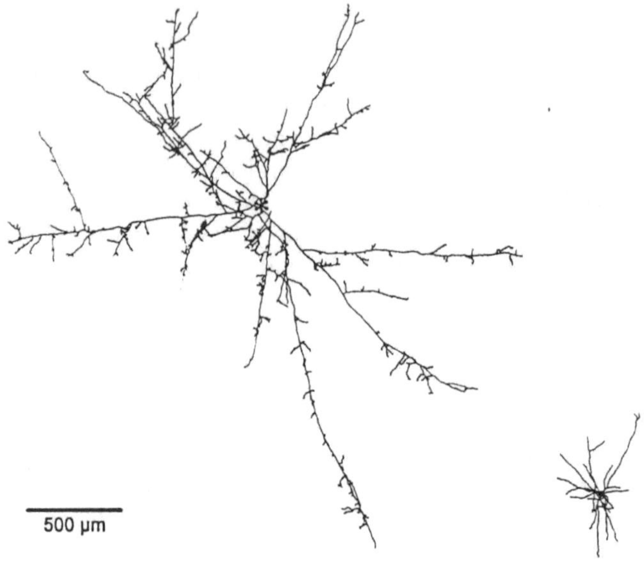

Fig. 5.6–1. Large biocytin-labeled basket cell in area 17 reconstructed from horizontal sections (Kisvárday and Eysel 1994). Soma and dendritic field are shown in the *lower right corner*. Position of the soma is indicated by an *asterisk*

The axonal boutons mapped from two large basket cells in area 17 increased with distance from the soma forming a plateau in boutonal frequency between 200 and 900 μm (Fig. 2B). The probability to encounter iso-, oblique-, or cross-orientation selectivity (always in relation to the orientation preference of the cell in the center) was derived from a map obtained by intrinsic signal recordings (Bonhoeffer and Grinvald 1991). We projected the probability functions in the same way as the relative boutonal frequency one-dimensionally onto the distance scale (Fig. 2C). The resulting curves indicate regions of iso-, oblique-, and cross-orientations and show the highest probability to meet cross-orientations at distances of about 600 μm from any point in area 17, while iso-orientations are met close to the origin and again at a distance of about 1200 μm; oblique orientations appear at shorter distance than cross-orientations and maintain constant probabilty for distances above 400 μm. The product of relative boutonal frequency and orientation probability is used to estimate the possible inhibitory impact of large basket cells on iso-, oblique, and cross-oriented cells in cat area 17 based on purely statistical data and not assuming any specificity of connections (Fig. 2D). According to this computation the inhibitory influence of large basket cells could be most prominent at cells of similar orientation preference at a distance of about 200 μm with a second lower peak

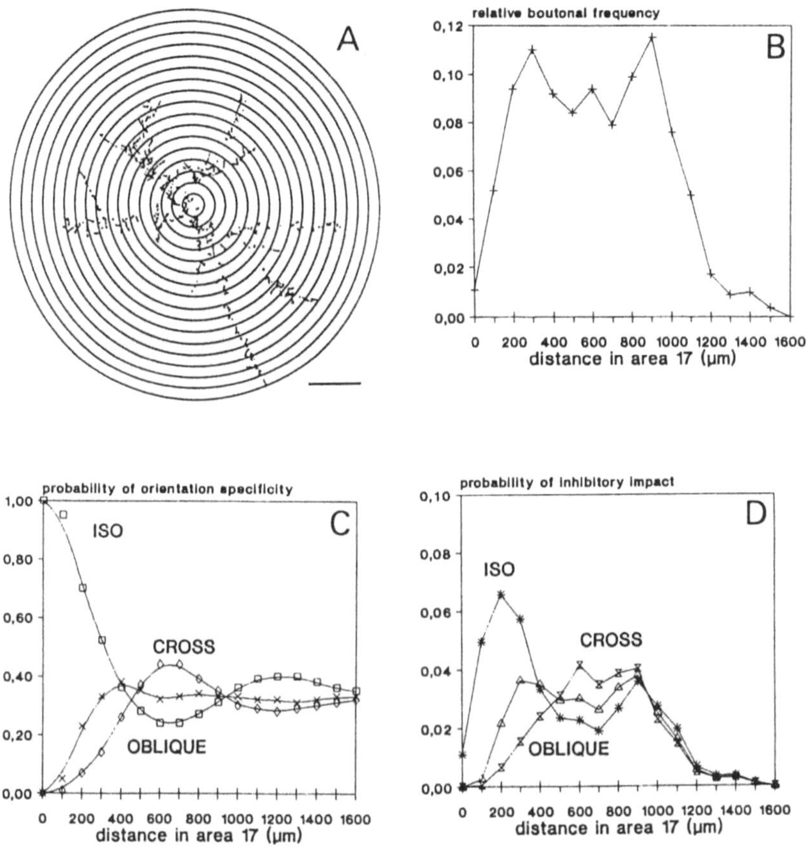

Fig. 5.6–2. Boutonal distribution (**A**), relative boutonal frequency (**B**), probability of orientation specificity (**C**) and computed probability of inhibitory impact (**D**) in terms of distance in area 17. **A** The pattern of synaptic boutons of the cell shown in Fig. 1. The concentrical rings indicate 100 μm steps in distance from the soma. **B** Relative frequency of boutons with respect to distance from the soma (absolute number counted within rings in **A** divided by the total number of boutons). Two large basket cells with a total of 6776 boutons were analyzed. **C** Computation of probability to meet iso-, cross-, or oblique orientation preferences at given distances from soma (map of Bonhoeffer and Grinwald 1991). **D** Probability of inhibitory impact of the two large basket cells at cells with different preferred orientation (derived from **B** and **C** by multiplication)

close to 900 μm. A substantial influence can also be expected at cells with cross-orientation between 600 and 900 μm, while inhibition of obliquely oriented cells can take place in a broad region between 300 and 900 μm. The regions with relatively high possible impact on oblique and cross oriented cells overlap completely.

Functional Topography. In a combined study with biocytin labeling and single-electrode mapping in the appropriate depth in layer III of area 17 we could show that basket cells indeed do not connect — as pyramidal cells — preferentially to a given orientation specificity. Instead, large basket cells connect to a broad range of orientation preferences including iso- and cross-orientations (Eysel and Kisvárday 1991).

Specificity. While abundantly sending synaptic contacts to different types of excitatory cells (pyramidal and spiny stellate), large basket cells display an interesting feature of specificity in relation to other GABAergic cells; among these cells they contact other large basket cells (Kisvárday et al. 1993).

5.6.4 Physiological Aspects of Lateral Inhibition

Statistics. The possible contributions of lateral inhibition can be derived from results obtained with local inactivation of visual cortex applying GABA at sites laterally remote from a recorded cell. The observed effects included losses or changes in directionality (Fig. 3A,B) as well as weakening of orientation tuning (Fig. 3C,D) and were interpreted as indicating elimination of both iso- and cross-inhibitory effects (Eysel et al. 1988, 1990; Crook et al. 1991). With a purely statistical approach we computed the number of GABA units (multiplication of GABA ejection current with ejection time) that were necessary to reduce directional tuning or orientational tuning at different distances from a given target cell in area 17 (Wörgötter and Eysel 1991). For those cases showing specifically increased responses the results were interpreted in the following way: lateral mechanisms resulting in a suppression of responses to nonpreferred orientations can be expected to originate from distances close to 0.5 mm (distance corresponding to half a hypercolum) while long-range iso-orientation inhibition (inferred from changes in direction selectivity) preferentially arises from a distance corresponding to the width of one hypercolumn (Wörgötter and Eysel 1991). According to the computations shown in Fig. 2D this interpretation corresponds to the possible lateral inhibitory influences of basket cells on iso- and cross-oriented target cells. It should be mentioned that the method of lateral remote inactivation does not allow the study of influences originating from 300 μm or closer to a target cell since from these distances the direct action of GABA at the target cell is superimposed to any effects of remote inactivation. The kind and strength of effects of remote local inactivation were compared for areas 17 and 18 (Crook et al. 1991), and no significant differences were found between the two input regions of cat visual cortex.

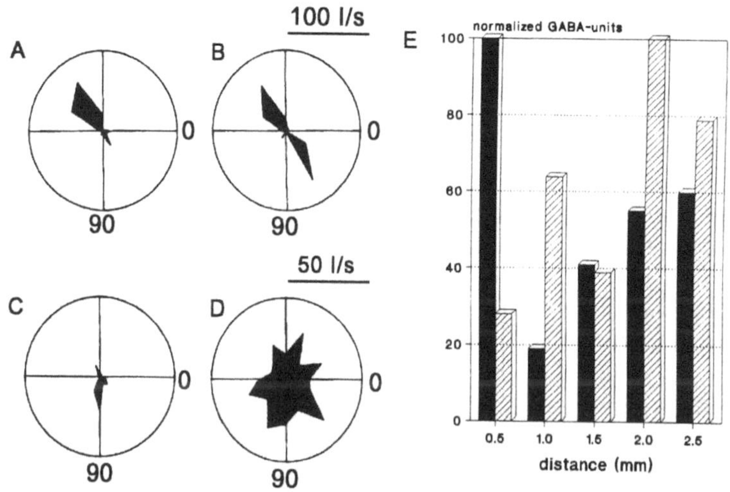

Fig. 5.6–3. Changes in direction selectivity (**A,B**), orientation tuning (**C,D**), and dependence of effects on distance from recording site (**E**). **A,B** Direction selective cell in cat area 17 loses direction specificity while maintaining orientation tuning. Microiontophoretical application of GABA with 60 nA 700 μm posterior from the recording site. The effect was completely reversible. (Modified from Eysel et al. 1988). **C,D** Highly orientation and direction selective simple cell reacts to GABA inactivation at 4 surrounding sites (each 500 μm away from recording site) with loss of orientation tuning (Eysel et al. 1990). **E** GABA units (product of ejection current and time) needed to elicit losses of direction selectivity (*black bars*) or orientation tuning (*hatched bars*) due to increased responses in nonpreferred directions or nonpreferred orientations, respectively. (Modified from Wörgötter and Eysel 1991)

Specificity. Further experiments performed in area 18 of the cat showed that the effects of inactivation are not related primarily to distance but to the actual specificities observed at the inactivation site. In fact, the effects of inactivation can be predicted if one knows the preferred orientation at the inactivation site: inactivation of sites with cross-orientation preference leads to a loss of orientation specificity at laterally remote cells by increasing responses to previously non-optimal orientations while not necessarily affecting responses to stimulation with optimal orientation in the preferred axis of motion (Crook and Eysel 1992). On the other hand, inactivation of sites with iso-orientation preference changes only directional responses, and when taking directional preferences at inactivation sites into consideration, losses or increases of direction selectivity can be predicted (Crook et al. 1992). For example, a cell can loose its direction selectivity when a remote site (distance 400–700 μm) with opposite direction selectivity is inactivated. Such sites seem to send inhibition that suppresses the response opposite to the preferred direction in the target cell.

5.6.5 Discussion

The visual cortex of the cat seems to contain two systems for extended monosynaptic lateral signal processing — an excitatory network originating from pyramidal cells and connecting patches with similar orientations and an inhibitory network of large basket cells connecting to a broad range of orientations.

Our anatomical and physiological findings strongly support the existence and functional significance of specific, laterally directed inhibitory mechanisms that evoke non-iso- as well as iso-orientation effects. While in one recent study using whole-cell patch recordings only iso-orientation inhibition was observed (Ferster and Jagadeesh 1992), another study using the same method very recently has shown both kinds of inhibition (Volgushev et al. 1993). Our morphological analysis combined with functional mapping as well as the statistical treatment of the boutonal distribution in relation to surface organization of orientation specificity demonstrate that large basket cells are good candidates for monosynaptically mediating both observed extremes of inhibitory effects. The involved lateral distances of 200–1000 μm indicate that close to the representation of the area centralis iso- and cross-orientation inhibition of large basket cells can be elicited from inside the classical receptive field. This is consistent with recent findings showing cross-orientation inhibition elicited from inside the excitatory receptive field (DeAngelis et al. 1992).

Computations performed with ringlike structures of lateral inhibition (circular inhibition) have shown a sharpening of orientation tuning when the ring of inhibitory cells have the diameter of half a hypercolumn (Wörgötter and Koch, 1991; Wörgötter et al. 1991). The distribution of boutons described here for layer III basket cells has a plateau centered around the distance of half a hypercolumn. Taking into account the predictions of the computer model, sharpening of orientation tuning seems indeed to be one possible functional role of the resulting lateral inhibition.

Basket cells are connected to cells with a variety of different preferred orientations, and the actual inhibitory effect exerted depends on the orientation of a stimulus, on the preferred orientation of basket and target cells, and on the specificity, strength, and distribution of connections. On the basis of the observed connectivity we propose a multifunctional role for the lateral inhibitory system of large basket cells in cat visual cortex: contribution to sharpening of orientation tuning with broad-band inhibition (including cross-orientation) and generation of directional preferences by means of iso-orientation inhibition.

Acknowledgments. The support of the Deutsche Forschungsgemeinschaft (Ey 8/17) and the European Communities (SC1 0329 C) is gratefully acknowledged.

References

Benevento LA, Creutzfeldt OD, Kuhnt U (1972) Significance of intracortical inhibition in the visual cortex. Nature 238: 124–126

Bonhoeffer T, Grinvald A (1991) Iso-orientation domains in cat visual cortex are arranged in pinwheel-like patterns. Nature 353: 429–431

Creutzfeldt OD, Kuhnt U, Benevento LA (1974) An intracellular analysis of visual cortical neurones to moving stimuli: responses in a cooperative neuronal network. Exp Brain Res 21: 251–274

Crook JM, Eysel UT (1992) GABA-induced inactivation of functionally characterized sites in cat visual cortex (area 18): effects on orientation tuning. J Neurosci 12: 1816–1825

Crook JM, Eysel UT, Machemer HF (1991) Influence of GABA-induced remote inactivation on the orientation tuning of cells in area 18 of feline visual cortex: a comparison with area 17. Neurosci 40: 1–12

Crook JM, Kisvárday ZF, Eysel UT (1992) Local Mechanisms of cortical direction selectivity: GABA-inactivation combined with injections of [^{3}H]-nipecotic acid. Soc Neurosci Abstr 18: 1033

DeAngelis GC, Robson JG, Ohzawa I, Freeman RD (1992) Organization of suppression in receptive fields of neurons in cat visual cortex. J Neurophysiol 68: 144–163

Eysel UT (1992) Lateral inhibitory interactions in areas 17 and 18 of visual cortex. In: Mize, Marc, Sillito (eds) GABA in the retina and central visual system. Progress Brain Res 90: 407–422

Eysel UT, Kisvárday ZF (1991) Specificity of lateral inhibitory connections in cat visual cortex. Soc Neurosci Abstr 17: 116.

Eysel UT, Wörgötter F, Pape HC (1987) Local cortical lesions abolish lateral inhibition at direction selective cells in cat visual cortex. Exp Brain Res 68: 606–612

Eysel UT, Muche T, Wörgötter F (1988) Lateral interactions at direction selective striate neurones in the cat demonstrated by local cortical inactivation. J Physiol 399: 657–675

Eysel UT, Crook JM, Machemer HF (1990) GABA-induced remote inactivation reveals cross-orientation inhibition in the cat striate cortex. Exp Brain Res 80: 626–630

Ferster D, Jagadeesh B (1992) EPSP-IPSP interactions in cat visual cortex studied with in vivo whole cell patch recording. J Neurosci 12: 1262–1274

Gabbott PLA, Somogyi P (1986) Quantitative distribution of GABA-immunoreactive neurons in the visual cortex (area 17) of the cat. Exp Brain Res 61: 323–331

Gilbert CD, Wiesel TN (1979) Morphology and intracortical projections of functionally characterized neurones in the cat visual cortex. Nature 280: 120–125

Gilbert CD, Wiesel TN (1983) Clustered intrinsic connections in cat visual cortex. J Neurosci 3: 1116–1133

Gilbert CD, Wiesel TN (1989) Columnar specificity of intrinsic horizontal and corticocortical connections in cat visual cortex. J Neurosci 9: 2432–2442

Kisvárday ZF (1992) GABAergic network of basket cells in the visual cortex. In: Mize, Marc, Sillito (eds) GABA in the retina and central visual system. Progress Brain Res 90: 385–405

Kisvárday ZF, Eysel UT (1992) Cellular organization of reciprocal patchy networks in layer III of cat visual cortex (area 17). Neurosci 46: 275–286

Kisvárday ZF, Eysel UT (1994) Functional and structural topography of horizontal inhibitory connections in cat visual cortex. Europ J Neurosci (in press)

Kisvárday ZF, Beaulieu C, Eysel UT (1993) Network of GABAergic large Basket cells in cat visual cortex (area 18). Implications for lateral disinhibition. J Comp Neurol 327: 398–415

Martin KAC, Whitteridge D (1984) Form, function and intracortical projections of spiny neurones in the striate visual cortex of the cat. J Physiol (Lond) 353: 463–504

Morrone M, Burr DC, Maffei L (1982) Functionals implications of cross-orientation inhibition of cortical visual cells. I. Neurophysiological evidence. Proc R Soc Lond B 216: 335–354

Sillito AM (1975) The contribution of inhinitory mechanisms to the receptive field properties of neurones in the striate cortex of the cat. J Physiol (Lond) 250: 305–329

Sillito AM (1977) Inhibitory processes underlying the directional specificity of simple, complex and hypercomplex cells in the cat's visual cortex. J Physiol (Lond) 271: 699–720
Sillito AM (1979) Inhibitory mechanisms influencing complex cell orientation selectivity and their modification at high resting discharge levels. J Physiol (Lond) 289: 33–53
Sillito AM (1984) Functional considerations of the operation of GABAergic inhibitory processes in the visual cortex. In: Jones EG, Peters A (eds) Cerebral cortex. Functional properties of cortical cells, vol 2. Plenum Press, New York, pp 91–117
Somogyi P (1986) Seven distinct types of GABA-immunoreactive neuron in the visual cortex of cat. Soc Neurosci Abstr 12: 583
Volgushev M, Pei X, Vidyasagar TR, Creutzfeldt OD (1993) Excitation and inhibition in cortical orientation selectivity revealed by whole cell recordings in vivo. Vis Neurosci 10: 1157–1169
Wörgötter F, Eysel UT (1991) Topographical aspects of intracortical excitation and inhibition contributing to orientation specificity in area 17 of the cat visual cortex. Europ J Neurosci 3: 1232–1244
Wörgötter F, Koch C (1991) A detailed model of the primary visual pathway of the cat: comparison of afferent excitatory and intraqcortical inhibitory connection schemes for orientation selectivity. J Neurosci 11: 1959–1979
Wörgötter F, Niebur E, Koch C (1991) Isotropic connections generate functional asymmetrical behavior in visual cortical cells. J Neurophysiol 66: 444–459

Chapter 6

Plasticity - Memory - Attention

6.1 Introduction: Functional Dynamics and Plasticity in the Neocortical Sensory Representation

T.R. VIDYASAGAR
Division of Neuroscience, John Curtin School of Medical Research,
Australian National University, Canberra, Australia

Neurons in the primary visual cortex and the visual association areas are known to respond to certain specific features of visual stimuli. The stimulus attributes of these cortical neurons vary from the simple cell of the striate cortex that codes stimulus orientation at a specific visual locus to neurons coding features such as faces in the superior temporal sulcus and various complex stimulus dimensions in the inferotemporal (IT) area. However, the functions of these neurons cannot be determined by the sensory input alone. There are widespread lateral connections in the cortex (Gilbert and Wiesel 1979) as well as extensive interconnections between different cortical areas (Van Essen et al. 1992). It has been a recurrent theme of Otto Creutzfeldt (1983, 1990a,b) that sensation and its neuronal counterparts have to be understood in the context of the sensory environment, the meaning and significance of the sensory stimulus and the attentional and motivational states of the animal. Thus, response to a sensory stimulus is a dynamic trace that is under a plethora of influences at any one time. Further, the response is also plastic, in that the response characteristics may change and be reorganized with time, especially after damage to one or the other level of the sensory system.

6.1.1 Functional Gating of Sensory Responses

Speaking at the occasion of honoring the memories of Richard Jung, Donald MacKay and Hans-Lukas Teuber, Otto Creutzfeldt (1990a) remarked about the factors that may gate the responses of cortical neurons to sensory stimuli: " ... the corresponding weighting functions need an evaluation system by which the output signals from different parts of the visual cortex are given a certain 'value' or meaning as defined by the purpose of the subject and its actions."

One of the cortical areas where gating of sensory inputs may occur is the posterior parietal cortex, which has long been implicated in attentional processes in both monkeys (Mountcastle et al. 1975; Colby et al. 1993) and humans (Critchley 1953). Mountcastle and co-workers had found that neurons in the area showed activity related to saccades to attended targets. More recently, Goldberg and his group found that when the monkey intends to make a saccade, the visual space as represented by the receptive fields of parietal neurons is remapped to fit the co-ordinates of the next fixation point even before the onset of the saccade. Now, in his presentation, Goldberg explores the neural correlates and sites of voluntary and involuntary attentional mechanisms. As you read these words, it is the voluntary attentional system that is in operation, whereas someone suddenly bursting into the room will cause you to be distracted through the involuntary mechanism. Experiments by Goldberg and colleagues suggest that the posterior parietal cortex may be the final staging area for both mechanisms, with the reciprocal connections between the posterior parietal area and the posterior cingulate area, mediating the involuntary component, and those between the posterior parietal area and the prefrontal cortex, the voluntary component. These findings are consistent with clinical syndromes associated with brain damage. When both mechanisms are affected as in parietal damage, there would be a general failure of contralateral attention, as in classical parietal neglect (Critchley 1953). If the posterior cingulate involuntary system alone is affected, there could be a dominance of the ipsilateral voluntary attentional system leading to neglect of the ipsilateral visual field, as in one patient with a medial retrosplenial lesion (Heilman et al. 1990).

Voluntary attention is also involved in searching for a particular object that is embedded in an array of irrelevant stimuli in the sensory scene. The underlying neuronal mechanism includes both memory and attentional processes. Thus, a stored representation of the object is necessary as well as a way of selecting that object from many others. It has been observed by Moran and Desimone (1985) that in area V4 of the monkey, cell responses to the preferred visual stimulus could be suppressed if the monkey were to attend to a different stimulus within the receptive field of the cell. In his paper, Desimone reports about responses in the inferotemporal cortex that could well be the neural basis of visual search. It appears that responses of some IT cells to the optimum visual stimulus are determined by whether or not the animal was looking for it, i.e., the response is enhanced if it were the cue (sample) in a delayed match-to-sample task and suppressed if it were not. It would be fascinating to unravel the neural inputs and circuitry that underly the above responses in the IT.

What sort of gating occurs further down the stream of sensory processing? Some lesion studies had suggested that the parahippocampal and hippocampal areas, which receive the information from the IT cortex as well as from a host of other neocortical areas, may be involved in memory functions (e.g., Squire 1987). However, Creutzfeldt and co-workers (Creutzfeldt 1990b; Vidyasagar et al. 1991; Salzmann et al. 1993) found that in these regions, responses to sensory stimuli were rarely related to the purely mnemonic elements of the task. They tend to be gated strongly by more complex factors such as motivation and expectation and by the general behavioral significance of the stimuli. Otto Creutzfeldt's abiding interest in the way neurons mediate higher cognitive functions led him to perform a pioneering series of recordings from conscious humans. Along with the neurosurgical team of George Ojemann in Seattle, which has been performing open temporal lobe surgery for treatment of epilepsy, Creutzfeldt recorded single units from the lateral temporal lobe during tests related to language, cognitive tasks and memory. They form one of the first recordings of how higher cognitive functions in the human brain influence the activity of single nerve cells and have thrown up a host of hypotheses (see Chap. 9.3).

6.1.2 Plasticity in Sensory Systems

If a cortical sensory system receives altered sensory inputs over a period of time, plastic changes are likely to occur. These have been documented not only during development (for review see Rauschecker and Marler 1987), but also in adult animals and humans (for review see Kaas 1991). After focal deafferentation, either by peripheral lesions or artificially induced functional scotomas, the areas of the deactivated cortex can be driven by stimuli exciting the topographically neighboring parts of the field. Thus, Gilbert reports that receptive fields within the scotomatous cortex expand severalfold in diameter. His group has found that the reorganizing mechanisms are not precortical, but intrinsic to the cortex, possibly mediated through the plexus of long-range horizontal connections. They have also observed that the reorganization occurs over several time scales and may involve various mechanisms ranging from proliferation of excitatory connections to loss or adaptation of inhibitory connections. Following artificial scotoma, changes may occur within minutes and even seconds, whereas the long-term effects may take months.

In their presentation, Kaas and Florence review the literature regarding the rapid and slower plastic changes that occur in sensory maps after afferent deactivation. Kaas and his colleagues have found substantial reorganization of the topographic map of the visual cortex within hours after peripheral deafferentation. Such reorganization of the cortical representation occurs also in the somatosensory and auditory systems. After deactivation of somatosensory inputs, changes are apparent also in the individual response properties of single neurons (Calford and Tweedale 1991), showing, for example, increase in the size of the receptive fields. While most of these massive reorganizations occur in the cortex, there are also some instances of limited subcortical reorganizations (Garraghty and Kaas 1991; Nicoletis et al. 1993).

In the auditory cortex, the spatial representation does not reflect the topographical localization of the source, but rather cochlear topography and thus the frequency spectrum. Scheich reports that during certain types of classical conditioning involving aversive stimuli, rapid plastic changes in the frequency representation on the gerbil's auditory cortex can be seen. He suggests that these changes are co-operative phenomena occurring in the cortex and that a population effect of the units leads to the shift in the frequency bands of the AAF and AI auditory fields. As in other sensory areas, such reorganizations may involve long-range horizontal connections.

When neurons in one part of a cortical sensory surface are assigned to a different part of the surface, as happens in the reorganizations mentioned above, some questions arise. Do such reassignments lead to misperceptions and errors? Or does the recruitment of a larger cortical surface area to a particular processing circuit lead to improved performance? Indeed, partial deafferentation does lead to misperceptions. Ramachandran et al. (1992) have found perceptual correlates of the massive cortical reorganization by showing that in patients with upper limb amputation, stimuli applied to the face, chest or axilla are referred to the phantom limb. This suggests that cortical areas corresponding to the amputated limb have become responsive to other stimuli. They have also found that such reorganization is modality specific and topographically ordered. These results closely correspond to those of Pons et al. (1991) in the monkey.

As Kaas points out, reduced levels of activity reaching the cortical surface of a sensory map lead to a situation where through horizontal connections, the deactivated area becomes a permissive target for adjacent circuits. On the other hand, normal levels of activity make them prohibitive targets. This principle was implicit in the early experiments of monocular deprivation by Wiesel and Hubel, which demonstrated a dynamic competition between the two eyes for cortical targets during development. It appears that a similar dynamic competition may exist between contiguous areas within a cortical sensory surface of the adult animal.

One can extend the question of plasticity within a modality to plasticity across modalities. Thus, in regions of the brain where inputs from different modalities converge, such as the superior colliculus, primate cortical area 7 or any of the polysensory areas of the temporal lobe, deactivation in one modality may lead to recruitment of the deafferented neurons by other modalities. This has indeed been seen in the superior colliculus (Vidyasagar 1978; Rauschecker and Harris 1983) and parietal area 7 (Hyvärinen 1982), when the deprivation is done during early development. This raises the question of whether the increased representation of one modality is reflected as higher performance or whether there is disruption in a system that is based upon interdependence between modalities. It was seen in guinea-pigs (Withington-Wray et al. 1990) that after visual deprivation, the development of the collicular auditory map was seriously disrupted. However, the intermodal effects after deactivation of a modality in the adult animal are yet to be investigated.

The experiments reported in this section are beginning to give us an understanding of the way sensory responses in the brain are under the dynamic influence of context and the changing demands imposed by injury and other perturbations of the system. Research in this area is entering an exciting phase and, as Otto Creutzfeldt would have put it, has the potential to impart "meaning" to neural mechanisms.

References

Calford MB, Tweedale R (1991) Immediate expansion of receptive fields of neurons in area 3b of macaque monkeys after digit denervation. Somatosens. Motor Res 8: 249–260

Colby CL, Duhamel J-R, Goldberg ME (1993) The analysis of visual space by the lateral intraparietal area of the monkey: the role of extraretinal signals. In: Hicks TP, Molotchnikoff S, Ono T (eds) Progress Brain Res 95: 307–316

Creutzfeldt OD (1983) Cortex cerebri. Springer, Berlin Heidelberg New York

Creutzfeldt OD (1990a) Brain, perception and mind. A tribute to Richard Jung, Donald Kay and Hans-Lukas Teuber. In: Spillmann L, Werner JS (eds) Visual perception: the neurophysiological foundations. Academic Press, San Diesgo, pp 449–463

Creutzfeldt OD (1990b) Neuronal responses in human and monkey temporal lobe during memory tasks: compartmentalization of memory functions. In: Iwai E (ed) Vision, memory and the temporal lobe. Elsevier, New York, pp 187–204

Critchley M (1953) The parietal lobes. Arnold, London

Garraghty PE, Kaas JH (1991) Functional reorganization in adult monkey thalamus after peripheral nerve injury. Neuroreport 2: 747–750

Gilbert CD, Wiesel TD (1979) Morphology and intracortical projections of functionally identified neurons in cat visual cortex. Nature 280: 120–125

Heilman KM, Bowers D, Watson RT, Day A, Valenstein E, Hammond E, Duara R (1990) Frontal hypermetabolism and thalamic hypometabolism in a patient with abnormal orienting and retrosplenial amnesia. Neuropsychologia 28: 161–169

Hyvärinen J (1982) The parietal cortex of monkey and man. Springer, Berlin Heidelberg New York

Kaas JH (1991) Plasticity of sensory and motor maps in adult mammals. Ann Rev Neurosci 14: 137–167

Mountcastle VB, Lynch JC, Georgopoulos A, Sakata H, Acuna C (1975) Posterior parietal association cortex of the monkey: command functions for operations within extrapersonal space. J Neurophysiol 38: 871–908

Moran J, Desimone R (1985) Selective attention gates visual processing in the extrastriate cortex. Science 229: 782–784

Nicoletis MAL, Lin RCS, Woodword DJ, Chapin JK (1993) Peripheral block of ascending cutaneous information induces immediate spatio-temporal changes in thalamic networks. Nature 361: 533–536

Pons TP, Garraghty PE, Ommaya AK, Kaas JH, Taub E, Mishkin M (1991) Massive cortical reorganization after sensory deafferentation in adult macaques. Science 252: 1857–1860

Ramachandran VS, Stewart M, Rogers-Ramachandran DC (1992) Perceptual correlates of massive cortical reorganization. Neuroreport 3: 583–586

Rauschecker JP, Harris LR (1983) Auditory compensation of the effects of visual deprivation in the cat's superior colliculus. Exp Brain Res 50: 69–83

Rauschecker JP, Marler P (1987) Imprinting and cortical plasticity. Wiley, New York

Salzmann E, Vidyasagar TR, Creutzfeldt OD (1993) Functional comparison of neuronal properties in the primate posterior hippocampus and parahippocampus (area TF/TH) during different behavioural paradigms involving memory and selective attention. Behav Brain Res 53: 133–149

Squire LR (1987) Memory and Brain. Oxford University Press, New York

Van Essen DC, Anderson CH, Felleman DJ (1992) Information processing in the primate visual system: an integrated systems perspective. Science 255: 419–423

Vidyasagar TR (1978) Possible plasticity in the rat superior colliculus. Nature 275: 140–141

Vidyasagar TR, Salzmann E, Creutzfeldt OD (1991) Unit activity in the hippocampus and the parahippocampal temporobasal association cortex related to memory and complex behaviour in the awake monkey. Brain Res 544: 269–278

Withington-Wray DJ, Binns KE, Keating MJ (1990) The maturation of the superior collicular map of auditory space in the guinea-pig is disrupted by developmental visual deprivation. Eur J Neurosci 2: 682–692

6.2 Circuitry and Functional Dynamics of Adult Visual Cortex

C.D. GILBERT AND T.N. WIESEL
Department of Neurobiology, The Rockefeller University, New York

Adapted from a review published in Cerebral Cortex (Gilbert 1993)

A fundamental understanding of the mechanisms of cortical processing of sensory information requires an examination of the relationships of cortical circuitry, cortical functional architecture, and response properties of single cells. An overriding aim of visual cortical physiology is to understand the lateral integration of information across visual space, from local line segments to global percepts. This was a recurrent theme of Otto Creutzfeldt's work, in which he examined the response of cells within the context of natural visual scenes. In this review, we present evidence that the substrate of lateral integration may be found in the plexus of horizontal connections seen in visual cortex. These connections not only endow cells with a sensitivity to the context within which features are presented but may also confer upon the visual system an ability to adapt to changes in the sensory environment.

6.2.1 Cortical Circuitry and Projection Patterns

The cortical circuit can be divided into its excitatory components, generated by pyramidal and spiny stellate cells, and its inhibitory components, formed by smooth stellate cells. The excitatory cells form a series of connections running across the cortical layers. The set of connections in cat primary visual cortex is summarized in Fig. 1: Input from the lateral geniculate nucleus arborizes within layer 4 and the upper part of layer 6. Layer 4 spiny stellate cells project up to the superficial layers of cortex, superficial layer pyramidal cells project down to layer 5, layer 5 projects to layer 6, and layer 6 cells project back up to layer 4, closing a loop of intrinsic excitatory connections (Gilbert and Wiesel 1979, 1981). In addition to these spiny neurons there is a population of smooth stellate cells, comprising approximately 20% of cortical neurons, which mediate inhibition within the cortex. At each stage in the cortical circuit the spiny cells are tapped for output to other areas, superficial layer cells projecting to other cortical areas, layer 5 cells projecting to the superior colliculus and layer 6 cells projecting to the lateral geniculate nucleus (Gilbert and Kelly 1975).

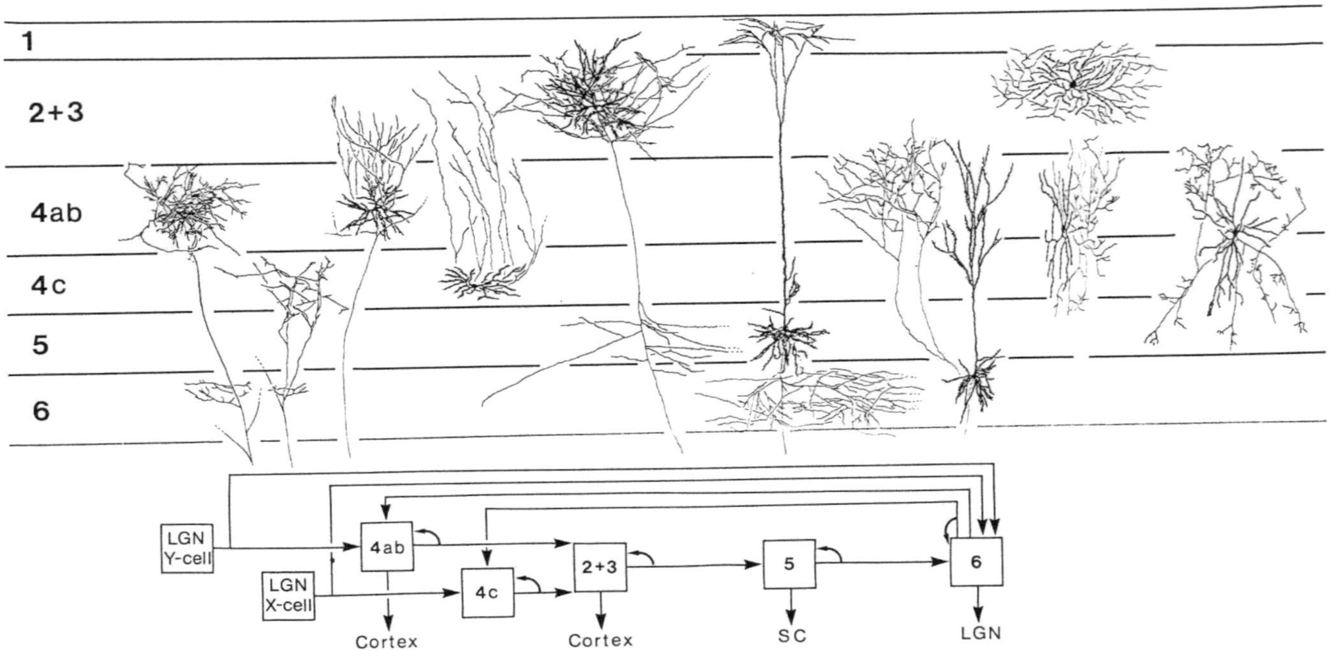

Fig. 6.2–1. Schematic diagram of the intracortical connections of the cat's striate cortex, determined by intracellular injections of horseradish peroxidase (Gilbert and Wiesel 1979). The spiny stellate and pyramidal cells are responsible for excitatory connections, and their intracortical and efferent connections are summarized in the block diagram (*below*). Smooth stellate cells, several types shown at *upper right*, mediate inhibitory interactions in the cortex. (Gilbert 1983; reproduced, with permission, from the Annual Review of Neuroscience, Vol. 6; copyright 1983 by Annual Reviews Inc)

The functional properties of cells in any cortical layer are well suited to the functions of the structures to which they project, and they reflect the characteristics of the inputs to the layer in which they reside (Gilbert 1977). Layer 4 cells, as well as cells in the superficial part of layer 6, receive input from the lateral geniculate nucleus and have simple receptive field properties, supporting the idea that the simple cell represents the first stage in visual cortical processing (Hubel and Wiesel 1962). Cells in layer 2+3 have complex receptive fields, with small diameters and end-inhibited, which is appropriate for the higher resolution and analysis of form that higher cortical areas mediate. Cells in layer 5 have larger receptive fields, tend to be directionally selective, and are sensitive to movement of small objects within the receptive field. Since these cells project to the superior colliculus, which helps direct the movement of the eyes to track moving objects, they are likely to provide the colliculus with information about the direction of object movement. Cells in layer 6 form a major feedback projection to the lateral geniculate nucleus, but the role of this projection remains largely unknown. They also project to layer 4 and play an important role in the generation of the property of end-inhibition (Bolz and Gilbert 1986), which together with orientation selectivity plays an important role in the analysis of curvature and shape (Hubel and Wiesel 1962).

6.2.2 Long-Range Horizontal Connections

In any vertical column cells have overlapping receptive fields with similar orientation specificity and eye preference (Hubel and Wiesel 1962). From considerations of functional architecture it seemed reasonable that there would be a systematic series of connections between layers, relating cells with similar functional properties. It was therefore quite surprising to find extensive horizontal connections, travelling parallel to the cortical surface, relating sites representing disparate locations in the visual field. These are the axon collaterals of cortical pyramidal cells (Gilbert and Wiesel 1979, 1983, 1989; Rockland and Lund 1982, 1983; Martin and Whitteridge 1984). An example of one of these cells, with its characteristic horizontally projecting clustered collateral arbor, is shown in Fig. 2.

The scale of the horizontal connections can be understood from electrophysiological studies mapping the size and position of receptive fields in vertical and tangential electrode penetrations. In any one column of the primary visual cortex, cells have overlapping receptive fields. Taking together the receptive field area and the scatter in receptive field position, the receptive fields of all the cells in the column cover a tiny fraction of the visual field. A rough rule of thumb governing topographic order in this area is that there is no overlap in the receptive fields of cells separated by a dis-

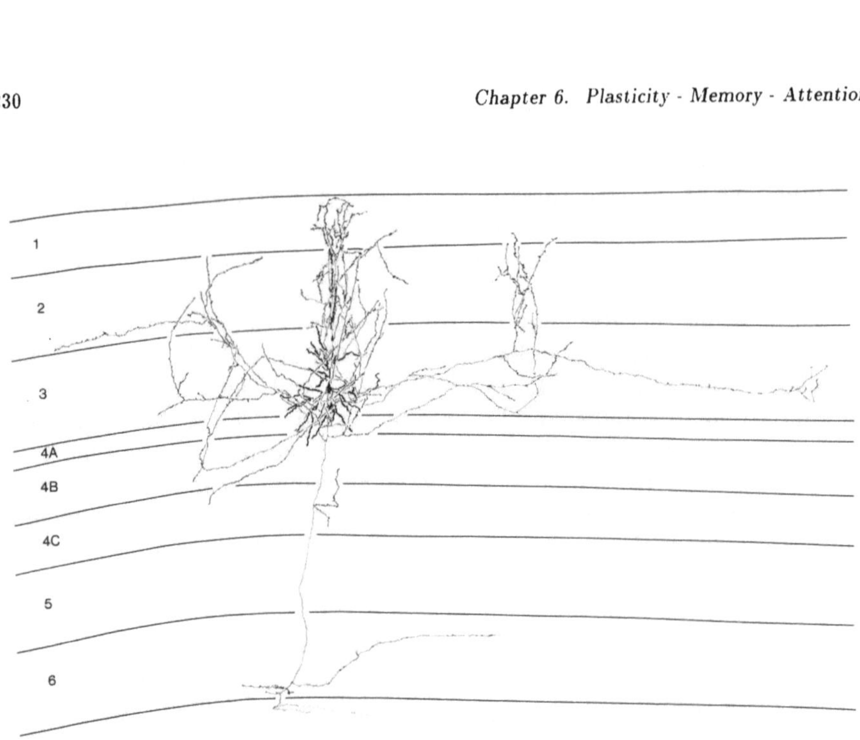

Fig. 6.2–2. Example of a horizontally projecting pyramidal cell in macaque striate cortex. The cell is a layer 3 pyramidal cell whose axon extended for more than 4 mm parallel to the cortical surface. The dendrite is seen as the *thicker, darker lines* in the center, and is studded with dendritic spines. The axon gives off several clusters of collaterals within the superficial layers, as well as projecting out of the immediate cortical area. Scale marker, 100 μm. (Adapted from McGuire et al. 1991; reprinted from Gilbert 1992; copyright Cell Press)

tance of 1.5 mm, taking into account the receptive field size and scatter for cells within a cortical column (a distance corresponding to two complete cycles of orientation columns, or two "hypercolumns;" Hubel and Wiesel 1974). Thus horizontal connections spanning 6–7 mm allow communication between cells with widely separated receptive fields. This represents puzzling evidence for cells' ability to integrate information over a larger part of visual space than that covered by their receptive fields, and calls into question the very definition of receptive field. The explanation for this seeming contradiction between cortical topography and receptive field structure is that the definition of the receptive field is stimulus dependent, and that a cell's response can be modulated by stimuli lying outside the classical receptive field. Put another way, a cell's response to a complex visual stimulus cannot be fully predicted from its response to a simple stimulus, such as a single short line segment.

Although the horizontal connections are very widespread, they are quite specific in terms of the functional properties of the target cells. Rather than contacting all cells within a certain radius, the axon collaterals of the horizontally projecting cells are distributed in discrete clusters. The clustering implies a possible relationship to the functional architecture of the cortex: the tendency of cells with similar functional properties to be grouped into columns of similar functional specificity. In the primary visual cortex, cells with common orientation specificity and ocular dominance are distributed in this fashion. Several lines of evidence show that the clustering of the horizontal axon collaterals allows them to mediate communication between columns of similar orientation preference. The spacing between clusters is roughly the same distance as that required to run through a full cycle of orientation columns (or one hypercolumn for orientation, which is about 750 μm wide). A physiological technique, known as cross-correlation analysis, demonstrates the functional relationship of cells comunicating via the horizontal connections. A cross-correlation is a histogram of differences in the spike times for a pair of cells. Cells that are connected, or that share a common input, show a peak in this histogram at a particular delay. Looking across a population of cells, cross-correlation analysis shows that cells in columns of similar functional specificity have correlated firing even when separated by distances as great as 2 mm (Ts'o et al. 1986; Ts'o and Gilbert 1988).

An anatomical technique, combining retrograde tracing with 2-deoxyglucose autoradiography, confirms the functional specificity of the horizontal connections. Injecting a retrograde tracer, such as rhodamine-filled latex microspheres, reveals the distribution of cells projecting to the injection site. After such an injection, cells are labeled across 8 mm of cortex, which represents an area roughly an order of magnitude larger than the receptive fields of the recipient cells. Within the labeled area the cells are distributed in discrete clusters, reminiscent of the pattern of axon collaterals seen in intracellular injections. When the retrograde tracing technique is combined with the 2-deoxyglucose technique, one can compare the distribution of the labeled cells with the distribution of orientation columns. These experiments show that the horizontal connections run between columns of similar orientation specificity (Gilbert and Wiesel 1989).

The widespread connections seen within each cortical area are also reflected in the highly convergent corticocortical connections between successive stages in the visual pathway. The connections between different cortical areas have a similar distribution to the intrinsic horizontal connections, in that they are widespread, and that cells give rise to clusters of terminals in the subsequent cortical area and, conversely, a site in a given cortical area receives input from clusters of cells in the antecedent cortical area. The picture that emerges is a sequence of convergent pathways mediating integration over progressively larger parts of visual space. Even in primary visual cortex, however, because of the visuotopic order of this area, the convergence of input from large areas of cortex and hence from relatively large parts of the visual field enables cells' stimulus specificities to be influenced by the context within which a local feature is presented.

6.2.3 Contextual Influences on Receptive Field Properties

Given the picture of the horizontal plexus, its distribution, relationship to functional architecture, and synaptic physiology, one is left with the question of their functional role in influencing receptive field properties and the perceptual consequences of these influences. An increasing body of evidence now shows that the size and orientation specificity of receptive fields can be changed by context and the presence of contours or textures outside the receptive field. These changes can be induced in various ways. The concurrent presence of the surrounding contours can alter the functional specificity (or "filter characteristics") of a cell, indicating a differential response between simple and complex stimuli. Even more surprisingly, stimulation in the receptive field surround can induce lasting changes in the responsiveness of the cell to the original simple stimulus, even after the surrounding pattern is removed. These effects may be related to observations of visual psychophysics, which have long established that the perceived attributes of a local feature can be influenced by the context in which that feature is presented.

Our first observation in this regard was made in the domain of orientation. When a line is presented within the receptive field in isolation, one can determine the orientation specificity of the cell by measuring its response to lines of different orientations, deriving an orientation tuning curve. This tuning curve can be altered when lines are presented in the surround, and the nature of the influence depends on the relationship between the orientation of the surround lines and the optimum orientation of the cell. With certain surround orientations the tuning curve may be either inhibited or potentiated, and with others one sees other effects including shifts in optimum orientation and changes in the bandwidth of the tuning curve. Some of the changes can be quantitatively compared to measurements of shifts in perceived orientation induced by context, a phenomenon known as the tilt illusion (Gibson and Radner 1937; Westheimer 1990). To make such a comparison one needs a population model in which the estimate of orientation is based on the relative firing levels of a set of cells, each cell tuned to a different orientation. One such model makes a vector representation of the cells; each cell is represented by a vector pointing in the direction of its optimum orientation and whose length represents its level of firing. Since the cells within the ensemble have overlapping tuning curves, any given stimulus activates many cells within the ensemble, and the estimate of orientation can be derived as the vector sum of the entire ensemble. Using such a model, we have found that several of the observed effects of contextual lines, including iso-orientation inhibition, shifts in the tuning curves toward the orientation of the surrounding lines, and changes in the bandwidth of orientation tuning could all account for the perceptual changes (Gilbert and Wiesel 1990). Further experiments are required to ascertain which of these changes most consistently account for the psychophysical measurements.

A second example of the context dependency of receptive field properties is in the domain of receptive field size. Receptive field properties may not only be modulated by the presence of contours in the area surrounding the receptive field, but the surround influences can alter receptive field structure over time. When an area several times the diameter of the receptive field is occluded, and the surrounding area is stimulated with lines or random dots, the receptive fields within the occluded area expand severalfold in diameter. When stimuli are then placed within the receptive field, the field collapses to its original size, and by alternately putting in and removing the mask from the conditioning stimulus, the field can go through several cycles of expansion and shrinkage (Fig. 3, Pettet and Gilbert 1992).

The mechanism of this reorganization is found in processes that are intrinsic to the cortex and are not found at antecedent levels in the visual pathway. One piece of evidence in support of this is the fact that it shows interocular transfer: if the conditioning stimulus, the artificial scotoma, is presented to one eye, one finds expansion of the receptive field in the other eye (Volchan and Gilbert 1992). This suggests that the source of the expansion are cells with binocular receptive fields, or at least ones receiving binocular input. The next step is to ascertain the synaptic mechanism of the change. Evidence of changes in synaptic weight that is correlated with the receptive field expansion comes from cross-correlation analysis. Ordinarily, cells with similar receptive field properties, such as common orientation preference, show correlated firing. A cross-correlogram provides a measure of the strength of connectivity between cells. If one records from a pair of cells with adjacent receptive fields, the strength of correlation, and hence the synaptic weight, increases as the receptive fields expand and the amount of overlap increases (Das and Gilbert 1993). It has been shown in in vitro recordings in cortical slices that the strength of the horizontally evoked synaptic potentials can be strengthened by use (Hirsch and Gilbert 1993), and the supposition is that the receptive field plasticity observed with conditioning with an artificial scotoma may have a corresponding increase in synaptic weights of excitatory connections, perhaps the horizontal connections, in the cortical circuit.

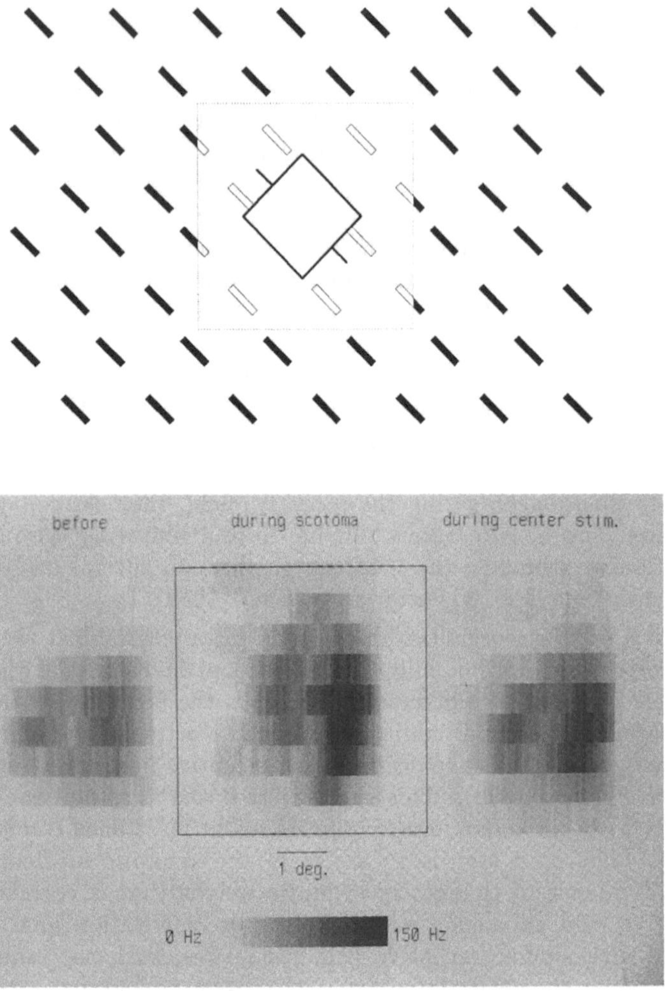

Fig. 6.2–3. Effect of an artificial scotoma on receptive field size. Just as the retinal lesions are capable of producing a considerable expansion in receptive field size within minutes after making the lesion (Gilbert and Wiesel 1992), mimicking the lesion by occluding a small part of the visual field also causes receptive fields located within the occluded area to expand (Pettet and Gilbert 1992). *Above*, the conditioning stimulus: a pattern of lines moving outside the receptive field (*square with a solid outline*, with the orientation specificity indicated by the two lines protruding from the square). The lines disappear when they move within the masked area (the mask is indicated by a *dotted line*, but is not explicitly drawn in the stimulating pattern; the stimulating lines are shown as *blackened rectangles*, and their disappearance indicated by *open rectangles*). After 10 min of conditioning the receptive field expands. *Below* this, is illustrated in a two-dimensional response profile of the cell, with the *darker portions* showing the greater response. The size and position of the occluder is indicated by the *outline in the center*, and the enlarged receptive field is indicated within the outline. Stimulating the center of the receptive field causes it to collapse in size, as indicated in the *lower right*. The receptive field can be caused to alternately expand and contract by a sequence of surround stimulation followed by center stimulation. (Adapted from Pettet and Gilbert 1992, Figs. 1 and 4, reprinted from Gilbert 1992; copyright Cell Press)

Other demonstrations of sensitivity for context include response to illusory contours, which has been shown in primate area 18 (von der Heydt and Peterhans 1989), response of supposedly "monocular" cells in the cortical representation of the optic disc to stimulation of the other eye by lines passing through and extending beyond the optic disc, a form of perceptual completion (Fiorani et al. 1992), and sensitivity to movement outside of the receptive field, altering cells' responses to moving stimuli within the receptive field (Allman et al. 1985; Tanaka et al. 1986; Gulyas et al. 1987).

There is evidence for perceptual correlates of contextual influences and receptive field mutability. It has long been known that one's perception of the attributes of localized features, such as brightness or color and orientation and position, can be influenced by the context within which these features are presented (Westheimer et al. 1976; Westheimer and McKee 1977; Butler and Westheimer 1978; Badcock and Westheimer 1985; Westheimer 1986). Under certain circumstances, context can induce the perception of brightness, color, lines or textures at points in the image where there is no physical stimulus present. These phenomena are referred to as perceptual fill-in or illusory contours (Yarbus 1957; Krauskopf 1961; Kanizsa 1979; Crane and Piantanida 1983; Ramachandran and Gregory 1991; Paradiso and Nakayama 1991). The visual system also tends to integrate along contours, contributing to the salience of a contour composed of a series of similarly oriented line segments (Field et al. 1993). This principle is related to the idea of "good continuation," one of the *Gestalt* laws of perceptual grouping, and may be related to visual segmentation. Which of these observations can be related to the context dependency and mutability of receptive fields seen in striate cortex remains to be established, but at least in the domains of orientation and position there appears to be a strong link.

We studied spatial localization around an artificial scotoma in human psychophysical experiments (Kapadia et al. 1994). In experiments analogous to the physiological studies we asked subjects to determine the position of short line segments in the vicinity of an artificial scotoma. Lines located near the boundary of the scotoma appear to be shifted toward the center of the scotoma. The perceived shift may be related to expansion of receptive fields within the scotoma: if each cell represents a "labeled line" for a particular visuotopic location, for example, the center of their receptive fields, when a group of cells expand their fields, stimuli located near the scotoma boundary cause activation of more cells located near the center of the scotoma than outside the scotoma, resulting in an estimate of visual field position closer to the center of the scotoma. If the psychophysical experiments do reflect the same processes as the physiological observations, they provide a useful measure of the time course of the changes; the positional shifts occur within one second of conditioning with the artificial scotoma.

6.2.4 Long-Term Changes in Cortical Topography

When the visual system receives changes in sensory experience for an extended period of time, even more dramatic alterations in receptive fields and cortical functional architecture can be seen. This is most dramatically shown when one makes lesions of the periphery. Initially, when sensory input is removed by retinal lesions, an area of cortex is left silenced, devoid of functioning input. By placing focal laser lesions in homologous positions in the two eyes, one can remove visual input from a cortical area of a defined extent. Over a period of a few months the silenced area recovers function, and it develops a representation of a different part of the retina than it represented before the lesion. The result is an enlarged representation of the area of retina outside the lesion (Fig. 4, Gilbert and Wiesel 1992). For individual cortical cells the receptive fields shift from positions within the lesioned part of the retina to loci surrounding the lesion. In cortical terms the size of the shift is on the order of 3–4 mm, similar in extent to the horizontal connections. In receptive field terms this of course depends on the retinotopic location of the lesion, but in primates, at about 5° eccentricity, the receptive field shift is on the order of 6° (Gilbert and Wiesel 1992).

Although the magnitude of the shift corresponds closely to the extent of horizontal connections, one must establish whether the reorganization is occurring within the cortex or at antecedent steps in the visual pathway and ascertain which sets of connections are responsible for propagating visual information from outside to within the cortical scotoma. We approached the first question by mapping the scotoma in the lateral geniculate nucleus at a time when the cortical scotoma had filled in. These experiments show that little reorganization occurs in the lateral geniculate nucleus, retaining a large area that cannot be activated by visual stimuli (Gilbert and Wiesel 1992). The size of this area corresponds to the normal representation of the lesioned retina, indicating that most of the reorganization seen in cortex is intrinsic to the cortex (Darian-Smith et al. 1992). Within the cortex, several classes of connections are potentially responsible: thalamic afferents, horizontal intrinsic connections arising from pyramidal cells, and feedback connections from higher cortical areas. The thalamic input is more restricted in its lateral spread than the horizontal connections; thus by itself would be insufficient to account for the degree of reorganization observed. Sprouting of thalamocortical connections could, however, cause the effect. When one maps the extent of the thalamocortical input to cortex by injection of multiple fluorescent retrograde tracers, there appears to be no such in-

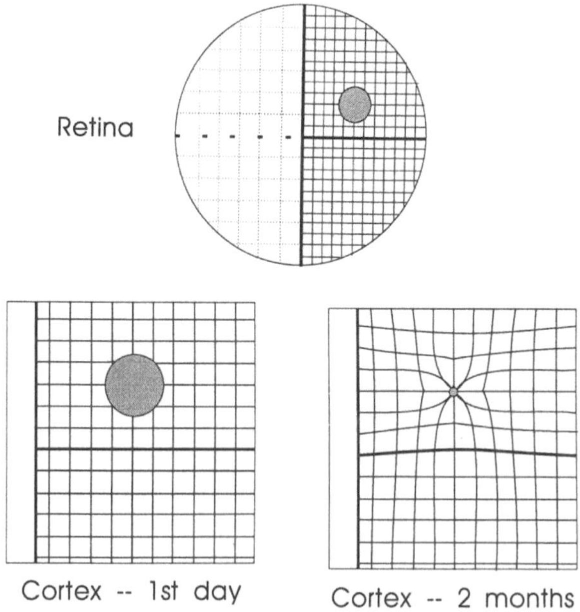

Fig. 6.2–4. Effect of making retinal lesions on cortical topography. The visuotopic axes are represented as the Cartesian grid superimposed on the retina, and the corresponding maps on the cortex in surface view. A lesion, made with a diode laser, destroys the photoreceptor layer in a restricted part of the retina (*shaded area, top*), effectively removing visual input from the cortical region representing that retinal area (*lower left*). To obtain an idea of the dimensions involved, in the primate, a lesion subtending 5° of visual field, centered about 4° in the periphery, silences an area of cortex 10 mm in diameter. Over a period of two months the topography of the cortex is reorganized (*lower right*), with a decreased representation of the silenced input, and an increased representation of the perilesion retina (Gilbert and Wiesel 1990, 1992; Kaas et al. 1990; Heinen and Skavenski 1991; reprinted from Gilbert 1992; copyright Cell Press)

crease in its lateral spread. This leaves us with the horizontal connections and feedback connections, which are normally long enough to account for the observed reorganization.

Even if the horizontal connections do not need to become more widespread to allow for the reorganization, they do appear to become strengthened in order to cause the shift in receptive field position. The strengthening might involve various mechanisms, from potentiation of excitatory connections to adaptation of inhibitory connections. Although potentiation of horizontal connections can occur, this strengthening takes place after a few seconds of conditioning and is not something normally requiring months to bring about. It is plausible that changes of this sort might be involved with the receptive field changes observed with an artificial scotoma, since these occur in minutes. Something else might differentiate the short- from the long-term changes, for example, morphological changes such as synaptic proliferation. At this stage we have preliminary evidence indicating that there might be an increase in the number of axon collaterals in the distant axon collateral clusters of cells projecting from outside the cortical scotoma toward the center of the scotoma (Darian-Smith and Gilbert 1993).

Although the full scale reorganization of cortical topography following retinal lesions takes a few months to occur, one can observe dramatic changes within minutes after making the lesions. Cells whose receptive fields are originally located near the boundary of the scotoma greatly increase their receptive field size, expanding into the part of visual field outside the lesioned area, with topographical shifts on the order of 1–2 mm (Gilbert and Wiesel 1992). Similarly, when a lesion is made in one eye, the cortex is silenced, but small areas of inactivated cortex, about 2 mm in diameter, can recover visual input a few hours after enucleation of the unlesioned eye (Chino et al. 1992). These findings might reflect the fact that under ordinary circumstances cells are capable of integrating information over a large part of visual space. Usually the inputs from outside the classic receptive field serve to modulate the response of the cell, but under the appropriate pattern of stimulation they can be boosted to a suprathreshold level. Removing the input that contributes to the receptive field center, for example, might allow the more peripheral inputs to be potentiated.

In summary, the functional consequences of the horizontal connections can be seen over several time scales. The longest term effects can mediate recovery following lesions of the central nervous system. Shorter term changes, occurring within minutes, either after retinal lesions or with conditioning with an artificial scotoma, reflecting a normal process of adaptation to changes in sensory experience. The functional role of this adaptation may, in primary visual cortex, be a normalization or calibration of the system to analysis of various visual attributes. In higher order cortical areas the same mechanisms may be applied to acquisition and storage of visual percepts. It may be that changes occur on an even shorter time scale, as suggested by the psychophysical experiments. If so, receptive fields may adjust their specificity dynamically, as one assimilates and analyzes each view. Any model of visual processing will have to take into account the fact that receptive field properties are dynamic, subject to the context within which features are presented, and capable of being shaped by visual experience. Importantly, this mutability starts at the earliest cortical stages of visual processing, including primary visual cortex.

References

Allman JM, Miezin F, McGuinnes E (1985) Direction- and velocity-specific responses from beyond the classical receptive field in the middle temporal visual area (MT) Perception 14: 105–126

Badcock DR, Westheimer G (1985) Spatial location and hyperacuity: the centre-surround localization function has two substrates. Vision Res 25: 1259–1269

Bolz J, Gilbert CD (1986) Generation of end-inhibition in the visual cortex via interlaminar connections. Nature 320: 362–364

Butler T, Westheimer G (1978) Interference with stereoscopic acuity: spatial, temporal, and disparity tuning. Vision Res 18: 1387–1392

Chino YM, Kaas JH, Smith III EL, Langston AL, Cheng H (1992) Rapid reorganization of cortical maps in adult cats following restricted deafferentation in retina. Vision Res 32: 789–796

Crane HD, Piantanida TP (1983) On seeing reddish green and yellowish blue. Science 221: 1078–1079

Darian-Smith C, Gilbert CD (1993) Role of leterally projecting intrinsic fibers in topographic reorganization of visual cortex in adult cats. Abst Soc Neurosci 19: 868

Darian-Smith C, Gilbert CD, Wiesel TN (1992) Cortical reorganization following binocular focal retinal lesions in the adult cat and monkey. Abst Soc Neurosci 18: 11

Field, DJ, Hayes A, Hess RF (1993) Contour integration by the human visual system: evidence for a local "association field." Vision Res 3: 173–193

Fiorani Jr M, Rosa MGP, Gattass R, Rocha-Miranda CE (1992) Dynamic surrounds of receptive fields in primate striate cortex: a physiological basis for perceptual completion? PNAS 89: 8547–8551

Gibson JJ, Radner M (1937) Adaptation, after-effect and contrast in the perception of tilted lines. J Exp Psychol 20: 453–467

Gilbert CD (1977) Laminar differences in receptive field properties of cells in cat primary visual cortex. J Physiol 268: 381–421

Gilbert CD (1983) Microcircuitry of the visual cortex. Ann Rev Neurosci 6: 217–247

Gilbert CD (1992) Horizontal integration and cortical dynamics. Neuron 9: 1–13

Gilbert CD (1993) Circuitry, architecture, and functional dynamics of visual cortex. Cerebral Cortex 3: 373–386

Gilbert CD, Kelly JP (1975) The projections of cells in different layers of the cat's visual cortex. J Comp Neur 163: 81–106

Gilbert CD, Wiesel TN (1979) Morphology and intracortical projections of functionally identified neurons in cat visual cortex. Nature 280: 120–125

Gilbert CD, Wiesel TN (1983) Clustered intrinsic connections in cat visual cortex. J Neurosci 3: 1116–1133

Gilbert CD, Wiesel TN (1989) Columnar specificity of intrinsic horizontal and corticocortical connections in cay visual cortex. J Neurosci 9: 2432–2442

Gilbert CD, Wiesel TN (1990) The influence of contextual stimuli on the orientation selectivity of cells in primary visual cortex of the cat. Vision Res 30: 1689–1701

Gilbert CD, Wiesel TN (1992) Receptive field dynamics in adult primary visual cortex. Nature 356: 150–152

Gulyas B, Orban GA, Duysens J, Maes, H (1987) The suppressive influence of moving texture background on responses of cat striate neurons to moving bars. J Neurophysiol 57: 1767–1791

Heinen SJ, Skavenski AA (1991) Recovery of visual responses in foveal V1 neurons following bilateral foveal lesions in adult monkey. Exp Brain Res 83: 670–674

Hirsch J, Gilbert CD (1993) Long-term changes in synaptic strength along specific intrinsic pathways in the cat's visual cortex. J Physiol 461: 247–262

Hubel DH, Wiesel TN (1962) Receptive fields, binocular interaction and functional architecture in the cat's visual cortex. J Physiol 160: 106–154

Hubel DH, Wiesel TN (1974) Uniformity of monkey striate cortex: a parallel relationship between field size, scatter and magnification factor. J Comp Neurol 158: 295–306

Kaas JH, Krubitzer LA, Chino YM, Langston AL, Polley EH, Blair N (1990) Reorganization of retinotopic cortical maps in adult mammals after lesions of the retina. Science 248: 229–231

Kanizsa G (1979) Organization in vision. Essays on gestalt perception. Praeger, New York

Kapadia M, Gilbert CD, Westheimer G (1994) A quantitative measure for short-term cortical plasticity in human vision. J Neurosci (in press)

Krauskopf J (1961) Heterochromatic stabilized images: a classroom demonstration. Am J Pyscho 80: 632–637

Martin KAC, Whitteridge D (1984) Form, function and intracortical projections of spiny neurnes in the striate visual cortex of the cat. J Physiol 353: 463–504
McGuire BA, Gilbert CD, Rivlin PK, Wiesel TN (1991) Targets of horizontal connections in macaque primary visual cortex. J Comp Neur 305: 370–392
Paradiso MA, Nakayama K (1991) Brightness perception and filling-in. Vision Res 31: 1221–1236
Pettet MW, Gilbert CD (1992) Dynamic changes in receptive field size in cat primary visual cortex. PNAS 89: 8366–8370
Ramachandran VS, Gregory TL (1991) Perceptual filling in of artificially induced scotomas in human vision. Nature 350: 699–702
Rockland KS, Lund JS (1982) Widespread periodic intrinsic connections in the tree shew visual cortex. Brain Res 169: 19–40
Rockland KS, Lund JS (1983) Intrinsic laminar lattice connections in primate visual cortex. J Comp Neurol 216: 303–318
Tanaka K, Hikosaka K, Saito H, Yukie M, Fukada Y, Iwai E (1986) Analysis of local and wide-field movements in the superior temporal visual areas of the macaque monkey. J Neurosci 6: 134–144
Ts'o D, Gilbert CD (1988) The organization of chromatic and spatial interactions in the primate striate cortex. J Neurosci 8: 1712–1727
Ts'o D, Gilbert CD, Wiesel TN (1986) Relationships between horizontal and functional architecture in cat striate cortex as revealed by cross-correlation analysis. J Neurosci 6: 1160–1170
Volchan E, Gilbert CD (1992) Cortical mechanisms of receptive field expansion in the cat striate cortex. Abst Soc Neurosci 18: 209
Von der Heydt R, Peterhans E (1989) Mechanisms of contour perception in monkey visual cortex. I. Lines of pattern discontinuity. J Neurosci 9: 1731–1748
Westheimer G (1986) Spatial interaction in the domain of disparity signals in human stereoscopic vision. J Physiol 370: 619–629
Westheimer G (1990) Simultaneous orientation contrast for lines in the human fovea. Vision Res 30: 1913–1921
Westheimer G, McKee SP (1977) Spatial configurations for visual hyperacuity. Vision Res 17: 941–949
Westheimer G, Shimamura K, McKee S (1976) Interference with line orientation sensitivity. J Opt Soc Am 66: 332–338
Yarbus AL (1957) The perception of an image fixed with respect to the retina. Biophysics 2: 683–690

6.3 Plasticity of Sensory Maps in Adult Mammals

J.H. Kaas and S.L. Florence
Department of Psychology, Vanderbilt University Nashville, Tennessee

6.3.1 Introduction

One of the common and productive approaches toward understanding the operations of the nervous system is to consider mechanical and computer analogs and view the brain as hard-wired or fixed in organization once it has been built in development. Of course many investigators (e.g., Mountcastle

1957; Creutzfeldt 1983) have also long recognized that such a view is an oversimplification that neglects the dynamic nature of neuronal function: the moment-to-moment changes in activity reflecting context, alertness, attention, physiological state, motivation, and learning. Such dynamic aspects of brain activity are now studied more commonly as methods have evolved for investigating modulations of the more robust and stable characteristics of the nervous system. However, in addition to these rapid and often highly reversible modulations in the nervous system, there are also slower alterations that require hours to months to emerge. Thus, the orderly representations of sensory surfaces that exist throughout the brain and occupy so much of cortex, are not completely stable over time. Representations can be altered in internal structure so that neurons acquire new receptive fields in patterns that in many ways resemble the changes that have been demonstrated repeatedly for the highly mutable developing nervous system. Such changes were not thought to occur in the adult nervous system, and early reports of adult plasticity were treated with skepticism. Only recently has the documentation of adult plasticity been so extensive that the concept is widely accepted. This acceptance brings us to consider several obvious secondary questions. (a) Where does plasticity occur? Are all brain structures capable of similar levels of change or are some more plastic than others? (b) How does plasticity occur? What are the mechanisms of change? (c) What are the functional consequences of plasticity? How do they relate to sensory and motor skills and performance? The answers to these questions are incomplete, and we are far from a consensus. Nevertheless, it seems productive to review conclusions supported by present evidence as a guide for further research and theorizing.

6.3.2 Rapid Alterations in Sensory Representations

Our major goal in this review is to address issues related to slowly emerging changes in sensory systems. Nevertheless it is important to note that many of the manipulations that produce slow changes also produce rapid modifications. Thus, amputation of body parts, section or deactivation of sensory nerves, retinal lesions, and electrical stimulation of the cortex all produce immediate changes in the location and extent of receptive fields for neurons in sensory representations in cortex (e.g., Metzler and Marks 1979; Calford and Tweedale 1991; Chino et al. 1992; Recanzone et al. 1992b; Nicolelis et al. 1993). These changes may result from the sudden loss of afferent drive of lateral inhibition throughout the sensory systems (Dykes et al. 1984), the potentiation of existing synapses as a result of bursts of activity (Brown et al. 1988, 1990), and the activation of widespread modulating systems (see Dykes 1990). Some of these same mechanisms may mediate rapid changes in receptive fields during experience and learning (e.g., Scheich et al. 1992; Weinberger et al. 1993; Diamond et al. 1993) and the presentation of stimuli outside the classical receptive field (see Allman et al. 1985; Pettet and Gilbert 1992).

6.3.3 Slowly Emerging Changes in Map Structure

Many of the changes that occur in sensory representations as a result of manipulations of inputs emerge slowly over a period of weeks to months. As an example from an early study, Merzenich et al. (1983a,b) cut the median nerve to the hand of monkeys, thereby depriving cortical maps of the body of activating inputs from the glabrous skin of the thumb, index and third fingers and the adjacent pads of the palm. The immediate consequence of cutting the nerve was a large and nearly complete deactivation of the deprived parts of the hand representation in primary somatosensory cortex (area 3b). However, over a period of weeks, the responsiveness of the deprived portions of cortex gradually recovered, and neurons throughout the deprived zone responded to stimuli on innervated parts of the hand, mainly the back of the hand and digits, which are innervated by the radial nerve. Because of the magnitude and the long time course of the change, the results were of special interest. Clearly more than short-term dynamic modulation of receptive field structure was involved, and a host of additional mechanisms, including the growth and formation of new connections, seemed possible.

Such early research on the somatosensory system led to a number of related studies, and it is now clear that partial removals of receptor surfaces are followed by reorganizations of the cortical representations of those surfaces in the visual (Kaas et al. 1990; Gilbert and Wiesel 1992; Chino et al. 1992; Heinen and Skavenski 1991) and auditory (Robertson and Irvine 1989; Schwaber et al. 1993) systems as well. In addition, reorganizations have been demonstrated after nerve section or amputations in primary somatosensory cortex of a number of different mammals, including rats, bats, raccoons, and cats (see Kaas 1991 for review). Thus, plasticity of somatosensory cortex is not a special feature of monkeys. Furthermore, reorganizations of somatosensory maps have been demonstrated in area 1, area 3a, and the second somatosensory area, S-II of monkeys, suggesting that all cortical maps are capable of change (Merzenich et al. 1983a, b; Pons et al. 1988). However, the reorganization of different cortical areas raises the issue of where changes are taking place. Since sensory systems are hierarchical, changes at one level are undoubtedly reflected at higher levels. Additionally, feedback connections may modify lower levels as a result of changes at higher levels.

While, in principle, major reorganizations of sensory maps could be relayed from subcortical structures to cortex, reorganizations of cortical maps are typically thought to occur in cortex, in part because many of the dynamic changes in the nervous system reflect cortical processes. Thus, it came as a surprise to us when we used microelectrodes to map the ventroposterior nucleus of monkeys weeks after median nerve section and discovered that the nucleus was completely reactivated in a manner comparable to that seen in cortex (Garraghty and Kaas 1991b). These experiments indicated that all of the changes previously seen in cortex after median nerve section (Merzenich et al. 1983a) could be a result of changes at the level of the thalamus or even below. Our current hypothesis, in fact, is that the reorganization takes place in the cuneate nucleus of the lower medulla rather than the thalamus. This hypothesis is not based on an unconventional supposition that subcortical structures are more plastic than cortical areas but rather on some unexpected observations on the anatomical organization of the cuneate nucleus. The cuneate nucleus, as other parts of the trigeminal-dorsal column nuclear complex, is characterized by a modular structure with clusters of cells separated by surrounding bands of fibers. When we injected tracers into the skin of different parts of the hand of monkeys, it became apparent that different clusters receive inputs from different digits and pads of the palm. However, the same clusters that receive inputs from glabrous skin receive inputs from the hairy skin of the back of the hand (Florence et al. 1991). Thus, small-scale local changes in synaptic effectiveness of inputs at the level of the cuneate nucleus could produce extensive somatotopic reorganizations at the levels of the thalamus and cortex. However, whether the reorganization occurs at the level of the cuneate nucleus or not, we now feel that it is important to consider the differences in the organization of sensory maps at different levels, since maps differ in the way in which they represent sensory surfaces. Thus, local changes in some maps, when relayed to other maps, could give the misleading appearance of major, widespread, reorganization.

Other experiments give the impression that subcortical plasticity is rather limited in extent, and that even primary cortical areas may not fully reactivate after partial deactivations. Somatotopic reorganizations at spinal cord, dorsal column nuclei, and thalamic levels of the somatosensory system have been described (see Snow and Wilson 1991), but the reported changes have either involved short distances in subcortical tissue, $200\,\mu$m or so (e.g., Wall and Egger 1971), or distances that are inconsistent with the results of more recent studies showing only limited modification (see Wilson and Snow 1987). Moreover, Eysel and coworkers (Eysel et al. 1981; Eysel 1982) have shown that there is only very limited reactivation (less than $200\,\mu$m) of parts of the lateral geniculate nucleus of cats deactivated by photocoagulation lesions of the retina. Thus, it appears that when large portions of subcortical stations in the somatosensory and visual systems are totally deactivated, only the margins of the deactivated zones recover responsiveness.

The reported magnitudes of reorganizations in primary somatosensory areas after partial deprivations are quite variable for uncertain reasons. The large recoveries noted after median nerve section (Merzenich et al. 1983a), and even median nerve plus ulnar nerve section (Garraghty and Kaas 1991a), can be accounted for by partial overlap of peripheral nerve inputs from the back of the hand with those from the front of the hand in the cuneate nucleus. However, this explanation cannot account for reorganization after amputations or from radial nerve section combined with ulnar or median nerve section, both of which remove inputs from both the back and front of the hand. Amputation of a single digit in monkeys is followed by recovery of responsiveness of the deprived part of area 3b, with the skin of the adjacent intact digits providing the driving input (Merzenich et al. 1984). However, removal of two or more digits is followed by only partial reactivation, with some unresponsive cortex remaining. Likewise, section of the radial and ulnar nerves is followed by only limited reactivation of the deprived zone of area 3b, with a large region of cortex persistently unresponsive (Garraghty et al. 1992). Under other conditions, however, more extensive reorganizations of area 3b of monkeys can occur. We have studied monkeys with accidental loss of a hand, or even forearm (Florence et al. 1993b). In some of these monkeys, considerable recovery occurs. For example, in a monkey with a long-standing amputation of the hand, the cortex related to the hand in area 3b was almost fully responsive to stimuli on the wrist and forearm (Fig. 1). Only a small portion of the deprived cortex remained unresponsive. An even more dramatic reactivation of cortex was observed in monkeys with a total lack of sensory input from the forearm as a result of long-standing section of these inputs at the level of the dorsal roots (Pons et al. 1991). In these monkeys all cortex normally devoted to the forelimb was responsive to the face. Clearly the amount of reactivation in area 3b of monkeys can be variable, and there is no simple relationship between the amounts of deprived cortex and the amount of cortex reactivated. Nevertheless, greater recoveries appear to be related to more severe deprivations and to longer recovery periods.

Two additional observations may be relevant for understanding mechanisms of change. First, some of the changes are highly reversible. In experiments on the long-term effects of nerve section, the cut nerves are ligated to prevent regeneration. Cut nerves otherwise regenerate, although incompletely and with many errors so that original skin targets are often missed. When the cut median nerve in monkeys was allowed to regenerate (Wall et

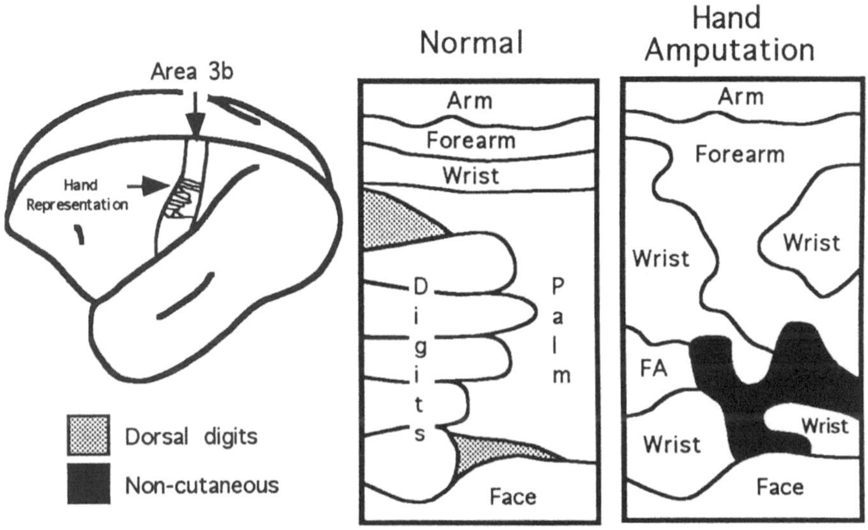

Fig. 6.3–1. Demonstration of the organization of primary somatosensory cortex (area 3b) in normal adult owl monkeys and after long-standing amputation of the hand. In normal monkeys the hand representation is large relative to the total extent of area 3b and is occupied predominantly by an orderly representation of each of the digits. After amputation at the level of the wrist the majority of the deprived zone in cortex becomes activated by inputs from the wrist and forearm. The relative location of area 3b on the dorsolateral brain surface is shown to the left. *FA*, Forearm. (Based on Florence et al. 1993b)

al. 1986), the regenerated inputs recaptured only part of the original median nerve territory in area 3b, and the map of the regenerated input was somatotopically disordered. However, when the median nerve was crushed and allowed to regenerate (Wall et al. 1983), the somatotopic maps of the skin subserved by median nerve were completely restored after regeneration, including details of the map specific to individual monkeys (maps were obtained before nerve crush and after regeneration in the same monkeys). Crushed nerves, unlike cut nerves, regenerate with great accuracy. These experiments suggest that the original connectional framework, the framework that emerged during development, is prepotent over the newly acquired connectional framework that emerges during cortical reorganization. Furthermore, at least in primary cortex, there is little or no correction for the somatotopic disorder produced by disordered nerve regeneration.

Presently the generality of conclusions about the reversibility of map changes is uncertain because we have information only on the reversibility of slowly developing reorganization from one type of cortical change, one that is mediated largely or completely at subcortical levels. Thus, further studies on the reversibility of cortical map modifications are needed.

A second observation that provides important clues about mechanisms of adult plasticity is based on the evidence for serial processing in somatosensory cortex of monkeys (Pons et al. 1987; Garraghty et al. 1990). Area 3b directly, and indirectly via projections to areas 1 and 2, provides the activating input to the second somatosensory area, S-II, so that a lesion of 3b and adjacent cortex deactivates S-II. However, if only part of the input to S-II is removed by ablating the hand representation in 3b, S-II reorganizes over weeks so that the deprived hand zone of S-II becomes responsive to the foot (Pons et al. 1988). It is not clear why this cortex is reactivated by inputs from the foot rather than other skin such as on the face, but the results provide compelling evidence that reorganizations can be based entirely on changes in cortex, since it is unlikely that the cortical lesion produced subcortical modifications that were then somehow relayed to S-II.

6.3.4 Mechanisms of Change

Currently, there is great uncertainty about how cortical and subcortical maps reorganize, but a brief review of the possibilities may usefully guide future research. Most changes appear to reflect the substitution of more active inputs for less active inputs, perhaps sometimes as a result of new patterns of correlated activity and synapse selection according to Hebbian rules (see Bear et al. 1987; Brown et al. 1990). Immediate and rapid changes in map microstructure can be mediated by (a) the changes in afferent driving that directly alter the balances of excitatory and inhibitory influences in neural networks and (b) the activation of widespread neuromodulatory systems. Amputations, for example, could both remove inputs that drive lateral inhibitions, thereby allowing subthreshold influences from remaining skin to be expressed in deprived neurons, and produce a burst of injury-related activity that would activate widespread neuromodulating systems, thereby potentiating subthreshold excitatory activity so that receptive field fringes are expressed. Presumably, modifications produced by widespread activation would dissipate over minutes to hours. In addition, the new balance of excitatory and inhibitory effects may drift as a result of slow, self-regulatory cellular mechanisms to reduce or increase the magnitude of immediate changes.

Bursts of activity, when correlated with other normally subthreshold influences can result in long-term potentiation of the subthreshold influences, apparently via the regulation of *N*-methyl-D-aspartate (NMDA) receptors, but these effects also presumably dissipate over a period of hours to days (Brown et al. 1988; 1990). Long-term potentiation, as part of a Hebbian protocol for modifying synaptic strength, may account for many of the map changes produced by correlation paradigms that develop over minutes to hours and persist over hours to days (e.g., Diamond et al. 1993).

The time courses of the more slowly developing changes in map structure are compatible with a host of self-regulatory mechanisms related to gene expression and the production of subcellular products. For example, deprived zones of cortex sometimes express lower levels of the inhibitory neurotransmitter, γ-aminobutyric acid (GABA; e.g., Welker et al. 1989; Garraghty et al. 1991; Rausell et al. 1992). Low levels of activity in the deprived cortex could provide a feedback signal that reduces the production of the inhibitory neurotransmitter, thereby allowing previously subthreshold excititory inputs to be expressed. As one might expect, chronically high levels of activity produced by maintained stimulation can increase the expression of GABA (Welker et al. 1992). Reduced activity could also signal the increased production of excitatory neurotransmitters and modulators. In addition, several types of neuronal growth could be promoted by reduced activity, including the formation of synapses, extension of dendrites, and the local extension of axon arbors. Such changes would be difficult to detect, but there is evidence that spinal cord damage induces synaptic turnover in cortex (Ganchrow and Bernstein 1981).

The large-scale changes in map organization may be mediated by even greater amounts of neural growth. It is probably very important to restrict new growth in the mature nervous system in order to maintain proper function, and indeed new growth more clearly characterizes the developing than the adult brain (e.g., McKinley and Kruger 1988; Killackey and Dawson 1989). Thus, there is little or no sprouting of retinogeniculate axon arbors of adult cats into parts of the geniculate deafferented by retinal lesions (Eysel 1982), and intact peripheral nerve axons do not sprout into spinal cord territories vacated by cutting the dorsal roots (e.g., Rodin and Kruger 1984). On the other hand, crushing a peripheral nerve appears to induce a central as well as a peripheral growth state in the regenerating neurons, so that growth occurs into spinal cord territories vacated by dorsal root section or nerve damage (e.g., Woolf et al. 1992) and within the spinal cord territory of the crushed nerve even when no vacant territory exists (Fig. 2; Florence et al. 1993a). These and other related experiments show that long-distance growth of axons can occur in the adult nervous system, and such growth could account for some of the more profound changes in cortical maps. For instance, there are no known preexisting connections that could account for the activation of the entire forearm region of 3b in monkeys by inputs from the face (Pons et al. 1991). Thus, Pons et al. (1991) hypothesized the growth and formation of new connections. The experiments of Florence et al. (1993a) are particularly pertinent to this theory, since axon growth occurred in tissue where inputs had reduced activity, as a result of the nerve crush, but no vacated territory. The growth did not extend into adjacent parts of the spinal cord that retained normal levels of activity. Thus, injury, and perhaps other factors such as reduced activity, may induce axon growth in the central nervous system, and neurons or neural tissue with reduced levels of activity may be permissive targets, while neurons or tissue with normal or increased levels of activity may be prohibitive targets.

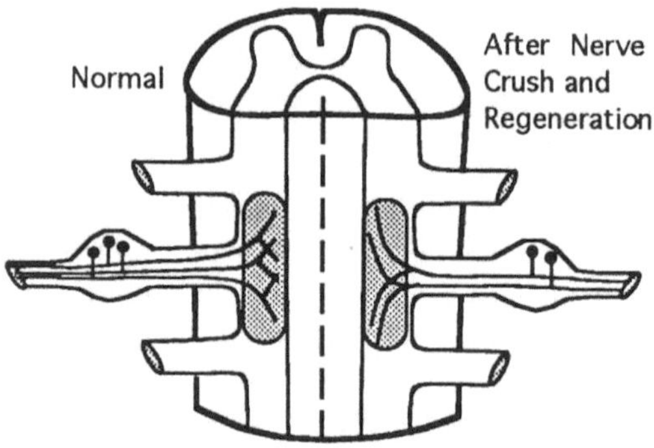

Fig. 6.3–2. Schematic drawing of the changes in peripheral nerve projections to the dorsal horn of the spinal cord of macaque monkeys after median nerve crush and regeneration. In normal monkeys (depicted to the left), axonal projections from the portion of the hand innervated by the median nerve each terminate in a restricted focus of the spinal cord within the median nerve territory in the dorsal horn (indicated by stippling). After median nerve crush and subsequent regeneration (shown to the right) central projections span a rostrocaudally more extensive zone in the dorsal horn, presumably as a result of sprouting of the central axon arborizations. Nonetheless, the median nerve projections do not encroach into regions innervated by other, intact sensory nerves where activity levels are unchanged by the median nerve crush. Rostral is to the top; the dashed line indicates the midline. (Based on Florence et al. 1993a)

6.3.5 What Are the Perceptual and Behavioral Consequences of Map Reorganization?

The reassignment of neurons in cortical maps from one to other parts of the sensory surface could produce either improved abilities or mislocalizations and errors. The argument for improved performance rests on the assumption that adding more neurons to the processing circuit for a sensory input enhances the capacity of that circuit, and there is some evidence to support this view. Most notably, Recanzone et al. (1992a) have presented evidence that monkeys trained to make a tactile discrimination of vibration frequency of a probe on a specific finger both improved over time and ultimately had a larger cortical representation of that finger. In a similar manner deafferentations often produce an expanded cortical representation of adjacent normally innervated skin, and skin next to a denervated zone may demonstrate hypersensitivity or increased tactile acuity (see Wall and Kaas 1985).

On the other hand, there are reasons to suppose that adding neurons to a network may not always improve performance. Performance may be limited by the density of the receptor array, and performance may already be maximal. In addition, the new neurons may simply add noise or error to the circuit, since the outputs to effector circuits may be relatively fixed. In fact, partial deafferentations commonly result in misperceptions in humans (see Wall and Kaas 1985). In recent dramatic demonstrations of injury-induced misperceptions, Ramachandran et al. (1992) reported that patients with forelimb amputations often feel tactile stimuli on the face as being both on the face and on the missing forelimb. This observation is particularly intriguing in that forelimb deafferentations in monkeys can result in reorganized somatosensory cortex so that stimuli on the face activate both the face portion and the forelimb portion of area 3b (Pons et al. 1987). Thus, abnormal activation of forelimb cortex by inputs from the face may not improve tactile abilities for the face but instead may produce misperceptions. While reorganizations may result in either improvements or errors, there are reasons to conclude that brain reorganizations often lead to improvements. Partial lesions of cortical maps in humans and monkeys (e.g., Newsome et al. 1985) commonly result in immediate reductions in performance followed by improvements over days to weeks. In these cases performance is not limited by receptor density but by reductions in the numbers of relevant neurons in the central circuits. Thus, any reassignment of neurons from other circuits to the impaired circuits could lead to enhanced performance.

References

Allman J, Miezin F, McGuinness E (1985) Stimulus specific responses from beyond the classical receptive field: neurophysiological mechanisms for local-global comparisons in visual neurons. Ann Rev Neurosci 8: 407–430

Bear MF, Cooper LN, and Ebner FF (1987) A physiological basis for a theory of synapse modification. Science 237: 42–48

Brown TH, Chapman PF, Kairiss EW, Keenan CL (1988) Long-term synaptic potentiation. Science 243: 724–728

Brown TH, Kairiss EW, Keenan CL (1990) Hebbian synapses: biophysical mechanisms and algorithms. Ann Rev Neurosci 13: 475–511

Calford MB, Tweedale R (1991) Immediate expansion of receptive fields of neurons in area 3b of macaque monkeys after digit denervation. Somatosens Motor Res 8: 249–260

Chino YM, Kaas JH, Smith EL III, Langston AL, Cheng H (1992) Rapid reorganization of cortical maps in adult cats following restricted deafferentation in retina. Vision Res 32: 789–796

Creutzfeldt OD (1983) Cortex cerebri. Leistung, strukturelle und funktionelle Organisation der Hirnrinde. Springer, Berlin Heidelberg New York

Diamond ME, Armstrong-James M, Ebner FF (1993) Experience-dependent plasticity in adult rat barrel cortex. Proc Natl Acad Sci USA 90: 2082–2086

Dykes, RW (1990) Acetylcholine and neuronal plasticity in somatosensory cortex. In: Steriade M, Biesold D (eds) Brain cholinergic systems. Oxford University Press, New York, pp 294–313

Dykes RW, Landry P, Metherate R and Hicks TP (1984) Functional role of GABA in cat primary somatosensory cortex: shaping receptive fields of cortical neurons. J Neurophysiol 52: 1066–1093

Eysel UT (1982) Functional reconnections without new axonal growth in a partially denervated visual relay nucleus. Nature 299: 442–444

Eysel UT, Gonzalez-Aquilar F and Mayer U (1981) Time-dependent decrease in the extent of visual deafferentation in the lateral geniculate nucleus of adult cats with small retinal lesions. Exp Brain Res 41: 256–263

Florence SL, Wall JT and Kaas JH (1991) Central projections from the skin of the hand in squirrel monkeys. J Comp Neurol 311: 563–578

Florence SL, Garraghty PE, Carlson M, Kaas JH (1993a) Sprouting of peripheral nerve axons in the spinal cord of monkeys. Brain Res 601: 343–348

Florence SL, Jain N, Beck PD, Kaas JH (1993b) Reorganization of somatosensory cortex and changes in the peripheral nerve innervation patterns in spinal cord and brain stem after amputation of the hand in monkeys. Soc Neurosci Abstr 19: 1706

Ganchrow D, Bernstein JH (1981) Bouton renewal patterns in rat hindlimb cortex after thoracic dorsal funicular lesions. J Neurosci Res 6: 525–37

Garraghty PE, Kaas JH (1991a) Large-scale functional reorganization in adult monkey cortex after peripheral nerve injury. Proc Natl Acad Sci USA 88: 6976–6980

Garraghty PE, Kaas JH (1991b) Functional reorganization in adult monkey thalamus after peripheral nerve injury. Neuroreport 2: 747–750

Garraghty PE, Pons TP, Kaas JH (1990) Ablations of areas 3b (SI proper) and 3a of somatosensory cortex in marmosets deactivate the second and parietal ventral somatosensory areas. Somatosens Motor Res 7: 125–135

Garraghty PE, Lachica EA, Kaas JH (1991) Injury-induced reorganization of somatosensory cortex is accompanied by reductions in GABA staining. Somatosens Motor Res 8: 347–354

Garraghty PE, Hanes DP, Florence SL, Kaas JH (1992) The extent of cortical reorganization after nerve injury is limited by the content of the deprivation. Soc Neurosci Abstr 18: 1548

Gilbert CD, Wiesel TN (1992) Receptive field dynamics in adult primary visual cortex. Nature 356: 150–152

Heinen SJ, Skavenski AA (1991) Recovery of visual responses in foveal V1 neurons following bilateral foveal lesions in adult monkey. Exp Brain Res 83: 670–674

Kaas JH (1991) Plasticity of sensory and motor maps in adult mammals. Ann Rev Neurosci 14: 137–167

Kaas JH, Krubitzer LA, Chino YM, Langston AL, Polley EH, Blair N (1990) Reorganization of retinotopic cortical maps in adult mammals after lesions of the retina. Science 248: 229–231

Killackey HP, Dawson DR (1989) Expansion of the central hindpaw representation following fetal forelimb removal in the rat. Eur J Neurosci 1: 210–221

McKinley PA, Kruger L (1988) Nonoverlapping thalamocortical connections to normal and deprived primary somatosensory cortex for similar forelimb receptive fields in chronic spinal cuts. Somatosensory Res 5: 311–323

Merzenich MM, Kaas JH, Wall J, Nelson RJ, Sur M, and Felleman D (1983a) Topographic reorganization of somatosensory cortical areas 3b and 1 in adult monkeys following restricted deafferentation. Neurosci 8: 33–55

Merzenich MM, Kaas JH, Wall JT, Sur M, Nelson RJ and Felleman DJ (1983b) Progression of change following median nerve section in the cortical representation of the hand in areas 3b and 1 in adult owl and squirrel monkeys. Neurosci 10: 639–665

Merzenich MM, Nelson RJ, Stryker MP, Cynader MS, Schoppman A, and Zook JM (1984) Somatosensory cortical map changes following digit amputation in adult monkeys. J Comp Neurol 224: 591–605

Metzler J, Marks PS (1979) Functional changes in cat somatic sensory-motor cortex during short-term reversible epidural blocks. Brain Res 177: 379–383

Mountcastle VB (1957) Modality and topographic properties of single neurons of cat's somatic sensory cortex. J Neurophysiol 20: 408–434

Newsome WT, Wurt RH, Dursteler MR, and Mikami A (1985) Deficits in visual motion processing following ibotenic acid lesions of the middle temporal visual area of the macaque monkey. J Neurosci 5: 825–840

Nicolelis MAL, Lin RCS, Woodword DJ, Chapin JK (1993) Peripheral block of ascending cutaneous information induces immediate spatio-temporal changes in thalamic networks. Nature 361: 533–536

Pettet MW, Gilbert CD (1992) Dynamic changes in receptive-field size in cat primary visual cortex. Proc Natl Acad Sci USA 89: 8366–8370

Pons TP, Garraghty PE, Friedman DP, Mishkin M (1987) Physiological evidence for serial processing in somatosensory cortex. Science 237: 417–420

Pons TP, Garraghty PE, Mishkin M (1988) Lesion-induced plasticity in the second somatosensory cortex of adult macaques. Proc Nat Acad Sci USA 85: 5279–5281
Pons TP, Garraghty PE, Ommaya AK, Kaas JH, Taub E, Mishkin M (1991) Massive cortical reorganization after sensory deafferentation in adult macaques. Science 252: 1857–1860
Ramachandran VS, Stewart M, Rogers-Ramachandran DC (1992) Perceptual correlates of massive cortical reorganization. Neuroreport 3: 583–587
Rausell E, Cusick CG, Taub E, and Jones EG, (1992) Chronic deafferentation in monkeys differentially affects nociceptive and nonnociceptive pathways distinguished by specific calcium-binding proteins and down-regulates γ-aminobutyric acid type A receptors at thalamic levels. Proc Natl Acad Sci USA 89: 2571–2575
Recanzone GH, Merzenich MM, Jenkins WM, Kamil AG, Dinse HR (1992a) Topographic reorganization of the hand representation in cortical area 3b of owl monkeys trained in a frequency-discrimination task. J Neurophysiol 67: 1031–1056
Recanzone GH, Merzenich MM, Dinse HR (1992b) Expansion of the cortical representation of a specific skin field in primary somatosensory cortex by intracortical microstimulation. Cerebral Cortex 2: 181–196
Robertson D, Irvine DRF (1989) Plasticity of frequency organization in auditory cortex of guinea pigs with partial unilateral deafness. J Comp Neurol 282: 456–71
Rodin BE, Kruger L (1984) Absence of intraspinal sprouting in dorsal root axons caudal to a partial spinal hemisection. A horseradish peroxidase transport study. Somatosens Res 2: 171–192
Scheich H, Simonis C, Ohl F, Thomas H, Tillein J (1992) Learning-related plasticity of gerbil auditory cortex: feature maps versus meaning maps. In: Gonzalez-Lima TH, Finheust DT, Scheich H (eds) Advances in metabolic mapping techniques for brain imaging of behavioral and learning functions. NATO ASI Series D 68: 447–474
Snow PJ, Wilson P (1991) Plasticity in the somatosensory system of developing and mature mammals. Progress in sensory physiology, vol II. Springer, Berlin Heidelberg New York, pp 482
Schwaber MK, Garraghty PE and Kaas JH (1993) Neuroplasticity of the adult primate auditory cortex following cochlear hearing loss. Amer J Otology 14: 252–258
Wall JT, Kaas JH (1985) Cortical reorganization and sensory recovery following nerve damage and regeneration. In: Cotman CW (ed) Synaptic plasticity. Guilford Press, New York, p 231–259
Wall JT, Felleman DJ, and Kaas JH (1983) Recovery of normal topography in the somatosensory cortex of monkeys after nerve crush and regeneration. Science 221: 771–73
Wall JT, Kaas JH, Sur M, Nelson RJ, Felleman DJ, Merzenich MM (1986) Functional reorganization in somatosensory cortical areas 3b and 1 of adult monkeys after median nerve repair: possible relationships to sensory recovery in humans. J Neurosci 6: 218–233
Wall PD, Egger MD (1971) Formation of new connections in adult rat brains after partial deafferentation. Nature 232: 542–45
Weinberger NM, David R, Lepun B (1993) Long-term retention of learning-induced receptive-field plasticity in the auditory cortex. Proc Natl Acad Sci USA 90: 2394–2398
Welker E, Soriano E, Van der Loos H (1989) Plasticity in the barrel cortex of the adult mouse: effects of peripheral deprivation on GAD-immunoreactivity. Exp Brain Res 74: 441–452
Welker E, Rau SB, Dorfi Van der Loos J, Melzer P, Van der Loos H (1992) Plasticity in the barrel cortex of the adult mouse: effects of chronic stimulation upon deoxyglucose uptake in the behaving animal. J Neurosci 12: 153–170
Wilson P, Snow PJ (1987) Reorganization of the receptive fields of spinocervical tract neurons following denervation of a single digit in the cat. J Neurophysiol 57: 803–13
Woolf CJ, Shortland P, Coggeshall RE (1992) Peripheral nerve injury triggers central sprouting of myelinated afferents. Nature 355: 75–78

6.4 Mapping of Stimulus Features and Meaning in Gerbil Auditory Cortex

H. SCHEICH[1], C. SIMONIS[2], F. OHL[1], H. THOMAS[2], AND J. TILLEIN[2]
[1]Institute of Neurobiology (IfN), Magdeburg, Germany
[2]Institute of Zoology, Technical University, Darmstadt, Germany

The basic functional organization of gerbil auditory cortex was mapped in parallel with unit recording and with the fluoro-2-deoxyglucose mapping (FDG) technique. Among at least seven subfields in this cortex the primary auditory cortex (AI) and the anterior auditory field (AAF) showed prominent tonotopic organization with parallel dorsoventral iso-frequency contours (electrophysiology) in correspondence to FDG labeled frequency band laminae. Aversive tone conditioning paradigms reshaped response curves of single units and thus produced spatial shifts of tone representation in the tonotopic maps of AI and AAF. This suggests that spectral features as well as meaning of sounds are represented in auditory cortex.

6.4.1 Introduction

The ongoing search for neural mechanisms of learning and memory is directed chiefly toward two goals: (1) the elucidation of elementary processes of plasticity at the synapse and (2) the identification of brain systems which select and control what is to be stored away from the incessant stream of information which reaches the brain. In the first context, the most successful approach so far has been the tracing of chains of events at modifiable synapses, i.e., from electrical to molecular and morphological, especially in the small nervous system of Aplysia (review Abrams and Kandel 1988) and in the mammalian hippocampus (review Gustafsson and Wigström 1988). In the realm of systems analysis, great progress has been made in singling out components of the limbic system, notably hippocampus and amygdala, with their general roles for cognitive and emotional aspects of learning (Mishkin and Petri 1983; Aggleton 1992), but also basic circuits for special learning tasks have been described in the vertebrate brain including sites of prominent synaptic plasticity. These cover, among others, the circuit for nictitating membrane conditioning in rabbits (Thompson et al. 1983) and auditory filial imprinting in chicks (Scheich et al. 1991).

Very little progress has been made in the analysis of what one could call the core of learning and memory processes, namely the architecture of the sensory memory in the brain. It has remained obscure where and how the host of information is stored which is acquired through experience during an individual's life and which represents the account to draw upon for experience-guided behavior. It appears to be a common belief that in the mammalian brain, the prime locus of storages is the neocortex and most speculations focus on so-called association areas as large and flexible entities for the accomodation of learned contextual information (Van Hoesen 1982; Squire 1987).

The findings presented in this paper gain their perspective from the use of the fluoro-2-deoxyglucose mapping (FDG) technique (Gonzalez-Lima and Scheich 1984a,b, 1986; Heil and Scheich 1985; Scheich et al. 1993), a simplification of the original method proposed by Sokoloff et al. (1977). The FDG mapping of functional activity has provided the information not easily obtained with other methods that widespread but defined changes of metabolic activity occur in several fields of gerbil auditory cortex during auditory conditioning.

Therefore, the first inference drawn from our data will be that before the possible involvement of association areas in learning, a primary sensory cortex can already be a site of learning-induced plasticity. In other words, primary auditory cortex seems to be not only a structure for auditory information processing per se, but also for the formation of what would be called a "memory trace".

The other and more specific finding, namely changes of spatial representation of auditory stimuli, has implications for general theories of sensory processing and stimulus recognition. Physical parameters of sensory stimuli (features) are spatially represented in orderly maps and believed to be identified by "reading" these maps. A change of this spatial representation when a stimulus parameter becomes meaningful by learning, as we can demonstrate here, would then conflict with stimulus identification. We shall make some propositions below how this conflict could be resolved. In any event, as a consequence of this finding, it might be reasonable to consider sensory cortical fields not simply as "feature maps", but at the same time as "meaning maps".

6.4.2 Basic Functional Organization of Auditory Cortex: Electrophysiology

As there were no previous studies, the tonotopic organization of gerbil primary auditory cortex (AI) and surrounding fields was analyzed in our laboratory using standard microelectrode mapping techniques (Scheich 1991a, b; Thomas et al. 1993; Scheich et al. 1993). The tonotopic (i.e., the best frequency (BF)-related) organization of AI covers the described hearing range of gerbils (Fig. 1). High BF were represented rostrally and low frequen-

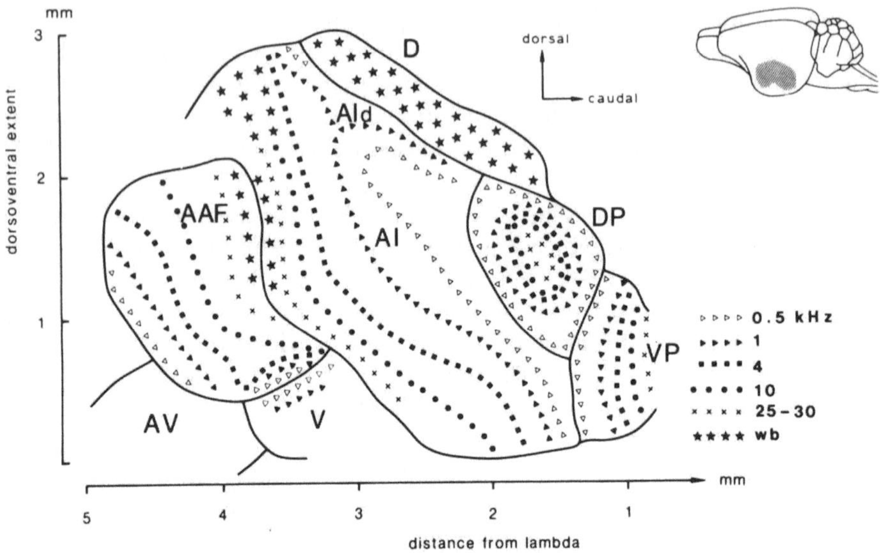

Fig. 6.4–1. Summary map of tonotopic organization of multiple fields in the left auditory cortex of the Mongolian gerbil. Approximate course and distances of iso-frequency contours schematized from data obtained in best frequency (BF) unit mapping eperiments. Areas marked with *asterisks* contain wide-band units responding with little variation to frequencies between 1 and 30 kHz. *AAF*, anterior auditory field; *AV*, anterior ventral field; *AI*, primary auditory cortex; *D*, field D (see text); *AId*, transition area between AI and D; *V*, ventral field; *DP*, dorsoposterior field; *VP*, ventroposterior field. (Modified from Scheich 1991b)

cies caudally in terms of iso-frequency contours with a roughly dorsoventral orientation.

There are additional tonotopic maps adjacent to AI. Rostral to AI a smaller field with a complete tonotopic gradient, but reversed to the one in AI, was mapped (mirror image representation) and termed anterior auditory field (AAF). BF in a range from 0.1 to 43.0 kHz were found in AI and AAF. Representation of lower BF up to 8 kHz had higher spatial resolution. Unit responses obtained in AI as well as in AAF were strong with usually narrow tuning and short latencies.

Caudal to AI, two additional small tonotopically organized fields, a dorsoposterior field (DP) and a ventroposterior field (VP) were identified. In both fields, low BF were located rostrally adjacent to the low-frequency representation in AI. High-frequency representations were found caudally in VP. In DP, BF were concentrically organized, with high BF located in the center. Responses to tone bursts within DP and VP were relatively weak, with longer latencies and broader tuning compared to AI and AAF.

Penetrations dorsocaudal to AI revealed an area of broadly tuned neurons without tonotopic organization, interpreted as a secondary belt region (field D). In the transition area between AI and D, responses were similar to those found in the low frequency range of AI and AAF, but without a straightforward tonotopy (AId). Ventral to the high-frequency border between AAF and AI, a rapid decrease to low BF leads to the assumption of an additional auditory field termed ventral field (V). Responses were similar to those of DP and VP. The location of AI and AAF, the two core fields, appeared to be within koniocortex in Nissl-stained sections, while the remaining fields lay outside.

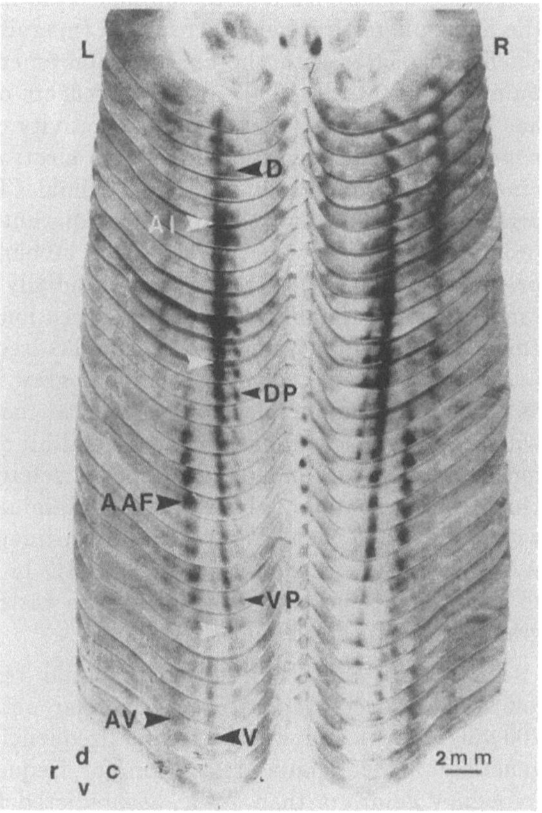

Fig. 6.4–2. Reconstruction of fluoro-2-deoxyglucose (FGD)-labeled auditory cortical fields in the gerbil. Serial horizontal section autoradiographs of left (*L*) and right (*R*) hemispheres after 2 s stimulation with 1 and 2 kHz alternating tones. The 1 and 2 kHz frequency band laminae in each of the different fields are fused to one very dark band as a result of small frequency "jumps" of one octave. The strong labeling facilitates identification of the fields. The spacing of the sections exaggerates the dorsoventral extent of auditory cortex, but allows us to follow the radial extent of labeling through the cortical depth in each section. Note spatial correspondence of the two data sets by comparing left hemisphere of this figure with electrophysiological map in Fig. 1. *D*, dorsal field; *DP*, dorsoposterior field; *AAF*, anterior auditory field; *VP*, ventroposterior field; *AV*, anterior ventral field; *V*, ventral field; *d*, dorsal; *r*, rostral; *v*, ventral; *c*, caudal

6.4.3 Basic Functional Organization of Auditory Cortex: Deoxyglucose Labeling

Gerbil auditory cortex fields were also mapped tonotopically with the FDG mapping technique (Scheich et al. 1993), using a more effective tracer and a simplified protocol of the original quantitative method described by Sokoloff et al. (1977). Narrow-band frequency-modulated (FM) tone bursts and pairs of alternating pure tone bursts (both at about 75 dB sound pressure level, SPL; 2 bursts/s) served as stimuli. Details of the stimulation methods used for gerbils are found in Scheich et al. (1993) and Caird et al. (1991). In short, adult awake gerbils in a small cage were exposed to the acoustic stimuli for 90 min in an anechoic chamber after an i.p. injection of 18 μCi ^{14}C-2-FDG. Serial horizontal sections of the brains were exposed to Kodak NMB X-ray film for two weeks. Topological measurements of labeled frequency band (FB) laminae were made using reconstructions of the hemispheres from enlarged prints (magnification, $\times$ 7.5). FDG uptake chiefly correlates with the energy requirements of the membrane sodium pump (Mata et al. 1980) and thus reflects ongoing electrical activity. It has been shown that peak metabolic activity corresponds to the location of neurons with the corresponding BF in electrophysiological and FDG combination experiments in the same animal (Theurich et al. 1984). The primary field (AI) and the rostrally adjacent smaller field (AAF) showed prominent frequency-specific stripes of labeling with a variable width (approximately 200 μm). Stripes reached radially across the cortical layers and tangentially in roughly dorsoventral direction (Fig. 2). FB laminae in AI and AAF shifted as a function of stimulus frequency relative to each other, i.e., the distance of the two FB laminae grew larger with lower frequencies, as expected from the electrophysiological maps. FB laminae also shifted with respect to a landmark in the lateral hippocampus which was found to form a reliable and independent spatial reference across individual brains. Quantitative spatial analysis of the FB laminae locations (Fig. 3) obtained with different frequencies revealed mirror-imaged tonotopic maps of AI and AAF, as in the electrophysiological study. In AI, the progression from low to high frequencies was from caudal to rostral (Fig. 4) and in AAF the gradient was reversed (not shown).

In the AI map (Fig. 4), the spatial resolution for frequencies below 8 kHz (speech range) showed similar octave intervals (approximately 200 μm/octave) and was larger for frequencies below 1 kHz. AI showed generally higher spatial resolution for frequencies as well as longer isofrequency contours than AAF, as predicted from the electrophysiological map (Figs. 1,2). Differential metabolic FDG responses to the described stimuli allowed us to distinguish, in addition to AI and AAF, a ventral (V), an anterior ventral (AV), a dorsoposterior (DP), a ventral posterior (VP), and a dorsal field (D; Fig. 2), corresponding to the electrophysiological map (Fig. 1).

The FDG patterns provided standard tonotopic maps of AI and AAF averaged across many individuals and, in addition, topological data of all

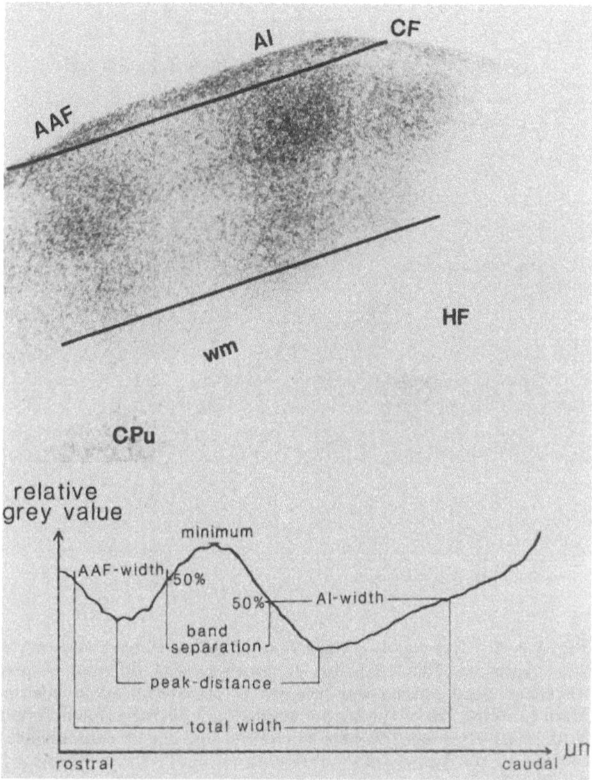

Fig. 6.4–3. Densimetric computer analysis of auditory cortex labeling in one horizontal section. *Top inset*, Autoradiograph with spatial window. Pixels are averaged in vertical columns between the two reference lines, which results in the profile at the *bottom*. Various measures such as values and locations of peak labeling and defined half-widths of frequency band (FB) laminae are indicated. *AAF*, anterior auditory field; *AI*, primary auditory cortex. *CF*, caudal auditory fields; *CPu*, caudate-putamen; *HF*, hippocampal formation; *wm*, white matter of cortex. The anterior tip of hippocampus was used as a spatial reference

fields in conjunction with reliable landmarks of gerbil auditory cortex. Functional FDG labeling cannot replace an electrophysiological approach due to missing temporal resolution. However, the technique allows us to collect in a comprehensive way reliable data sets of representation, i.e., of the geometry of evoked auditory activity patterns in cortical fields. Therefore, the technique also appeared to be suitable for studying learning-induced activity. Prominent learning-induced changes in FDG uptake have previously been shown in the auditory and the limbic system of rats using a classical aversive conditioning paradigm (Gonzalez-Lima and Scheich 1984, 1986a,b). The use of the FDG technique for learning studies appears to

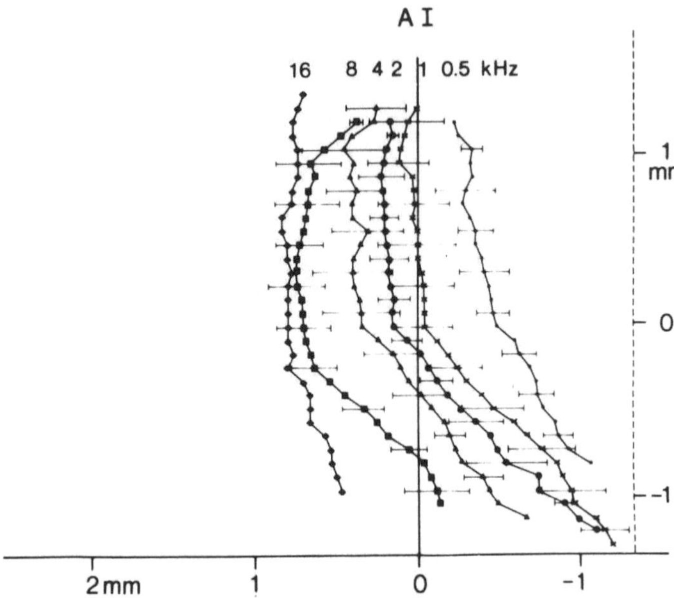

Fig. 6.4–4. Average tonotopic map of gerbil primary auditory cortex (AI) obtained from fluoro-2-deoxyglucose (FDG) labeling experiments with different frequencies. The peak labeling in each frequency band lamina was referred to the rostrocaudal position of a reliable landmark in the brain (anterior tip of the hippocampus) and averaged for different cases and the two hemispheres. Note relatively small standard deviations and the correspondence of the dorsal aspect of the 1 kHz contour to the hippocampus reference ($x = 0$). For comparison with electrophysiological map of AI, see Fig. 1

be favorable in general, as for a number of stereological and energy reasons FDG accumulation occurs predominantly in presynaptic terminals and dendrites, thus reflecting primarily circumsynaptic electrical activity (Heil and Scheich 1986).

6.4.4 Learning-Induced Changes of Fluoro-2-Deoxyglucose Uptake

The following learning studies in gerbil auditory cortex were designed to test whether aversive conditioning changes the frequency representation in fields AI and AAF (Simonis and Scheich, in prep.). Animals in a first experiment were injected with 18 μCi FDG prior to classical conditioning in a small box using a 1 kHz tone as conditioned stimulus (CS) followed by a foot shock as unconditioned stimulus (US). The 1 kHz tone (10 s, 70 dB SPL) was periodically frequency modulated (100 Hz modulation depth, 3 Hz modulation frequency) and was followed by the foot shock upon termination. A total of 100 pairs were given within a period of 45–60 min. Control animals received 100 tones and shocks in pseudorandomized order.

Quantitative topological analysis of labeling in autoradiographs was performed using an image processor (Fig. 3). Densimetric values of pixels were averaged across the radial and dorsoventral extent of FB laminae with a resolution of about 10 μm, which resulted in profiles of labeling with sharp peaks. The peak to peak distance between the labeled FB laminae in the supragranular layers of AI and AAF was significantly larger in the paired animals with respect to unpaired controls (H-test, $p \leq 0.01$). This corresponds to the visual impression of a larger separation of the two bands, as shown in the montage of Fig. 5. Peak shifts were in the range of 100–200 μm. In a tonotopic frame of reference of the two mirror-imaged fields AI and AAF, an increased peak distance of labeling would correspond to a shift of maximum activity towards lower frequencies.

In a second experiment, animals were trained in a shuttle box to avoid the foot shock after a 1 kHz tone. In the following FDG session (recall session) the animals received only occasional foot shocks when they made mistakes. Controls were previously habituated to only the tones in the shuttle box and received no shocks (not shown). This control was considered to correspond to the ideal of a "meaningless tone situation" better than animals being confronted with the tone and the box for the first time. The peak to peak distance between AI and AAF FB laminae was not different in this avoidance-conditioned group compared to the controls. However, the total width of the labeled FB laminae across the tonotopic gradients in AI and in AAF in the avoidance conditioned group was significantly larger than in the controls ($p \leq 0.001$).

The results of these two studies suggest that the populations of neurons which are maximally activated during associative learning and during recall of learned information are partially different. Both populations are also partially different from the population maximally activated by the meaningless tones.

To show that the described effects are frequency specific and not due to some general increase of cortical excitability by any frequency, animals were trained with a differential conditioning paradigm. The two frequencies (CS+ = 1 kHz and CS– = 10 kHz) were presented randomly and the animals had to respond to the CS+ by avoiding a foot shock, whereas the CS– was not paired with a shock. For a second group, the pairing was reversed. A habituated control group was stimulated with the same frequencies without shocks. For both experimental groups after training, we found in a subsequent FDG session a relatively more intense and broader CS+ representation, while the representation of CS– was relatively depressed. Due to limited resolution resulting from the representation of two frequencies

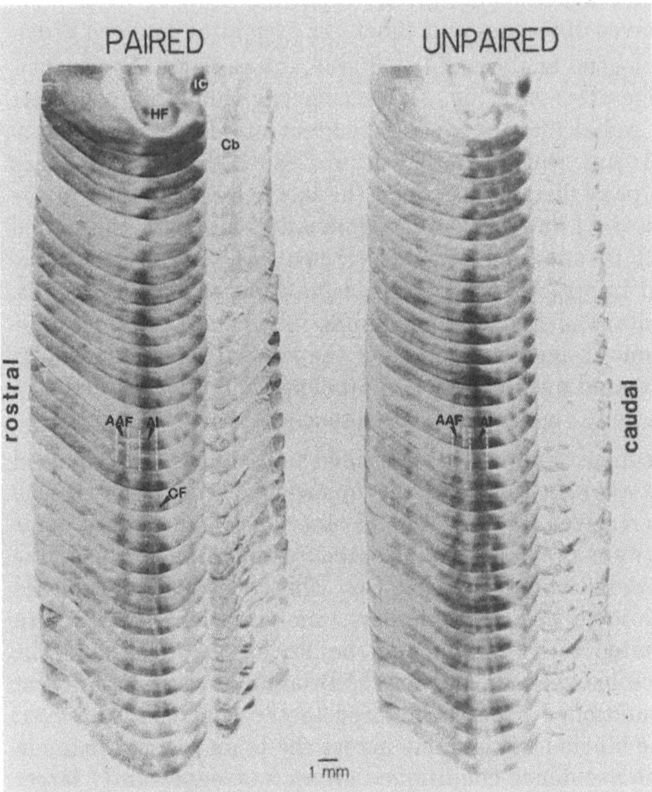

Fig. 6.4–5. Fluoro-2-deoxyglucose (FDG) labeling of gerbil auditory cortex during aversive classical conditioning to 1 kHz tone bursts (100 stimuli in 45 min). In comparison to paired presentation of tones and foot shocks in the conditioned animals, the controls received the same number of the two stimuli during the FDG session, but in an unpaired fashion. The most conspicuous difference between the two groups of animals is the change in the distance between the two labeled 1 kHz frequency band laminae in primary auditory cortex (AI) and anterior auditory field (*AAF*), which is larger in the conditioned cases. Thus, classical conditioning changes tonotopic representation toward lower frequencies. *HF*, hippocampal formation; Cb, cerebellum; CF, caudal auditory fields; IC, inferior colliculus

in each of the two fields (in principle four FB laminae), spatial shifts of labeling were difficult to substantiate.

The results together lead to the following preliminary conclusions: tone conditioning appears to produce primarily spatial changes of activity patterns in fields of auditory cortex. In differential tone conditioning experiments due to an intra-animal comparison of FB laminae of two frequencies, small differential changes of activity levels could also be substantiated. Spatial changes are not simply expansions of the conditioned frequency representation but shifts of peak labeling in one or the other direction across the tonotopic gradient. Thus, spatial representation of frequencies in cortical tonotopic maps does not seem to be rigid, but can be influenced to some degree by learning. The relative spatial shifts (distance between AI and AAF laminae) are large (100–300 μm). Note that octave intervals correspond to approximately 200 μm in the 1 kHz range of field AI (Fig. 3). Thus, if both populations of neurons in AI and AAF contribute to the shift by an equal amount, it corresponds to almost half an octave.

6.4.5 Electrophysiological Changes After Classical Conditioning

A parallel approach to learning was taken in awake, chronically prepared gerbils with single- and multi-unit recordings in AI (chamber technique) before and after classical conditioning (Ohl 1991). Gerbils with painless cranial fixation sat quietly in a housing tube in multiple 3 h recording sessions distributed over some weeks. They tolerated mild electrocutaneous stimulation at the tail as an unconditioned stimulus (US), which produced small tail flicks and bradycardia. The frequency-receptive field (FRF) of each unit was analyzed before conditioning with tone bursts of appropriate frequencies at approximately 70 dB SPL free-field stimulation at the contralateral ear. Then, during further stimulation with these tones, one of the frequencies was paired 50 to 100 times with the US while the others remained unpaired. Subsequently, the FRF was analyzed again.

As a typical non-frequency-specific effect, both spontaneous and evoked spike rate of units usually dropped after conditioning. If the BF of a given unit had been used as a CS frequency, no remarkable influences were seen later with respect to the overall shape of the FRF. Specific and reliable changes of FRF configuration occured with CS frequencies located on the slopes of FRF. It was not the response to the CS nor to the BF, however, which maximally changed, but responses to frequencies adjacent to the CS. These "neighborhood effects" consisted of relative increases of response strength symmetrically or more often asymmetrically around the conditioning frequency. This is shown for three representative units in Fig. 6. A number of units showed more complicated conditioned changes in their FRF involving relatively large response changes to frequencies more distant to the CS frequency. The described neighborhood effect, however, was a consistent feature of the FRF reorganization, independent of the magnitude or direction of the response change to the CS frequency (Fig. 6). About 63% of the tested neurons could be conditioned in the described way.

It seems clear that the changes are not due to individual properties of any given conditioned neuron but rather to a cooperative effect in a larger network, either cortical or subcortical. Firstly, the frequency-specific condi-

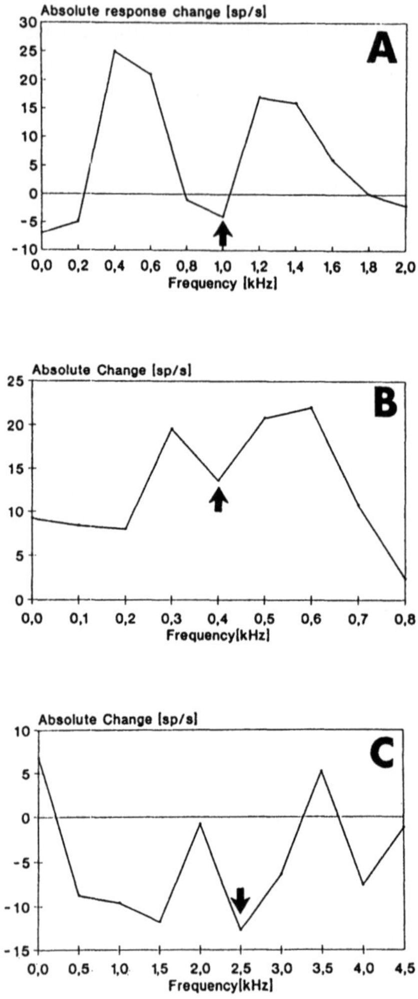

Fig. 6.4–6. A-C Conditioning induced changes of tone evoked iso-intensity frequency responses of three units in the primary auditory cortex (AI). Response changes, in spikes/s, are plotted as function of tone frequency. At 0.0 kHz, the changes of spontaneous activity are included in the graphs. The *arrows* mark the conditioning frequency (CS). As can be seen in all three cases, changes of response strength may occur at all frequencies in the response area and not only to the CS frequency. The changes at the CS+ frequency can either be very small (**A**) or manifest themselves as a significant increase (**B**) or decrease (**C**) of evoked spike rate. In all three cases, however, the overall change of the iso-intensity response is such that responses to CS-adjacent frequencies become enhanced relative to the CS+ frequency

tioning effect does not depend on the individual tuning characteristics of a neuron, but rather on the conditioning frequency. In other words, retuning of a neuron is not in a simple way proportional to the unconditioned frequency responses in the FRF. More specifically, there is no response change or a slight depression of response to the conditioned frequency itself. Thus, the influence on frequencies adjacent to the CS must involve other, presumably neighboring and connected neurons driven by the CS.

The electrophysiological results are suitable to explain, heuristically, the FDG results of conditioning and both together provide some insight in how learning changes cortical representation of sounds. Obviously, frequency conditioning does not alter very much the tuning of those units with a BF corresponding to the CS, i.e., these units appear to remain in the ensemble of conditioned neurons. Instead, tonotopically neighboring units are incorporated into the conditioned ensemble to a varying degree. The incorporation effect would not necessarily result in a simple increase of labeling and widening of a labeled FB lamina but, depending on the asymmetry of the enhancement of response strength around the conditioning frequency, to rather complex spatial shifts of the distribution of labeling. The fine pattern of such redistributions of activity in large neuronal ensembles may depend on the type of conditioning and on context variables, as suggested by the FDG results. It is appealing to speculate that the meaning of sounds is stored by such cooperative effects which shift the weight of activity within and beyond the limits of previously responsive neuronal ensembles.

6.4.6 Discussion

It has been reasonably well established that the energy requirements of the membrane sodium pump are the closest correlate of 2-deoxyglucose (2DG) accumulation in nervous tissue (Mata et al. 1980). Taking this into account, stereological considerations suggest that in brain structures, the largest accumulation of 2DG (or FDG) should theoretically occur in the neuropil fraction and not in the perikarya. This is due to the fact that, for instance, in a unit volume of tissue consisting only of perikarya, the ratio of excitable membrane surface to intracellular fluid content is much smaller than in the same unit volume consisting only of small neuropil processes. The relatively large intracellular compartments of perikarya must lead to a strong dilution of the glucose uptake (here FDG) which is to fuel membrane pumping. To be realistic, these considerations require, in addition, correction factors in any given brain area, because maximum changes in the membrane potential and thus in ion pumping activity are presumably unique for perikarya, dendrites, axon terminals, and synaptic boutons of different neuron populations. Furthermore, the stereological mix of small and large intracellular compartments is also unique for each brain structure. In any event, clearcut cases such as structures which contain a laminar segregation of synaptic neuropil and perikarya, predictably show the weakest FDG accumulation over the perikarya (Heil and Scheich 1986; Gonzalez-Lima and Scheich 1986b). In summary, various observations and inferences suggest that the perisynaptic neuropil, i.e., axon terminals, boutons, and dendrites, contribute the bulk of FDG accumulation in the brain. This, in turn, favors certain functional interpretations and limits others. One is that 2DG or FDG methods are particularly suitable for the study of learning-induced changes in the brain as the elementary mechanisms of learning, i.e., changes in the synaptic strengths are well reflected by these methods. Another is that relatively little information might be derived from the study of local labeling with respect to the postsynaptic spike output of modified synaptic connections. In other words, it is chiefly the input activity and dendritic activity of local neurons which is reflected. In sensory maps of cortex, however, this principle might be compromised. First, within radial cortical columns local activity is always strongly influenced by vertical feedback, which blurs the distinction between input and output activity. Secondly, due to the orderly functional organization in a map, the locus of neuronal activity provides the most important information on stimulus properties. This spatial aspect, therefore, is the link between pre- and postsynaptic activities.

The general purpose of this article is to demonstrate that the FDG mapping technique reveals learning-induced alterations in sensory maps; more specifically, it catches spatial aspects of changed neuronal activity which would be very difficult or impossible to substantiate with electrophysiological mapping techniques. Even with fine-grain single-unit recording, the spatial resolution in cortex is usually not better than 100 μm, and this only in single (one-dimensional) electrode tracks. With FDG autoradiography in each plane of sectioning, a two-dimensional coherent picture with a resolution of 50 μm is easily obtained (Heil and Scheich 1986). With densimetric tools and spatial averaging, as used here, the resolution goes far beyond this limit. Therefore, the method has virtues which, in principle, are beyond the reach of electrophysiological techniques, namely in the analysis of spatial activity changes in neuropil. There is a whole set of complicated questions which concern electrical activity in neuropil, i.e., in the intertwining and largely overlapping fields of dendritic and axon terminal ramifications and their synapses. As one approach to this aspect various attempts have been made to increase the resolution of 2DG methods to a "cellular" level, mainly by reducing diffusion in preparations (Benson et al. 1985; Buchner and Buchner 1982; McCashland and Woolsey 1988a,b). While this type of analysis at the microcompartmental level of neurons has its own justifications, it must be separated from the level of analysis which is of higher relevance here, i.e., for activity in networks. There is a concept, different from that of cellular analysis, behind the interest in neural networks which advocates activity measurements in larger spatial entities. This involves on the one hand the finding of neuronal assemblies, e.g., in visual cortex (Singer et al. 1991), and on the other hand insights obtained from modeling of neural nets (Carpenter and Grossberg 1991). Both suggest that single neuron properties are largely the result of cooperativity in an assembly of interconnected neurons, similar to the cooperative effects illustrated by our electrophysiological results of conditioning in auditory cortex. The implication of this is that the distribution of the weights of connections is the most relevant measure in order to understand cooperativity. The FDG mapping technique essentially allows us to focus on this aspect if several requirements are fulfilled and if one keeps in mind the fact that the method is a *relative* measure of activity in neuropil.

Our results on learning-induced retuning of neurons in gerbils seem to be generally in line with results of different conditioning procedures in other mammalian auditory cortices (Diamond and Weinberger 1989; Edeline et al. 1990; Weinberger and Diamond 1987, 1990). In conjunction with the FDG data, however, single-unit plasticity is put on a more comprehensive scale of observation: large and widespread changes in spatial representation of sounds occur in tonotopic maps which involve most, if not all, of the spatial extent of spectral representation along FB laminae.

It has already been shown that tonotopic organization is not rigid, either during development (King and Moore 1991) or after partial cochlear lesions (Robertson and Irvine 1989). As the cochlear place code still matures during development (Rubel and Ryals 1983) and compensatory inputs to cortical territories are to be expected when the original inputs are inactivated, the above effects appear somewhat less surprising than the similar effects shown here as a result of simple conditioning in a fully mature and organized system.

The shifts of maximum activation across the tonotopic gradient observed with conditioning are in the range of, at least, half an octave and are found along most of the dorsoventral extent of FB laminae (several millimeters). Therefore, very large numbers of neurons must be involved in forming new functional ensembles. How this conflicts with the spectral processing and identification of other sounds remains to be determined. In principle, a strategy of large, but not completely overlapping representations of spectrally similar sounds could be the way cortex solves this problem, if an additional requirement is fulfilled. There should be a relatively wide meshwork of non-plastic and rigidly responding neurons which guarantee tonotopic fidelity and recognition of physical sound properties (features). This condition is properly fulfilled, as in our electrophysiological study 37% of the neurons were not conditionable by the described methods. Many neurons within the meshwork could then be flexibly involved to shape the representation of sounds with different meanings. Thus, auditory (and other sensory) cortical maps may be a compromise between "feature maps" and "meaning maps".

References

Abrams TW, Kandel ER (1988) Is contiguitiy detection in classical conditioning a system or a cellular property? Learning in Aplysia suggests a possible molecular site. Trends Neurosci 11: 128–135

Aggleton JP (1992) (ed) The amygdala. Wiley, New York

Benson TE, Burd GD, Greer CA, Landis DMD, Shepherd GM (1985) High-resolution 2-deoxyglucose auto-radiography in quick-frozen slabs of neonatal rat olfactory bulb. Brain Res 339: 67–78

Buchner E, Buchner S (1982) Functional neuroanatomical mapping in insects by 3H-2-deoxy-D-glucose at electron microscopical resolution. Neurosci Lett 28: 235–240

Caird D, Scheich H, Klinke K (1991) Functional organization of auditory cortical fields in the Mongolian gerbil (Meriones unguiculatus). Binaural 2-deoxyglucose patterns. J Comp Physiol [A] 168: 13–26

Carpenter GA, Grossberg S (1991) (eds) Pattern recognition by self-organizing neural networks. MIT Press, Cambridge, Massachussetts

Diamond DM, Weinberger NM (1989) Role of context in the expression of learning-induced plasticity of single neurons in auditory cortex. Behav Neurosci 103: 471–494

Edeline J-M, Massioui N-E, Dutrieux G (1990) Frequency-specific cellular changes in the auditory system during acquisition and reversal of discrimative conditioning. Psychobiology 18: 382–393

Gonzalez-Lima F, Scheich H (1984) Neural substrates for tone-conditioned bradycardia demonstrated with 2-deoxyglucose. I. Activation of auditory nuclei. Behav Brain Res 14: 213–233

Gonzalez-Lima F, Scheich H (1986a) Neural substrates for tone-conditioned bradycardia demonstrated with 2-deoxyglucose. II. Auditory cortex plasticity. Behav Brain Res 20: 281–293

Gonzalez-Lima F, Scheich H (1986b) Classical conditioning of tone-signaled bradycardia modifies 2-deoxyglucose uptake patterns in cortex, thalamus, habenula, caudate-putamen and hippocampal formation. Brain Res 363: 239–256

Gustafsson B, Wigström H (1988) Physiological mechanisms underlying long term potentiation. Rends Neurosci 11: 156–162

Heil P, Scheich H (1985) Quantitative analysis and two-dimensional reconstruction of the tonotopic organization of the auditory field L in the chick form 2-deoxyglucose data. Exp Brain Res 58: 532–543

Heil P, Scheich H (1986) Effect of unilateral and bilateral cochlea removal on 2-deoxyglucose patterns in the chick auditory system. J Comp Neurol 252: 279–301

King AJ, Moore DR (1991) Plasticity of auditory maps in the brain. Trends Neurosci 14: 31–37

Mata M, Fink DJ, Grainer H, Smith CB, Davidson J, Savaki H, Schwartz WJ, Sokoloff L (1980) Activity dependent energy metabolism in rat pituitary primarily reflects sodium pump activity. J Neurochem 34: 213–215

McCashland JS, Wollsey TA (1988a) New high-resolution 2-deoxyglucose method featuring double labeling and automated data collection. J Comp Neurol 278: 543–554

McCashland JS, Wollsey TA (1988b) High resolution 2-deoxyglucose mapping of functional cortical columns in mouse barrel cortex. J Comp Neurol 278: 555–569

Mishkin M, Petri HL (1983) Memories and habits: Some implications for the analysis of learning and retention. In: Butters N, Squire LR (eds) The neuropsychology of memory. Gulford, New York, pp 287–296

Ohl F (1991) Elektrophysiologische Korrelate plastischer Prozesse im auditorischen Cortex der Mongolischen Wüstenrennmaus (Meriones unguiculatus). Diploma Thesis, Technische Hochschule Darmstadt

Robertson D, Irvine DRF (1989) Plasticity of frequency organization in auditory cortex of guinea pigs with partial unilateral deafness. J Comp Neurol 282: 456–471

Rubel EW, Ryals BM (1983) Development of the place principle: acoustic trauma. Science 219: 512–514

Scheich H (1991a) Representational geometrics of telencephalic auditory maps in birds and mammals. In: Finlay BL, Innocenti G, Scheich H (eds) The neocortex, ontogeny and phylogeny. NATO Series, New York, pp 119–136

Scheich H (1991b) Auditory cortex: comparative aspects of maps and plasticity. Curr Opin Neurobiol 1: 236–247

Scheich H, Wallhäußer-Franke E, Braun K (1991) Does synaptic selection explain auditory imprinting? In: Squire LP, Weinberger NM, Lynch G, McGaugh JL (eds) Memory: organization and locus of change. Oxford University Press, Oxford, pp 114–159

Scheich H, Heil P, Langner G (1993) Functional organization of auditory cortex in the Mongolian gerbil (Meriones unguiculatus) II. Tonotopic 2-deoxyglucose mapping. Eur J Neurosci 5: 898–914
Singer W, Gray CM, Engel AK, König P, Artola A, Bröcker W (1991) Formation of cortical cell assemblies. In: Sejnowski PJ, Stevens CF (eds) Cold Spring Harbour symposium on quantitative biology, vol LV. Cold Spring Harbour Lab Press, Cold Spring Harbour, pp 939–952
Sokoloff L, Reivich H, Kennedy C, Des Rosiers MH, Patlak CS, Pettigrew KD, Sakurada O, Shinohara, N (1977) The (14C)deoxyglucose method for the measurement of local cerebral glucose utilization. Theory, procedure and normal valures in the conscious and anesthetized albino rat. J Neurochem 28: 897–916
Squire LR (1987) Memory and brain. Oxford University Press, Oxford
Theurich M, Müller CM, Scheich H (1984) 2-Deoxyglucose accumulation parallels extracellarly recorded spike activity in the avian auditory neostriatum. Brain Res 322: 157–161
Thomas H, Tillein J, Heil P, Scheich H (1993) Functional organization of auditory cortex in the Mongolian gerbil (Meriones unguiculatus). I. Electrophysiological mapping of frequency representation and distinction of fields. Eur J Neurosci 5: 882–897
Thompson RF, Clark GA, Donegan NH, Lavond DG, Madden J, Mamounas LA, Mank MD, McCormick DA (1983) Neuronal substrates of basic associative learning. In: Butters N, Squire LR (eds) The neuropsychology of memory. Gulford, New York, pp 424–442
Van Hoesen GW (1982) The parahippocampal gyrus. Trends Neurosci 5: 345–350
Weinberger NM, Diamond DM (1987) Physiological plasticity in auditory cortex: rapid induction by learning. Prog Neurobiol 29: 1–55
Weinberger NM, Diamond DM (1990) Retuning of auditory cortex: a preliminary model. Concepts Neurosci 1:91–132

6.5 Cortical Mechanisms for Voluntary and Involuntary Attention: Posterior Cingulate and Lateral Intraparietal Areas in the Monkey

M.E. Goldberg[1], S.Y. Musil[1], C.L. Colby[1], J.-R. Duhamel[1], and C.R. Olson[2]
[1]Laboratory of Sensorimotor Research, National Eye Institute, Bethesda, Maryland
[2]Department of Anatomy, University of Maryland, Baltimore, Maryland

6.5.1 Introduction

In the nineteenth century, William James pointed out that attention can be divided into voluntary and involuntary processes. Voluntary attention enables the subject to select an object from the environment for analysis. The object itself need not be particularly salient. An example of voluntary attention is the processing of the words in this article (James 1878). Involuntary attention allows a previously unexpected, salient object from the environment to grab hold of the analysis mechanism. For example, if the telephone on your desk were suddenly to burst into flames, your attention would be drawn to it despite your attempts to keep reading this article.

Previous studies of the neurophysiology of attention have dealt primarily with voluntary attention (see Colby 1991 for a review). Goldberg and Wurtz discovered that the activity of a visual neuron in the superficial layers of superior colliculus could be enhanced when a monkey made a saccade to a stimulus in the neuron's receptive field. They postulated that this enhancement of activity was the substrate of visual attention and that attentional mechanisms acted like a gain control on the visual activity of single neurons (Goldberg and Wurtz 1972). Mountcastle and his group (Mountcastle et al. 1975; Lynch et al. 1977) showed that the posterior parietal cortex had activity associated with saccadic eye movements to attended targets. This activity was shown also to be an enhancement of the visual response (Robinson et al. 1978) that occurred even when, in the context of a peripheral attention task, eye movements were expressly forbidden (Bushnell et al. 1981). Thus, a brain mechanism is known whose properties are consistent with voluntary attention. Much less work has been done on involuntary attention. For example, its anatomical substrate in the primate is unknown, and neuronal properties consistent with it have not been described in monkeys. In this report, we demonstrate properties of neurons in the posterior cingulate cortex of the monkey which are consistent with a role in involuntary attention. We contrast the properties of the posterior cingulate (CGp) with those of the lateral intraparietal area (LIP), the area of the posterior parietal cortex thought to be most intimately involved with attentional processes (Colby et al. 1993a).

6.5.2 Lateral Intraparietal Area and Voluntary Attention

The posterior parietal cortex can be subdivided into a number areas on the basis of cytoarchitecture, connections, and physiology (Andersen et al. 1985, 1990; Colby et al. 1988, 1993b). An excellent candidate for an attentional area is LIP, which has connections with the frontal eye field and the intermediate layers of the superior colliculus. Neurons in LIP are visually responsive and discharge before visually guided and memory-guided saccades (Gnadt and Andersen 1988; Barash et al. 1991; Goldberg et al. 1990; Colby et al. 1993a).

We studied the activity of neurons in LIP while a monkey performed a number of visuomotor tasks, in order to ascertain the relationship of the neuron's activity to the monkey's behavior. Rhesus monkeys were prepared for chronic neurophysiological recording by the surgical implantation of head-holding devices, recording cylinders, and subconjunctival scleral search coils, using sterile surgical technique and general anesthesia induced by ketamine and maintained by isofluorane inhalation or barbiturate perfusion. Postoperative analgesia (banamine) was provided as necessary. Monkey training and physiological data collection were done using a PDP 11 computer system that has been described previously (Goldberg 1983; Colby et al. 1993b). All animal protocols were approved by the National Eye Institute Animal Care and Use Committee and were within the NIH Guidelines for the Care and Use of Laboratory Animals.

Neurons in LIP discharge in response to visual stimuli (Fig. 1). This is ascertained by studying the response of the neuron while the monkey performs a fixation task (Wurtz 1969). The monkey looks at a spot of light projected on a tangent screen in a dimly lit room. It earns a reward by signaling when the spot of light dims by releasing its hand from a contact rod. A small visual stimulus (a dim red spot 0.4 log units above background) appears in the receptive field, but the monkey does not have to respond to the stimulus in any way. The trials are performed in blocks, so the monkey can rely upon the stimulus' irrelevance. By changing the task, we can make the stimulus relevant: in the peripheral attention task the monkey looks at the same fixation point and the same peripheral stimulus appears, but now the monkey must release its hand from the rod when the peripheral stimulus dims. The monkey may not make a saccade to the stimulus: if its eye leaves a small window around the fixation point the trial is aborted, and the monkey has no possibility of earning a reward (Bushnell et al. 1981). The monkey now must attend to the stimulus, and for neurons in LIP the response to this visual stimulus is enhanced relative to the response of the same neuron to the same stimulus in the fixation task. Again, the trials are run in blocks. The enhancement occurs when the stimulus appears and not necessarily when the monkey responds to the stimulus about a second later (Fig. 1). A similar enhancement occurs when the stimulus is the target in a delayed saccade task. In this task, the monkey looks at the fixation point and the target appears and disappears while it fixates. Subsequently, the fixation point disappears and the monkey now must make a saccade in darkness to the location where the target had been (Hikosaka and Wurtz 1983). If the monkey makes the correct eye movement, the target, now near the fovea, reappears, and the trial continues. The same neuron also bursts before the saccade in the dark (Fig. 1). In order to distinguish between a presaccadic response and a reactivation of the visual response (Fischer and Boch 1985) we studied the activity of the neuron in a learned saccade task (Bruce and Goldberg 1985). In this task, trials requiring visually triggered saccades to a briefly flashed target are pseudorandomly alternated with trials in which the central fixation point disappears but no target is flashed. The monkey is rewarded for making the same saccade that he has been recently making, even though no visual

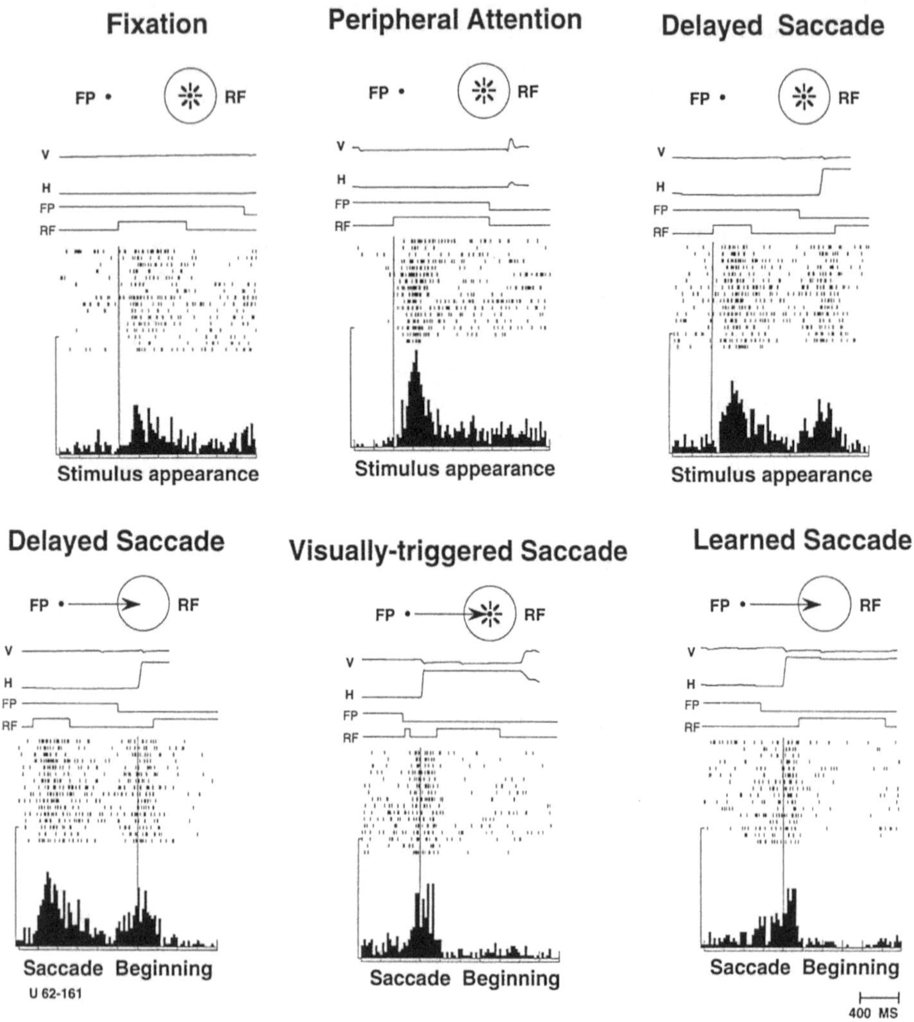

Fig. 6.5–1. Responses of a neuron in the lateral intraparietal area (LIP) in five standard tasks. In this and the following figures, each panel shows a single eye movement trace from one trial *(V*, vertical; *H*, horizontal), the timing of the onset and offset of the fixation point *(FP)* and receptive field *(RF)* stimuli, response rasters from 16 individual trials, and the associated histogram made from these rasters. The *calibration bar* at the left of each histogram indicates 100 spikes/s. The label at the top of each panel indicates the type of trial illustrated, and the label at the bottom indicates the event on which the rasters are aligned. The time of this event is shown by the *continuous vertical line* extending through both the histogram and the rasters. For the delayed saccade task, the same rasters are shown twice, first aligned on the onset of the visual stimulus (*top row*) and then aligned on the beginning of the saccade *(bottom row)*. This cell has an enhanced visual response in both the peripheral attention and the delayed saccade tasks (*top row*). It also discharges at the time of the saccade, whether or not a visual stimulus appeared in the course of the trial (*bottom row*). (From Goldberg et al. 1990)

stimulus appears. The cell gives a weak but clear presaccadic response in this condition. Thus, neurons in LIP discharge in response to visual stimuli, but they also discharge before saccades under circumstances where no visual stimulus has recently appeared. They give enhanced responses to attended stimuli whether those stimuli are the targets of saccades or saccades are expressly forbidden. It is difficult to implicate them in the control of saccades or the analysis of the visual world. However, in all of the above situations, the discharge occurs when the animal shifts its attention to a spatial location. The attention is voluntary and can be described with James' words: the monkey "attends to the object for the sake of some remote interest which the effort will serve."

6.5.3 The Nature of the Spatial Signal in LIP

If LIP neurons signal a shift of voluntary attention to a spatial locus, then the question arises as to how that spatial location is coded. We think of space as being egocentrically coded, but a consistent regular map of supraretinal space has never been described. Gaze-independent visual responses have been described in premotor areas (Rizzolatti et al. 1987) and in somatosensory visual bimodal areas (Colby et al. 1993b), but these are special visual representations having to do with the predominantly somatosensory perio-oral area or specific limb movements. Andersen and Mountcastle (1983) showed that the excitability of visual neurons in the posterior parietal cortex varied as a function of the position in the eye of the orbit. These neurons resemble units in the hidden layer of a network that converts retinal locus and eye position into absolute target location (Zipser and Andersen 1988). Zipser and Andersen suggested that the brain calculates absolute target position in space using a network whose output is so distributed as to render the spatial signal inapparent at the level of a single neuron.

A different strategy for the representation of spatial attention is to describe the coordinates of its displacement, rather than its absolute location. This is equivalent to saying that attended loci are represented relative to the center of gaze. It is well known that a purely retinal representation of visual space is inadequate. For example, the oculomotor system can make spatially accurate saccades when, because of an intervening saccade, there is a dissonance between the retinal location of a target and the amplitude and direction of the saccade necessary to acquire that target (Hallett and Lightstone 1976). Neurons in LIP can compensate for intended eye movements and respond to stimuli that will be brought into their receptive fields by an impending saccade (Fig. 2). This phenomenon is a transient shift in the visual excitability of the neuron. It requires that the impending sac-

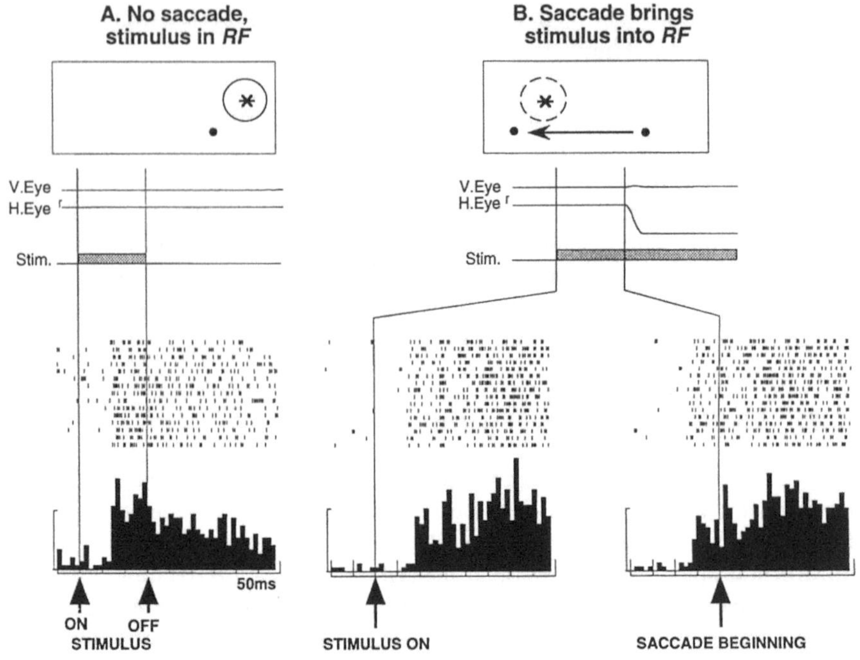

Fig. 6.5–2. Effect of impending saccade on visual responsiveness. Diagrams for each condition show the fixation point (*dot*), visual stimulus (*star*), receptive field (*circle*) and saccade (*arrow*). Rasters and histogram are aligned on the event indicated by the *long vertical line. Tick marks* at bottom indicate 50 ms intervals. The neuron does not fire in relation to the saccade alone, nor to a stimulus outside of the receptive field when the monkey is not preparing a saccade. **A** Visual response to stimulus in receptive field in fixation task. **B** Visual response to a stimulus that is initially outside of the receptive field. The visual stimulus and the new fixation target appear simultaneously. *Left raster* aligned on stimulus appearance, *right raster* on beginning of saccade. Note that discharge precedes saccade. (Adapted from Duhamel et al. 1992)

cade brings the stimulus into the receptive field. If the monkey makes the saccade when an appropriate stimulus is not present, the cell is silent. It does not discharge in response to the stimulus flashed at a spatial location outside the receptive field unless an impending saccade will bring that spatial location into the receptive field. This predictive shift of the receptive field into the coordinates of the next fixation enables the parietal cortex to maintain a retinocentric, yet spatially accurate representation of visual space.

6.5.4 Posterior Cingulate Cortex and Involuntary Attention

The posterior cingulate cortex (CGp) has been proposed to be involved in a number of cognitive functions including emotion (Papez 1937) and attention (Mesulam 1981). However, single neuron recording in the awake, behaving monkey has not previously been used to analyze the nature of the posterior cingulate contribution to those processes. Because of strong reciprocal connections with posterior parietal cortex (Cavada and Goldman-Rakic 1989), we decided to study CGp using tasks similar to those that had revealed attentional modulation in the posterior parietal cortex. The recording sites in CGp are shown in Fig. 3A.

When we tested CGp neurons in visually guided saccade tasks, we discovered that the neurons fail to respond to the small saccade targets, despite the fact that the monkey must have attended to those targets. Over half, however, did discharge in a tonic fashion after the saccade (Fig. 3B). Neurons were broadly tuned for stimulus direction, but discharge was usually strongest after contralateral saccades. Neuronal discharge was determined by a number of factors: orbital position (Fig. 4A), saccade direction (Fig. 4B), and saccade amplitude (Fig. 4C). The activity of most neurons was influenced by all three factors, but a few were sensitive exclusively to orbital position or to saccade amplitude and direction.

Despite the fact that no neurons were driven by the onset of the saccade target (a small dim spot), most neurons were indeed visually sensitive. This was indicated by the fact that the postsaccadic activity of most neurons was attenuated in the dark (Fig. 5A). Thus reafferent visual stimulation must contribute to the postsaccadic response. To assess further the nature of visual responsiveness in CGp, we stimulated neurons with a variety of stationary and moving stimuli. We found that many neurons responded to large textured patterns (e.g., a checkerboard stimulus subtending 16°) over extensive receptive fields often encroaching on three or four quadrants of the visual field (Fig. 5B). The same neurons did not respond to small dim spots even when these were objects of voluntary attention (Fig. 5C). Most CGp neurons simply did not respond to stimuli with a diameter less than 4° (Olson et al. 1993; Musil et al. 1992). The neurons are indifferent to stimulus pattern, orientation, or spatial frequency.

The preference of neurons in CGp for large, bright stimuli suggests that they are monitoring the environment for salient objects. The stimuli which drive these neurons resemble those which James described as the objects of involuntary attention: "a sense-impression, very intense, voluminous, or sudden ... [with] a directly exciting quality ... bright things, moving things."

6.5.5 Conclusions

The classic syndrome of parietal damage in human (Critchley 1953) and monkey (Lynch and McLaren 1989) is neglect. Neglect reflects a failure of both voluntary and involuntary attentional mechanisms. The data on en-

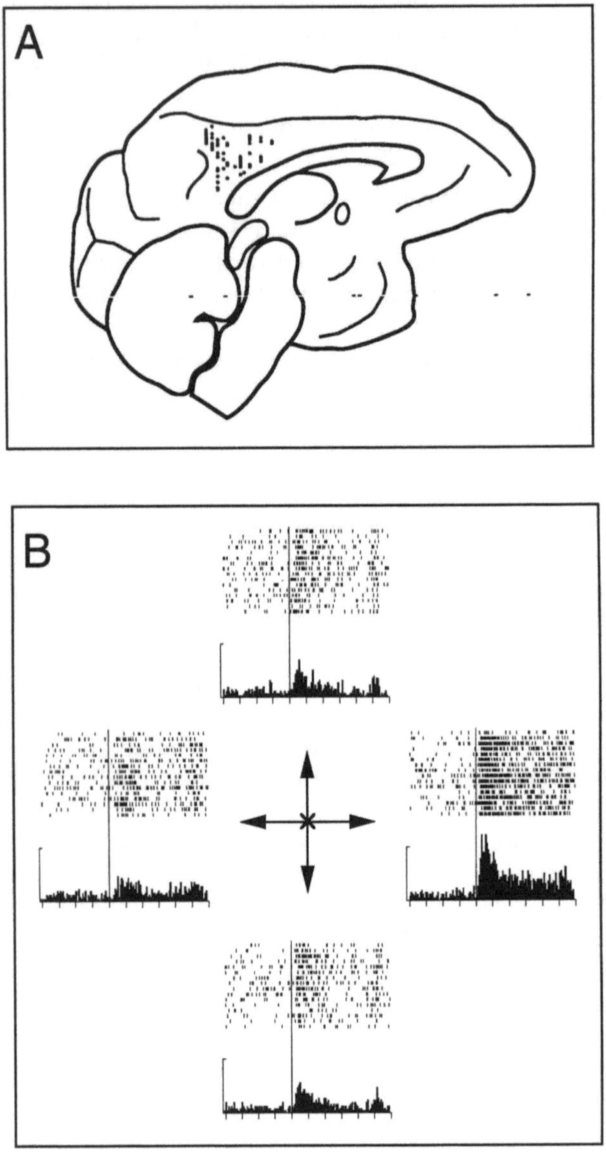

Fig. 6.5–3. A Recording sites at which neurons exhibited postsaccadic activity are projected onto a medial view of the left hemisphere of a macaque monkey. **B** Activity of a posterior cingulate cortex (CGp) neuron during a visually guided saccade task. In each raster display and histogram, activity is aligned on the onset of the saccadic eye movement. Neuronal activity increased after saccadic eye movements and was especially high following rightward (contraversive) eye movements

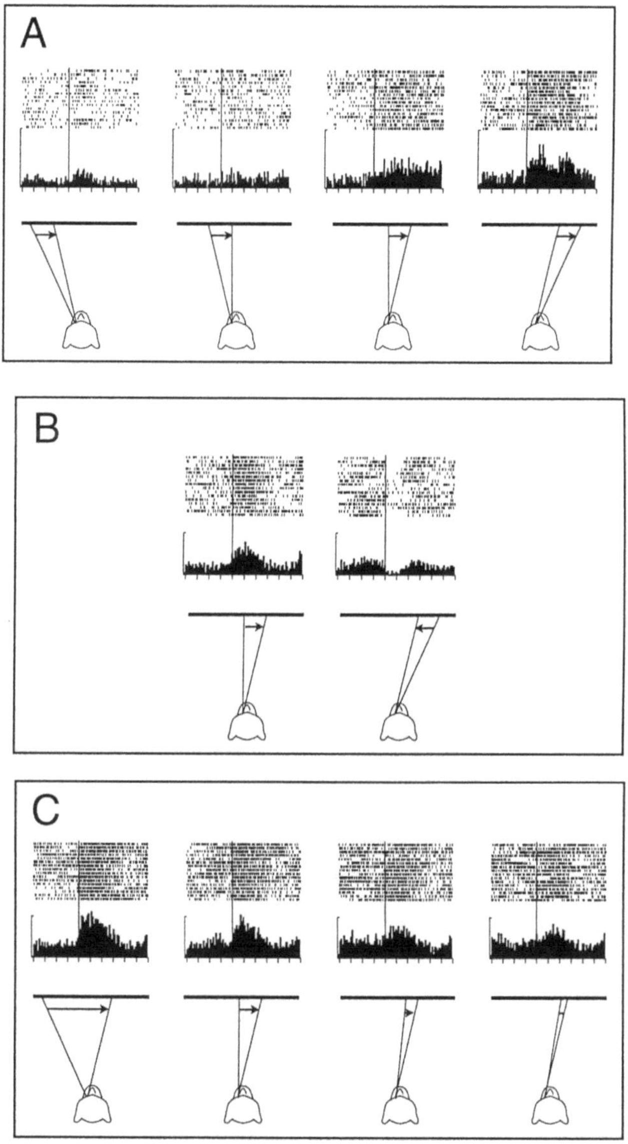

Fig. 6.5–4. **A** Activity of a posterior cingulate cortex (CGp) neuron sensitive to orbital position. Saccades of identical amplitude (16°) and direction (rightward) were initiated from four different starting points on pseudorandomly interleaved trials. Note that activity became greater as the eye was deviated farther to the right. **B** Activity of a neuron sensitive to saccade direction. Saccades of 16° amplitude were directed to the same end-point from diametrically opposed starting points on pseudorandomly interleaved trials. Note that activity was greater following rightward saccades. **C** Activity of the same neuron demonstrating sensitivity to saccade amplitude. On interleaved trials, saccades of different amplitude (32°, 16°, 8° and 4°) were executed in the same direction to an identical end point. Note that postsaccadic activity was greatest after the largest eye movements

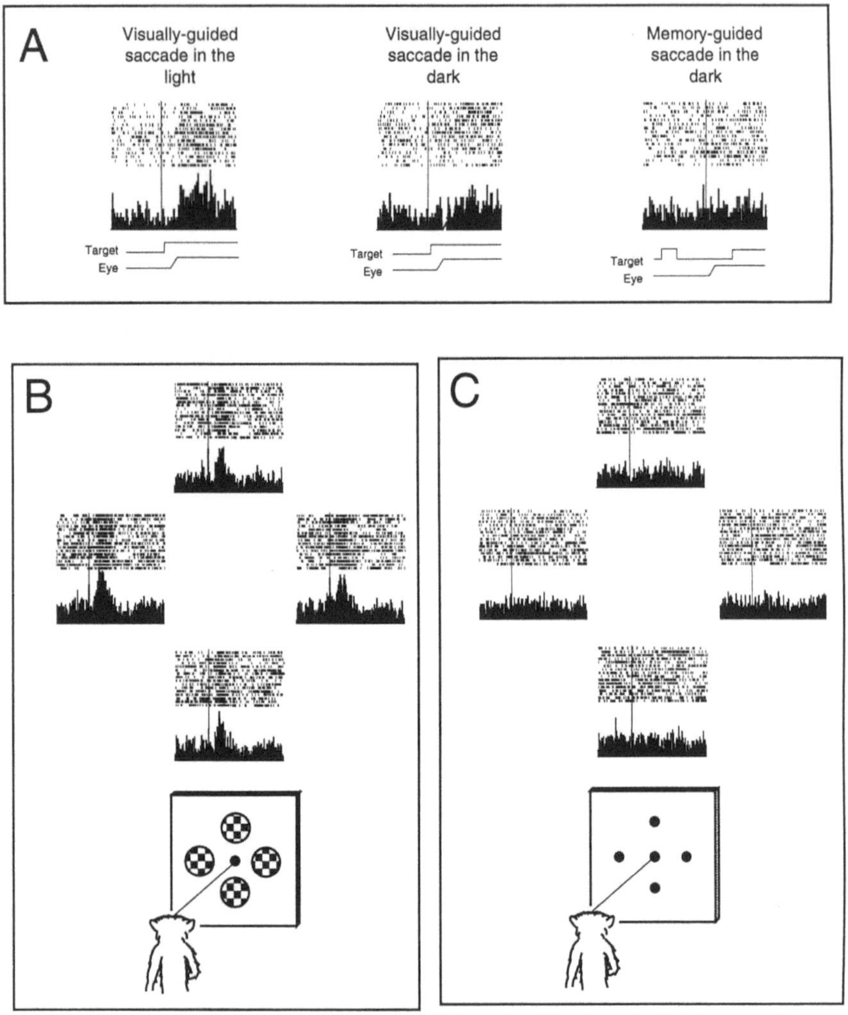

Fig. 6.5–5. **A** Activity of a posterior cingulate cortex (CGp) neuron following visually guided 16° rightward saccades made under three conditions in separate blocks of trials. *Left*, the target and the environment were visible at the time of the eye movement. *Middle*, the target was visible, but the room was in darkness at the time of the eye movement. *Right*, the eye movement was executed in total darkness to the remembered position of a target extinguished before movement onset. **B** Response of a CGp neuron to a large (16°) textured stimulus. Trials are aligned on the onset of the visual stimulus. **C** Activity of the same neuron during performance of a peripheral attention task. Raster and histogram displays are aligned with respect to the time of onset (*vertical line*) of the peripheral spot. Note that the neuron did not fire either in response to the onset of the peripheral stimulus or in conjunction with the covert shift of attention to it

hancement presented here support the view that parietal cortex is involved in voluntary attention. A role in involuntary attention is suggested by the finding that some visual neurons in the posterior parietal cortex respond better to large, bright stimuli (Robinson et al. 1978). We therefore propose that parietal cortex in general and LIP in particular serve as a final staging area for both voluntary and involuntary attentional processes.

Signals related to involuntary and voluntary attention may arise from distinct cortical systems projecting to parietal cortex. We suggest that the involuntary attentional component comes from CGp. This speculation is supported by our observation that CGp neurons respond to physically salient visual stimuli. It is also consonant with neuropsychological findings reported recently by Heilman and his group (Heilman et al. 1990). They describe a curious case of a human with a medial retrosplenial lesion, an area which may include the human homolog of CGp. This patient had mild ipsilateral neglect, associated with a relative increase in activity of the ipsilateral prefrontal cortex on positron emission tomography (PET) scan. The authors argue that the ipsilateral neglect arose from hyperfunction of the frontal mechanism scanning the contralateral field. This is what one would expect if the involuntary attentional, distractive mechanism of the contralateral field were shut down, shifting the balance toward a voluntary attentional program. A plausible candidate for the source of the voluntary attentional component is prefrontal cortex. Responsiveness to distracting stimuli is a hallmark of patients with prefrontal damage (Fuster 1989). This deficit may reflect a weakening of the voluntary attentional system in the presence of an intact involuntary system.

References

Andersen RA, Mountcastle VB (1983) The influence of the angle of gaze upon the excitability of the light-sensitive neurons of the posterior parietal cortex. J Neurosci 3: 532–548

Andersen RA, Asanuma C, Cowan M (1985) Callosal and prefrontal associational projecting cell populations in area 7a of the macaque monkey: a study using retrogradely transported fluorescent dyes. J Comp Neurol 232: 443–455

Andersen RA, Asanuma C, Essick G, Siegel RM (1990) Corticocortical connections of anatomically and physiologically defined subdivisions within the inferior parietal lobule. J Comp Neurol 296: 65–113

Barash S, Bracewell RM, Fogassi L, Gnadt JW, Andersen RA (1991) Saccade-related activity in the lateral intraparietal area. II. Spatial properties. J Neurophysiol 66: 1109–1124

Bruce CJ, Goldberg ME (1985) Primate frontal eye fields. I. Single neurons discharging before saccades. J Neurophysiol 53: 603–635

Bushnell MC, Goldberg ME, Robinson DL (1981) Behavioral enhancement of visual responses in monkey cerebral cortex. I. Modulation in posterior parietal cortex related to selective visual attention. J Neurophysiol 46: 755–772

Cavada C, Goldman-Rakic PS (1989) Posterior parietal cortex in rhesus monkey: II. Evidence for segregated corticocortical networks linking sensory and limbic areas with the frontal lobe. J Comp Neurol 287: 422–445

Colby CL (1991) The neurophysiology of attention. J Child Neurol 6: 88–118

Colby CL, Gattass R, Olson CR, Gross CG (1988) Topographic organization of cortical afferents to extrastriate visual area PO in the macaque: a dual tracer study. J Comp Neurol 238: 1257–1299
Colby CL, Duhamel JR, Goldberg ME (1993a) The analysis of visual space by the lateral intraparietal area of the monkey: the role of extraretinal signals. In: Hicks TP, Molotchnikoff S, Ono T (eds) Progress Brain Res, 95: 307–316
Colby CL, Duhamel J-R, Goldberg ME (1993b) Ventral intraparietal area of the macaque: anatomic location and visual response properties. J Neurophysiol 69: 902–914
Critchley M (1953) The parietal lobes. Arnold, London
Duhamel J-R, Colby CL, Goldberg ME (1992) The updating of the representation of visual space in parietal cortex by intended eye movements. Science 255: 90–92
Fischer B, Boch R (1985) Peripheral attention versus central fixation: modulation of the visual activity of prelunate cortical neurons of the rhesus monkey. Brain Res 345: 111–123
Fuster JM (1989) The prefrontal cortex, 2nd edn. Raven, New York
Gnadt JW, Andersen RA (1988) Memory related motor planning activity in posterior parietal cortex of macaque. Exp Brain Res 70: 216–220
Goldberg ME (1983) Studying the neurophysiology of behavior: methods for recording single neurons in awake behaving monkeys. In: Barker JL, McKelvy JF (eds) Methods in cellular neurobiology, vol 3. Wiley, New York, pp 225–248
Goldberg ME, Wurtz RH (1972) Activity of superior colliculus in behaving monkey. II. The effect of attention on neuronal responses. J Neurophysiol 35: 560–574
Goldberg ME, Colby CL, Duhamel J-R (1990) The representation of visuomotor space in the parietal lobe of the monkey. Cold Spring Harbor Symp Quant Biol 55: 729–739
Hallett PE, Lightstone AD (1976) Saccadic eye movements to flashed targets. Vision Res 16: 107–114
Heilman KM, Bowers D, Watson RT, Day A, Valenstein E, Hammond E, Duara R (1990) Frontal hypermetabolism and thalamic hypometabolism in a patient with abnormal orienting and retrosplenial amnesia. Neuropsychologia 28: 161–169
Hikosaka O, Wurtz RH (1983) Visual and oculomotor functions of monkey substantia nigra pars reticulata. III. Memory-contingent visual and saccade responses. J Neurophysiol 49: 1268–1284
James William (1878) Psychology: briefer course. In: Writings 1878–1899. Library of America, New York, pp 1–443
Lynch JC, McLaren JW (1989) Deficits of visual attention and saccadic eye movements after lesions of parieto-occipital cortex in monkeys. J Neurophysiol 61: 74–90
Lynch JC, Mountcastle VB, Talbot WH, Yin TCT (1977) Parietal lobe mechanisms for directed visual attention. J Neurophysiol 40: 362–389
Mesulam M-M (1981) A cortical network for directed attention and unilateral neglect. Ann Neurol 10: 309–325
Mountcastle VB, Lynch JC, Georgopoulos A, Sakata H, Acuña C (1975) Posterior parietal association cortex of the monkey: command functions for operations within extrapersonal space. J Neurophysiol 38: 871–908
Musil SY, Olson CR, Goldberg ME (1992) Visual response properties of single neurons in posterior cingulate cortex of rhesus monkey. Soc Neurosci Abstr 17: 442
Olson CR, Musil SY, Goldberg ME (1993) Posterior cingulate cortex and visuospatial cognition: properties of single neurons in the behaving monkey. In: Vogt BA, Gabriel M (eds) Neurobiology of cingulate cortex and limbic thalamus: a comprehensive handbook. Birkhäuser, Boston, pp 366-380
Papez JW (1937) A proposed mechanism of emotion. Arch Neurol Psychiatry 38: 725–743
Rizzolatti G, Gentilucci M, Luppino L, Matelli M, Ponzoni-Maggi S (1987) Neurons related to goal-directed motor acts in inferior area 6 of the macaque monkey. Exp Brain Res 67: 220–224
Robinson DL, Goldberg ME, Stanton GB (1978) Parietal association cortex in the primate: sensory mechanisms and behavioral modulations. J Neurophysiol 41: 910–932
Wurtz RH (1969) Visual receptive fields of striate cortex neurons in awake monkeys. J Neurophysiol 32: 727–742
Zipser D, Andersen RA (1988) A back-propagation programmed network that stimulates response properties of a subset of posterior parietal neurons. Nature 331: 679–684

6.6 Neural Mechanisms for Memory-Guided Visual Search

R. DESIMONE[1], L. CHELAZZI[1], E. MILLER[1], AND J. DUNCAN[2]
[1]Laboratory of Neuropsychology, National Institute of Mental Health, Bethesda, Maryland
[2]MRC Applied Psychology Unit, Cambridge, United Kingdom

For Otto Creutzfeldt, a key concept in understanding the functions of the sensory cortex was "action" (Creutzfeldt 1993). In his view, visual processing is tightly integrated with behavioral output systems. This quickly brings us to one of the major problems in visuomotor integration, namely how particular objects in the crowded visual world are selected for their behavioral relevance.

Although spatial attention may be the most commonly studied selection mechanism in neurophysiology, objects are frequently selected for attention and action based on non-spatial features (shape, color, etc.). When searching a crowd for a particular person's face, the person's location is not necessarily known in advance and the target face does not necessarily pop out automatically on the basis of a featural difference such as luminance or motion. In this type of visual search, an internal representation of the target object, stored in short- or long-term memory, must be used to guide selection. There are two general classes of models for how non-spatial selection is accomplished.

According to "serial search" accounts, focal attention is directed to each object in the scene, one (or perhaps a small group) at a time (e.g., Bergen and Julesz 1983; Treisman 1988). Each selected object is evaluated against the stored representation of the target object in memory, until a match is found. The greater the number of objects in the scene, the longer it takes to scan the array and the longer it takes to find the target.

By contrast, according to "parallel search" accounts, all the objects in the scene are processed in parallel in visual cortex, within some type of competitive network (Duncan and Humphreys 1989; Bundesen 1990). Within this network, neural populations coding the target features are biased, or primed, by the representation of the target stored in memory, sometimes termed the "attentional template." Depending on target-nontarget similarity, some of the priming signal may spill over onto populations of cells coding the nontargets; however, the fact that the bias is greater for the target will ultimately cause the correct population of cells to win. The greater the number of objects in the scene, the smaller the signal-to-noise ratio (or the longer the competition takes to resolve) and the longer the averaging needed to find the target. According to this scheme, increasing the size of the search array or increasing the similarity between target and nontargets increases search time for much the same reason that it takes longer to find a dim luminance target on a uniform background than it does to find a bright one.

6.6.1 Visual Memory

Clearly, visual memory plays a critical role in both of these schemes for non-spatial selection. Recent physiological studies have revealed at least three major classes of memory mechanisms in inferior temporal (IT) cortex. First, the responses of many IT neurons are sensitive to simple stimulus repetition (Gross et al. 1979; Baylis and Rolls 1987; Mikami and Kubota 1980; Miller et al. 1991a, 1991b, 1993; Riches et al. 1991; Eskandar et al. 1992; Li et al. 1993). When current stimuli are similar to those seen in either the immediate or long-term past, IT neuronal responses are often suppressed. Thus, stimuli that are novel or not recently seen will be processed preferentially by these cells, a property we have termed "adaptive mnemonic filtering" (Miller et al. 1991b, 1993). This property presumably contributes to the fact that novel stimuli tend to "pop out" of visual displays (Wang et al. 1992), much the same way that a red apple pops out from a green field. The adaptive mechanism appears to operate passively, or automatically (Miller and Desimone 1994), and could turn out to be a feature of several visual areas.

Many times, though, we need to search for a particular familiar object and suppress orienting to novel ones. Consistent with this, some IT cells give enhanced responses to just the one stimulus the monkey must search for in a temporal sequence of stimuli. It is as though these cells are biased, or primed, to detect the one item the monkey is looking for. Thus, there is a second memory mechanism in IT cortex that is not sensitive to simple stimulus repetition, but rather is linked to specific stimuli held in working memory. Preliminary evidence suggests that one source of this priming signal could be ventral prefrontal cortex (Fuster et al. 1985; Miller and Desimone 1994; Wilson et al. 1993; Chelazzi et al. 1993b).

Finally, many IT cells have maintained activity linked to items held in working memory, in the absence of any external stimulus (Fuster and Jervey 1981; Miyashita and Chang 1988; Fuster 1990; Miller et al. 1993; Chelazzi et al. 1993a). For example, if a monkey is required to remember stimulus A for a few seconds, some IT cells that normally respond to A would have a higher maintained activity during the retention interval than would cells that would not normally respond to A. Unlike in prefrontal cortex, though, this maintained activity appears to be easily disrupted by any intervening stimuli during the delay (Miller et al. 1993; Miller and Desimone 1994; Chelazzi et al. 1993b).

6.6.2 Memory-Guided Search

The mechanism for the type of visual search that underlies finding a face in a crowd requires the interaction of memory and attention. There must be a means both to hold the object of search in memory and to use this stored object to resolve competition among the array elements. We investigated the neural basis of search in the anterior ventral portion of IT cortex (Chelazzi et al. 1993a), the same portion in which we found the three types of memory-related activity described above.

In this task, monkeys were briefly presented with a complex picture (the cue) at the center of gaze to hold in memory. The cue was always either a "good" stimulus that elicited a strong response from the cell or a "poor" stimulus that elicited little or no response. After a blank delay period of 1.5–3.0 s the good and the poor stimulus were both presented simultaneously as choice stimuli, at an eccentricity of 4°–5° from the center of gaze. The monkey was required to make a saccadic eye movement to the target stimulus that matched the cue, ignoring the nonmatching stimulus (the "distractor"). That is, the monkey had to find the target based on the features of the cue that it had previously stored in memory. The spatial locations of the target and distractor were varied randomly, so the monkey could not find the target based on prior knowledge of its location. Because of the large receptive fields of IT neurons, the cue was virtually always contained within the receptive field, as were all target and distractor locations within the contralateral field. We will consider the results with the choice array confined to the contralateral field first.

As expected from previous studies of memory in IT, the cue typically initiated activity that persisted through the following delay period among the neurons that were tuned to the cue's features. That is, many cells had a higher level of maintained activity in the delay following the good cue (7.9 spikes/s on average) than following the poor cue (5.6 spikes/s on average). To explore how the target might be selected, we compared responses to identical choice arrays preceded by either the good or poor cue. This is equivalent to comparing responses to choice arrays in which either the good stimulus or poor stimulus was the target. In absolute terms, the firing rate during the initial phase of the response was somewhat higher following the good cue than following the poor cue, because the activity during the delay interval often persisted into the initial phase of the choice array response. Relative to the preceding delay activity, however, the initial response (averaged across the population of cells) to the array was about the same, regardless of which choice stimulus was the target. That is, the choice array tended to activate cells tuned to either stimulus in the array.

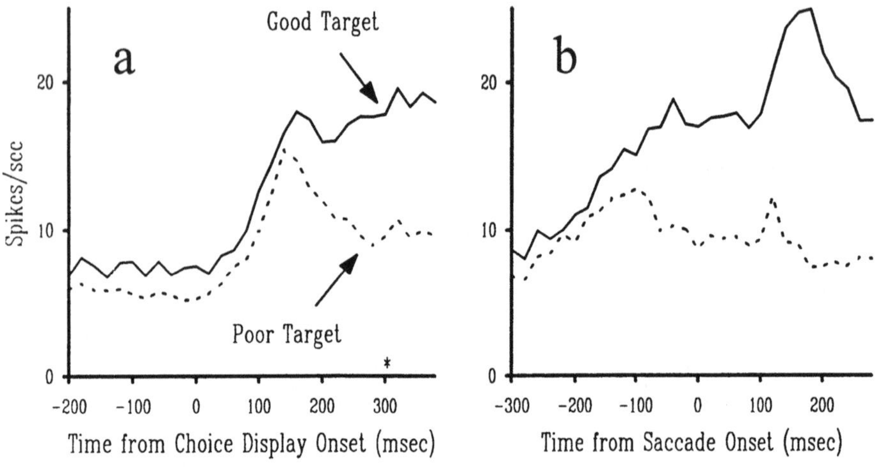

Fig. 6.6–1. Average responses to identical choice arrays, in which either the good stimulus or poor stimulus (as determined individually for each recorded cell) was the target. This is equivalent to taking responses to identical choice arrays preceded by good or poor stimulus as cue. The relative locations of target and distractor within the contralateral visual field varied randomly from trial to trial, and the graphs are averaged over all stimulus locations. Responses are averaged from the entire sample of 58 inferior temporal (IT) cells that responded to the choice arrays. **a** Responses time-locked to onset of the choice array. The average saccadic latency (indicated by the *asterisk*) was 306 ms after array onset. **b** Same data, time-locked to the onset of the saccade. Starting approximately 100 ms before the eye movement (time 0), responses appeared to be "captured" by the target, so that neuronal activity would reflect only the target's properties. (Adapted from Chelazzi et al. 1993a)

By contrast, the late phase of the response changed dramatically depending on whether the animal was about to make an eye movement to the good or poor stimulus. When the target was the good stimulus, the response remained high. However, when the target was the poor stimulus for the recorded cell, the response to the good distractor stimulus was suppressed even though it was still within the receptive field (Fig. 1). On average, suppression of the distractor response began about 200 ms after the onset of the choice array or about 90–120 ms before the start of the eye movement. The cells responded as though 100 ms or so before the eye movement, the target stimulus "captured" the response of the cells, so that neuronal activity in IT cortex reflected only the target's properties. We found similar results when the choice array size was increased to five elements, the maximum tested.

Previous studies had shown that when animals attend to stimuli at one location and ignore stimuli at another, responses to the stimuli at the ignored location are suppressed (Moran and Desimone 1985). This is the first evidence that IT responses are also gated when selection is guided by object features. Once the target is selected, however, there may be some spatial information about the target available in IT cortex, in spite of the large receptive fields. The receptive fields of many IT neurons have a "hot spot" at the fovea, which extends asymmetrically into the upper or lower contralateral visual field. Thus, the response of many IT cells to the target varied somewhat depending on its location in the contralateral field.

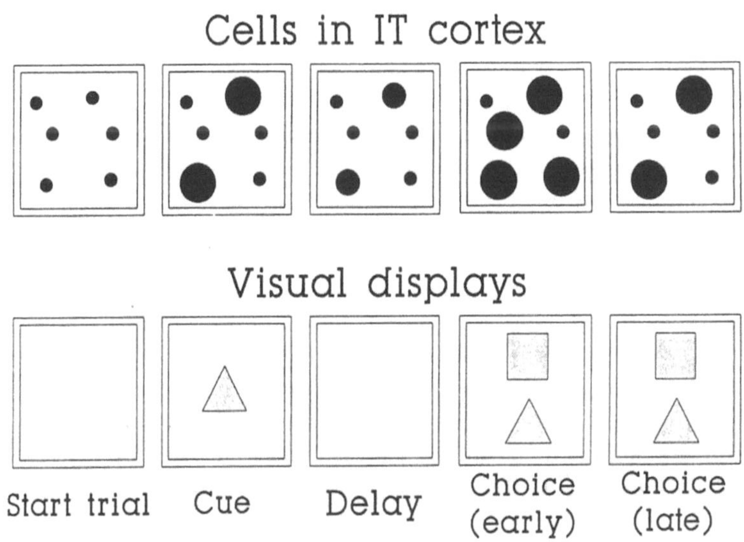

Fig. 6.6–2. Schematic representation of target selection within a population of cells in inferior temporal (IT) cortex. The *upper* diagrams represent activity within a population of cells. Each *filled circle* represents a cell, and the diameter of the circle represents the cell's firing rate. The *lower* diagrams represent the stimulus displays during the task. The cue activates those cells in IT cortex tuned to its features, and these cells maintain a higher level of firing during the delay interval. When the choice array first appears, it activates IT cells tuned to the features of either stimulus. The average activation (relative to the preceding delay activity) in IT cortex in this case is independent of which stimulus is the target, although individual cells may show either enhancement or suppression due to the engagement of memory mechanisms in IT cortex. Later, within about 100 ms of saccade onset, activity within the population is limited to those cells tuned to the features of the target. Cells tuned to the distractor are suppressed, possibly due to competitive interactions within IT cortex. Cells tuned to the target may be biased, or primed, to win this competition due to either the persisting activity from the delay or the action of memory mechanisms in the early portion of the choice array response. After the target is selected, oculomotor systems direct the eyes to the correct location. (Adapted from Chelazzi et al. 1993a)

We appear to have narrowed the target selection process down to a critical 200 ms interval following stimulus onset (Fig. 2). During this interval, we find all of the mnemonic effects on IT responses that we found in our studies of memory. First, the responses of some cells to the choice array are suppressed when their preferred stimulus is the target, due to the action of the adaptive mnemonic filter mechanism (the good stimulus was presented twice in the trial, both as cue and target). Second, other cells show enhanced responses when their good stimulus is the target, similar to what we found in our working memory experiments (Miller and Desimone 1994). The preceding cue seems to "prime" these cells to detect the target stimulus. When neuronal responses are averaged across the entire population of cells, the enhanced responses of these cells tend to be balanced out by the suppressed responses of the adaptive cells. Finally, following the good cue, many cells showed elevated activity in the initial phase of response to the choice array, due to persisting mnemonic activity from the delay.

If IT cells coding the different array elements are in competition with each other during presentation of the choice array, the cue-selective delay activity and target response enhancement may weight the competition towards the cells whose preferred stimulus is the target. According to this scheme, the responses of IT cells selective for the distractor stimulus are suppressed by the action (direct or indirect) of those IT cells selective for the target stimulus, consistent with the "parallel search" model described earlier in this chapter. This possibility is supported by the fact that three fourths of the cells with significant target effects also showed significant cue-selective activity during the delay. Furthermore, the target effects were much reduced when target and distractor were located across the midline, in opposite hemifields, a finding that is consistent with prior studies of spatial attention in IT cortex (Sato 1988; Moran and Desimone, unpublished data). The IT cells maximally activated by target and distractor in this configuration were located in opposite hemispheres and, therefore, may have been less able to interact competitively with each other. Nonetheless, we cannot yet rule out the possibility that target selection in IT cortex requires interaction with a separate attentional control system.

What is the consequence of target selection in IT cortex? We believe that target information in IT cortex is ultimately sent to oculomotor systems in order to direct the eyes to the appropriate target. More generally, visual memory mechanisms in IT cortex may select the stimuli that are attended and acted upon by motor systems, a conclusion that probably would not have surprised Otto Creutzfeldt.

Acknowledgments. We gratefully acknowledge the advice and support of Mortimer Mishkin in all phases of the work. The work was supported in part by the Human Frontiers Science Program Organization.

References

Baylis GC, Rolls ET (1987) Responses of neurons in the inferior temporal cortex in short term and serial recognition memory tasks. Exp Brain Res 65: 614–622

Bergen JR, Julesz B (1983) Parallel versus serial processing in rapid pattern discrimination. Nature 303: 696–698

Bundesen C (1990) A theory of visual attention. Psych Rev 97: 523–547

Chelazzi L, Miller EK, Duncan J, Desimone R (1993a) A neural basis for visual search in inferior temporal cortex. Nature 363: 345–347

Chelazzi L, Miller EK, Lueschow A, Desimone R (1993b) Dual mechanisms of short-term memory: ventral prefrontal cortex. Soc Neurosci Abstr 19: 975

Creutzfeldt OD (1993) Cortex cerebri. Performance, structural and functional organization of the cortex (English translation). Mary Creutzfeldt, Göttingen

Duncan J, Humphreys GW (1989) Visual search and stimulus similarity. Psych Rev 96: 433–458

Eskandar EN, Richmond BJ, Optican LM (1992) Role of inferior temporal neurons in visual memory: I. Temporal encoding of information about visual images, recalled images, and behavioral context. J Neurophysiol 68: 1277–1295

Fuster JM (1990) Inferotemporal units in selective visual attention and short-term memory. J Neurophysiol 64: 681–697

Fuster JM, Jervey JP (1981) Inferotemporal neurons distinguish and retain behaviorally relevant features of visual stimuli. Science 212: 952–955

Fuster JM, Bauer RH, Jervey JP (1985) Functional interactions between inferotemporal and prefrontal cortex in a cognitive task. Brain Res 330: 299–307

Gross CG, Bender DB, Gerstein GL (1979) Activity of inferior temporal neurons in behaving monkeys. Neuropsychologia 17: 215–229

Li L, Miller EK, Desimone R (1993) The representation of stimulus familiarity in anterior inferior temporal cortex. J Neurophysiol 69: 1918–1929

Mikami A, Kubota K (1980) Inferotemporal neuron activities and color discrimination with delay. Brain Res 182: 65–78

Miller EK, Desimone R (1994) Parallel neuronal mechanisms for short-term memory. Science (in press)

Miller EK, Gochin PM, Gross CG (1991a) Habituation-like decrease in the responses of neurons in inferior temporal cortex of the macaque. Vis Neurosci 7: 357–362

Miller EK, Li L, Desimone R (1991b) A neural mechanism for working and recognition memory in inferior temporal cortex. Science 254: 1377–1379

Miller EK, Li L, Desimone R (1993) Activity of neurons in anterior inferior temporal cortex during a short-term memory task. J Neurosci 13: 1460–1478

Miyashita Y, Chang HS (1988) Neuronal correlate of pictorial short-term memory in the primate temporal cortex. Nature 331: 68–70

Moran J, Desimone R (1985) Selective attention gates visual processing in the extrastriate cortex. Science 229: 782–784

Riches IP, Wilson FA, Brown MW (1991) The effects of visual stimulation and memory on neurons of the hippocampal formation and the neighboring parahippocampal gyrus and inferior temporal cortex of the primate. J Neurosci 11: 1763–1779

Sato T (1988) Effects of attention and stimulus interaction on visual responses of inferior temporal neurons in macaque. J Neurophysiol 60: 344–364

Treisman A (1988) Features and objects: the fourteenth Bartlett Memorial Lecture. Q J Exp Psychol [A] 40: 201–237

Wang Q, Cavanagh P, Green M (1992) Familiarity and pop-out in visual search. Assoc Res Vis Ophth Abstr 33: 1262

Wilson FAW, Ó Scalaidhe SP, Goldman-Rakic PS (1993) Dissociation of object and spatial processing domains in primate prefrontal cortex. Science 260: 1955–1958

Chapter 7

Functional Aspects I

7.1 Introduction: Functional Aspects of Neocortical Sensory Areas

P. Heggelund
Department of Neurophysiology, University of Oslo, Norway

This section will address recent advances in our understanding of the function of neocortical areas related to perception and sensory control. The neural basis of perception is a topic in which Otto Creutzfeldt was strongly involved ever since his early days in Richard Jung's laboratory in Freiburg, and it is a topic to which he made numerous important scientific contributions. His deep interest in this subject was manifested in the key position that neural mechanisms of perception had in the research activity in his department, first in Munich and later in Göttingen. The main emphasis in this section is on visual areas, as it was in Creutzfeldt's own work. The presentations span central questions ranging from functional architecture of the primary visual cortex to higher order functions related to visual attention, object representation, and visuomotor integration.

The section starts with two papers on functional architecture of the visual area V1 and V2. These areas are both organized into functional subsystems that are assumed to be involved in processing of distinct types of visual information. In V1, cells dominated by the same eye are grouped together in ocular dominance columns. Cells responding to the same optimal stimulus orientation tend to be grouped together in clusters termed iso-orientation domains or orientation columns. A third subsystem, the cytochrome oxidase blobs, contain a high proportion of color selective cells. The subsystems are connected by long-range lateral intrinsic projections that may extend horizontally over several millimeters. Three subsystems have also been described in V2. These are stripe compartments (thin, thick, and pale) running parallel to the surface and extending through the cortical layers.

It has recently become possible to study the organization of the subsystems by using optical recording techniques in combination with established electrophysiological and anatomical techniques. This approach was used by Amiram Grinvald and Rafi Malach to study the horizontal extention of the subsystems in V1. They showed that iso-orientation domains are organized radially such that the various iso-orientation domains at a given cortical locus form a pinwheel-like structure. The majority of the pinwheels are centered along the midline of the ocular dominance columns. The cytochrome oxidase blobs also tend to be centered along the midline of the ocular dominance columns, but the location of their centers is independent of the location of the pinwheel centers. Lateral intrinsic connections between ocular dominance columns usually link columns dominated by the same eye. In the orientation domain, the long-range connections tend to link columns with similar optimal orientation, although this like-to-like relationship was less precise than for the ocular dominance system.

Daniel Ts'o and colleagues have studied organization within subsystems in V1 and V2. The blobs in V1 may be color selective, containing red-green or yellow-blue opponent cells. Some blobs seem to lack color opponent cells and instead have broadband, center-surround cells. Bridges of dense cytochrome oxidase staining that may occur between blobs contain color selective cells, showing that they are part of the color system in V1. Horizontal connections between blobs seem to run between blobs of matching color opponency. In V2, the stripes do not seem to be homogeneous, but rather collections of functional clusters. For example, in the thin stripes, which contain color selective cells, there are patches dominated by red-green opponent cells as well as patches dominated by yellow-blue opponent cells. In the retinotopic organization of the stripe compartments, each stripe system seems to have its own retinotopic representation such that there is a break in the representation of the visual field at the transition from one stripe to the next. The connectivity between V1 and V2 seems to be characterized by a high degree of convergence and divergence. The connectivity between modules in V1 and V2 is such that cell pairs tended to be correlated only when the receptive field properties of the cells are matched.

Visual information is processed in more than 30 cortical areas that are connected by more than 300 pathways, as discussed by David Van Essen and his colleagues. From an asymmetry in the laminar pattern of the connectivity, it is possible to establish a hierarchy between the areas within which several streams of processing can be distinguished. The response properties of cells in the various areas are less well characterized. In some of the higher visual areas, the cells respond to complex stimuli including natural objects such as faces and hands. An interesting question is how complex stimulus selectivities are generated from the more simple selectivities found in V1. Another striking difference between cells in higher visual areas and V1 is that cells in V1 have small receptive fields, whereas cells in higher visual areas may have large receptive fields within which they can respond selectively to stimuli over large ranges of stimulus size and position. Thus, another interesting set of questions are related to the mechanisms used to generate such scale and position invariance that may be important for our ability to recognize objects independent of distance and viewing angle. Van Essen and colleagues suggest that these mechanisms may be related to directed visual attention. They propose a model for directed visual attention that involves dynamic routing of information flow through successive stages of the visual hierarchy where both cortical and subcortical structures are involved in the regulation of the routing.

An important complex of problems for our understanding of perception is related to how the brain constructs representations of the environment. These problems are adressed by Wolf Singer. It is assumed that the representations are distributed as activity in assemblies of interacting neurons where different neurons in an assembly may code for different aspects of the object. The interacting neurons can be in the same or in different cortical areas. The individual neurons participate at different times in different assemblies and thus reduce the number of cells needed to represent different objects. It is suggested that the neurons in a given assembly are distinguished by temporal synchronous firing patterns. In this way it may be possible for the brain to distinguish cell groups giving information about different aspects of the same object from cell groups giving information about different objects. Experiments designed to test predictions from this assembly hypothesis may indicate that feature selective cells in visual cortex synchronize the firing when stimulated by the outlines of the same object. The synchronization probability seems to reflect some of the well-known *Gestalt* laws of perceptual organization (e.g., proximity, similarity, continuity, common fate). At least part of the synchronization seems to be due to cortico cortical association connections. These connections may be shaped through use-dependent modifications during postnatal development.

Visual representations may be integrated with motor information to generate movements. One essential step in this sensorimotor integration is planning of the movements. Richard Andersen presents evidence for the hypothesis that the step from visual representation to intention to make movements is related to neural activity in the posterior parietal lobe. Neural activity in a small area in the posterior parietal cortex called the lateral intraparietal area (LIP) seems to be related to the planning of saccadic eye movements. Another essential step in the sensorimotor integration is related to coordinate transformation. The visual information represented in retinotopic coordinates must be transformed to coordinates of muscle for movements. These coordinate transformations seem to take place in LIP and area 7a of the posterior parietal lobe where representations of space intermediate between the representations in retinotopic and motor coordinates are found. Intermediate representations of space in head-, body-, and perhaps also world-centered coordinates seem to exist. Such representations may be generated by combining visual information with information about eye and head position from the somatosensory and the vestibular system. These representations do not seem to be explicit at the single cell level, but are found in a population code.

Our understanding of the functional aspects of cortical areas related to sensory systems other than vision is less detailed than our understanding of the visual areas. Particularly little is known about the cortical representation of the vestibular system. In the last paper in this section, Otto-Joachim Grüsser and his colleagues show that the cortical processing of vestibular information in primates is distributed over different areas, as in the processing of visual information. At least three different cytoarchitectonic areas are involved. In addition to vestibular input, cells in these areas also receive input from other cortical association areas. The inputs from the other cortical association areas transmit somatosensory, optokinetic, gaze motor, and somatomotor information.

7.2 Functional Architecture and Connection Rules in Primary Visual Cortex of Macaque Monkey

A. GRINVALD AND R. MALACH
Department of Neurobiology, The Weizmann Institute of Science, Rehovot, Israel

In this chapter we begin by summarizing the relationships between cytochrome oxidase (CO) blobs, ocular dominance columns and iso-orientation domains, subsystems of primary visual cortex of macaque monkey, underlying visual perception. Next, we discuss the relationship between these functional subsystems and the lateral intrinsic circuitry as revealed by tracer injections. More details have been described elsewhere (Bartfeld and Grinvald 1992; Malach et al. 1993).

The primary visual cortex of macaque monkey is organized in a columnar fashion: neurons sharing similar response properties are clustered together forming columns which often traverse the entire cortical depth from pia to the white matter. High resolution maps of the three subsystems mentioned above were acquired by optical imaging based on activity-dependent intrinsic signals. The most prominent organizational feature of orientation preference was a radial arrangement, forming a pinwheel-like structure. More than 80% of these pinwheels were centered along the midline of ocular dominance columns. The iso-orientation contours of adjacent pinwheels crossed borders of ocular dominance columns at approximately right angles. Pinwheels with the same or opposite directions of orientation preference change were smoothly connected with each other. In exactly the same cortical area, the cytochrome oxidase blobs were also mapped, using cytochrome oxidase histology. Like the centers of pinwheels, the centers of blobs also lie along the midline of ocular-dominance columns. However, the centers of pinwheels did not coincide with the centers of blobs; these two subsystems are spatially independent.

The rules underlying the relationship between these functional subsystems and the lateral intrinsic circuitry have remained a hotly debated issue despite an intensive research effort. To resolve this second issue we employed a combined approach using the methods of tract tracing and optical imaging on the same patches of cortex. First, high resolution maps of the cortical functional columns were revealed by in vivo optical imaging. Having these maps in hand, we then placed extracellular injections of the sensitive anterograde tracer biocytin into selected functional domains. We found that long-range intrinsic connections tend to form patches which correspond to same-eye dominance columns. Interestingly, binocular dominated cortical sites were selectively connected among themselves and avoided strictly monocular regions. In the orientation domain, long-range intrinsic connections tended to link columns sharing similar orientation preferences. However, this selectivity was more of a statistical trend, with no less than 30±10% of the interconnected patches manifesting different functional preferences. These patterns of connections between patches provide cortical neurons with a rich diversity of response properties and provide the neuronal substrata for functional plasticity.

7.2.1 Introduction

Mountcastle (1957) in the cat somatosensory cortex and Hubel and Wiesel (1968, 1974, 1977) in the cat and monkey visual cortices have found that cortical cells with similar response properties are frequently clustered together, forming columns (modules) which often traverse the entire cortical depth, from the pia to the white matter. However, the exact organization of different functional modules parallel to the cortical surface still remains an unresolved question for all areas of the mammalian cortex. Over the last three decades, a functional segregation has been demonstrated for three types of visual response properties in monkey primary visual cortex. The first, and best understood, is the segregation of the inputs from the right eye and from the left one. These segregated inputs form the ocular dominance columns, which run nearly parallel to one another in slabs that are 400 μm wide (Levay et al. 1975). A second system of segregated neurons is referred to as the iso-orientation domains (or orientation preference bands): each domain contains cells that respond best to a given stimulus orientation (Hubel et al. 1978). The third, and least understood, subsystem contains neurons that are thought to be selective for other attributes of the visual stimulus, such as color (Livingstone and Hubel 1984; Ts'o and Gilbert 1988) and spatial frequency (Tootell et al. 1988). These cells are located in the CO blobs (Livingstone and Hubel 1984; Ts'o and Gilbert 1988; Tootell et al. 1988; Horton and Hubel 1981). Hubel and Wiesel (1977) and Livingstone and Hubel (1984) have proposed a schematic "ice-cube" model for describing the functional architecture of the repeating hypercolumns that contain all three subsystems. However, the exact layout and the relationships between blobs, iso-orientation domains, and ocular dominance columns were not determined until recently. This was done by optical imaging (Blasdel 1992; Bartfeld and Grinvald 1992). Previous works (Mountcastle 1957; Hubel and Wiesel 1968, 1974, 1977; Levay et al. 1975; Hubel et al. 1978; Livingstone and Hubel 1984; Ts'o and Gilbert 1988; Tootell et al. 1988; Horton and Hubel 1981) relied on electrophysiological, anatomical, and histological techniques to reveal the organization of each type of subsystem. Optical imaging based on voltage-sensitive dyes (Tasaki et al. 1968; Salzberg et al. 1973; Grinvald et al. 1988) is a particularly attractive technique for exploring these relations. Blasdel and Salama (1986) provided a striking demonstration of the usefulness of optical imaging to study functional architecture in the primate visual cortex. While they thought their imaging was based on voltage-sensitive dyes, we have developed an alternative optical imaging approach (Grinvald et al. 1986; Frostig et al. 1990; Ts'o et al. 1990), based on activity-dependent, intrinsic optical changes (Frostig et al. 1990; Hill and Keynes 1949; Chance et al. 1962; Jobsis et al. 1977; Grinvald et al. 1991) without dyes.

In parallel with studies of the functional architecture, neuroanatomical studies have revealed that intrinsic horizontal connections form an elaborate network linking neighboring cortical sites (Livingstone and Hubel 1984; Rockland and Lund 1982, 1983; Gilbert and Wiesel 1979, 1983). These horizontal connections are most extensive in superficial cortical layers. They extend to about 2–3 mm in area V1 (Amir et al. 1993; King et al. 1989; LeVay 1988a) and are thought to be mostly excitatory (McGuire et al. 1984; LeVay 1988b). These connections are characterized by a tendency to project selectively to preferred sites, forming distinct axonal arborizations of 200–300 μm in diameter (Rockland and Lund 1983; Amir et al. 1993). We term these axonal arborizations "patches". The exact relationship between the axonal patches and the functional domains of ocular dominance and orientation preference is quite complex and has eluded a conclusive resolution despite major research efforts. Thus, the selectivity of intrinsic connections to eye dominance (Livingstone and Hubel 1984; Ts'o and Gilbert 1988) and orientation (Gilbert and Wiesel 1989; Matsubara et al. 1985) remains an open question.

7.2.2 Materials and Methods

Monkeys (*Macaca fascicularis*) were initially anesthetized with ketamine HCl (20 mg/kg, intramuscularly), followed by sodium pentothal anesthesia (10–20 mg/kg per h, administered i.v. and supplemented as needed). To monitor the state of the anesthetized monkeys, the electroencephalogram, electrocardiogram, temperature, and expired CO_2 were observed throughout the experiment to assure adequate depth.

Imaging procedure. To minimize pulsation, a sealed, stainless steel chamber was mounted over the exposed cortex. A slow scan charge-coupled device (CCD) camera was mounted above the optical chamber. The cortical image was projected on the CCD camera using a tandem lens arrangement (macroscope), which provides high light collection capability and a shallow depth of focus. Thus, by focusing 300 μm below the cortical surface, artifacts originating from surface blood vessels were largely eliminated. The CCD camera provided digitized images of the cortex (eccentricity 6°–8°, close to the V1/V2 border). Drifting gratings (6° by 4°) of eight orientations were used for each 2 s stimulus; we collected 5–10 frames over a period of 3 s. The functional maps obtained by optical imaging based on intrinsic signals provide information about the organization of the upper cortical layers down to a depth of about 600 μm, as discussed elsewhere (Grinvald et al. 1988; Frostig et al. 1990; Ts'o et al. 1990).

Data Analysis. Functional maps can be obtained by comparing the cortical images obtained when the animal views various stimuli. To remove the baseline image and correct for the effects of uneven illumination and other common-mode noise originating from the microvasculature, each of these images are divided by the "blank" image, obtained when the animal viewed a blank screen. The resulting image provides a "single (stimulus) condition" map. To reduce microvascular artifacts even more, in our studies the sum of all images was used instead of the blank. (These images were obtained by using many stimulus conditions activating the cortex uniformly, and their sum is referred to as a "cocktail blank.") Additional technical details and definitions of the various types of functional maps derived from optical imaging have been provided elsewhere (Ts'o et al. 1990; Bonhoeffer and Grinvald 1991).

Histological Preparations. The animals were then killed by pentobarbital overdose and perfused transcardially (phosphat-buffered solution (PBS) followed by glutaraldehyde 2.5%, paraformaldehyde 0.5% fix). The cortex was flattened (Tootell and Silverman 1985) and cut tangentially at 60–80 μm. Alternate sections were stained for CO and biocytin. Sections were inspected and photographed under bright and dark field illumination (Amir et al. 1993).

Biocytin Injections Targeted at Optically Identified Functional Domains. After the functional maps were obtained, cortical sites of interest received iontophoretic injections of the tracer biocytin (2% in tris pH 7.6, 4 μA/4 min, DC current). Ten injections were fully analyzed in relation to the optical maps. The tracer biocytin allows a precise definition of its effective tracer-uptake zone by localizing, under high magnification, labeled neuronal somata at the site of injection. Injection site diameters varied from 100 to 300 μm. Following the injections, animals were kept alive under anesthesia for another 8–12 h with continuous monitoring of their physiological states and levels of anesthesia.

Matching of Optical Maps with Histological Sections. The precise match between the optical maps obtained in vivo and the histological material is the most critical step in combining the two data sets. This goal was accomplished by relying on the vascularization pattern as a matching guide. The pattern of the superficial blood vessels was photographed during the optical imaging session and then recovered by carefully sectioning the tissue in order to preserve this pattern in the histological material. The pattern of blood vessels taken under the in vivo and postmortem conditions was then superimposed using simple linear transformations, i.e., magnification, translation, and rotation, (only) of the entire image. Further refinement of the matching was obtained by introducing slight linear shifts of one image, so that the histologically revealed CO blobs were positioned at the precise centers of all the ocular dominance bands imaged optically. The match we obtained between these two sets was excellent over the entire imaged cortical area (less $3 \times 2\,\mathrm{mm}$ for each map), confirming the reliability of this matching procedure. This latter matching, by relying on the blobs, has certain advantages, since it works even if the axis perpendicular to the cortical layers does not perfectly coincide with the optical axis used for the in vivo imaging.

7.2.3 Results

To optically image orientation domains, i.e., the upper layers of the orientation columns, anesthetized monkeys viewed drifting gratings of a given orientation. Using a CCD camera, optical images of the exposed cortex were taken separately for each stimulus orientation. These eight maps obtained with eight stimulus orientations, separated by 22.5°, showed a mosaic-like arrangement for each of the partially overlapping "iso-orientation" domains. The patchy domains of the various orientation preferences are arranged like flower petals, partially overlapping. To provide a detailed picture of the interrelationship among the various iso-orientation domains, we used an analytical procedure similar to the one pioneered by Blasdel and Salama (Blasdel and Salama 1986; Blasdel 1988) that vectorially adds the strength of the response for each of the eight stimulus orientations on a pixel-by-pixel basis. The resultant vector for each pixel is defined by its angle and magnitude, and the angle provides the orientation preference at each cortical site. Several types of displays can be used to emphasize various aspects of the architecture of orientation domains, three of them will be used here.

7.2.3.1 The Pinwheel-Like Organization of Orientation Preference

In a "discrete angle" map (e.g., Fig. 1C of Bartfeld and Grinvald 1992), each color codes for a particular range of orientation preferences. Thus, each color corresponds to an orientation domain represented with a resolution of 22.5°. The discrete angle map is useful in showing the direction of iso-orientation contours across the cortical surface. However, it cannot distinguish whether orientation preference changes smoothly or discretely across the cortical surface. The same data, displayed using a continuous color scale, indicates that orientation preference indeed changes smoothly and continuously across the cortex, except at a few points of singularity (Fig. 1D of Bartfeld and Grinvald 1992). This map is reminiscent of orientation preference maps previously obtained by Blasdel and Salama (1986).

Inspection of more than 60 angle maps indicated that the most prominent feature of orientation domains was their radial, pinwheel-like organization (Braitenberg and Braitenberg 1979). It is clear that the orientation domains are ordered radially around a point of singularity and that each orientation is represented just once. To show the gradual change in orientation preference, the same pinwheel can also be depicted using a continuous color scale. Such presentation shows that in a pinwheel, the orientation preference of cortical neurons changes smoothly in a radial manner around the point of singularity. These optical maps and previous results suggest that the term "iso-orientation slabs (bands)" should probably be revised, because discrete iso-orientation bands were not revealed even by fine electrical recordings (Hubel and Wiesel 1974, 1977; Livingstone and Hubel 1984; Blasdel and Salama 1986; Bonhoeffer and Grinvald 1991). "Angle magnitude" maps show both the preferred angle, color-coded as before, and the magnitude of the response. The magnitude of the resultant vector (Blasdel and Salama 1986; Ts'o et al. 1990; Blasdel 1988) can be represented by the brightness of the color. Inspection of angle magnitude maps revealed that the magnitudes of the resultant vectors were small at the centers of pinwheels. This leads to a general darkening of the colors as the center of a pinwheel is approached, and a dark halo appears surrounding the singularity point. Inspection of the eight orientation maps resulting from the various stimulus orientations showed that the responses of a population of cells in the vicinity of the center of a pinwheel were all large. To examine whether this reflected a significant reduction of orientation tuning of single cells or was a property of a heterogeneous population, we explored the electrical response properties of single cells. Three- to five-unit recordings were obtained at seven singularity points, previously mapped by optical imaging in vivo. In more than 75% of such recordings, we found that the cells at or near the singularity point (within 50 μm) were sharply tuned. Thus, the darkening near the singularity point may be due to a mixture of cells with different orientation preferences in this small region.

7.2.3.2 Relationships Between Pinwheels and Ocular Dominance Columns

The ocular dominance maps in the same patch of visual cortex were then evaluated. The resulting map shows dark and light parallel bands. These correspond to the right eye and left eye ocular dominance columns, respectively. To show their relationship to the pinwheels, we first drew thin lines on the map along the borders of ocular dominance columns. These borders were then superimposed on the map of orientation preference. From this superposition, a novel relationship between these two functional domains became apparent: most pinwheels were nearly centered on one of the ocular dominance columns. In maps obtained from three animals which included 141 pinwheel centers, 81% were located along the midline of an ocular dominance column (the midline was taken as a tenth of the width of a column). Most of remaining pinwheel centers were located close to ocular dominance borders as one might expect (see Discussion).

A second novel relationship was that the iso-orientation contours tend to cross borders of ocular dominance columns at right angles. Even in regions where ocular dominance columns are highly curved, the principle of orthogonal crossing was maintained. Hubel and Wiesel (1974) predicted that orientation bands would be orthogonal to borders of ocular dominance columns, but this was not demonstrated in subsequent studies (Hubel et al. 1978; Blasdel and Salama 1986). To quantify these observations, the distribution of angles at which the iso-orientation contours cross borders of ocular dominance columns was analyzed; 76% of the crossing angles were between 75° and 90°. Another long standing question has been whether all orientation domains are equally represented in the cortex. We found that on average, all orientations were in fact equally represented.

7.2.3.3 Relationship Between Pinwheels and Blobs

It was also of interest to determine how the third subsystem of cells, residing in CO blobs, is related to the first two subsystems. One possibility, proposed by Gotz (1988), is that the blobs are positioned exactly at the centers of pinwheels. However, when the locations of the blobs revealed by postmortem CO histology were compared with the locations of the pinwheel centers in exactly the same patch of cortex, the pinwheel centers usually did not coincide with centers of blobs.

The angle maps apparently show that continuous changes in orientation preference across the cortical surface (i.e., the pinwheels) were by and large not perturbed by the blobs. However, these angle maps do not provide any information about orientation tuning strength. To clarify whether the locations of the blobs coincided with cortical regions having poor (Livingstone and Hubel 1984; Ts'o and Gilbert 1988) or sharp orientation tuning, the orientation tuning in blobs and interblob regions was determined directly from the optical data. A comparison between orientation selectivity of blobs and interblob regions was made; 78% of the pixels that corresponded to blobs exhibited poor orientation selectivity (less than half the maximal selectivity index). In contrast, cortical neurons that were mostly in interblob regions were more sharply tuned. Our single-unit recordings did not encounter many neurons (less than 15%) with blob-like response properties within 50 μm of the seven singularity points we investigated. It showed, once again, that the centers of pinwheels did not coincide with blobs. In addition, we noticed that our results are fully consistent with the long, tangential penetrations described by Livingstone and Hubel (1984). In their data, the locations of blobs and locations of sudden and large jumps in orientation preference often did not coincide.

7.2.3.4 Lateral Connections Between Cortical Regions that Exhibit Monocular Responses

To explore the relationship between the system of long-range horizontal connections and the ocular dominance columns, we first revealed the ocular dominance architecture by in vivo imaging and then targeted biocytin injections at specific sites of interest within the optically imaged columns. A first hint as to the relationship of the horizontally connected patches to the ocular dominance columns was obtained by superimposing the pattern of the biocytin-labeled patches on the system of blobs revealed in neighboring sections stained for CO. It has been established that rows of CO blobs are located at the centers of ocular dominance columns (Horton and Hubel 1981), so they can be used as indirect marker for the location of these columns. The results showed that injections at the center of a row of CO blobs produced patches that tended to "skip" the neighboring row, suggesting a tendency of long-range lateral connections to connect to the same eye columns, and avoided the opposite eye columns. Occasionally, one could also observe patches that clearly violated this rule, e.g., the patch residing in between the CO rows (presumably in a more binocular region). In this case, the injection site overlapped a CO blob and the patches produced by this injection targeted mostly other CO blobs, in apparent confirmation of the report that CO blobs are selectively interconnected (Livingstone and Hubel 1984). To obtain a more accurate description of the relationship between lateral connections and ocular dominance columns, we relied on the optical maps. As expected from the relationship to the CO patterns, the connections were found mostly clustered in the same eye column as that of the injection site and showed a tendency to skip the contralateral eye column. A quantitative estimate of this phenomenon was performed in which the eye preference of cortical neurons within the patches of lateral connections was compared to that found at the injection site. The histograms we obtained show the distribution of pixel values in the optical map either at the injection site or at the patches as a function of ocularity. The results show that 87% of the pixels at the injection site had the same ocularity index as that found at the center of the injection site. In the patches, the distribution was wider: only 65% of the pixels had the same ocularity index as that found at the center of the injection site. Despite this wider distribution, none of the patches linked regions dominated exclusively by the opposite eye, further indicating that the lateral projections were far from being random. Thus, the intrinsic lateral connections appear to be quasi-specific with respect to ocularity.

7.2.3.5 Lateral Connections Between Binocular Regions

To test whether binocular regions are selectively interconnected, we targeted injections, in three cases, at binocular regions, i.e., at transition zones between right and left eye columns. Interestingly, the patches produced by these injections showed a clear tendency to reside in binocular regions, avoiding the highly monocular cortical regions of both eyes. This was also reflected in the quantitative analysis comparing the distribution of the degree of ocularity at the injection site with that of the patchy projections. In this experiment, only 5% of the injected label spread into monocular regions (labeled right and left) and similarly, only 10.5% of the patches overlapped with highly monocular regions.

7.2.3.6 Lateral Connections Among Orientation Domains

To examine how orientation domains are interconnected, we first obtained high-resolution optical maps of orientation preference. Injections in these domains revealed that the majority of patches tended to be connected to orientation domains similar to that of the injection site. A quantitative analysis of this relationship showed that at this injection site, neurons exhibited preferential responses to orientations of 180$\pm$22.5, whereas at the patches only 66% had the same orientation tuning, indicating some level of divergence of connections to sites that exhibit different orientation preferences.

To obtain a more complete picture of the orientation specificity of connected patches, biocytin-labeled connections were superimposed on a high-resolution orientation-preference angle map obtained by optical imaging. Quantitative analysis of these optical maps and the histology were derived from four angle maps obtained from experiments in which the injection sites contained a narrow range of orientations; the orientation distributions within the injection sites and patches peaked around the same orientation index, confirming the qualitative impression that matched orientation columns tended to be linked. However, the distribution of orientation preferences at the patches was wider than that found at the injection sites.

To examine whether the deviating patches were specifically targeting poorly tuned cortical regions, we repeated the analysis, but excluded all the regions that were revealed by the magnitude maps to be poorly tuned. The results of this analysis remained essentially unchanged, deviating by less than $\pm 1.5\%$. There was little, if any, overlap between these injection sites or the patches and the CO blobs. To conclude, long-range connections show a tendency to link similar orientation domains with a divergence of about $30 \pm 10\%$.

7.2.3.7 The Short-Range Connections

Inspection of all the injection sites revealed they were all surrounded by a massive "halo" of axons and dendritic arbors within a distance of about 400 μm in diameter. These connections are not merely fibers on their way to the patches, but carry numerous synaptic contacts uniformly distributed along their length, as shown in detail elsewhere (Amir et al. 1993). The dendritic arbors of cortical neurons showed a tendency to cross freely through diverse orientation and ocular dominance domains. Similarly, short-range axonal connections, unlike the long-range patches, were highly distributed and did not appear to selectively connect to particular functional domains. On close examination, numerous individual fibers carrying synaptic boutons were followed from the site of injection, crossing a diverse set of orientation domains along their way.

7.2.4 Discussion

7.2.4.1 The Relationship Between Three Subsystems of the Functional Architecture of Macaque Primary Visual Cortex

The Pinwheel-Like Architecture of Orientation Preference. The pinwheels seen here are smoothly connected to each other across the borders of ocular dominance columns. Connected pinwheels frequently had opposite directions of orientation preference changes, i.e., clockwise versus counterclockwise (Gotz 1988). However, we could not find a consistent rule for the pairing of neighboring pinwheels. It appears that during development, some points centered on some patches of ocular dominance columns behave like a seed for a crystal around which the orderly petals of various orientation domains form pinwheels. Such a developmental process, probably a dynamic one, requires only short range interactions of up to 400 μm. In contrast, maintaining an orderly arrangement for clockwise and counterclockwise pinwheels requires a much greater interaction range and a formidable remodeling of synaptic connections during development.

Is Monkey Primary Visual Cortex Composed of a Regular Mosaic of Repeating Fundamental Modules? Here we define a fundamental module as the smallest cortical area that includes a complete representation of each of the three subsystems. This term is probably related to "hypercolumns" (Hubel and Wiesel 1977), a term that has been used by many authors to describe various types of repeating modules and therefore has become somewhat ambiguous. An intriguing question has been whether the cortex is built of repeating fundamental modules. A nearly perfect mosaic of repeating modules was observed here only for a submodule, composed of a pinwheel centered on a patch of an ocular dominance column. Such "monocular" pinwheels are continuous with one another along and across ocular dominance boundaries. Given that a pinwheel can process visual information derived from all orientations detected by one eye, one might suspect that two adjacent patches of complementing ocularity, each containing one pinwheel and portions of a few blobs, form the fundamental processing module in primary visual cortex. Many such fundamental modules were indeed found; however, at present we cannot define a fundamental module in a way that would result in a regular mosaic of repeating units as defined above.

Multiple Mosaics of All Subsystems as an Alternative to a Single Mosaic of Fundamental Modules. It appears either that we failed to see the regular mosaic-like arrangement of fundamental modules hidden in our data or that such an arrangement does not exist in the primary visual cortex of the macaque monkey. What we see in our data is a mosaic-like arrangement of blobs and mosaic-like arrangements for each of the partially overlapping orientation domains. Hubel and Wiesel (1977) showed that during tangential penetrations, the electrode must be advanced approximately 2 mm across the cortical surface (approximately 2 hypercolumns) to locate cells that do not have overlapping receptive fields. The size of each fundamental module we detected was approximately 800 μm by 400 μm (eccentricity of 6°–8° close to the V1/V2 border). Thus, at least four fundamental modules (2×2) processed some information from a single point of a retinal image in parallel (in the upper cortical layers). In such a large area, we observed a redundancy of the individual domains in the regular mosaics of blobs and of orientation domains. This organization of subsystems can serve at least two important objectives. First, the redundancy can functionally compensate for imperfections in the mosaic-like layout of fundamental modules. A regular mosaic of fundamental modules is actually not required, because regular mosaics of all the individual subsystems are adequately represented. Second, this redundancy provides an abundant neuronal substrate for local computations and population coding of the complex visual scene thoroughly analyzed by the primate visual cortex.

7.2.4.2 The Relationship Between the Functional Architecture and Patches of Intrinsic Lateral Connections in Macaque Primary Visual Cortex

A Binocular Subsystem. Several studies have indicated that ocular dominance in upper cortical layers varies continuously from monocular regions dominated by the right eye through binocular regions to monocular regions dominated by the left eye and so on. One interpretation of these results is that the ocular dominance system is built up by two sets of columns, one for each eye. An alternative organization, fully consistent with the above findings, is an ocularity system that consists of at least three sets that smoothly blend into one another, two monocular sets and a binocular system (LeVay 1988a). It is interesting to note that the connections found in the present study seem to confirm the latter suggestion by revealing that binocular regions possess a distinct set of interconnections. The present results, however, do not imply that these subsystems are discrete, exhibiting sharply defined borders.

Specificity and Precision of Long-Range Intrinsic Connections. Our results demonstrate a consistent trend in the relationship between intrinsic connections and the underlying functional architecture; functional domains sharing similar response properties tended to be linked to each other. In that respect, our results support the conclusions reported for connections among orientation domains by Gilbert and Wiesel (1979) in cat area 17 and argue against the notion that orthogonal orientation domains are interconnected (Matsubara et al. 1985). For the ocular dominance system, our results are compatible with the suggestion, based on cross-correlation studies, that monkey ocular dominance columns of the same eye are interconnected (Ts'o and Gilbert 1988). Quantitative analysis of the data revealed that the targeting of functional domains by the lateral connections occasionally diverged to dissimilar functional domains. Thus, to achieve a high degree of precision in such a system, some form of population coding may be required to "average" out of the connectional deviations. However, we do not rule out the alternative possibility that the "deviating patches" target, in a precise manner, a subset of neurons which belong to a different, yet undetermined, columnar subsystem that partially overlaps the ocular dominance and orientation domains imaged here.

In conclusion, the balance between specificity and diversity manifested by the long-range horizontal connections produces a potential for a vast spectrum of neuronal response properties, while preserving the precision and orderly organization at the level of entire neuronal populations.

This anatomical substrate may be involved in an array of functions ranging from computing interpolations, enhancing contrast sensitivity, and mediating interactions of responses beyond the classical receptive field and plastic changes in cortical response properties in the primary visual cortex of the macaque.

Acknowledgements. We thank Drs. D.Y. Ts'o, D. Malonek, C.D. Gilbert, and T.N. Wiesel for their contributions, and Drs. R. Tootell, E. Kaplan, L.C. Katz, S. Leytus, S. Barash and E. Achissar for their useful comments. This research was supported by a grant from the Israeli Academy of Arts and Sciences to R. M. and by Margaret Enoch, The Schilling and the Green Foundations to A. G.

References

Amir Y, Harel M, Malach R (1993) Cortical hierarchy reflected in the organization of intrinsic connections in macaque monkey visual cortex. J Comp Neurol (in press)

Bartfeld E, Grinvald A (1992) Relationships between orientation-preference pinwheels, cytochrome oxidase blobs, and ocular dominance columns in primate striate cortex. Proc Natl Acad Sci USA 89: 11905–11909

Blasdel GG (1988) In: Lund JS (ed) Sensory processing in the mammalian brain: neural substrates and experimental strategies. Oxford University Press, Oxford, pp 242–268

Blasdel GG (1992) Differential imaging of ocular dominance and orientation selectivity in monkey striate cortex. J Neurosci 12: 3139–3161

Blasdel GG, Salama G (1986) Voltage-sensitive dyes reveal a modular organization in monkey striate cortex. Nature 321: 579–585

Bonhoeffer T, Grinvald A (1991) Iso-orientation domains in cat visual cortex are arranged in pinwheel like patterns. Nature 353: 429–431

Braitenberg V, Braitenberg C (1979) Geometry of orientation columns in the visual cortex. Biol Cybern 33: 179–186

Chance B, Cohen P, Jobsis F, Schoener B (1962) Intracellular oxidation-reduction states in vivo. Science 137: 499–508

Frostig R, Lieke EE, Ts'o DY, Grinvald A (1990) Cortical functional architecture and local coupling between neuronal activity and the microcirculation revealed by in vivo high-resolution optical imaging of intrinsic signals. Proc Natl Acad Sci USA 87: 6082–6086

Gilbert CD, Wiesel TN (1979) Morphology and intracortical projections of functionally identified neurons in cat visual cortex. Nature 280: 120–125

Gilbert CD, Wiesel TN (1983) Clustered intrinsic connections in cat visual cortex. J Neurosci 3: 1116–1133

Gilbert CD, Wiesel TN (1989) Columnar specificity of intrinsic horizontal and corticocortical connections in cat visual cortex. J Neurosci 9: 2432–2442

Gotz KG (1988) Cortical templates for the self-organization of orientation-specific d- and l-hypercolumns in monkeys and cats. Biol Cybern 58: 213–223

Grinvald A, Lieke E, Frostig RD, Gilbert CD, Wiesel TN (1986) Functional architecture of cortex revealed by optical imaging of intrinsic signals. Nature 324: 361–364

Grinvald A, Frostig R, Lieke EE, Hildesheim R (1988) Optical imaging of neuronal activity. Phys Rev 68: 1285–1366

Grinvald A, Frostig RD, Siegal RM, Bartfeld E (1991) High resolution optical imaging of neuronal activity in awake Monkey. Proc Natl Acad Sci USA 88: 11559–11563

Hill DK, Keynes RD (1949) Opacity changes in stimulated nerve. J Physiol 108: 278–281

Horton CH, Hubel DH (1981) Regular patchy distribution of cytochrome oxidase staining in primary visual cortex of macaque monkey. Nature 292: 762–764

Hubel DH, Wiesel TN (1968) Receptive fields and functional architecture of monkey striate cortex. J Physiol 195: 215–243

Hubel DH, Wiesel TN (1974) Uniformity of monkey striate cortex: a parallel relationship between field size, scatter, and magnification factor. J Comp Neurol 158: 267–294
Hubel DH, Wiesel TN (1977) Functional architecture of macaque monkey visual cortex. Proc R Soc Lond B 198: 1–59
Hubel DH, Wiesel TN, Stryker MP (1978) Anatomical demonstration of orientation columns in macaque monkey. J Comp Neurol 177: 361–380
Jobsis FF, Keizer JH, LaManna JC, Rosental MJ (1977) Reflectance spectrophotometry of cytochrome aa3 in vivo. J Appl Physiol 43: 858–872
King MA, Lois PM, Hunter BE, Walker DW (1989) Biocytin: a versatile anterograde neuroanatomical tract-tracing alternative. Brain Res 497: 361–367
LeVay S (1988a) The patchy intrinsic projections of visual cortex. In: Hicks TP, Benedek G (eds) Progress in brain research vol 75. Elsevier, Amsterdam
LeVay S (1988b) Patchy intrinsic projections in visual cortex, area 18, of the cat: morphological and immunocytochemical evidence for an excitatory function. J Comp Neurol 269: 265–274
Levay S, Hubel DH, Wiesel TN (1975) The pattern of ocular dominance columns in macaque visual cortex revealed by a reduced silver stain. J Comp Neurol 159: 559–576
Livingstone MS, Hubel DH (1984) Anatomy and physiology of a color system in the primate visual cortex. J Neurosci 4: 309–380
Malach R (1992) Dendritic sampling across processing streams in monkey striate cortex. J Comp Neurol 315: 303–312
Malach R, Amir Y, Harel M, Grinvald A (1993) Novel aspects of columnar organization revealed by optical imaging and in-vivo targeted biocytin injections in primate striate cortex. Proc Natl Acad Sci USA (in press)
Matsubara J, Cynader M, Swindale NV, Stryker MP (1985) Intrinsic projections within visual cortex: evidence for orientation-specific local connections. Proc Natl Acad Sci USA 82: 935–939
McGuire BA, Hornung JP, Gilbert CD, Wiesel TN (1984) Patterns of synaptic input to layer 4 of striate cortex. J Neurosci 4: 3021–3033
Mountcastle B (1957) Modality and topographic properties of single neurons of cat's somatic sensory cortex. J Neurophysiol 20: 408–434
Rockland KS, Lund JS (1982) Widespread periodic intrinsic connections in the tree shrew visual cortex. Science 215: 1532–1534
Rockland KS, Lund JS (1983) Intrinsic laminar lattice connections in primate visual cortex. J Comp Neurol 216: 303–318
Salzberg BM, Davila HV, Cohen LB (1973) Optical recording of impulses in individual neurones of an invertebrate central nervous system. Nature 246: 508–509
Tasaki I, Watanabe A, Sandlin R, Carnay L (1968) Changes in fluorescence, turbidity and birefringence associated with nerve excitation. Proc Natl Acad Sci USA 61: 883–888
Tootell RBH, Silverman MS, Hamilton SL, Switkes E, De Valois RL (1988) Functional anatomy of macaque striate cortex II: retinotopic organization. J Neurosci 8: 1610–1624
Ts'o DY, Gilbert CD (1988) The organization of chromatic and spatial interactions in the primate striate cortex. J Neurosci 8: 1712–1727
Ts'o DY, Frostig RD, Lieke EE, Grinvald A (1990) Functional organization of primate visual cortex revealed by high resolution optical imaging. Science 249: 417–420
Tootell RB, Silverman MS (1985) Two methods for flat-mounting cortical tissue. J Neurosci Methods 15: 177–190

7.3 The Organization, Connectivity and Interactions of Color Processing in the Visual Cortex

D.Y. Ts'o and A.W. Roe
Division of Neuroscience, Baylor College of Medicine, Houston, Texas

Color perception is one of the many products extracted by the visual cortex from our rich visual world. As such, it provides a powerful cue for visual recognition and segmentation. The visual cortex has evolved a complex machinery for making the most of the limited spectral information afforded by the eye. These mechanisms underlying color perception are poorly understood. Nevertheless, we can discern some of the principles and architecture of color processing through studies of the functional organization, connectivity and response properties of the color-selective cells of the visual cortex. Also of particular relevance are the relationship and interactions between the color pathways and the neural components responsible for processing the other aspects of visual perception, such as orientation, movement and depth. In particular, the classical studies of Hubel and Wiesel established the columnar organization of such properties as ocular dominance and orientation selectivity, and helped formulate a modular description of the organization of visual cortex (Hubel and Wiesel 1968, 1974, 1977). One fundamental demonstration arising from these studies is the interlacing of several functional maps, e.g., ocular dominance and orientation, within the same cortical structure. This arrangement neatly facilitates the processing of multiple dimensions of visual attributes for a given region in visual space.

Recent findings have begun to piece together the relationship between these early studies on functional organization and the architecture of color processing, both in primary visual cortex (V1) and of extrastriate visual areas. An important discovery was the patchy staining of visual cortex for the mitochrondrial enzyme cytochrome oxidase (Wong-Riley 1979). In V1, this staining procedure, in tangential section, reveals a lattice of oval patches or blobs that physiological studies suggest are involved in color processing (Fig. 1). In other visual areas, cytochrome oxidase histology shows quite different staining patterns. For example, in V2, instead of blobs one sees a series of darkly stained bands or stripes. Electrophysiological studies show that these stripes contain clusters of cells selective for color or disparity. These and other lines of evidence indicate that although each visual area is highly organized, the specific properties that are organized and the geometry of the organization are different from area to area. Furthermore, response properties that are well-organized (e.g., retinotopy, ocular domi-

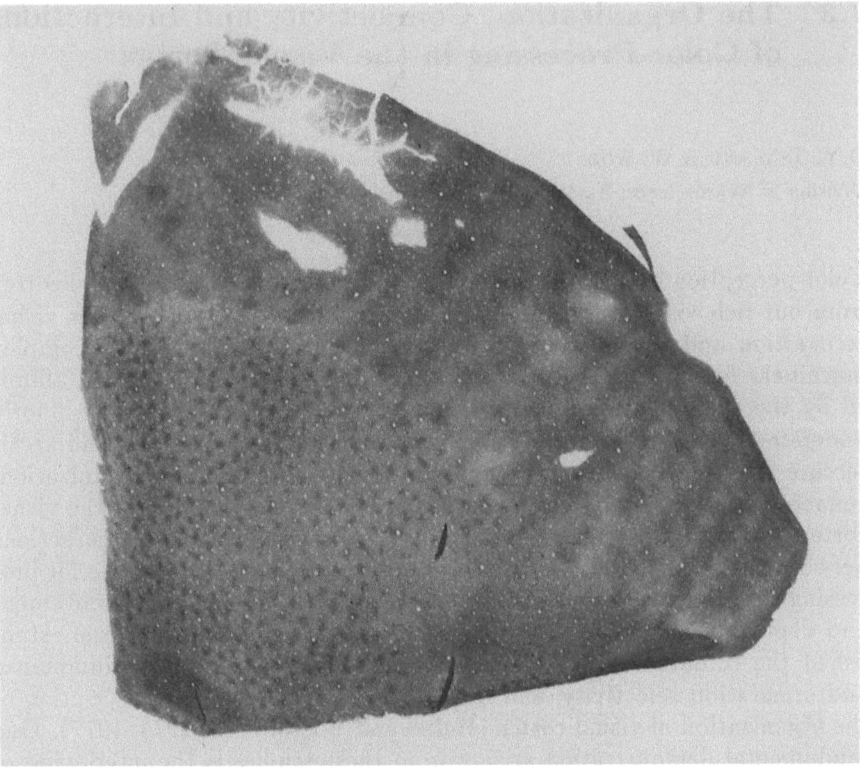

Fig. 7.3–1. Cytochrome oxidase histology of V1 and V2. A tangential section from visual cortex of the squirrel monkey, showing the cytochrome oxidase-rich blobs of V1 (the polka dotted pattern). The lightly staining regions between neighboring blobs are known as the interblobs. In V2 is seen the thick, thin and pale stripes

nance and orientation in V1), in earlier cortical stages, often become less well or not-at-all organized in higher areas (e.g., V2, MT). Other properties, on the other hand, though present at early stages (e.g., disparity or directionality in V1), are not well-organized until higher areas. This observation suggests that there may be some theoretical or developmental limit on the number of properties that can be organized in the cortical structure. Most importantly, this shifting of the organizing factors from area to area further reinforces the notion that the properties that are well-organized in an area indicate or even dictate the role of that area in visual processing.

In this chapter we will discuss the functional properties and organization of color cells in visual area V1 and V2, and their relationship to other visual cells. In addition, we will examine the segregation between these different functional compartments and the degree to which these compartments interact.

7.3.1 Functional Properties and Organization in V1

Physiological studies in V1 have found a wide range of response properties, several of which have been demonstrated to be systematically represented. In addition to a precise retinotopy, V1 was found to contain a pronounced columnar organization for ocular dominance and orientation. Also introduced by the classical work of Hubel and Wiesel was the notion of a hypercolumn, a complete set of columns of one type, representing, for example, all orientations, arranged in an orderly manner. Interwoven with the columns of one functional property are the columns of other functional properties, such that a complete set of hypercolumns for all properties form the basic module of cortical organization required to process a given region in visual space. Based on electrophysiological studies, the original "icecube model" of Hubel and Wiesel (1977) depicted an orthogonal arrangement of ocular dominance and orientation hypercolumns.

Functional Organization of Color. More recently, an additional important feature of the modular organization of the monkey striate cortex was revealed by the finding of a patchy labeling of striate cortex to a stain developed for the mitochrondrial enzyme cytochrome oxidase (Wong-Riley 1979; Horton and Hubel 1981; Humphrey and Hendrickson 1983; Horton 1984). High concentrations of this enzyme are thought to indicate long-term elevated metabolic activity. The regions of dense cytochrome oxidase reactivity, in tangential sections of macaque striate cortex, are centered on ocular dominance columns and laid out as a matrix of oval patches or blobs, approximately $150 \times 200\,\mu m$ each. Optical imaging studies have further showed that the blobs coincide with centers of monocularity (Ts'o et al. 1990a; Fig. 2).

Livingstone and Hubel (1984a) found that the blobs contained cells with receptive fields that were monocular, unoriented and often color selective. Many of these cells could be categorized into groups or types that were similar to types found in the lateral geniculate nucleus (LGN; Wiesel and Hubel 1966). However, several blob color cell types did not correspond to LGN types. One prominent example were cells described as double color-opponent, having spatially opponent fields with color opponency in both center and surround. Subsequent studies (Ts'o and Gilbert 1988) showed that the majority of these cells did not have true double color-opponent properties (Daw 1968), since their surround was always suppressive and the response to stationary isoluminant color contrast was poor. Accordingly, this class of color cell was named "modified type II," suggesting that their suppressive surround represents a modification of the LGN type II field. Color is then yet another functional dimension that is integrated into the cortical matrix. These data prompted a revision of the icecube model that

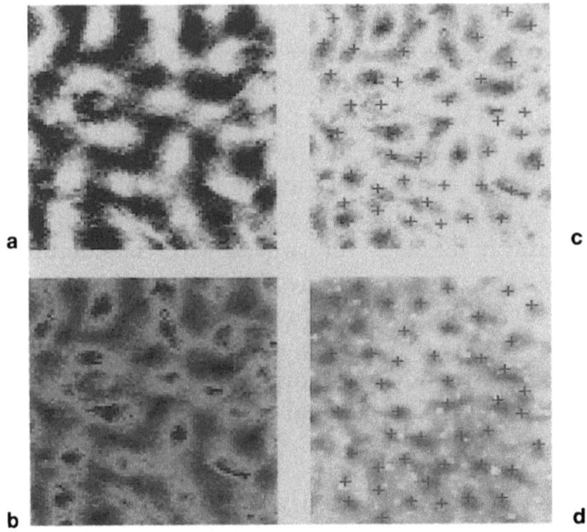

Fig. 7.3–2. The imaging of cytochrome oxidase-rich blobs by examining regions of highly monocular responses. **A** The ocular dominance map for a portion of V1. This map was color-coded (**B**) such that left-eye dominated activity is seen in red, right-eye in green and the extremes of left or right-eye dominated activity in blue. These blue regions lie in the middle of both ocular dominance bands and correspond to centers of high monocularity. **C** A re-coding of the map in **B**, producing a map of monocularity, where the darker regions are monocular and the whiter regions are binocular. This map was then compared to the post mortem cytochrome oxidase histology **D**. Cross marks were placed on the blobs seen in the histology **D**. This pattern of crosses was then transferred to the map of monocularity **C**. Comparison of the monocularity map with the cytochrome oxidase histology, showing excellent correspondence between the centers of monocularity and the blobs

incorporates the blobs into the centers of each ocular dominance column (Livingstone and Hubel 1984a).

Color-Opponent-Specific Blobs. The view of the organization of the blob color system has been further refined by the finding that blobs are color-opponency-specific. In addition to confirming (at least in the superficial cortical layers) a patchy or columnar organization of color in V1 (see also Gouras 1974; Michael 1981), Ts'o and Gilbert (1988) reported that single vertical electrode penetrations within blobs encounter color cells of only one color-opponency, either red versus green or blue versus yellow (see also Dow and Vautin 1987). Cells recorded in neighboring vertical penetrations within a single blob share the same color-opponency. These findings suggest that individual blobs are dedicated to the processing of one color-opponency system. In addition, blobs were found that did not contain any color cells,

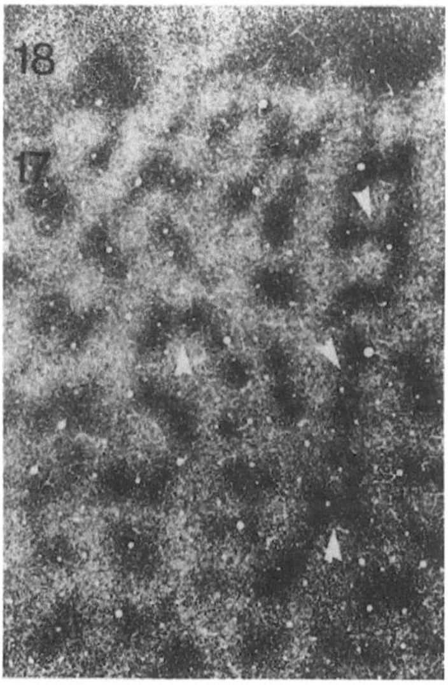

Fig. 7.3–3. Cytochrome oxidase histology of the blobs and bridges of V1 (area 17) of the macaque monkey. The blobs are located in the center of the ocular dominance columns, which run perpendicular (roughly bottom to top in this figure) to the V1/V2 (area 17/18) border. The bridges *(white arrows)* span neighboring blobs, both across opposite ocular dominance columns and along the same column. Also apparent is the pairing of ocular dominance columns such that the blobs of a given column are joined via bridges with the blobs of only one of its two neighboring ocular dominance columns. This pattern produces an appearance of ladders or railroad tracks, in which the bridges appear as the rungs or ties

but only contained broadband center-surround (type III) cells. This finding may be relevant in interpreting the role of the blobs in nocturnal primates such as the owl monkey that have very poor color vision.

The color-specific blobs across cortex do not share equal representation in V1. Based on multiple electrode penetrations into the perifoveal cortical representation of V1, Ts'o and Gilbert (1988) found a 3:1 ratio of red/green to blue/yellow blobs, paralleling biases found in the retina (5:2; see Gouras 1968; DeMonasterio and Gouras 1975; Schiller and Malpeli 1978) and the LGN (12:1; see Wiesel and Hubel 1966; Dreher et al. 1976; Krüger 1977). Furthermore, blue/yellow blobs seem to cluster together, suggesting a non-uniform or patchy distribution of blue cone inputs to cortex.

Bridges. An intriguing observation first made by Horton (1984) is that the ocular dominance columns often seem paired such that the blobs of a given left eye dominance band are connected by bridges to the blobs of only one neighboring right eye dominance band, and not the other (Fig. 3). This pattern of pairing gives the appearance in cytochrome oxidase histology, of railroad tracks or ladders, with the bridges appearing as ties or rungs, crossing two ocular dominance bands (the rails of the track). Horton speculated on the origins of this arrangement may be tied to a separation of the ocular dominance columns during development.

Recordings in the bridges suggest that they, like the blobs, also contain unoriented cells that are often color selective (Ts'o and Gilbert 1988; Landisman and Ts'o 1992). Thus the bridges are apparently an additional component of the V1 color system. Binocular, unoriented, color selective cells were found in bridges between blobs of opposite ocular dominance. This view has been confirmed by optical imaging studies designed to reveal the color-selective regions of cortex (Landisman et al. 1991; Landisman and Ts'o 1992) that directly show that the bridge regions are color-selective. Bridges have also been found spanning neighboring blobs of different color-opponency. Color response properties in such bridges are often found to be neither red/green nor blue/yellow, but mixed in spectral selectivity.

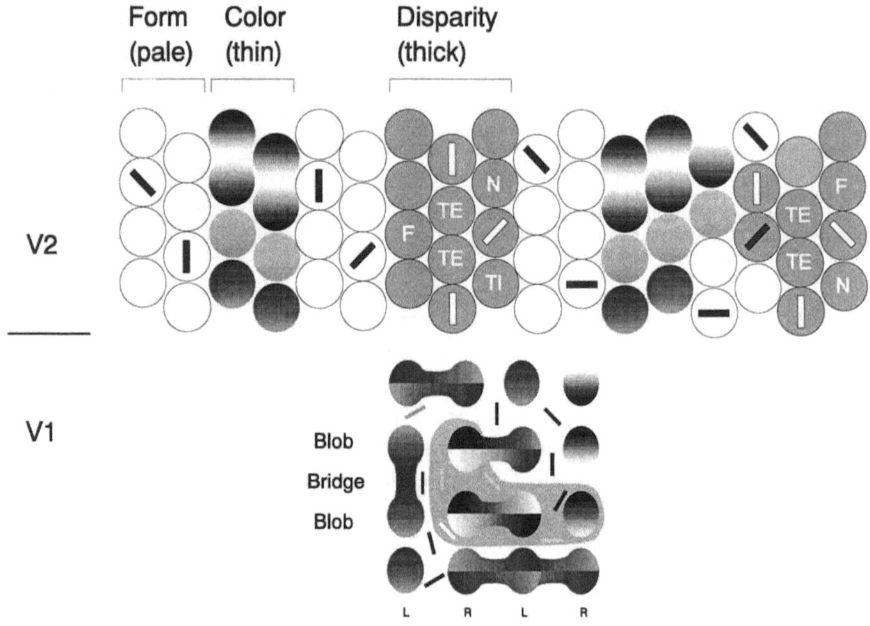

Fig. 7.3–4. Model of V1 and V2 functional organization. In V1 *(bottom half of figure)*, the red/green blobs outnumbered the blue/yellow blobs by 3:1. Bridges of color cells can be seen between neighboring blobs, running both along and across the ocular dominance columns. Color-oriented cells lie at the edges of the blob/interblob borders. At a coarse level, V2 consists of alternating thick, pale and thin stripes. Each stripe is actually composed of multiple patches or subcompartments with a particular functional specialization. V2 patches in the color stripes are selective for different color properties, while patches in the disparity stripes are responsive to a range of disparities, from near to far. Among oriented regions in pale and disparity stripes, there are regions with highly uniform orientation selectivity and columnar organization, as well as regions in which orientation is poorly organized and not columnar. There are regions in V2 where color and disparity stripe have joined together (depicted *on the right*). At the border of these regions, cells are tuned for both color and disparity

Interposed between the interblob and the blob regions are the "periblob" regions, in which color-oriented cells are often found (Dow and Vautin 1987; Ts'o and Gilbert 1988). Some initial studies using cross-correlation analysis indicate that color-oriented cells receive input from color blob cells with matching color specificity (Ts'o and Gilbert 1988). The distribution and receptive field properties of color-oriented cells and preliminary cross-correlation data all suggest that these cells represent an interplay between the blob and interblob regions or alternatively, between the form and color pathways.

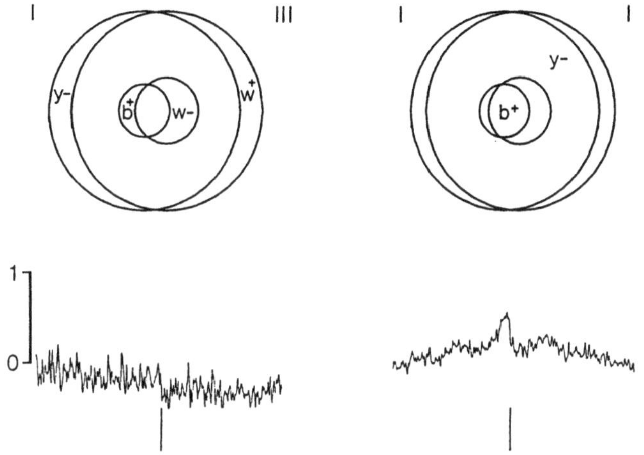

Fig. 7.3–5. *Left*, Blue-ON yellow-OFF type I to type III OFF-center correlogram. *Right*, Blue-ON yellow-OFF type I to Blue-ON yellow-OFF type I cell. Correlograms were normalized by the baseline correlation. The shape and pattern of the peak indicates an excitatory monosynaptic connection from one type I to the other. Spike counts: Na, 4378; Nb, 19370; contribution, 13.2%; effectiveness, 3.0% (see Ts'o et al. 1986 for methods)

Connectivity of Color Processing. Since the blob regions represent a major component of the functional and cytoarchitectonic organization of the monkey striate cortex, it is natural to ask how the blobs might interact with the non-blob regions. Livingstone and Hubel (1984b), using focal extracellular injections of horseradish peroxidase (HRP), demonstrated connections between adjacent blobs, and also between neighboring non-blob (interblob) regions. No connections between blob and interblob regions were found. This segregation of connections was confirmed and extended with results using cross-correlation analysis which showed that blob-blob connections existed between cells with matching color opponency (Fig. 5) and that interblob-interblob connections existed between cells with matching orientation preference (Ts'o and Gilbert 1988). Preliminary anatomical results, using focal injections at blob sites whose color opponency had been previously determined, seem to support this arrangement (Ts'o 1989). Cross-correlation studies have also shown that lateral connections also contribute to the construction of specific color-receptive field properties. For example, type II cells (cells with color-opponent centers but no surrounds) provide monosynaptic input to modified type II cells (cells with color-opponent centers and broad-band surrounds) and type I cells (basic center-surround color cells) and modified type II cells contribute directly to oriented color cells in V1 (Ts'o and Gilbert 1988). These findings then suggest that lateral connections within the cortex have a like-to-like pattern, both within the orientation domain and the color domain. Undoubtedly similar rules of connectivity may be found within other functional groupings.

Initial studies have shown that many of these cortical patterns of cross-correlation are not static, but rather change their strength and signature with changes in state or stimulus context. This type of finding, which also has been observed in the corticocortical interactions between V1 and V2 (Ts'o et al. 1993) as well as in other cortical areas (Aertsen et al. 1991; Ahissar et al. 1992) emphasizes the dynamic nature of cortical processing.

7.3.2 Functional Properties and Functional Organization in V2

Inputs to V2. Unlike the visual system in the cat, the second visual area (V2) in the primate receives its visual inputs primarily from V1. Direct geniculate input to V2 is sparse and scattered, and arises from the interlaminar zones of the LGN (Bullier and Kennedy 1983). Reversible inactivation of V1 by cortical cooling renders cells in V2 visually unresponsive (Girard and Bullier 1989; Schiller and Malpeli 1977), suggesting that V1 is the primary source of visual drive for V2.

Stripes in V2. In contrast to V1, cytochrome oxidase staining in the second visual area, or V2, reveals a pattern of bands or stripes of dense labeling, separated by pale interstripes (Tootell et al. 1983; Livingstone and Hubel 1982, 1984a; Humphrey and Hendrickson 1983; DeYoe and Van Essen 1985; Horton 1984; Shipp and Zeki 1985). The V2 stripes are functionally distinct and contain different populations of functional cell types. In contrast to V1, there is no apparent organization to ocular dominance in V2 and most V2 cells are binocularly responsive. Cells in the thin stripes are color-selective, some of which are similar to the modified type II cells seen in V1. Interstripe regions are characterized by oriented, non-color-selective fields that commonly exhibit end-stopping. Thick stripe cells are usually oriented, lack end-stopping, and are often disparity sensitive (DeYoe and Van Essen 1985; Hubel and Livingstone 1987; Ts'o et al. 1990b). In addition to these physiological findings, the pattern of V2 corticocortical connections seen in anatomical tracer studies suggests that thin, pale, and thick stripes are the V2 components of the color, form and motion/disparity pathways, respectively. Some reports, however, suggest a less clear-cut relationship between V2 cell properties and the V2 stripe organization. Studies using achromatic and isoluminant chromatic gratings to quantitatively characterize V2 cells have found a less striking segregation of cell types (Levitt and Movshon 1990). These differences in findings may be related to differences in the methods of cell classification or approaches to correlating physiology with anatomy.

Stripe Substructure. Within a single stripe in V2, there is some evidence to indicate clustering of cell properties such as color, disparity, and orientation (DeYoe and Van Essen 1985; Tootell and Hamilton 1989; Ts'o et al. 1989). Other functional methods used to image stripe organization such as 2-deoxyglucose labeling (Tootell et al. 1983; Tootell and Hamilton 1989) and optical imaging of intrinsic signals (Ts'o et al. 1990a,b, 1993) clearly indicate clustering of functional activity along the length of the stripes. In thin stripes, for example, physiological recordings encounter patches of red-green opponent dominated patches as well as blue-yellow dominated patches. Disparity-tuned thick stripes cells, as mentioned previously, tend to fall into near cell-dominated, far cell-dominated, tuned excitatory, and tuned inhibitory clusters (Ts'o et al. 1990b, 1991). In pale stripes, physiological recordings frequently encounter a cluster of similarly oriented cells adjacent to another cluster of oriented cells selective for a quite different orientation. It is clear that V2 stripes are, in fact, not homogeneous structures, but rather collections of functional modules (Fig. 4).

Higher Order Receptive Field Types in V2. Several types of higher order receptive field types have been described in V2. Two types of color cells described are the spot cell and the color-oriented cell. Spot cells (Baizer et al. 1977; Hubel and Livingstone 1987) are unoriented color-selective cells which respond optimally to a small spot. However, unlike a standard center-surround cell, the spot cell has a spatial independence such that a small spot (e.g., $0.25°$) is effective over a relatively large portion (e.g., $4° \times 4°$) of the visual field (Hubel and Livingstone 1985, 1987). These spot cells are not found in V1 and may represent a further elaboration in V2 of the modified type II cells of V1. One possible circuit would involve convergence from modified type II cells with matching color opponency and both eye dominances and with receptive field centers scattered over the large area corresponding to the spot cell's field (typical 4–8 times larger than a modified type II cell at the same eccentricity). Such convergence would require input either intrinsically from the modified type II cells within V2 or via corticocortical connections from V1 modified type II cells over a relatively wide area of cortex.

Another color cell type found in V2 is the color-oriented cell. While it is still unclear how oriented color cells are organized in either V1 or V2, preliminary evidence suggests that in V1 they are localized in the border regions between the color-selective blobs and the orientation-selective interblobs (Ts'o and Gilbert 1988). In V2, they are similarly located at such border zones, i.e., thin/pale borders. It is possible that oriented color cells in V2 receive input directly from color-oriented cells in V1 (and thus comprise a separate oriented color cell pathway) or, alternatively, they may be constructed de novo via convergent input from unoriented color cells in V1 or V2 (Roe and Ts'o 1992).

An organization apparently absent in V1 that is quite prominent in V2 is the organization for disparity. Disparity cells, defined by their selectivity for stimuli presented at a specific retinal disparity, are common to the thick stripes of V2. Most remarkable is the predominance of obligatory binocular cells, disparity cells which respond vigorously to stimulation of both eyes when stimuli are present at the optimal disparity, but are absolutely silent to monocular stimulation of any type. Studies describing near cells (crossed disparity) and far cells (uncrossed disparity) in V2 suggest a possible clustered organization of near cells and of far cells in the V2 thick stripes (Ts'o et al. 1990b, 1991). Another example of a mixed property cell occurring at a stripe border region is the color disparity cell. Seen in both cytochrome oxidase stained sections and in optical images of functional organization in V2 are the regions where thick and thin stripes appear to merge, resulting in the absence of an intervening pale stripe. These color/disparity borders are often characterized by patches of color disparity cells, and may result from the convergence of the two functional compartments (Ts'o et al. 1990b, 1991).

Form processing also gains in complexity in V2. Unlike standard orientation-selective cells, some V2 cells, termed "subjective contour" cells (von der Heydt and Peterhans 1989; Peterhans and von der Heydt 1989), respond to stimulus patterns which are perceived to contain contours not actually present in the stimulus. These investigators proposed a model for the construction of the V2 subjective contour cell from the convergence of appropriately positioned end-inhibited cells in V1 and/or V2.

Topography in V2. Yet another distinguishing feature of V2 is its retinotopic organization. It was previously known that receptive fields in V2 are larger than those in V1 and retinotopy in V2 is less precise than that in V1 (Gatass et al. 1981). With the discovery of cytochrome oxidase and the stripe organization it revealed in V2, a new issue arose concerning visuotopic mapping in V2. In V1, there is a point-to-point mapping of the visual field onto the cortex such that roughly a square millimeter of cortex, which spans a left eye and a right eye ocular dominance column, represents a specific locus in visual space. In V2, each thick-pale-thin-pale cycle spans approximately 4 mm of cortex. Therefore, roughly speaking, at any single isoeccentricity each cycle of thin-pale-thick-pale stripes potentially receives input from a cortical area spanning 4–5 ocular dominance hypercolumns, or approximately the width of 10 blobs. How does the region of visual space which is represented across several ocular dominance columns map onto a set of three functionally distinct stripes? Evidence from receptive fields recorded in long tangential penetrations across the stripes in V2 suggest that there is a multiple and discontinuous mapping of the visual field in V2 (Roe and Ts'o 1993). In other words, any region of visual space is represented at least in triplicate across the thin-pale-thick stripes in V2. An electrode traveling across any single stripe will encounter a continuous progression of receptive fields, followed by a jump back in receptive field progression at the stripe border, followed by another continuous progression representing the same region of visual space, in a different functional domain. Therefore, several ocular dominance columns of V1 must map in a topographic manner onto each of the color, form, and disparity stripes across the V1/V2 border. Results from anatomical studies (Livingstone and Hubel 1984a) are consistent with this prediction.

In conclusion, visual areas V1 and V2 are segregated into functional compartments specialized for such properties as orientation, color, and disparity. Not only is this segregation manifest in the clustering of cells with differing functional properties, but each major functional compartment in V2 apparently constructs its own representation of the visual field. The segregation is far from complete, however, and much interaction occurs between neighboring compartments. Whether the purpose of compartmentalization is simply for developmental or anatomical convenience or for facilitating specific computations within a given functional domain or both, it is likely that the organization of multiple feature domains within a cortical area reflects the processing strategies employed and is an important clue as to the role of that cortical area.

References

Aertsen A, Vaadia E, Abeles M, Ahissar E, Bergman H, Karmon B, Lavner Y, Margalit E, Nelken I, Rotter S (1991) Neural interactions in the frontal cortex of a behaving monkey: signs of dependence on stimulus context and behavioral state. J Hirnforsch 32: 735–743

Ahissar E, Vaadia E, Ahissar M, Bergman H, Arieli A, Abeles M (1992) Dependence of cortical plasticity on correlated activity of single neurons and on behavioral context. Science 257: 1412–1415

Baizer JS, Robinson DL, Dow BM (1977) Visual responses of area 18 neurons in awake, behaving monkey. J Neurophysiol 40: 1024–1037

Bullier J, Kennedy (1983) Projection of the lateral geniculate nucleus onto cortical area V2 in the macaque monkey. Exp Brain Res 53: 168–172

Daw NW (1968) Color-coded ganglion cells in the goldfish retina: extension of their receptive fields by means of new stimuli. J Physiol (Lond) 251: 167–196

DeMonasterio FM, Gouras P (1975) Functional properties of ganglion cells of the rhesus monkey retina. J Physiol (Lond) 251: 167–196

DeYoe EA and Van Essen DC (1985) Segregation of efferent connections and receptive field properties in visual area V2 of the macaque. Nature 317: 58–61
Dow BM, Vautin RG (1987) Horizontal segregation of color information in the middle layers of foveal striate cortex. J Neurophysiol 57: 712–739
Dreher B, Fukada Y, Rodieck R (1976) Identification, classification and anatomical segregation of cells with X-like and Y-like properties in the lateral geniculate nucleus of old-world primates. J Physiol (Lond) 258: 433–452
Gatass R, Gross CG, Sandell JH (1981) Visual topography of V2 in the macaque. J Comp Neurol 201: 519–539
Girard P, Bullier J (1989) Visual activity in area V2 during reversible inactivation of area 17 in the macaque monkey. J Neurophysiol 62: 1287–1302
Gouras P (1968) Identification of cone mechanisms in monkey ganglion cells. J Physiol (Lond) 199: 533–547
Gouras P (1974) Opponent-color cells in different layers of foveal striate layers. J Physiol (Lond) 238: 583–602
von der Heydt R, Peterhans E (1989) Mechanisms of contour perception in monkey visual cortex. I. Lines of pattern discontinuity. J Neurosci 9: 1731–1748
Horton JC (1984) Cytochrome oxidase patches: a new cytoarchitectonic feature of monkey visual cortex. Phil Trans Royal Soc Lond 304: 199–253
Horton JC, Hubel DH (1981) Regular patchy distribution of cytochrome oxidase staining in primary visual cortex of macaque monkey. Nature 292: 762–764
Hubel DH, Livingstone MS (1985) Complex-unoriented cells in a subregion of primate area 18. Nature 315: 325–327
Hubel DH, Livingstone MS (1987) Segregation of form, color, and stereopsis in primate area 18. J Neurosci 7: 3378–3415
Hubel DH, Wiesel TN (1968) Receptive fields and functional architecture of monkey striate cortex. J Physiol (Lond) 195: 215–243
Hubel DH, Wiesel TN (1974) Sequence regularity and geometry of orientation columns in the monkey striate cortex. J Comp Neurol 158: 267–293
Hubel DH, Wiesel TN (1977) Functional architecture of macaque monkey visual cortex. Proc R Soc Lond 198: 1–59
Humphrey AL, Hendrickson AE (1983) Background and stimulus-induced patterns of high metabolic activity in the visual cortex (Area 17) of the squirrel and macaque monkey. J Neurosci 3: 345–358
Krüger J (1977) Stimulus dependent color specificity of monkey lateral geniculate neurones. Exp Brain Res 30: 297–311
Landisman CE, Ts'o DY (1992) Color processing in the cytochrome oxidase-rich blobs and bridges of macaque striate cortex. Soc Neurosci Abstr 18: 592
Landisman CE, Grinvald A, Ts'o DY (1991) Optical imaging reveals preferential labeling of cytochrome oxidase-rich regions in reponse to color stimuli in areas V1 and V2 of macaque monkey. Soc Neurosci Abstr 17: 1089
Levitt JB, Movshon JA (1990) Receptive fields and functional architecture of macaque V2. Invest Ophthalmol Vis Sci 31:89
Livingstone MS, Hubel DH (1982) Thalamic inputs to cytochrome oxidase-rich regions in monkey visual cortex. Proc Natl Acad Sci USA 79: 6098–6101
Livingstone MS, Hubel DH (1984a) Anatomy and physiology of a color system in the primate visual cortex. J Neurosci 4: 309–356
Livingstone MS, Hubel DH (1984b) Specificity of intrinsic connections in primate primary visual cortex. J Neurosci 4: 2830–2835
Michael CR (1981) Columnar organization of color cells in layer 4Cb of the monkey striate cortex. J Neurophysiol 46: 587–604
Peterhans E, von der Heydt R (1989) Mechanisms of contour perception in monkey visual cortex. II. Contours bridging gaps. J Neurosci 9: 1749–1763
Roe AW, Ts'o DY (1992) Functional connectivity between V1 and V2 in the primate. Soc Neurosci Abstr 18: 11
Roe AW, Ts'o DY (1993) Visual field representation within primate V2 and its relationship to the functional distinct stripes. Invest Ophthalmol Vis Sci 34: 812
Schiller PH, Malpeli JG (1977) The effect of striate cortex cooling on area 18 cells in the monkey. Brain Res 126: 366–369
Schiller PH, Malpeli JG (1978) Functional specificity of lateral geniculate nucleus laminae of the rhesus monkey. J Neurophysiol 41: 788–797

Shipp S, Zeki S (1985) Segregation of pathways leading from area V2 to areas V4 and V5 of macaque monkey. Nature 315: 322–325
Tootell RBH, Hamilton SL (1989) Functional anatomy of the second visual area (V2) in the macaque. J Neurosci 9: 2620–2644
Tootell RBH, Silverman MS, De Valois RL, Jacobs GH (1983) Functional organization of the second cortical visual area of primates. Science 220: 737–739
Ts'o DY (1989) The functional organization and connectivity of color processing. In: Lam DM, Gilbert CD (eds) Neural mechanisms of visual perception. Portfolio, The Woodlands, Texas, pp 87–115
Ts'o DY, Gilbert CD (1988) The organization of chromatic and spatial interactions in the primate striate cortex. J Neurosci 8: 1712–1727
Ts'o DY, Gilbert CD, Wiesel TN (1986) Relationships between horizontal interactions and functional architecture in the cat striate cortex as revealed by cross-correlation analysis. J Neurosci 6: 1160–1170
Ts'o DY, Gilbert CD, Frostig RD, Grinvald A, Wiesel TN (1989) Functional architecture of visual area 18 of macaque monkey. Soc Neurosci Abstr 15: 161
Ts'o DY, Frostig RD, Lieke EE, Grinvald A (1990a) Functional organization of primate visual cortex revealed by high resolution optical imaging. Science 249: 417–420
Ts'o DY, Gilbert CD, Wiesel TN (1990b) Functional architecture of color and disparity in visual area 2 of macaque monkey. Soc Neurosci Abstr 16: 203
Ts'o DY, Gilbert CD, Wiesel TN (1991) Orientation selectivity of and interactions between color and disparity subcompartments in area V2 of macaque monkey. Soc Neurosci Abstr 17: 1089
Ts'o DY, Roe AW, Shey J (1993) Functional connectivity within V1 and V2: patterns and dynamics. Soc Neurosci Abstr (in press)
Wiesel TN, Hubel DH (1966) Spatial and chromatic interactions in the lateral geniculate body of the rhesus monkey. J Neurophysiol 29: 1115–1156
Wong-Riley, MTT (1979) Changes in the visual system of monocularly sutured or enucleated cats demonstrable with cytochrome oxidase histochemistry. Brain Res 171: 11–28

7.4 Anatomical, Physiological, and Computational Aspects of Hierarchical Processing in the Macaque Visual Cortex

D.C. Van Essen[1], B. Olshausen[1], J. Gallant[1], W. Press[1], C. Anderson[1], H. Drury[1], G. Carman[2], and D. Felleman[3]

[1]Department of Anatomy and Neurobiology, Washington University School of Medicine, St. Louis, Missouri,
[2]The Salk Institute, LaJolla, California
[3]Department of Neurobiology and Anatomy, University of Texas, Medical School, Houston, Texas

Our understanding of the organization and function of cerebral cortex has progressed dramatically since the 1960s, when the efforts of a few pioneering physiologists and anatomists ushered in the modern era of cortical research (see reviews by Mountcastle 1967; Hubel and Wiesel 1977; Creutzfeldt

1993). Much of this endeavor has involved studies of nonhuman primates, especially the macaque monkey, and much of it has focused on the visual system as a modality that is particularly amenable to experimental analysis. This chapter reviews recent work from our laboratory, in which we have used a combination of anatomical, physiological, and computational approaches to study the functional organization of the visual system in the macaque. The anatomical analysis indicates that the visual system is organized as an intricate hierarchy containing dozens of different subdivisions, hundreds of different pathways, and numerous distinct stages of processing. The complexity of the anatomy has motivated efforts to develop new approaches to representing, analyzing, and communicating information about cortical organization, and these efforts will be discussed briefly. Our physiological and computational studies have concentrated on two interrelated aspects of pattern recognition and form analysis. One issue concerns the way in which visual receptive field properties become progressively more complex at successive stages of the cortical hierarchy. Another issue concerns the mechanisms underlying translation and scale invariance, a critical aspect of pattern recognition that is intimately linked to the process of directed visual attention.

7.4.1 Organization of Visual Cortex

Although classical neuroanatomists recognized only a few cortical areas implicated in vision, it is now evident that there are literally dozens of different visual areas in primates. In the macaque monkey, 32 distinct visual areas have been reported that can be distinguished by their connectivity, architecture, topography, and/or physiological characteristics (Felleman and Van Essen 1991). The locations of these areas can be represented on a three-dimensional reconstruction of the entire cerebral cortex, as illustrated by the different shaded regions in Fig. 1. The reconstruction is based on the contours of layer IV rather than the pial surface, thereby allowing partial visualization into the depths of various sulci (see legend for details). In this scheme, there are 9 areas largely or exclusively involved in visual processing in the occipital lobe, 10 in the parietal lobe, 11 in the temporal lobe, and 2 in the frontal lobe. Some of these areas have sharply defined borders and have been identified unequivocally by several independent criteria. However, for the majority of areas there are substantial uncertainties concerning their exact borders, and in some regions the uncertainty is even greater because alternative partitioning schemes have been suggested (e.g., Seltzer and Pandya 1980; Desimone and Ungerleider 1989).

Part of the difficulty in attempting to resolve these uncertainties and discrepancies arises from the lack of an adequately accurate and consistent

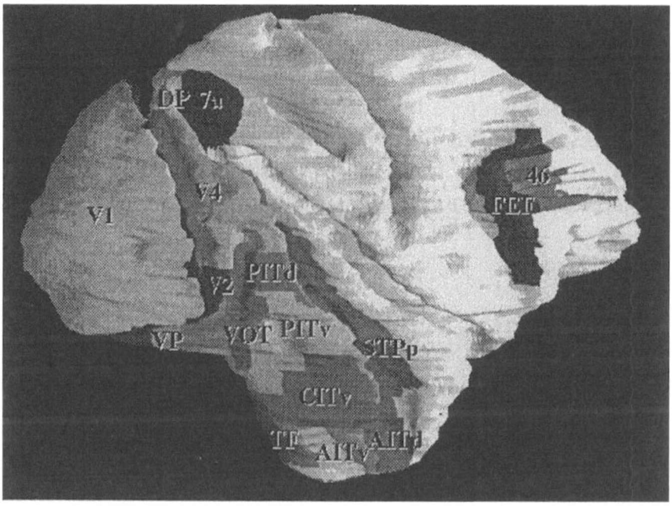

Fig. 7.4–1. This three-dimensional reconstruction shows a surface rendering of the right hemisphere of a macaque brain, including the locations of different visual areas as reviewed by Felleman and Van Essen (1991). The reconstruction is based on manually traced contours of layer IV of the cortex traced from a series of Nissl-stained horizontal sections spaced at 0.5 mm intervals. The contours were scanned into a Macintosh computer and transmitted to a Silicon Graphics IRIS graphics workstation. Contours were then traced, subsampled, and aligned relative to one another (with local contour adjustments made in order to compensate for distortions during histology). Corresponding nodes on adjacent sections were then linked by a semi-automated tessellation routine, leading to wire-frame representation which is based on 30 000 nodes in 84 sections. A surface rendering was then generated by creating a polygon mesh from the tessellation and rendering the polygons using Gouraud shading

way to represent and display experimental findings. Individual brains differ markedly in their size, gyral configuration, and in the size and exact location of different visual areas. These differences are often aggravated by inaccuracies and distortions in transferring areal boundaries onto drawings of the intact brain or onto manually prepared two-dimensional maps of the unfolded cortex (see Van Essen and Maunsell 1980). A major motivation underlying our creation of the computerized brain reconstruction shown in Fig. 1 is to provide an improved framework for addressing these difficulties. To attain this objective, though, a number of additional steps will be necessary. For example, it will be desirable to have displays that include the actual extent of the uncertainty limits for the borders between different areas. Another desirable option is to allow different partitioning schemes to be represented independently on the same three-dimensional brain reconstruction, in such a way that each scheme can either be viewed side by side simultaneously or can be viewed sequentially by switching from one scheme to the other on the same brain view. A third step will be to apply computerized unfolding techniques to the entire macaque cortex, so that the complete surface can be visualized in a format that minimizes distortions in surface area (see Carman 1990; Schwartz et al. 1989; Dale and Sereno 1993). Finally, there is a need for strategies to accurately transform data from the hemisphere in which they are acquired to a standard brain, by determining the optimal warping between the two structures (Miller et al. 1994). Collectively, these approaches should markedly enhance our ability to decipher the organization of the cortical mosaic.

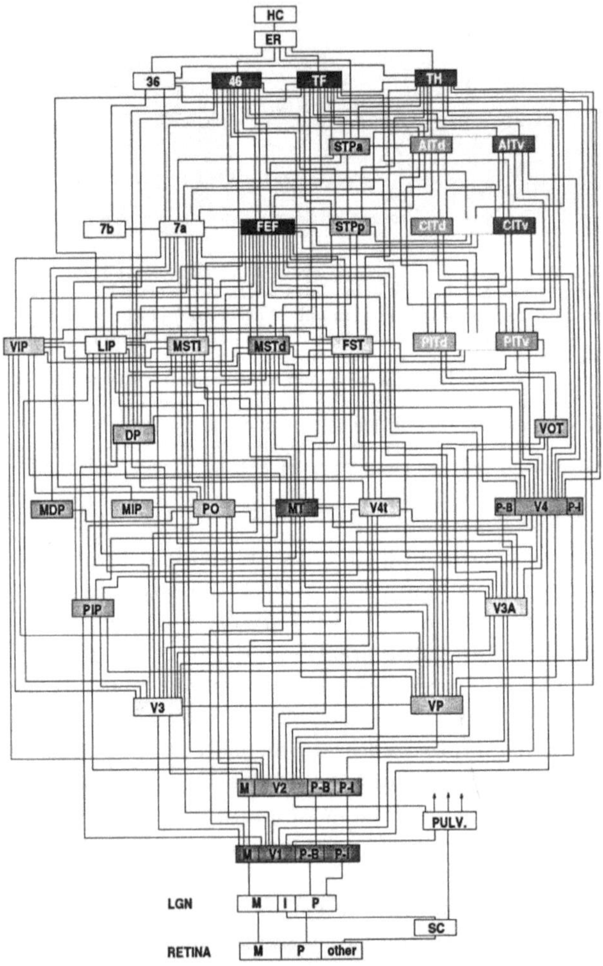

Fig. 7.4–2. A hierarchy of visual areas in the macaque, based on laminar patterns of anatomical connections. About 90% of the known pathways are consistent with this hierarchical scheme. Reprinted, with permission, from Van Essen et al. (1992)

Connections Among Cortical and Subcortical Centers. The 32 visual areas are extensively interconnected by 305 different pathways that have been reported by various laboratories. Most of these pathways are organized into reciprocal pairs, in which an asymmetry in laminar patterns of connectivity allows one direction to be assigned as ascending (forward) and the other direction as descending (feedback). Systematic application of this approach allows the formulation of an anatomically based hierarchical scheme (Felleman and Van Essen 1991), which is illustrated in Fig. 2. There are ten different levels in the cortical visual hierarchy, which starts with V1 and V2 at the bottom and extends to areas in the temporal, parietal, and frontal lobes at the top. Within this hierarchy there are multiple processing streams that can be distinguished by their connectivity, their physiological characteristics, and their contributions to specific aspects of behavior. Of particular relevance to this chapter is the distinction between high-level processing streams leading to the temporal versus the parietal lobe (Desimone and Ungerleider 1989). The ventral stream, leading to areas in inferotemporal cortex, has been implicated in tasks related to object recognition (the "what" stream), whereas the dorsal stream, leading to areas in posterior parietal cortex, has been implicated in analysis of spatial relationships between objects (the "where" stream). Both of these streams are implicated in the mechanisms of directed visual attention, but in quite different ways (see below).

Subcortical Connections. An additional level of complexity arises from the fact that each cortical area also has extensive connections with subcortical structures such as the superior colliculus and the pulvinar, some of which are reflected in the generic arrows emanating from the pulvinar in Fig. 2. The pulvinar contains multiple subdivisions whose relationship to different cortical areas is only partially understood. Part of the difficulty in evaluating these relationships arises from the complex three-dimensional organization of the pulvinar and the fact that its architectonically defined subdivisions are not in precise correspondence with subdivisions defined by topography or connectivity.

We have become particularly interested in the pulvinar because of its putative involvement in visual attention and the possibility that it plays a prominent role in regulating information flow through the cortex (Olshausen et al. 1993; see below). To facilitate the analysis of pulvinar organization and function, we have developed a graphically oriented database for analyzing its connections with different cortical areas. The aim is to provide easy access to results from individual anatomical studies and, in addition, to facilitate analyses that combine results from multiple studies (Press et al. 1993). As an example of the latter, Fig. 3 shows our preliminary re-

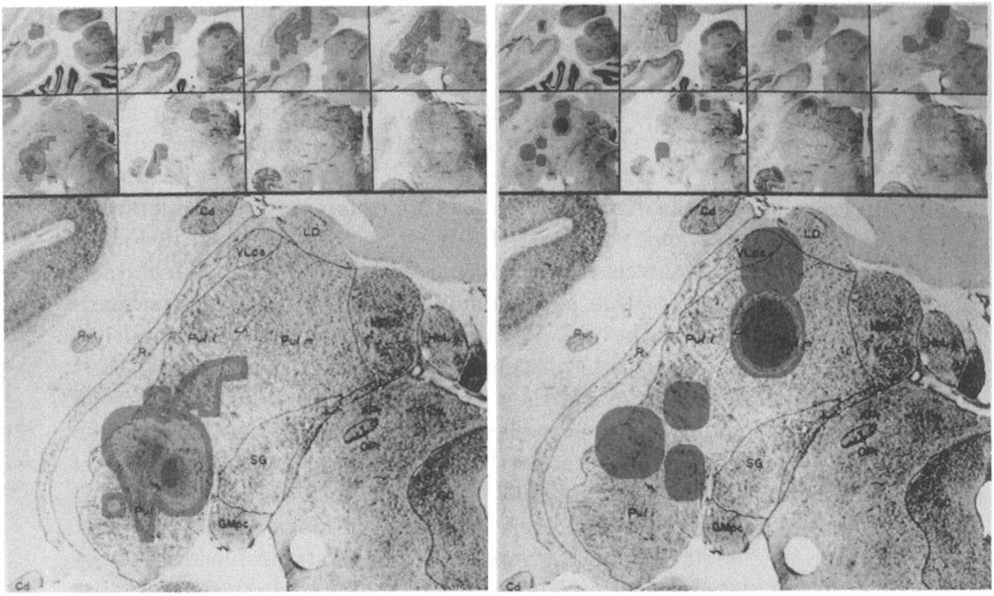

Fig. 7.4–3. **A,B** Graphical database of connectivity pattern between visual cortex and the pulvinar. Smoothed representation of the distribution of reported connections between area V4 and the pulvinar (**A**) and between area 7a and the pulvinar (**B**) are shown. The density of connections is based on illustrated case examples published by Trojanowski and Jacobson (1976), Benevento and Davis (1977), Lysakowski et al. (1988), and Baleydier and Morel (1992). The small images at the *top* are coronal sections from the atlas of Kusama and Mabuchi (1970). Clicking the mouse on any of these images brings it into the foreground window at the bottom for close inspection. Menu options for selecting different cortical areas, pulvinar subnuclei, or particular published studies are provided in adjoining windows that are not illustrated here

sults from an analysis that compares the connections to area V4 (Fig. 3A) with those to area 7a (Fig. 3B), based on studies by Trojanowski and Jacobson (1976), Benevento and Davis (1977), Lysakowski et al. (1988), and Baleydier and Morel (1992). The top rows are low resolution views of eight coronal sections through the thalamus from the published atlas of Kusama and Mabuchi (1970). The bottom panels are a higher resolution view of a single section that can be selected from this collection. V4 is primarily connected with the inferior pulvinar and with central and ventral portions of the lateral pulvinar, whereas area 7a is mainly connected with the dorsomedial nucleus of the pulvinar (Pdm) and with several portions of the lateral and medial nuclei. Thus, there is substantial overlap, but the bulk of the 7a versus V4 connections are to distinct pulvinar subregions.

7.4.2 Physiological and Computational Aspects of Hierarchical Processing

The hierarchical scheme for cortical organization described in the preceding section was based strictly on anatomical criteria, and it obviously needs to be tested using other approaches more closely related to cortical function. Two key issues arise from considering the differences in receptive field properties between cells at early and late stages of the ventral cortical stream, which, as already noted, has been implicated in form processing. Cells in V1 generally have very small classic receptive fields, respond well to simple stimuli such as bars and gratings, and show a high degree of selectivity along basic stimulus dimensions such as orientation, spatial frequency, length, width, and even curvature (see Van Essen 1985). In inferotemporal cortex, some cells appear to be selective for complex stimuli, including natural objects such as faces and hands (see Desimone 1991). Hence, one issue is how complex stimulus selectivities are generated from the low-level building blocks provided by V1. Equally importantly, inferotemporal neurons typically have very large receptive fields, and they display form selectivity over a wide range of stimulus sizes and positions. Thus, a second issue concerns the neurobiological mechanisms underlying position and scale invariance of individual cells (and of our perceptual system as a whole).

Complex Form Selectivity. To address the issue of complex receptive field properties, it is necessary to have a rich repertoire of visual stimuli available for a systematic and sensible exploration of the information space contained in natural images. As a step in this direction, we have recently begun using a family of stimuli that is: (a) motivated by computational theory and psychophysics; (b) mathematically defined and suitable for calculation of tuning curves along perceptually significant dimensions; and (c) well defined in spatial frequency and orientation composition, two characteristics known to be important at the level of V1. These stimuli, illustrated schematically in Fig. 4A, include conventional sine-wave gratings, hyperbolic gratings, and polar gratings (concentric, radial, and intermediate spiral patterns). We refer to the polar and hyperbolic gratings collectively as non-Cartesian stimuli and to the conventional gratings as Cartesian stimuli.

Using these stimuli, we studied 237 neurons in area V4 of anesthetized macaque monkeys (Gallant et al. 1992, 1993a). The most striking finding was that some cells responded very selectively to non-Cartesian stimuli, as illustrated in Fig. 4 using a display format in which response magnitudes are indicated by the size of the squares in each two-dimensional array. The cell in Fig. 4B responded best to concentric gratings of several different spatial frequencies, whereas the cell in Fig. 4C responded best to hyperbolic gratings of a particular orientation and spatial frequency range. For com-

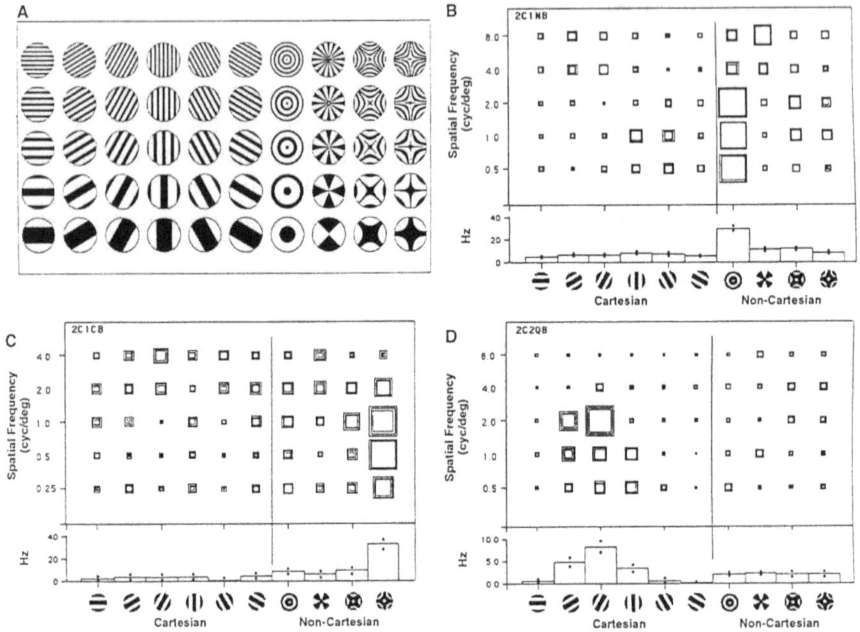

Fig. 7.4–4. **A** Stimulus set consisting of Cartesian and non-Cartesian (polar and hyperbolic) sinusoidal gratings. For clarity the gratings shown are square wave gratings, and the frequencies of high-frequency gratings have been reduced. *Rings* surrounding gratings represent borders of the stimuli, which were aligned as closely as possible with the "classic receptive field". Gratings used in experiments were quantized to 250 luminance levels and were displayed for 0.5 s on a neutral gray background. **B** Responses of a single V4 concentric cell to Cartesian and non-Cartesian gratings. The *abscissa* specifies grating type (six Cartesian grating orientations, concentric and radial gratings, and two hyperbolic grating orientations). The *ordinate* specifies spatial frequency. The marginal histogram gives the mean relative firing rates and standard errors of the means collapsed across spatial frequency. *Squares* in the two-way plot give normalized relative firing rates and standard errors. **C** Responses of a single V4 hyperbolic cell to Cartesian and non-Cartesian gratings. **D** Responses of a single V4 Cartesian cell to the same stimulus set. Reprinted, with permission, from Gallant et al. (1993a)

parison, Fig. 4D illustrates a cell that responded best to Cartesian gratings of appropriate orientation and spatial frequency.

About one-fifth of the cells studied in this way responded significantly better to the most effective non-Cartesian stimulus than to the most effective Cartesian stimulus. Few cells responded significantly better to Cartesian stimuli, especially in our recent tests, which have explored a wider range of non-Cartesian stimuli. The remaining cells in the population showed more complex tuning curves that typically included peaks in both Cartesian and non-Cartesian domains. However, a cluster analysis of their tuning properties indicates that most of these cells are closely related to non-Cartesian cells (Gallant et al. 1993b). Control experiments showed that selectivity for non-Cartesian stimuli was in general not an artifact of a simpler type of selectivity for length, width, or stimulus position. Some cells in V4 were so selective as to be essentially unresponsive to conventional bars or gratings. Since virtually all cells in V1 are reported to respond well to simple stimuli (Hubel and Wiesel 1968; Livingstone and Hubel 1984), it seems likely that selectivity for non-Cartesian stimuli is a property first established in extrastriate cortex (though not necessarily just in V4).

Previous studies in V4 have shown that cells can be selective for stimuli dimensions much smaller than the classical receptive field (Desimone and Schein 1987) and for complex, but less well-characterized, geometrical characteristics (Tanaka et al. 1986, 1991). Our results expand upon these findings by showing that many cells have well behaved tuning curves along a set of stimulus dimensions new to neurophysiology. The precise functional significance of non-Cartesian cells in V4 remains unclear, but we suspect that they play an important role in form perception. For example, one interesting possibility is that they represent important types of symmetry and structure contained in natural visual images.

Models for Translation and Scale Invariance. The positional and scale invariance mentioned above for neurons in inferotemporal cortex is a likely substrate for our ability to recognize complex objects independent of viewing distance and direction of gaze. These perceptual capabilities are all the more remarkable given that we are able to perform them in a rich natural environment containing many different objects (e.g., when we recognize a single face in a crowd of people). This capability is dependent upon the process of directed visual attention, whereby at any one instant our pattern recognition machinery is directed toward the analysis of a restricted region of the visual field (the window of attention) at a particular spatial scale. Attention is inherently a very dynamic process, as both the position and scale of attention can be switched on a time scale of about 100 ms.

One model for directed visual attention is based on the notion of dynamic routing of information flow through successive stages of the visual hierarchy (Olshausen et al. 1993). Figure 5A illustrates four major processing stages associated with visual attention in this model, and Fig. 5B shows plausible neurobiological substrates for mediating the different stages. The first stage involves selection of a target region within the image to which it is appropriate to direct attention, based on bottom-up cues of visual saliency and/or top-down cognitive influences. This is modeled as a winner-take-all mechanism physically based in the superior colliculus and/or posterior parietal cortex. The next stage involves the conversion of this winner-take-all output into a set of control signals to be used to regulate the flow of information through the cortical hierarchy. The pulvinar is a plausible candidate to serve this role, based in part on its extensive connections with different

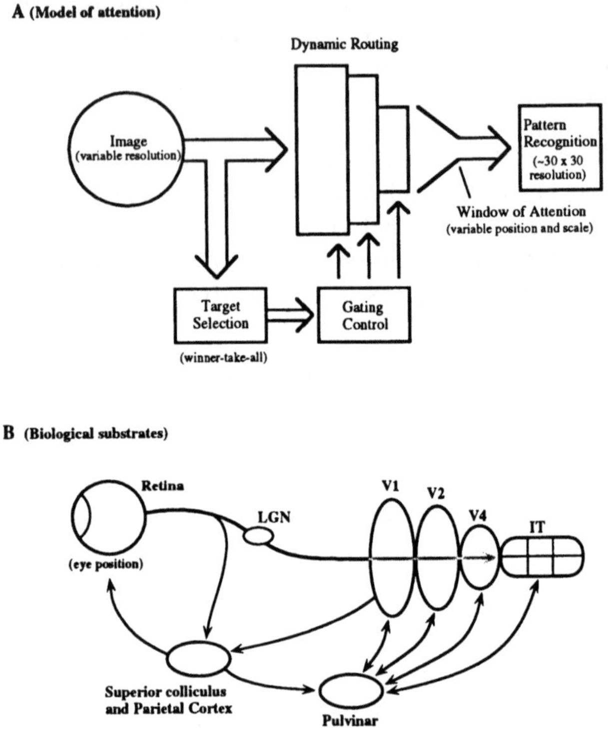

Fig. 7.4–5. A,B Schematic model of directed visual attention. The model involves four conceptually distinct stages illustrated schematically (**A**), and the proposed neurobiological substrates (**B**). *(1)* An initial step is to decide where, and at what spatial scale, attention should be directed. This *target selection* process can be driven by bottom-up (image-based) and/or top-down (volitional) factors, and we presume that it takes place in the posterior parietal cortex and in the superior colliculus. *(2)* This decision is implemented by generating a set of *control signals* that regulate how information flows through visual cortex. We suggested that the pulvinar is intimately involved in this control process. *(3)* At each major stage of the cortical hierarchy (visual areas *V1, V2, V4*, and inferotemporal area *IT*), ascending visual signal are *dynamically routed*, so that only the information from the desired window of attention reaches pattern recognition centers in the IT complex. This results in a transformation of the visual representation from retinal coordinated in V1 to object-based coordinates in IT. *(4)* Within IT, *pattern recognition* is carried out on the attended region using an associative memory that finds the best match to whatever patterns have been stored. *LGN*, lateral geniculate nucleus

cortical areas (see Fig. 3). The third stage involves the actual regulation of information flow through different cortical areas (especially V1, V2, and V4). This could be achieved if the pulvinar had a multiplicative or modulatory type of influence on the transmission of ascending corticocortical projections, such that only a subset of the converging inputs to each level are functionally enabled at any given moment. Appropriate coordination of information flow within as well as between levels would result in high-level pattern recognition centers in inferotemporal cortex receiving signals from only the desired portion of the retinal image and at the desired scale. The final stage of the process involves recognition of the appropriately shifted and scaled portion of the image inputs by matching to a stored template, presumably by some type of associative memory process.

A simplified version of this model has been successfully implemented in a computer simulation, which is capable of discriminating between characters presented at different positions and scales on a 22 $\times$ 22 input image (Olshausen et al. 1993). This demonstrates that the strategy is in fact capable of autonomous pattern recognition, and it provides a valuable substrate for exploring several challenging problems that must be faced in scaling up to more complex pattern recognition tasks. Obviously, though, it does not directly address the question of whether analogous processes operate in visual cortex. However, the model as outlined in Fig. 5 makes a number of strong predictions concerning the dynamic aspects of receptive field properties in visual cortex, the types of attention-related activity in the pulvinar, and the effects of pulvinar activity on the transmission of ascending corticocortical signals. The prediction of shifts in the position and scale of receptive fields is consistent with experimental observations reported by Moran and Desimone (1985) and is currently under more detailed investigation in our laboratory (Connor et al. 1993).

7.4.3 Concluding Remarks

Despite progress on a number of fronts, our understanding of cortical function clearly remains fragmentary and incomplete in many important respects. For example, within this chapter we have illustrated physiological data (non-Cartesian selectivity in area V4) that are in need of a computationally sound explanation and interpretation and we have discussed a computational model of visual attention that is only beginning to be put to a critical experimental test. These two lines of analysis have evolved largely independently, and they clearly need to be integrated into a more comprehensive framework. This framework needs to respect the constraints imposed by the underlying neuroanatomy, not only at the level of broad hierarchical relationships as reviewed in this chapter, but also at the level of the fine-grained connectivity matrix that links individual neurons within and between visual areas. Tackling this spectrum of issues will remain an exciting, albeit daunting, challenge for years to come.

Acknowledgments. This grant was supported by grants from the National Institutes of Health and the Office of Naval Research. We thank D. Bilitch for help with software development.

References

Baleydier C, Morel A (1992) Segregated thalamocortical pathways to inferior parietal and inferotemporal cortex in macaque monkey. Visual Neuroscience 8: 391–405

Benevento LA, Davis B (1977) Topographical projections of the prestriate cortex to the pulvinar nuclei in the macaque monkey: an autoradiographic study. Brain Research 30: 405–424

Carman GJ (1990) Mappings of the cerebral cortex. Thesis, California Institute of Technology

Connor CE, Gallant JL, Van Essen DC (1993) Effects of focal attention on receptive field profiles in area V4. Soc Neurosci Abstr 19: 974

Creutzfeldt OD (1993) Cortex cerebri. Performance, structural and functional organization of the cortex. (English translation) Mary Creutzfeldt, Göttingen

Dale AM, Sereno MI (1993) Improved localiation of cortical activity by combining EEG and MEG with MRI cortical surface reconstruction: a linear approach. J Cog Neurosci 5: 162–176

Desimone R (1991) Face-selective cells in the temporal cortex of monkeys. J Cog Neurosci 3: 1–8

Desimone R, Schein SJ (1987) Visual properties of neurons in aarea V4 of the macaque: sensitivity to stimulus form. J Neurophysiol 57: 835–868

Desimone R, Ungerleider LB (1989) Neural mechanisms of visual processing in monkeys. In: Boller F, Grafman J (eds) Handbook of neuropsychology, vol 2. Elsevier, New York, pp 267–299

Felleman DJ, Van Essen DC (1991) Distributed hierarchical processing in primate visual cortex. Cerebral Cortex 1: 1–47

Gallant JL, Braun J, Van Essen DC (1992) Position and phase independence of neural responses to non-conventional gratings in macaque area V4. Soc Neurosci Abstr 18: 592

Gallant JL, Braun J, Van Essen DC (1993a) Selectivity for polar, hyperbolic, and cartesian gratings in macaque visual cortex. Science 259: 100–103

Gallant JL, Connor CE, Rakshit S, Van Essen DC (1993b) Selectivity for polar and hyperboic gratings in macaque visual area V4: Multivariate classification and anatomical distribution. Soc Neurosci Abstr 19: 771

Hubel DH, Wiesel TN (1968) Receptive fields and functional architecture of monkey striate cortex. J Physiol (Lond) 195: 215–243

Hubel DH, Wiesel TN (1977) Functional architecture of macaque monkey visual cortex. Proc R Soc Lond B 198: 1–59

Kusama T, Mabuchi M (1970) Stereotaxic atlas of the brain of Macaca fuscata. University of Tokyo Press, Tokio

Livingstone MS, Hubel DH (1984) Anatomy and physiology of a color system in the primate visual cortex. J Neurosci 4: 309–356

Lysakowski A, Standage GP, Benevento LA (1988) An investigtion of collateral projections of the dorsal lateral geniculate nucleus and other subcortical structures to cortical areas V1 and V4 in the macaque monkey: a double label retrograde tracer study. Exp Brain Res 69: 651–661

Miller MI, Christensen GE, Amit Y, Grenander U (1994) A mathematical textbook of deformable neuro-anatomies. Proc Natl Acad Sci (USA) (in press)

Moran J, Desimone R (1985) Selective attention gates visual information processing in the extrastriate cortex. Science 229: 782–784

Mountcastle VB (1967) The problem of sensing and the neural coding of sensory events. In: Quarton GC, Melnechuk T, Schmitt FO (eds) The neurosciences. A study program. Rockefeller University Press, New York, pp 393–408

Olshausen BA, Anderson CH, and Van Essen DC (1993) A neurobiological model of visual attention and invariant pattern recognition based on dynamic routing of information. J Neurosci 13: 4700–4719

Press WA, Olshausen BA, Van Essen DC (1993) Analyzing connections between the pulvinar and visual cortex: an interactive graphical database. Soc Neurosci Abstr 19: 331

Schwartz EL, Shaw A, Wolfson E (1989) A numerical solution to the generalized mapmaker's problem. IEEE Trans Pattern Anal Mach Intell 11: 1005–1008

Seltzer B, Pandya DN (1980) Converging visual and somatic sensory cortical input to the intraparietal sulcus of the rhesus monkey. Brain Res 192: 339–351

Tanaka K, Saito H, Fukada Y, Moriiya M (1991) Coding visual images of objects in the inferotemporal cortex of the macaque monkey. J Neurophys 66: 170–189

Tanaka M, Weber H, Creutzfeldt OD (1986) Visual properties and spatial distribution of neurones in the visual association area on the prelunate gyrus of the awake monkey. Exp Brain Res 65: 11–37

Trojanowski JQ, Jacobson S, (1976) Areal and laminar distribution of some pulvinar cortical efferents in rhesus monkey. J Comp Neurol 169: 371–392
Van Essen DC (1985) Functional organization of primate visual cortex. In: Peters A, Jones EG (eds) Cerebral cortex, vol 3. Plenum, New York, pp 259–329
Van Essen DC, Maunsell JHR (1980) Two-dimensional maps of the cerebral cortex. J Comp Neurol 191: 255–281
Van Essen DC, Anderson CH, Felleman DJ (1992) Information processing in the primate visual system: an integrated systems perspective. Science 255: 419–423

7.5 The Putative Role of Synchrony in Cortical Processing

W. Singer
Max Planck Institute for Brain Research, Frankfurt am Main, Germany

7.5.1 Prolog

In 1966 Otto Creutzfeldt, then at the Max Planck Institute for Psychiatry in Munich, offered a series of lectures, together with the psychiatrist Paul Matussek, investigating possible relations between cognitive functions and brain mechanisms. These lectures focused on the problem of how processes distributed across the two cerebral hemispheres could be bound together to assure coherent perception and action. Creutzfeldt's proposal was that certain cognitive disorders might be due to disturbances of a postulated binding mechanism. At that time, Moruzzi's seminal work had already been very influential and neurophysiologists were aware that the ascending reticular arousal system (ARAS) and nonspecific thalamocortical projections played an important role in the control of telencephalic and, in particular, cortical processes. It had been established that these systems determine the frequency spectrum of the electroencephalogram (EEG) and assure simultaneous arousal of both hemispheres, and it has commonly been assumed that they establish phase coherence of EEG oscillations across different cortical areas and even hemispheres. Otto Creutzfeldt hypothesized that mechanisms which assured the phase locking of EEG oscillations might actually be involved in the coordination of distributed processes. It had also been known from the neuropsychological studies of Roger Sperry that patients whose telencephalic commissures had been severed suffered from the inability to coordinate the actions of the two hemispheres and to perceptually bind features of stimuli that extended across the visual hemifields. Consequently, Otto Creutzfeldt wondered to what extent the telencephalic commissures contributed to the maintenance of interhemispheric synchronization of EEG patterns. If they did, so his reasoning, this would be support of the possibility that synchronizing mechanisms might serve to coordinate spatially distributed processes in the brain. Creutzfeldt suggested to one of the students attending these lectures to investigate this possibility experimentally. The outcome was a thesis and a paper, written in German (Singer and Creutzfeldt 1969), which described interhemispheric synchronization, but failed to establish a synchronizing role of the callosum. Now, 25 years later, it has become a well-established notion that cortical processes are organized in a highly distributed way (for review see Ungerleider and Mishkin 1982; Felleman and van Essen 1991; Zeki et al. 1991), which accentuates the binding problem that Creutzfeldt had addressed in his lectures. Interestingly, it has again been proposed that synchronization of neuronal activity might serve as a mechanism for the coordination of distributed processes in the neocortex (Milner 1974; von der Malsburg 1985; Freeman 1975; Damasio 1990; Crick and Koch 1990; for a recent review see Singer 1993). In this chapter, experiments will be briefly reviewed which were designed to examine the putative role of synchrony in cortical processing.

7.5.2 Representations and the Binding Problem

Most perceptual objects can be decomposed into components, and the features of these components are generally not unique for a particular object. The individuality of objects results from the specific composition of elementary features and their relations rather than from the specificity of the component features. Hence, for a versatile representation of sensory patterns in the nervous system, three basic functions have to be accomplished: (1) elementary features need to be represented by neuronal responses, (2) responses to features constituting a particular object have to be distinguished and bound together in a flexible way, and (3) the specific relations between these features have to be encoded and preserved.

One way to achieve the grouping of features and to establish an unambiguous code for their specific relations is to connect the set of neurons which respond to the component features of a particular object to a higher-order neuron which will represent the object. If the thresholds of these higher-order neurons are adjusted so that each cell responds only to one particular combination of feature detectors, the responses of these higher-order neurons would provide an unambiguous description of the relations between the component features and hence would be equivalent to the representation of the pattern. In this scheme, the features of the object are bound together by convergence of fixed connections which link neurons representing component features with neurons representing the whole pattern. The relations between features are encoded by the specific architecture of these convergent connections.

Alternative proposals for the binding of component features are based on the assumption that representations of specific feature constellations consist of assemblies of a large number of simultaneously active neurons, which may be contained in a single cortical area but which may also be distributed over many cortical areas (Abeles 1991; Braitenberg 1978; Crick 1984; Edelman 1987, 1989; Edelman and Mountcastle 1978; Grossberg 1980; Hebb 1949; Palm 1982, 1990; Singer 1985, 1990; von der Malsburg 1985). The essential feature of assembly coding is that individual cells can participate at different times in the representation of different objects. The assumption is that just as a particular feature can be present in many different patterns, a neuron coding for this feature can be shared by many different representations. This reduces substantially the number of cells required for the representation of different objects and allows for considerably more flexibility in the generation of new representations.

Basic requirements for representing objects by such assemblies are: first, the responses of the cells responding to a visual scene need to be compared with one another and examined for possible "meaningful" relations; second, cells coding for features which can be related need to become organized into an assembly; third, neurons which have joined a particular assembly must be identifiable as members of this very assembly. Their responses need to be tagged so that they can be recognized as being related. It is commonly assumed that these organizing steps, the probing of possible relations, the formation of an assembly, and the labeling of responses are achieved in a single self-organizing process by selective reciprocal connections between the distributed neuronal elements.

The suggestion most closely related to the experimental data reviewed in this chapter is that assemblies should be distinguished by a temporal code (von der Malsburg 1985; von der Malsburg and Schneider 1986). A similar suggestion, although formulated less explicitly, was made previously by Milner (1974). The assumption is that the assembly-forming connections should establish temporal coherence on a millisecond time scale between the responses of the coupled cells. Thus, neurons having joined into an assembly would be identifiable as members of the assembly because their responses would exhibit synchronous episodes.

If assemblies are indeed distinguished by the temporal coherence of the responses of the constituting neurons, a set of predictions can be derived:

1. Spatially segregated neurons should synchronize their responses if activated by features that can be grouped together. This should be the case, for example, for features constituting a single object.
2. Synchronization should be frequent among neurons within a particular cortical area, but it should also occur across cortical areas.
3. The probability that neurons synchronize their responses both within a particular area and across areas should reflect some of the *Gestalt* criteria used for perceptual grouping.
4. Individual cells must be able to rapidly change the partners with which they synchronize their responses if stimulus configurations change and require new associations.
5. If more than one object is present in a scene, several assemblies should form. Cells belonging to the same assembly should synchronize their responses, while no consistent temporal relations should exist between the discharges of neurons belonging to different assemblies.
6. Synchronization probability should, at least in part, depend on the functional architecture of reciprocal corticocortical connections and should change if this architecture is modified.

7.5.3 Evidence for Temporal Relations

Systematic search for such stimulus-dependent synchronization phenomena and their anatomical substrate was initiated by the observation that adjacent neurons in the cat visual cortex can transiently engage in highly *synchronous* discharges when presented with their preferred stimulus (Gray and Singer 1987). Groups of neurons recorded simultaneously with a single electrode were found to discharge in synchronous bursts which followed one another at intervals of 15–30 ms. Typically, these sequences of synchronous firing last no more than a few hundred milliseconds and may occur several times during a single passage of the moving stimulus. If the clusters of synchronized discharges occur at regular intervals over a sufficiently long period of time, autocorrelograms computed from such response epochs often exhibit a periodic modulation (Gray and Singer 1987, 1989; Eckhorn et al. 1988; Gray et al. 1990; Schwarz and Bolz 1991; Livingstone 1991).

This phenomenon of local response synchronization has been observed with multiunit and field potential recordings in several independent studies in different areas of the visual cortex of anesthetized cats (areas 17, 18, 19, and the posteriomedial lateral suprsylvian area PMLS; Eckhorn et al. 1988, 1992; Gray and Singer 1989; Gray et al. 1990; Engel et al. 1991a), in area 17 of awake cats (Raether et al. 1989; Gray and Viana di Prisco 1993), in the optic tectum of awake pigeons (Neuenschwander and Varela 1990), and in the visual cortex of anesthetized (Livingstone 1991) and awake, behaving monkeys (Kreiter and Singer 1992).

Multi-electrode recordings have revealed that similar response synchronization can occur also between spatially segregated cell groups within the same visual area (Gray et al. 1989, 1992; Engel et al. 1990; König et al. 1993), between different cortical areas (Eckhorn et al. 1988, 1992; Engel et al. 1991a; Murthy and Fetz 1992; Nelson et al. 1992), and even across hemispheres (Engel et al. 1991b; Munk et al. 1992). Measurements in awake cats (Raether et al. 1989; Gray and Viana di Prisco 1993) and monkeys (Kreiter and Singer 1992; Murthy and Fetz 1992; Ahissar et al. 1992) have shown that this phenomenon of response synchronization is not confined to anesthesia, but is readily demonstrable in various neocortical areas (visual, auditory, somatosensory, and motor) of alert, behaving animals. These single-unit data are complemented by a large body of evidence derived from field potential and EEG recordings which all indicate that distributed groups of neurons can engage in synchronous activity (for a review of the extensive literature, see Singer 1993).

7.5.4 The Dependence of Response Synchronization on Stimulus Configuration

As outlined above, the hypothesis of temporally coded assemblies requires that the probabilities with which distributed cells synchronize their responses should reflect some of the *Gestalt* criteria applied in perceptual grouping. Another, related prediction is that individual cells must be able to change the partners with which they synchronize, whereby the selection of partners should occur as a function of the patterns used to activate the cells. Detailed studies in anesthetized cats and recently also anesthetized and awake monkeys have revealed that synchronization probability for remote groups of cells is determined both by factors within the brain as well as by the configuration of the stimuli (Engel et al. 1990, 1991a,b,c; Gray et al. 1989; Kreiter et al. 1992; König et al. 1993). In general, synchronization probability within a particular cortical area decreases with increasing distance between the cells. If cells are so closely spaced that their receptive fields overlap, the probability is high that their responses will exhibit synchronous epochs if evoked with a single stimulus. The latter condition requires that the orientation and direction preferences of the cell pairs are sufficiently similar or that their tuning is sufficiently broad to allow for coactivation by a single stimulus. As recording distance increases, synchronization probability becomes more and more dependent on the similarity between the orientation preferences of the neurons (Ts'o et al. 1986; Engel et al. 1990).

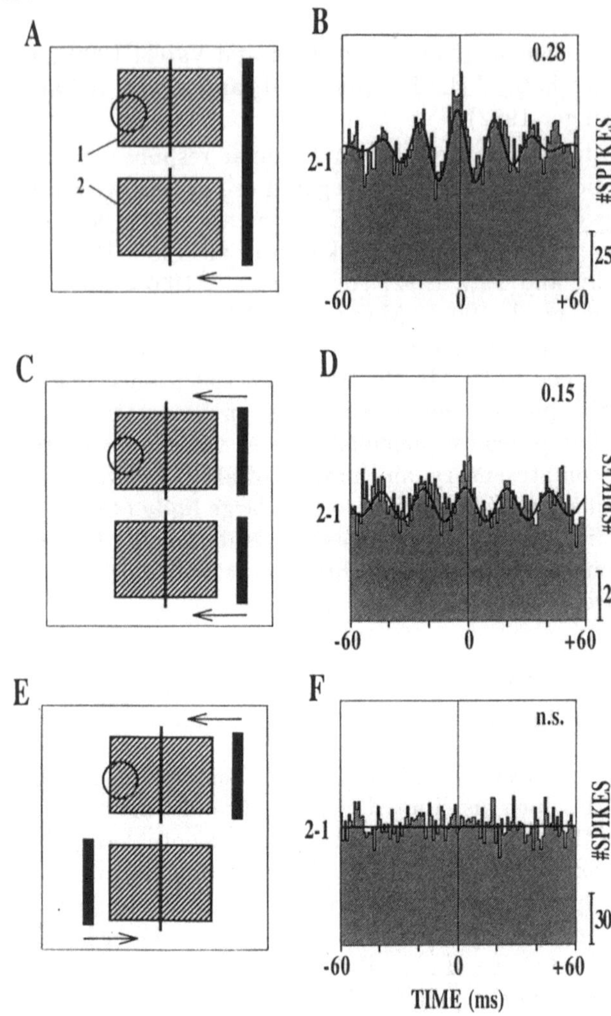

Fig. 7.5–1. Long-range synchronization is influenced by stimulus coherence. Multiunit activity was recorded from two sites in area 17 of cat visual cortex that were separated by 7 mm. The two cell groups preferred vertical orientations. **A,C,E** Plots of the receptive fields and stimulation conditions. The colinear arrangement of the fields allowed the comparison of three different stimulus paradigms: a long continuous light bar moving across both fields (**A**), two independent light bars moving in the same direction (**C**), and the same two bars moving in opposite directions (**E**). The circle represents the center of the visual field, and the *thick line* drawn across each receptive field indicates the preferred orientation. **B,D,F** The respective cross-correlograms obtained with each stimulus paradigm. Using the long light bar, the two oscillatory responses were synchronized with zero phase lag, as indicated by the prominent center *peak* and the strong modulation of the cross-correlogram with alternating peaks and troughs (**B**). If the continuity of the stimulus was interrupted, the synchronization became weaker (**D**), and it totally disappeared if the motion of the stimuli was incoherent (**F**). This change of the stimulus configuration affected neither the strength nor the oscillatory nature of the two responses (not shown). The graph superimposed on each of the correlograms represents a Gabor function that was fitted to the data to assess the strength of the modulation (Engel et al. 1990). The *number in the upper right corner* indicates the "relative modulation amplitude", a measure of correlation strength that was determined by computing the ratio of the amplitude of the Gabor function to its offset. *ns*, not significant. *Scale bars* indicate the number of spikes. (Adopted from Engel et al. 1992)

Concerning the dependence of synchronization probability on the configuration of visual stimuli, single linearly moving contours have so far been found to be most efficient. Gray et al. (1989) recorded multiunit activity from two locations in cat area 17 separated by 7 mm. The receptive fields of the cells were nonoverlapping, had nearly identical orientation preferences, and were spatially displaced along the axis of preferred orientation. This enabled stimulation of the cells with bars of the same orientation under three different conditions: (1) two bars moving in opposite directions, (2) two bars moving in the same direction, and (3) one long bar moving across both fields coherently. No significant correlation was found when the cells were stimulated by oppositely moving bars. A weak correlation was present for the coherently moving bars, but the long bar stimulus resulted in a robust synchronization of the activity at the two sites (Fig. 1). This effect occurred in spite of the fact that the overall number of spikes produced by the two cells and the oscillatory patterning of the responses were similar in the three conditions.

In a related experiment, Engel et al. (1991a) demonstrated that the synchronization of activity between cells in areas 17 and PMLS of the cat depends in a very similar way on the properties of the visual stimulus as the intra-areal synchronization.

These results indicate clearly that synchronization probability depends not only on the spatial segregation of cells and on their feature preferences, the latter being related to the cells' position within the columnar architecture of the cortex, but also and to a crucial extent on the configuration of the stimuli. So far, synchronization probability appears to reflect rather well some of the *Gestalt* criteria for perceptual grouping. The high synchronization probability of nearby cells corresponds to the binding criterion of "vicinity", the dependence on receptive field similarities agrees with the criterion of "similarity", the strong synchronization observed in response to continuous stimuli obeys the criterion of "continuity" and the lack of synchrony in responses to stimuli moving in opposite directions relates to the criterion of "common fate".

Experiments have also been performed in order to test the prediction that simultaneously presented, but different contours should lead to the organization of two independently synchronized assemblies of cells (Engel et al. 1991c; Kreiter et al. 1992). If groups of cells with overlapping receptive fields, but different orientation preferences are activated with a single moving light bar they synchronize their responses even if some of these groups are suboptimally activated (Engel et al. 1990, 1991c). However, if such a set of groups is stimulated with two independent spatially overlapping stimuli which move in different directions, they split into two independently synchronized assemblies, those groups joining the same synchronously active

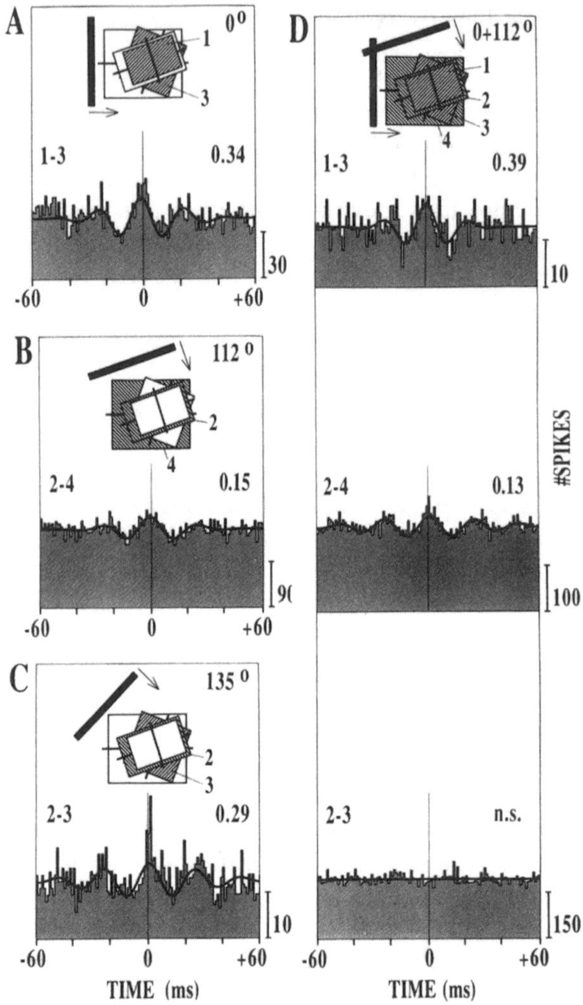

Fig. 7.5–2. A-D Stimulus dependence of short-range interactions. Multiunit activity was recorded from four different orientation columns of area 17 of cat visual cortex separated by 0.4 mm. The four cell groups had overlapping receptive fields and orientation preferences of 22° (group 1), 112° (group 2), 157° (group 3) and 90° (group 4), as indicated by the *thick line* drawn across each receptive field in (**A-D**). The figure shows a comparison of responses to stimulation with single moving light bars of varying orientation (*left*) with responses to the combined presentation of two superimposed light bars (*right*). For each stimulus condition, the *shading* of the receptive fields indicates the responding cell groups. Stimulation with a single light bar yielded a synchronization between all cells activated by the respective orientation. Thus, groups 1 and 3 responded synchronously to a vertically orientated (0°) light bar (**A**), groups 2 and 4 to a light bar at an orientation of 112° (**B**), and cell groups 2 and 3 to a light bar of intermediate orientation (**C**). Simultaneous presentation of two stimuli with orientations of 0° and 112°, respectively, activated all four groups (**D**).

assembly that have a preference for the same stimulus (Fig. 2). Thus, the two stimuli evoke simultaneous responses in a large array of spatially interleaved neurons, but these neurons are organized in two assemblies which can be distinguished because of the temporal coherence of responses within and the lack of coherence between assemblies. Cells representing the same stimulus exhibit synchronized response epochs, while no consistent correlations occur between the responses of cells that are evoked by different stimuli. The parameters of the individual responses such as their amplitude or oscillatory patterning were not affected by changes in the global configuration of the stimuli. Thus, it is not possible to tell from the responses of individual cells whether they were activated by a single contour or by two different stimuli. The only cue for this distinction was provided by the evaluation of synchronicity at a millisecond time scale.

7.5.5 The Synchronizing Connections and Their Experience-Dependent Selection

The hypothesis requires that synchronization probability depends to a substantial extent on interactions between the neurons whose responses actually represent the features that need to be bound together. Thus, corticocortical connections should contribute to the synchronization process. This postulate is supported by the finding that synchronization between cells located in different hemispheres is abolished when the corpus callosum is cut (Engel et al. 1991b; Munk et al. 1992). This is direct proof (a) that corticocortical connections contribute to response synchronization and (b) synchronization with zero phase lag can be brought about by reciprocal interactions between spatially distributed neurons despite considerable conduction delays in the coupling connections. Thus, synchrony is not necessarily an indication of common input, but may also be the result of a dynamic organization process which establishes coherent firing by reciprocal interactions.

The theory of assembly coding implies further that the criteria according to which particular features are grouped together reside in the functional

---

However, in this case, the groups segregated into two distinct assemblies, depending on which stimulus was closer to the preferred orientation of each group. Thus, responses were synchronized between groups 1 and 3, which preferred the vertical stimulus, and between 2 and 4, which preferred the stimulus oriented at 112°. The two assemblies were desynchronized with respect to each other, and so there was no significant synchronization between groups 2 and 3. The cross-correlograms between groups 1 and 2, 1 and 4, and 3 and 4 were also flat (not shown). Note that the segregation cannot be explained by preferential anatomical wiring of cells with similar orientation preference (Ts'o et al. 1986), because cell groups can readily be synchronized in all possible pair combinations in response to a single light bar. The correlograms are shown superimposed with their Gabor function. The number to the *upper right* of each correlogram indicates the relative modulation amplitude. *ns*, not significant. *Scale bars* indicate the number of spikes. (From Engel et al. 1992) architecture of the assembly-forming coupling connections. It is of particular interest, therefore, to study the development of the synchronizing connections, to identify the rules according to which they are selected, to establish correlations between their architecture and synchronization probabilities, and, if possible, to relate these neuronal properties to perceptual functions.

In mammals, corticocortical connections develop mainly postnatally (Callaway and Katz 1990; Innocenti 1981; Luhmann et al. 1986; Price and Blakemore 1985a) and attain their final specificity through an activity-dependent selection process (Callaway and Katz 1991; Innocenti and Frost 1979; Luhmann et al. 1990; Price and Blakemore 1985b). Recent results from strabismic kittens indicate that this selection is based on a correlation rule and leads to disruption of connections between cells which often exhibit decorrelated activity (Löwel and Singer 1992). Normally, these connections link cortical territories, irrespective of whether these are dominated by the same or by different eyes. In the strabismics, by contrast, the tangential intracortical connections come to link with high selectivity only territories served by the same eye. These anatomical changes in the architecture of corticocortical connections are reflected by altered synchronization probabilities. In strabismics, response synchronization no longer occurs between cell groups connected to different eyes, while it is normal between cell groups connected to the same eye (König et al. 1993).

These results have several important implications. First, they are compatible with the notion that tangential intracortical connections contribute to response synchronization (see above). Second, they agree with the postulate that the assembly-forming connections should be susceptible to use-dependent modifications and be selected according to a correlation rule. Third, the modifications of intracortical connections and synchronization probabilites add to the list of substrate changes that may be related to the specific perceptual deficits associated with early onset squint. Strabismic subjects usually develop normal monocular vision in both eyes, but they become unable to fuse signals conveyed by different eyes into coherent percepts, even if these signals are made retinotopically contiguous by optical compensation of the squint angle (von Noorden 1990). Thus, in strabismics, binding mechanisms appear to be abnormal or missing between cells driven from different eyes. The lack of corticocortical connections and the lack of response synchronization could be one of the reasons for this deficit in addition to the loss of binocular neurons.

7.5.6 Correlation Between Perceptual Deficits and Response Synchronization in Strabismic Amblyopia

Further indications for a relation between experience-dependent modifications of synchronization probabilities and functional deficits come from a recent study of strabismic cats who had developed amblyopia. Strabismus, when induced early in life, does not only lead to a loss of binocular fusion and stereopsis, but may also lead to amblyopia of one eye (von Noorden 1990). This condition develops when the subjects solve the problem of double vision not by alternating use of the two eyes, but by constantly suppressing the signals coming from the deviated eye. The amblyopic deficit usually consists of reduced spatial resolution and distorted and blurred perception of patterns. A particularly characteristic phenomenon in amblyopia is crowding, the drastic impairment of the ability to discriminate and recognize figures if these are surrounded with other contours. The identification of neuronal correlates of these deficits in animal models of amblyopia has remained inconclusive, because the contrast sensitivity and the spatial resolution capacity of neurons in the retina and the lateral geniculate nucleus were found to be normal. In the visual cortex, identification of neurons with reduced spatial resolution or otherwise abnormal receptive field properties remained inconclusive (for a discussion see Crewther and Crewther 1990; Blakemore and Vital Durand 1992). However, multielectrode recordings from striate cortex of cats exhibiting behaviorally verified amblyopia have revealed highly significant differences in the synchronization behavior of cells driven by the normal and the amblyopic eye, respectively. The responses to single moving bars that were recorded simultaneously from spatially segregated neurons connected to the amblyopic eye were much less well synchronized with one another than the responses recorded from neuron pairs driven through the normal eye (Roelfsema et al. submitted). This difference was even more pronounced for responses elicited by gratings of different spatial frequency. For responses of cell pairs activated through the normal eye, the strength of synchronization tended to increase with increasing spatial frequency, while it tended to decrease further for cell pairs activated through the amblyopic eye (Fig. 3). Apart from these highly significant differences between the synchronization behavior of cells driven through the normal and the amblyopic eye, no other differences were found in the commonly determined response properties of these cells. Thus, cells connected to the amblyopic eye, continued to respond vigorously to gratings whose spatial frequency had been too high to be discriminated with the amblyopic eye in the preceding behavioral tests. These results suggest that disturbed temporal coordination of responses such as reduced synchrony may be one of the neuronal correlates of the amblyopic deficit. Indeed, if synchronization

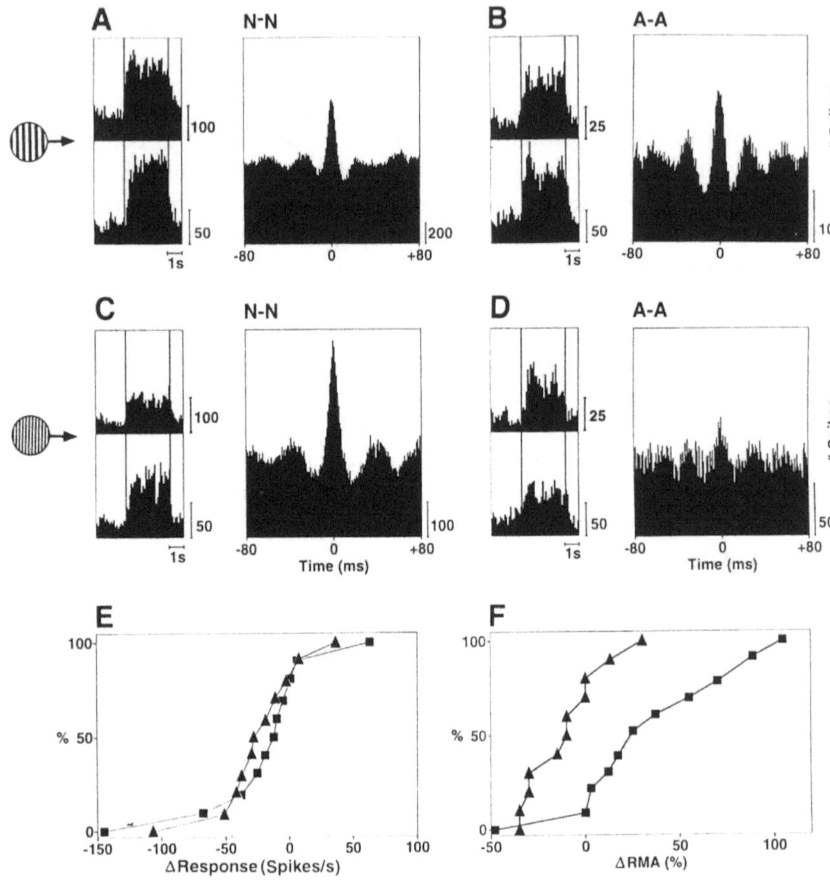

Fig. 7.5–3. Amplitudes and synchronization of responses to gratings of different spatial frequencies recorded from cats with strabismic amblyopia. **A-D** Responses to low (**A,B**) and high (**C,D**) spatial frequency gratings, recorded simultaneously from two cell groups driven by the normal eye (N-sites; **A,C**) and two cell groups driven by the amblyopic eye (A-sites; **B,D**), respectively. The *left* and *right* panels show the response histograms and the corresponding cross-correlograms, respectively. Note that response amplitudes decrease at the higher spatial frequency in both cases, while the relative modulation amplitude increases for the N-N pair, but decreases for the A-A pair. **E** Cumulative distribution functions of the differences between the *amplitude* of responses to low and high spatial frequency gratings of optimal orientation. *Squares*, N-sites (n=53); *triangles*, A-sites (n=35). *Abscissa*, responses to high spatial frequency minus responses to low spatial frequency gratings. Note the similarity of the two distributions (p>0.1). **F** Cumulative distribution functions of the differences between relative modulation amplitudes (ΔRMA) of cross-correlograms obtained for responses to high and low spatial frequency gratings of N–N pairs (*squares; n*=24) and A–A pairs (*triangles; n*=11). ΔRMA values (*abscissa*) were calculated by subtracting the relative modulation amplitude obtained with the low spatial frequency from that obtained with the high spatial frequency. The difference between the ΔRMA distributions of N–N pairs and A–A pairs is highly significant. (From Roelfsema et al. 1993)

of responses at a millisecond time scale is used by the system for feature binding and perceptual grouping, disturbance of this temporal patterning could be the cause for the crowding phenomenon, as this can be regarded as a consequence of impaired perceptual grouping. Furthermore, because of reduced synchrony, the responses elicited through the amblyopic eye are less salient for subsequent processing stages and this may explain why signals from this eye are so readily and consistently suppressed and excluded from conscious experience.

7.5.7 Epilog

The experimental results reviewed in this chapter are compatible with the hypothesis that synchronization of neuronal responses on a millisecond time scale may be exploited in cortical processing. In a distributed network with a high degree of convergence on individual cells, but low safety factor of individual connections, synchronization of discharges is an efficient strategy to increase the saliency of responses and may be more effective than merely raising discharge rates. As synchronization enhances with great selectivity the saliency of only those response episodes that contain coincident discharges, it can in principle be used to select with high spatial and temporal precision responses that need to be related with one another and passed on for further processing. One function of the system of corticocortical association fibers could thus consist in the adjustment of timing of the discharges in spatially distributed neurons. Eventually, this will of course also influence firing rates, but at levels of processing that are remote from motor output such shifting of temporal relations may not be associated with marked rate changes.

The evidence available so far indicates clearly that synchronization can occur with high temporal precision among widely distributed neurons, but we have only correlative data on putative relations between these synchronization phenomena and perceptual functions. While available data agree with predictions derived from the hypothesis that synchrony is used as a code for relations among distributed neurons, further, more direct investigations of this conjecture are now required. As synchrony can only be evaluated by simultaneously recording the activity of different cells, multi-electrode recordings in awake, behaving animals will be indispensible, and most likely it will also be necessary to develop new analytical methods to reliably detect transient temporal relations among the activities of widely distributed groups of neurons. Thus, the approach that Otto Creutzfeldt had in mind in 1966 is still pursued, and it is likely that there will be a renaissance of correlation studies in the future.

References

Abeles M (ed) (1991) Corticonics. Cambridge University Press, Cambridge
Ahissar E, Vaadia E, Ahissar M, Bergmann H, Arieli A, Abeles M (1992) Dependence of cortical plasticity on correlated activity of single neurons and on behavioral context. Science 257: 1412–1415
Blakemore C, Vital-Durand F (1992) Different neural origins for 'blur' amblyopia and strabismic amblyopia. Ophthal Physiol Opt 12: 83
Braitenberg V (1978) Cell assemblies in the cerebral cortex. In: Heim R, Palm G (eds) Lecture notes in biomathematics. 21. Theoretical approaches in complex systems. Springer, Berlin Heidelberg New York, pp 171–188
Callaway EM, Katz LC (1990) Emergence and refinement of clustered horizontal connections in cat striate cortex. J Neurosci 10: 1134–1153
Callaway EM, Katz LC (1991) Effects of binocular deprivation on the development of clustered horizontal connections in cat striate cortex. Proc Natl Acad Sci 88: 745–749
Crewther DP, Crewther SG (1990) Neural sites of strabismic amblyopia in cats: spatial frequency deficit in primary cortical neurons. Exp Brain Res 79: 615–622
Crick F (1984) Function of the thalamic reticular complex: the searchlight hypothesis. Proc Natl Acad Sci USA 81: 4586–4590
Crick F, Koch C (1990) Towards a neurobiological theory of consciousness. Sem Neurosci 2: 263–275
Damasio AR (1990) Synchronous activtion in multiple cortical regions: a mechanism for recall. Sem Neurosci 2: 287–296
Eckhorn R, Bauer R, Jordan W, Brosch M, Kruse W, Munk M, Reitboeck HJ (1988) Coherent oscillations: a mechanism for feature linking in the visual cortex? Biol Cybern 60: 121–130
Eckhorn R, Schanze T, Brosch M, Salem W, Bauer R (1992) Stimulus-specific synchronizations in cat visual cortex: multiple microelectrode and correlation studies from several cortical areas. In: Basar E, Bullock TH (eds) Induced rhythms in the brain. Birkhäuser, Boston, pp 47–80
Edelman GM (1987) Neural Darwinism: the theory of neuronal group selection. Basic Books, New York
Edelman GM (1989) The Remembered Present. Basic Books, Inc., New York
Edelman GM, Mountcastle VB (1978) The Mindful Brain. Mass. MIT Press, Cambridge
Engel AK, König P, Gray CM, Singer W (1990) Stimulus-dependent neuronal oscillations in cat visual cortex: Inter-columnar interaction as determined by cross-correlation analysis. Eur J Neurosci 2: 588–606
Engel AK, Kreiter AK, König P, Singer W (1991a) Synchronization of oscillatory neuronal responses between striate and extrastriate visual cortical areas of the cat. Proc Natl Acad Sci USA 88: 6048–6052
Engel AK, König P, Kreiter AK, Singer W (1991b) Interhemispheric synchronization of oscillatory neuronal responses in cat visual cortex. Science 252: 1177–1179
Engel AK, König P, Singer W (1991c) Direct physiological evidence for scene segmentation by temporal coding. Proc Natl Acad Sci USA 88: 9136–9140
Engel, AK, König P, Kreiter, AK, Schillen, TB, Singer, W (1992) Temporal coding in the visual cortex: new vistas on integration in the nervous system. Trends in Neurosci. 15: 218–226
Felleman DJ, van Essen DC (1991) Distributed hierarchical processing in the primate cerebral cortex. Cerebral Cortex 1: 1–47
Freeman WJ, ed (1975) Mass action in the nervous system. Academic Press, New York
Gray CM, Singer W (1987) Stimulus-specific neuronal oscillations in the cat visual cortex: a cortical functional unit. Soc Neurosci Abstr 13: 1449
Gray CM, Singer W (1989) Stimulus-specific neuronal oscillations in orientation columns of cat visual cortex. Proc Natl Acad Sci USA 86: 1698–1702
Gray CM, Viana di Prisco G (1993) Properties of stimulus-dependent rhythmic activity of visual cortical neurons in the alert cat. Soc Neurosci Abstr 19: 868
Gray CM, König P, Engel AK, Singer W (1989) Oscillatory responses in cat visual cortex exhibit inter-columnar synchronization which reflects global stimulus properties. Nature 338: 334–337
Gray CM, Engel AK, König P, Singer W (1990) Stimulus-dependent neuronal oscillations in cat visual cortex: Receptive field properties and feature dependence. Eur J Neurosci 2: 607–619
Gray CM, Engel AK, König P, Singer W (1992) Synchronization of oscillatory neuronal responses in cat striate cortex. Temporal properties. Vis Neurosci 8: 337–347
Grossberg S (1980) How does the brain build a cognitive code? Psychol Rev 87: 1–51
Hebb DO (1949) The organization of behavior. Wiley, New York

Innocenti GM (1981) Growth and reshaping of axons in the establishment of visual callosal connections. Science 212: 824–827
Innocenti GM, Frost DO (1979) Effects of visual experience on the maturation of the efferent system to the corpus callosum. Nature 280: 231–234
König P, Engel AK, Löwel S, Singer W (1993) Squint affects synchronization of oscillatory responses in cat visual cortex. Eur J Neurosci 5: 501–508
Kreiter AK, Singer W (1992) Oscillatory neuronal responses in the visual cortex of the awake macaque monkey. Eur J Neurosci 4: 369–375
Kreiter AK, Engel AK, Singer W (1992) Stimulus-dependent synchronization of oscillatory neuronal activity in the superior temporal sulcus of the macaque monkey. Eur Neurosci Assoc Abstr 15: 1076
Livingstone MS (1991) Visually evoked oscillations in monkey striate cortex. Soc Neurosci Abstr 17: 176
Löwel S, Singer W (1992) Selection of intrinsic horizontal connections in the visual cortex by correlated neuronal activity. Science 255: 209–212
Luhmann HJ, Martinez-Millan L, Singer W (1986) Development of horizontal intrinsic connections in cat striate cortex. Exp Brain Res 63: 443–448
Luhmann HJ, Singer W, Martinez-Millan L (1990) Horizontal interactions in cat striate cortex. I. Anatomical substrate and postnatal development. Eur J Neurosci 2: 344–357
Milner PM (1974) A model for visual shape recognition. Psychol Review 81: 521–535
Munk MHJ, Nowak LG, Chouvet G, Nelson JI, Bullier J (1992) The structural basis of cortical synchronization. Eur J Neurosci Suppl 51: 21
Murthy VN, Fetz EE (1992) Coherent 25- to 35-Hz oscillations in the sensorimotor cortex of awake behaving monkeys. Proc Natl Acad Sci USA 89: 5670–5674
Nelson JI, Salin PA, Munk MHJ, Arzi M, Bullier J (1992) Spatial and temporal coherence in cortico-cortical connections: a cross-correlation study in areas 17 and 18 in the cat. Vis Neurosci 9: 21–38
Neuenschwander S, Varela FJ (1990) Sensory-triggered oscillatory activity in the avian optic tectum. Soc Neurosci Abstr 16: 109
Palm G (1982) Neural assemblies. Springer, Berlin Heidelberg New York
Palm G (1990) Cell assemblies as a guideline for brain research. Concepts Neurosci 1: 133–137
Price DJ, Blakemore C (1985a) The postnatal development of the association projection from visual cortical area 17 to area 18 in the cat. J Neurosci 5: 2443–2452
Price DJ, Blakemore C (1985b) Regressive events in the postnatal development of association projections in the visual cortex. Nature 316: 721–724
Raether A, Gray CM, Singer W (1989) Intercolumnar interactions of oscillatoary neuronal responses in the visual cortex of alert cats. Eur Neurosci Assoc 12: 72.5
Roelfsema PR, König P, Sireteanu R, Singer W (1993) Squint-induced amblyopia is associated with reduced synchronization of cortical processes. Soc Neurosci Abstr 19: 867
Schwarz C, Bolz J (1991) Functional specificity of the long-range horizontal connections in cat visual cortex: a cross-correlation study. J Neurosci 11: 2995–3007
Singer W (1985) Activity-dependent self-organization of the mammalian visual cortex. In: Rose D, Dobson VG (eds) Models of the visual cortex. Wiley, Chichester, pp 123–136
Singer W (1990) Search for coherence: A basic principle of cortical self-organization. Concepts Neurosci 1: 1–26
Singer W (1993) Synchronization of cortical activity and its putative role in information processing and learning. Ann Rev Physiol 55: 349–374
Singer W, Creutzfeldt OD (1969) Die Bedeutung der Vorderhirnkommissuren für die Koordination bilateraler EEG-Muster. Exp Brain Res 7: 195–213
Ts'o D, Gilbert C, Wiesel TN (1986) Relationship between horizontal interactions and functional architecture in cat striate cortex as revealed by cross-correlation analysis. J Neurosci 6: 1160–1170
Ungerleider LG, Mishkin M (1982) Two cortical visual systems. In: Ingle J, Goodale MA, Mansfield RJW (eds) Analysis of visual behavior. MIT Press, Cambridge, pp 564–586
von der Malsburg C (1985) Nervous structures with dynamical links. Ber Bunsenges Phys Chem 89: 703–710
von der Malsburg C, Schneider W (1986) A neural cocktail-party processor. Biol Cybern 54: 29–40
von Noorden GK (1990) Binocular vision and ocular motility, theory and management of strabismus. Mosby, St. Louis
Zeki S, Watson JDG, Lueck CJ, Friston KJ, Kennard C, Frackowiak RSJ (1991) A direct demonstration of functional specialization in human visual cortex. J Neurosci 11: 641–649

7.6 Sensory-Motor Integration and the Posterior Parietal Cortex

R.A. ANDERSEN
Division of Biology, California Institute of Technology, Pasadena, California

Experiments outlined here suggest that the posterior parietal cortex performs visuomotor integration functions including motor planning and coordinate transformations. Neural activity related to planning saccades is found in the lateral intraparietal area (LIP), a recently discovered cortical area specialized for saccadic eye movements. The planning related activity encodes the movements that the animal intends to make. Coordinate transformations are performed in area LIP and 7a of the posterior parietal cortex and produce a representation of space in both these areas. These representations are not explicit at a single cell level but rather utilize a population code. The coordinate transformations are achieved by a specific operation which systematically combines visual and eye position signals to form "planar gain fields".

7.6.1 Introduction

A major aspect of sensory-motor integration is the planning of movements. At some point in the sensory-motor integration process, sensory signals give way to signals related to what the subject intends to do. This chapter will discuss evidence that this step from sensory representation to decisions to make movements utilizes the neural circuitry within the posterior parietal cortex. We will focus on a recently discovered small area within the posterior parietal cortex, the lateral intraparietal area (LIP), which appears to be specialized for saccadic eye movements. This area carries sensory information related to the targets for eye movements and signals related to the planning of eye movements.

Coordinate transformations are a second major aspect of sensory-motor integration and are required because sensory information is derived in the coordinates of the retina and must be transformed to the coordinates of muscles for movements. The experiments discussed here suggest there exist, within the posterior parietal cortex, intermediate and abstract representations of space interposed between the sensory input and the motor output. These intermediate representations are formed by combining visual information with body position information to form representations of space in head-, body-, and perhaps even world-centered coordinates.

7.6.2 Area LIP

The posterior parietal cortex plays an important role in the processing of eye movements. Bilateral lesions to the posterior parietal cortex in human patients result in the inability to will saccades, although spontaneous saccades are unaffected (Balint 1909). Electrical stimulation of the monkey posterior parietal cortex produces saccadic eye movements (Wagman 1964; Fleming and Crosby 1955), and lesions to this region in monkeys produce deficits in saccades (Lynch and McLaren 1989; Keating and Gooley 1988).

Mountcastle and his colleagues (1975) reported cells in the inferior parietal lobule (which encompasses approximately the posterior half of the posterior parietal cortex) which were selective for saccades, as well as neurons selective for smooth pursuit eye movements, reaching movements of the limbs, and fixation. They reported that there appeared to be a columnar organization for these functional properties. Lynch et al. (1977) later reported that these columns were not segregated by functional types into particular parts of the posterior parietal cortex, suggesting that repeating columns of each functional class are rather evenly distributed across the inferior parietal lobule. These results must be interpreted with some caution, however, since the data were pooled from several different brains and referenced to sulcal patterns which vary considerably from animal to animal. We reasoned that these different types of columns could be selectively labeled with anatomical tracers by assuming that they had connections with different brain structures (Andersen et al. 1985a). Thus, for instance, columns of cells with saccade related activity would likely project to the frontal lobe, in the region of the frontal eye fields. However, when we injected retrograde tracers into the frontal eye fields and adjoining dorsolateral prefrontal cortex, label was found predominantly in the lateral bank of the intraparietal cortex. We named this area the lateral intraparietal area (LIP) because it was located on the lateral bank of the intraparietal sulcus, lateral to the ventral intraparietal area (VIP) described earlier by Maunsell and Van Essen (1983). Subsequent recording experiments showed that most area LIP cells had activity related to eye movements, and a majority of these cells responded prior to the saccades (Gnadt and Andersen 1988; Andersen et al. 1990; Barash et al. 1991a,b). Other studies showed that reach activity was confined largely to area 7b (Hyvärinen 1981; Hyvärinen and Shelepin 1979; Robinson and Burton 1980a,b) and smooth pursuit activity to the medial superior temporal area MST (Newsome et al. 1988). Fixation activity typically varies with direction of gaze (Lynch et al. 1977; Andersen et al. 1987; Sakata et al. 1980), and conveys information about eye position (Andersen et al. 1987; Andersen 1989). These eye position-related activities are typically found in areas LIP and 7a. Thus, many of the functional types discovered by Mountcastle and colleagues (1975) are actually segregated into small cortical fields rather than interdigitated cortical columns. Area LIP is the cortical field where saccade related activity is found.

7.6.3 Visual and Saccade Related Activities in Area LIP

There was a brief period of controversy regarding whether parietal neurons had saccade related activity. When Mountcastle and colleagues (1975) first observed saccade responses, they proposed that the area issued general commands to make eye movements. Soon thereafter, Robinson, Goldberg and Stanton (1978) observed visual responses from parietal neurons and they challenged Mountcastle's command hypothesis, arguing that the cells were responding in a sensory fashion to the saccade targets as visual stimuli rather than in a motor fashion related to the eye movement. Using a memory saccade task which separated sensory from motor responses, we showed that posterior parietal neurons had both visual and saccade related activity (Andersen et al. 1987). These results suggested that it was more appropriate to consider posterior parietal cortex as being involved in sensorimotor integration, rather than as a strictly sensory or motor structure (Andersen 1987).

The visual and saccade activity in LIP has recently been studied in great detail by our laboratory and compared to visual and saccade activity in area 7a (Barash et al. 1991a,b). Saccade related responses in area LIP generally begin prior to eye movements, whereas most area 7a saccade responses are postsaccadic, beginning after initiation of eye movements. This observation, and the reduced activity related to planning eye movements in 7a compared to LIP (see below), led us to propose that area LIP participates in the planning of eye movements, whereas area 7a appears to subserve other functions. These studies also showed that the visual receptive fields and motor fields of LIP neurons generally were found to overlay one another.

7.6.4 "Memory" Related Activity in LIP

Recently, we found a memory related activity in area LIP using a task which required monkeys to make saccades to remembered locations in the dark (Gnadt and Andersen 1988; Andersen et al. 1990; Barash et al. 1991a,b). The cells remained active during the period in which the animal withheld its response while remembering the location of an extinguished saccade target (Fig. 1).

Using a double saccade task similar to the one developed by Mays and Sparks (1980), we were able to distinguish between whether the cells were coding the location of the sensory stimulus or whether the cells were coding the intention to make a saccade of a particular amplitude and direction. The

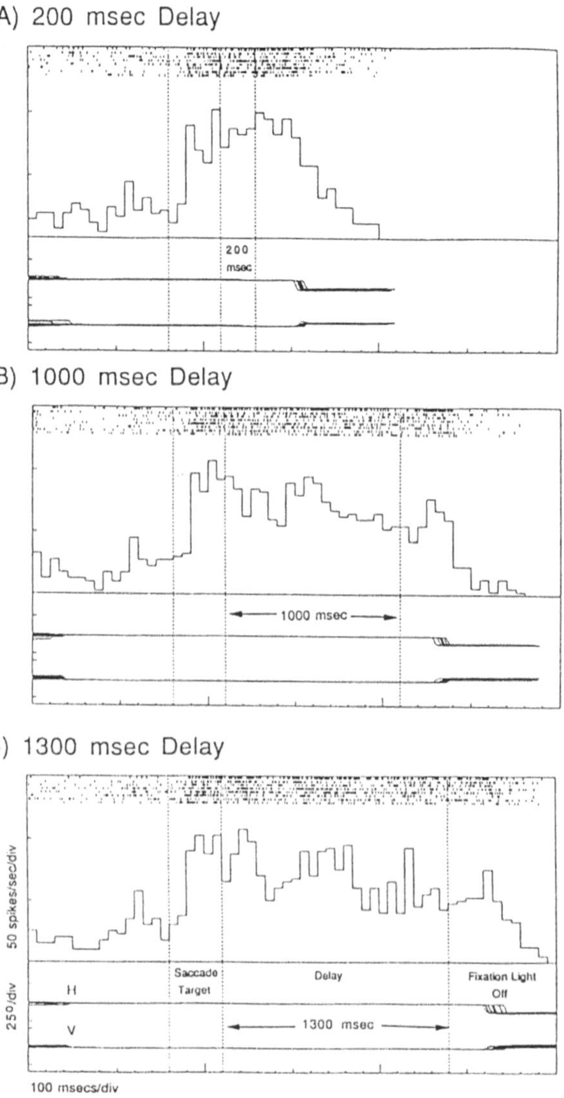

Fig. 7.6–1. **A-C** Memory saccade task with different delays demonstrating the memory character of the activity during the delay. Delays are 200 ms (**A**), 1000 ms (**B**), and 1300 ms (**C**). The *rasters* show the actual neural activity that is used to make the histograms. The period between the *first two dotted vertical lines* represents the time the saccade target is present and the period between the *second and third lines* is the delay period. The fixation light goes off coincident with the *third dotted vertical line.* Both horizontal *(H)* and vertical *(V)* eye position traces are shown. In this experiment the saccade target appeared 15° to the left. There is a vertical component in the leftward eye movement; this upward component for horizontal eye movements is common for saccades to remembered locations made in the dark. (Reprinted, with permission, from Andersen et al. 1990)

activity could be evoked even when the sensory stimulus did not fall in the receptive field but the eye movement was made into the cell's motor field, arguing for the latter alternative (Fig. 2). We interpreted these activities as being part of a motor plan which has been latched in and remained active during the waiting period. The memory activity of LIP cells is directionally tuned, and these memory fields coincide with the visual and saccade fields. These results suggest that the memory activity reflects the intention of the monkey to make the next saccade.

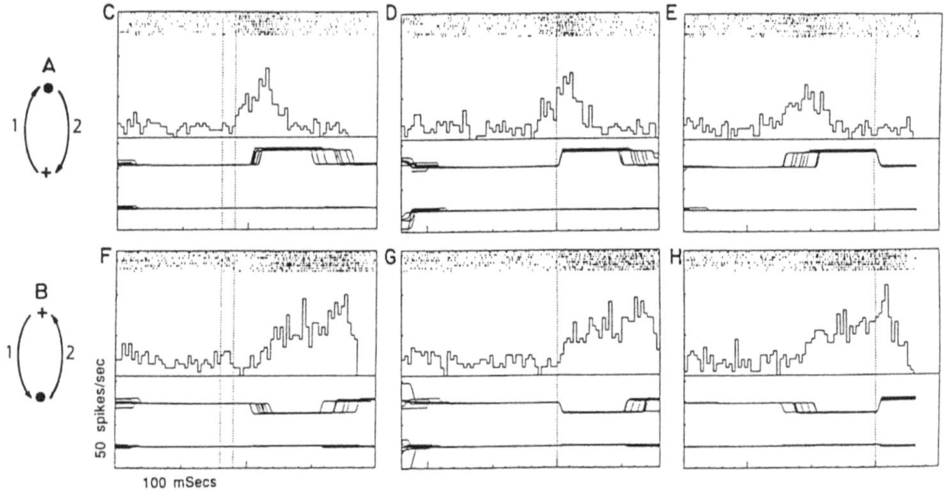

Fig. 7.6–2. Back-saccade paradigm. **A,B** Scheme of the two saccades in the task. The first saccade is to the (single) target, and the second saccade is made in the dark back to the location of the original fixation point. **C–H** Activity in the back-saccade task of an lateral intraparietal (LIP) neuron. The preferred direction of this neuron, for the LS (light-sensitive), M (memory), and S (saccade) phases, is upward. Hence, in the *top row*, the visual stimulation and the first movement are in the preferred direction, and the second movement is in the opposite, nonpreferred direction. In the *bottom row*, the visual stimulation and the first saccade are in the nonpreferred direction, but the second saccade is in the preferred direction. **C** and **F** are aligned on the sensory stimuli. The *first dotted vertical line*, denotes the offset of the fixation spot and the simultaneous onset of the target. The *second dotted line* represents the target offset. **D** and **G** are aligned on the beginning of the first saccade, and the *dotted line* denotes the time the first saccade begins. **E** and **H** are aligned on the beginning of the second saccade, and the *dotted line* denotes the time the second saccade begins. Shown in each panel, from the top, are the spike rasters, where each *horizontal trace* represents a trial, and each *tick within a line* marks the time of occurrence of a spike, the resulting histogram, and the horizontal and vertical eye position traces of the various trials, superimposed. (Reprinted, with permission, from Barash et al. 1991b)

To test the hypothesis that this activity is related to intention, we have trained monkeys in a change-in-plan task. In this task, an animal was required to make an eye movement to a remembered target after a delay (Bracewell et al. 1991). During the delay period of some trials, the target would flash on at a new location, requiring a change in the direction of the planned saccade. The activity of the cells would turn on and off consistently with the motor plan that the animal had to formulate, as required by the task. Remarkably, the plan to make an eye movement in a particular direction could be determined by examining the activity of the LIP cells without the animal emitting any behavior. To further test this intention hypothesis we trained monkeys to make saccades to the remembered locations of auditory targets (Bracewell et al. 1991). We found many of the cells to exhibit intended movement activity for both visual and auditory stimuli, consistent with the idea that the activity is related more to the motor plan than to the modality of the sensory stimulus. Lastly, in a memory double saccade task, we showed that the memory activity was only present for the next intended movement (Bracewell et al. 1991; Mazzoni et al. 1992). Two targets were flashed briefly, and the animal had to remember the location of both targets. If the second target fell in the visual receptive field of a cell, but both saccades were of a different amplitude and direction than the motor field of the cell, then the cells usually showed no response. Even when the task was configured so that the second saccade target fell in the visual receptive fields and the second saccade was made into the motor fields, the cells still did not become active until after the first saccade.

An important question is whether neural activity is related to attention or intention; for instance, Goldberg and colleagues (1990) have argued that the memory activity in area LIP is related to the animal's visual attention and not intention to make movements. A visual attention interpretation is consistent with the change in plan results, but is not consistent with the auditory memory results and requires that the interpretation be broadened to posit that the attentional activity in this area is multimodal. The memory double saccade results do not appear to be consistent with any simple attention hypothesis, since the animal must attend and memorize both visual targets, but most cells will have little or no memory activity for the visual targets in their receptive fields if the task does not require eye movements into their motor fields. A correlation of the memory activity with intention is the most straightforward interpretation for these data.

7.6.5 Representations of Visual Space in Extraretinal Coordinates

Combining eye and body position information can form several frames of reference, depending on the type of body position information that is used. A head-centered representation is a coordinate system framed with respect to the head and is formed by combining information about eye position and the location of a visual stimulus imaged on the retinas (Fig. 3). A body-centered coordinate representation is achieved by combining head, as well as eye and retinal, position information.

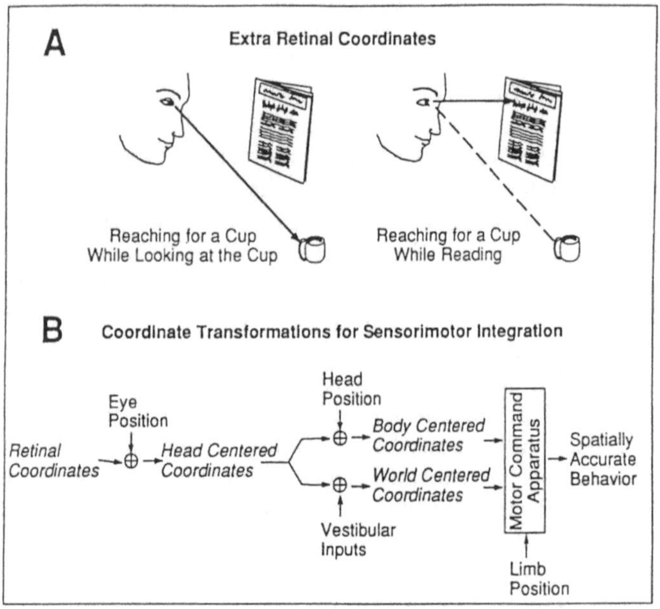

Fig. 7.6–3. **A** Demonstration of why representations of space in extraretinal coordinates are required for accurate motor behaviors. The term "extraretinal" refers to the encoding of visual stimuli in higher level coordinate frames than simple retinal coordinates. In the sketch on the *left* of the figure, a person is fixating the cup and it is imaged on the foveas, whereas on the *right* he or she is fixating the newspaper and the cup is imaged on a peripheral part of the retinas. In both cases, the subject is able to localize the cup with a reaching movement. Since different parts of the retinas are stimulated in the two conditions, information about eye position must also be available to accurately determine that the cup was at the same location in space. **B** How extraretinal coordinate frames can be computed from retinal coordinates. Visual stimuli are imaged on the retinas and are inputed to the brain in retinal coordinates. Eye position signals can be added to form representations in head-centered coordinates, and body-centered coordinates can be formed by also adding head position information. One way of forming world coordinates is to add vestibular signals, which code the location of the head in the world, to a head-centered coordinate frame. The figure shows these signals being added sequentially for illustrative purposes. It is presently not known whether there is a hierarchical organization of extraretinal coordinate frames in the brain or whether several of these signals come together at once to immediately form body and world coordinate frames. These signals, when combined with information about limb position derived from proprioceptive inputs, can encode accurate reaching movements. (Reprinted, with permission, from Andersen et al. 1993)

A world-centered coordinate frame (Fig. 3) can be achieved by combining vestibular signals with eye position and retinal position signals. The reason that it is world-centered is that the eye position information provides information about the location of the stimulus with respect to the head, and the vestibular information provides information about the location of the head in the world. Below we will provide data consistent with all three of these representations being present in areas 7a and LIP of the posterior parietal cortex. The studies outlined here provide a glimpse into the internal operations of the brain which form the basis of our spatial perceptions and actions.

7.6.6 Representation of Space in Head-Centered Coordinates

We examined how space is represented in the posterior parietal cortex by exploring the spatial receptive field properties of neurons in behaving monkeys. One might have imagined that locations in head-centered coordinates are encoded using receptive fields similar to retinal receptive fields, but anchored in head-centered, rather than retinal, coordinates. If this were the case, each time the eyes moved, the receptive field would change the location on the retina from which it derives its input, in order to code the same location in space.

Our early investigations of area 7a of the posterior parietal cortex showed, however, that locations in head-centered coordinates could be coded in an entirely different format (Andersen and Mountcastle 1983; Andersen et al. 1985b). The receptive fields of the neurons did not change their retinal locations when eye position changed. Rather the visual and eye position signals interacted to form "planar gain fields" in which the amplitude of the visual response was modulated by eye position (Andersen et al. 1985b; see Fig. 4). The gain fields are said to be planar because the amplitude of the response to stimulation of the same patch of retina varied linearly with horizontal and vertical eye position (Andersen et al. 1985b). We later found similar gain fields for eye position in area LIP (Andersen et al. 1990).

These results indicated that spatial locations are not represented explicitly at a single cell level using receptive fields in space. However, the location of a target in head-centered coordinates can still be easily determined if the activity of several neurons is examined together; in other words, the representation of space is *distributed* in this area. Figure 5 demonstrates why this representation is distributed. The contour plot of activity is made for the variables of location in head-centered space and eye position. When examined in this fashion, it can be seen that neurons are tuned to a particular location in head-centered space, but only for a limited range of eye positions. To derive a signal for location in head-centered space independent of eye position it is necessary to examine the activity of a subset of parietal neurons and thus the code is a distributed one.

Looking at the behavior of single cells as only components of a much larger, distributed network has been crucial in advancing our understanding

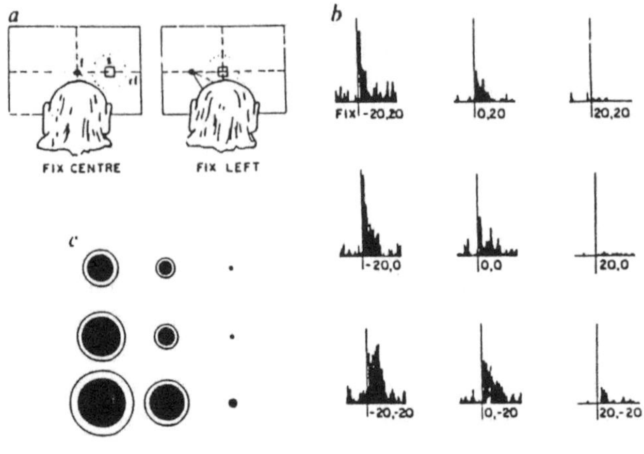

Fig. 7.6–4. **a** Experimental protocol for determining spatial gain fields, with the projection screen viewed from behind the monkey's head. To determine the effect of eye position, the monkey with head fixed, fixates on a point, *f*, at one of nine symmetrically placed locations on the projection screen. The stimulus, *s*, is always presented at the same retinal location, chosen as the maximum response zone of the retinal receptive field. The stimulus consists of 1°- or 6°-diameter spots flashed for 500 ms. Some cells were tested with 6°-diameter stimuli and others with 1°-diameter stimuli; however, stimulus size was not a relevant factor for this effect. Each measurement is repeated eight times. **b** Peristimulus histograms of a typical gain field determination. The nine histograms are located in the same relative positions as the fixations that produced them. The *vertical line* indicates the time of visual stimulus onset. **c** A graphic method for illustrating these data in which the diameter of the *darkened inner circle*, representing the visually evoked gain fields, is calculated by subtracting the background activity recorded 500 ms before the stimulus onset from the total activity during the stimulus. The annulus diameter corresponds to the background activity that is due to an eye position signal alone, recorded during the 500 ms before the stimulus presentation. (Reprinted, with permission, from *Nature* (Zipser and Andersen 1988); copyright 1988 Macmillan Magazines)

of how the brain computes locations in space. Neural networks trained to convert inputs of eye position and retinal position into locations in head coordinates at the output develop a distributed representation in the "hidden layer" interposed between the input and output layers (Zipser and Andersen 1988). This distributed representation appears to be the same as that found in areas 7a and LIP, with the "hidden" units exhibiting planar gain fields. A mathematical analysis of this network indicates that the planar gain fields are the basis of an algorithm for adding eye and retinal position vectors in a distributed network (Goodman and Andersen 1990). Thus the method of integrating these two signals is not random, but is systematic and requires the gain fields to be planar.

One of our neural network models for areas 7a and LIP was trained to produce output units with receptive fields in head-centered coordinates (Zipser and Andersen 1988). The middle layer of this model produced gain fields similar to those found in area 7a, suggesting that gain fields are an intermediate stage between retinal and spatial receptive fields. A possible

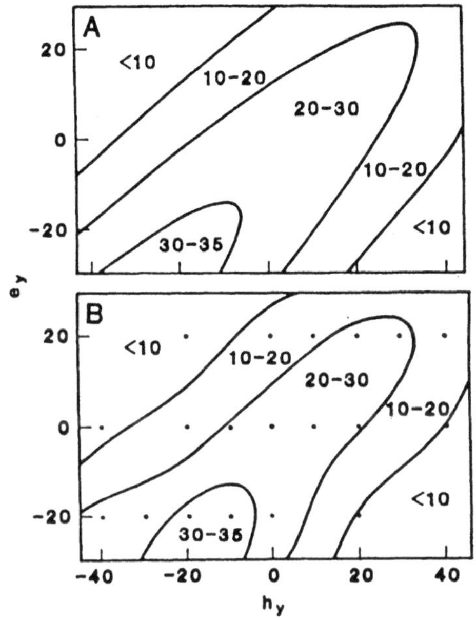

Fig. 7.6–5. **A** Computer simulation of the response (in spikes per second) of an area 7a neuron predicted by multiplying the vertical axis of a planar gain field by the vertical axis of a Gaussian receptive field. The results are represented on the contour plot with the stimulus head-centered coordinates (h_y) plotted along the abscissa and eye position (e_y) along the ordinate. **B** Contour plot of actual recording data for a cell with the same gain field and receptive field characteristics as the model neuron plotted in **A**. Each data point represents the mean evoked response to eight repetitions of the stimulus; the average standard error for these data points was 2 spikes/s. (Reprinted, with permission, from *Science* (Andersen et al. 1985); copyright American Association for the Advancement of Science)

objection to this model is that cells resembling its output (receptive fields in space) are not routinely found. However, we also trained a second network with an output representation similar to the activity found in oculomotor structures and motor centers in general. In this format, activity varies monotonically as a function of location with respect to the head. We have shown that such a network can be trained to make eye movements and have argued that receptive fields in space are an unnecessary encoding of spatial location (Andersen et al. 1990; Goodman and Andersen 1989). Instead, cells with planar gain fields appear to be an intermediate step in the transformation from visual to motor coordinates.

7.6.7 Body-Centered Coordinates and World-Centered Coordinates

The above experiments tested the interaction of eye position and retinal position signals for animals with their heads mechanically immobilized. As a result, head-centered representations could not be distinguished from body-centered or world-centered representations. With this in mind, we have recently examined the effect of head position on the visual response of cells in the posterior parietal cortex (Brotchie et al. 1991). Neural network simulations performed prior to the experiments suggested that posterior parietal neurons should have gain fields for head position as well as eye position if they are representing space in body-centered coordinates. Furthermore, the eye and head gain fields of individual parietal neurons should have the same gradients (two-dimensional slopes), even though the gradients of different cells may vary considerably. The recording experiments from areas 7a and LIP bore out these predictions. About half of the cells with eye position gain fields were also found to have head position gain fields. When both these gain fields were present in individual cells their slopes were the same, consistent with the models prediction. These results suggest that there may be two representations of space in the posterior parietal cortex, one in head-centered coordinates (units with gain fields for eye position) and the other in body-centered coordinates (units with gain fields for eye and head position). Recent recordings from our laboratory have shown that vestibular signals are also integrated with the various other signals (Snyder et al. 1993). When monkeys are rotated in the chair in the dark, many cells which show tonic activity related to eye position also have similar changes in activity for movement of the head relative to the room. Presumably these cells are receiving an integrated vestibular signal. Since these cells code both location of the eye in the head and location of the head in the world, they are coding the direction of gaze in the world. Another subset of cells has visual (retinal) receptive fields which have gain fields for eye position and chair rotation. The modulation of the visual response by a vestibular signal suggests that this population of cells may code locations of visual stimuli in world coordinates.

7.6.8 Biologically Plausible Learning Rule

One criticism of neural network models has been that the learning rule used for training the networks is unlikely to be used by the nervous system. In response to this criticism we have trained neural networks to perform the transformation from retinal- to head-centered coordinates using a reinforcement learning rule developed by Barto and Jordan (1987), which is more biologically plausible than backpropagation (Mazzoni et al. 1991a,b). We found the reinforcement-trained networks produced the same gain fields which are produced by the backpropagation-trained networks and which are found in the brain. This result suggests the algorithm discovered for computing the coordinate transformation is largely independent of the exact learning rule used to generate it. It also suggests that posterior parietal neurons could learn or adjust spatial representations using a learning paradigm that is more reasonable in terms of what is currently known about learning mechanisms in the central nervous system.

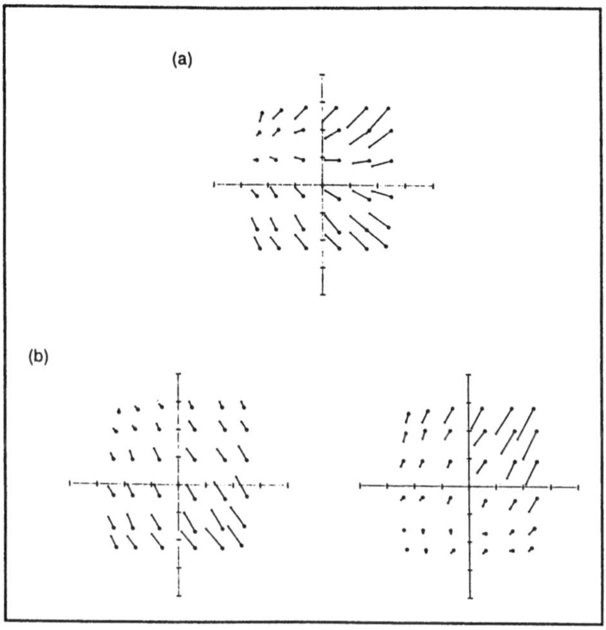

Fig. 7.6–6. **a** Eye movements recorded when two hidden units of a monotonic output network are stimulated simultaneously. **b** Eye movements recorded when each of the two hidden units is stimulated alone. Note that the result of simultaneous stimulation (**a**) is more or less the vector addition of the two saccade fields that result from stimulation of the individual units. (Reprinted, with permission, from *Journal of Cognitive Neuroscience* (Goodman and Andersen 1989); copyright MIT Press)

7.6.9 Microstimulation Experiments

We have also examined the effects of "microstimulating" the Zipser–Andersen model on eye movements (Goodman and Andersen 1989). Our model was connected to a simplified set of oculomotor muscles (four instead of six), and individual hidden units were maximally activated to simulate microstimulation. The most typical result of stimulating individual hidden units was the change in amplitude pattern seen with stimulation of area LIP (Fig. 6b). This model suggests that the change in amplitude pattern is indicative of a distributed representation of space in LIP. Because the representation of head-centered space is distributed, a single cell does not drive the eyes to a goal in space; rather, such behavior requires the activity of many LIP neurons. To demonstrate directly how these change in amplitude patterns could code single locations among a group of neurons, we showed that stimulation of two or more hidden units produced a pattern of eye movements which converge toward a single goal in head-centered space (Fig. 6a).

7.6.10 Conclusions

There has been substantial progress in research on the role of area LIP in the processing of saccades. This area appears to play an interesting, high-level role in the processing of eye movements. For instance, the memory activity in this area appears to be directly related to intention and motor planning.

The data reviewed here also shed light on the nature of abstract representations of space. Spatial representations are derived by integrating visual signals with information about eye position and head position. These signals are brought together in the posterior parietal cortex to form a specific, distributed representation of space which is typified by planar gain fields.

Acknowledgements. The research reviewed here was supported by the National Institutes of Health, the Office of Naval Research, and the McDonnell-Pew Program in Cognitive Neuroscience.

References

Andersen RA (1987) The role of the inferior parietal lobule in spatial perception and visual-motor integration. In: Plum F, Mountcastle VB, Geiger SR (eds) The handbook of physiology, Sect 1. The nervous system, vol 5. Higher functions of the brain part 2. American Physiological Society, Bethesda, pp 483–518

Andersen RA (1989) Visual and eye-movement functions of the posterior parietal cortex. Ann Rev Neurosci 12: 377–403

Andersen RA, Mountcastle VB (1983) The influence of the angle of gaze upon the excitability of the light sensitive neurons of the posterior parietal cortex. J Neurosci 3: 532–548

Andersen RA, Asanuma C, Cowan WM (1985a) Callosal and prefrontal associational projecting cell populations in area 7a of the macaque monkey: a study using retrogradely transported fluorescent dyes. J Comp Neurol 232: 443–455

Andersen RA, Essick GK, Siegel RM (1985b) Encoding of spatial location by posterior parietal neurons. Science 230: 456–458

Andersen RA, Essick GK, Siegel RM (1987) Neurons of area 7 activated by both visual stimuli and oculomotor behavior. Exp Brain Res 67: 316–322

Andersen RA, Bracewell RM, Barash S, Gnadt JW, Fogassi L (1990) Eye position effects on visual, memory and saccade-related activity in areas LIP and 7A of macaque. J Neurosci 10: 1176–1196

Andersen RA, Snyder L, Li C-S, Stricanne B (1993) Coordinate transformations in the representions of spatial information. Current Opinion in Neurobiology 3: 171–176

Balint R (1909) Seelenlähmung des "Schauens", Optische Ataxie, Räumliche Störung der Aufmerksamkeit. Psychiatr Neurol 25: 51–81

Barash S, Bracewell RM, Fogassi L, Gnadt JW, Andersen RA (1991a) Saccade-related activity in the lateral intraparietal area. I. Temporal properties. J Neurophysiol 66: 1095–1108

Barash S, Bracewell RM, Fogassi L, Gnadt JW, Andersen RA (1991b) Saccade-related activity in the lateral intraparietal area. II. Spatial properties. J Neurophysiol 66: 1109–1124

Barto AG, Jordan MI (1987) Gradient following without backpropagation in layered networks. Proc IEEE Int Conf on Neural Networks 2: 629–636

Bracewell RM, Barash S, Mazzoni P, Andersen RA (1991) Neurones in the macaque lateral intraparietal cortex appear to encode the next intended saccade. Soc Neurosci Abstr 17: 1282

Brotchie PR, Andersen RA (1991) A body-centered coordinate system in posterior parietal cortex. Soc Neurosci Abstr 17: 1282

Fleming JFR, Crosby EC (1955) The parietal lobe as an additional motor area: the motor effects of electrical stimulation and ablation of cortical areas 5 and 7 in monkeys. J Comp Neurol 103: 485–512

Gnadt JW, Andersen RA (1988) Memory related motor planning activity in posterior parietal cortex of macaque. Exp Brain Res 70: 216–220

Goodman S, Andersen RA (1989) Microstimulation of a neural-network model for visually guided saccades. J Cognitive Neurosci 1: 317–326

Goodman SJ, Andersen RA (1990) Algorithm programmed by a neural network model for coordinate transformation. Proc Int Joint Conf Neural Networks San Diego II: 381–386

Goldberg ME, Colby CL, Duhamel J-R (1990) The representation of visuomotor space in the parietal lobe of the monkey. Cold Spring Harbor Symp Quant Biol 55: 729–739

Hyvärinen J (1981) Regional distribution of functions in parietal association area 7 of the monkey. Brain Res 206: 287–303

Hyvärinen J, Shelepin Y (1979) Distribution of visual and somatic functions in the parietal associative area 7 of the monkey. Brain Res 169: 561–564

Keating EG, Gooley SG (1988) Disconnection of parietal and occipital access to the saccadic oculomotor system. Exp Brain Res 70: 385–398

Lynch JC, McLaren JW (1989) Deficits of visual attention and saccadic eye movements after lesions of parieto-occipital cortex in monkeys. J Neurophysiol 61: 74–90

Lynch JC, Mountcastle VB, Talbot WH, Yin TCT (1977) Parietal lobe mechanisms for directed visual attention. J Neurophysiol 40: 362–389

Maunsell JHR, Van Essen DC (1983) The connections of the middle temporal visual area (MT) and their relationship to a cortical hierarchy in the macaque monkey. J Neurosci 3: 2563–2586

Mazzoni P, Andersen RA, Jordan MI (1991a) A more biologically plausible learning rule for neural networks. Proc Nat Acad Sci USA 88: 4433–4437

Mazzoni P, Andersen RA, Jordan MI (1991b) A more biologically plausible learning rule than backpropagation applied to a network model of cortical area 7a. Cerebral Cortex 1: 293–307

Mazzoni P, Bracewell RM, Barash S, Andersen R (1992) Intended movement activity of area LIP neurons in delayed double saccade tasks. Soc Neurosci Abstr 18: 148

Mays LE, Sparks DL (1980) Dissociation of visual and saccade-related responses in superior colliculus neurons. Science 43: 207–232

Mountcastle VB, Lynch JC, Georgopoulos A, Sakata H, Acuna C (1975) Posterior parietal assoication cortex of the monkey: command function for operations within extrapersonal space. J Neurophysiol 38: 871–908

Newsome WT, Wurtz RH, Komatsu H (1988) Relation of cortical areas MT and MST to pursuit eye movements. I. Differentiation of retinal from extraretinal inputs. J Neurophysiol 60: 604–620

Robinson CJ, Burton H (1980a) Organization of somatosensory receptive fields in cortical areas 7b, retroinsular postauditory and granular insula of M. Fascicularis. J Comp Neurol 192: 69–92

Robinson CJ, Burton H (1980b) Somatic submodality distribution within the second somatosensory (SII), 7b, retroinsular, postauditory and granular insular cortical areas of of M. Fascicularis. J Comp Neurol 192: 93–108

Robinson DL, Goldberg ME, Stanton GB (1978) Parietal association cortex in the primate: sensory mechanisms and behavioral modulations. J Neurophysiol 41: 910–932

Sakata H, Shibutani H, Kawano K (1980) Spatial properties of visual fixation neurons in posterior parietal association cortex of the monkey. J Neurophysiol 43: 1654–1672

Snyder LH, Brotchie PR, Andersen RA (1993) World-centered encoding of location in posterior parietal cortex of monkey. Soc Neurosci Abstr 19: 770

Wagman, IH (1964) Eye movements induced by electric stimulation of cerebrum in monkeys and their relationship to bodily movements. In: Bender MB (ed) The oculomotor system. Harper and Row, New York, pp 18–39

Zipser D, Andersen RA (1988) A back-propagation programmed network that simulates response properties of a subset of posterior parietal neurons. Nature 331: 679–684

7.7 Comparative Physiological and Anatomical Studies of the Primate Vestibular Cortex

O.-J. GRÜSSER, W.O. GULDIN, S. MIRRING, AND A. SALAH-ELDIN
Department of Physiology, Freie Universität Berlin, Germany

7.7.1 Introduction

Compared to the extensive knowledge of the cortical visual and visuomotor system, impressively demonstrated at this conference, our scientific understanding of the cortical vestibular mechanisms is quite fragmentary. In this review we summarize the results of investigations on the cortical representation of the primate vestibular system performed in different primate species during the past 15 years (Grüsser et al. 1982, 1983, 1990a,b, 1992; Akbarian et al. 1988, 1992, 1993; Guldin et al. 1987, 1989, 1992). Since the cortical mechanisms related to vestibular stimulation in Old World monkeys (*Macaca fascicularis*, Java monkey) and New World monkeys (*Saimiri sciureus*, squirrel monkey, *Callithrix jacchus*, common marmoset monkey) are very similar, one can conclude that they must have common genetic determinants which developed before Gondwanaland divided into two continents, Africa and South America, 65 million years ago. As in other sensory modalities, the cortical representation of the vestibular system is realized by a distributed neuronal network, the "inner cortical vestibular circuit", which is composed of at least three different cytoarchitectonic areas. In addition to afferent vestibular signals, these areas receive inputs from other cortical association areas transmitting somatosensory, optokinetic, gaze motor, and somatomotor information.

7.7.2 Material and Methods

The studies were performed during the years 1980–1993 in 13 Java monkeys (*Macaca fascicularis*), 12 squirrel monkeys (*Saimiri sciureus*) and 8 marmoset monkeys (*Callithrix jacchus*).

Neurophysiological Recordings. After adequate conditioning of the animals to the monkey chair and the stimulation conditions, stereotaxic recordings were performed with conventional methods, whereby varnished tungsten microelectrodes or tungsten- or elgiloy-in-glass microelectrodes were used (for details see Grüsser et al. 1990a; Guldin et al. 1992). In addition to single neuron recordings, eye position was monitored by means of horizontal and vertical DC-electrooculography. Experiments were performed three times a week, on the average, and the experimental sessions lasted between 3 and 8 months in one animal with pauses of 1–4 weeks in between.

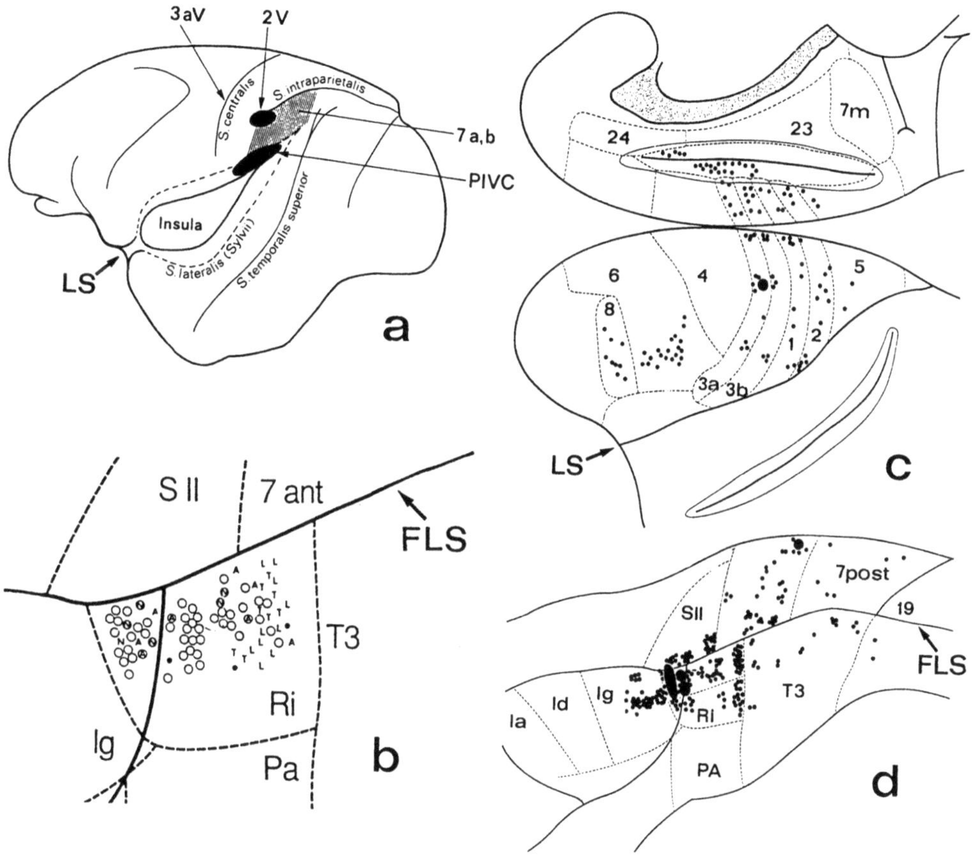

Fig. 7.7–1. **a** Schematic drawing of the location of the three cortical areas forming the inner cortical vestibular circuit in the macaque brain. The lateral sulcus is unfolded; the PIVC is located deep in the parietal operculum at the posterior end of the insula and the retroinsular region. **b** The PIVC of the New World monkey *Saimiri sciureus* in a view of the retroinsular region with an unfolded parietal and temporal operculum. The PIVC corresponded to area Ri located in the temporal part of the retroinsular region. *Circles*, location of nerve cells responding to vestibular and optokinetic stimulation; *N*, neurons responding to mechanical stimulation of neck receptors; *A*, to stimulation of the deep proprioceptors of the forearm; *L*, of the leg; *T*, of the trunk. **c,d** Distribution of HRP-labeled cells in the brain of a squirrel monkey. The injection was placed in the anterior part of area Ri as shown in **d**. *Small dots*, 1–5 HRP-labeled neurons found below the cortical surface within a 40 μm coronal section; *large dots*, more than 10 HRP-labeled neurons. **d** The parietal and the temporal operculum of the insula and the retroinsular area are unfolded. A similar distribution of HRP-labeled nerve cells was found in the corresponding contralateral cortex regions. *FLS*, fundus of the lateral sulcus. (**b-d** From Guldin et al. 1992)

Stimulation. The monkey chair was placed on a turntable with a semicircular rail system which allowed sinusoidal vestibular stimulation in a yaw, pitch, roll, or any other plane in between. The interaural axis of the head was the center of rotation. Steady tilt and horizontal off-axis rotation were possible (see Fig. 1 in Grüsser et al. 1990a). Optokinetic stimulation was achieved by either a vertically striped cylinder rotating around the animal or a rotating random light dot pattern projected to the wall of the cylindrical experimental room by means of a planetarium projector. These optokinetic stimuli could be moved in any direction either continuously or sinusoidally back and forth and could be coupled to the vestibular stimulation. Sinusoidal trunk rotation with the head unmoved in space was used to stimulate the neck receptors. In addition, other mechanical stimuli were applied to the skin, muscles, or joints.

Data Processing. Eye position, single nerve cell action potentials, horizontal (yaw) and vertical (pitch, roll) chair position, chair velocity, position and velocity of the optokinetic drum or the planetarium projector were stored on electromagnetic tape or directly on the hard disc of a digital computer after AD conversion. Conventional computer programs averaged the 8–40 responses to repeated stimulation to yield peristimulus time histograms (PSTHs).

Neuroanatomical Techniques. A 0.1–0.4 μl solution of tracers was administered to different sites of the monkey brain after identification of the injection site by single neuron recordings. Injections of the following retrograde tracers were made to either cortical structures of the vestibular system or brainstem vestibular nuclei: horseradish peroxidase (HRP), Wheatgerm-HRP, nuclear yellow, fast blue, and different types of labeled dextrans. Pressure injection was employed in most of the experiments; HRP was applied in some by microiontophoresis. Wheatgerm-HRP and dextran-rhodamines were used as anterograde tracers. After adequate delays the animals were sacrificed by deep anesthesia (>60 mg/kg pentobarbital intraperitoneally) and a transcardial perfusion with 0.5 liter of a physiological saline solution followed by 1–4 liter of phosphate-buffered 4% paraformaldehyde solution (pH 7.35). Further histological processing was by standard techniques (for details see Guldin et al. 1992). For the microscopic examination of the slides in search of HRP-filled cells a Zeiss light microscope was used, while a Zeiss fluorescence microscope with appropriate filters was used to study the fluorescence tracer marked nerve cells. A quantitative evaluation of the distribution of labeled nerve cells yielded "labellograms", a method described by Akbarian et al. (1992).

7.7.3 The Inner Cortical Vestibular Circuit

To date only responses of cortical vestibular nerve cells to dynamic changes in head-in-space position have been recorded in primates. The cortical representation of linear head acceleration and steady head-in-space position is still unknown. The cortical system processing head-in-space rotation consists mainly of three reciprocally interconnected areas: the parieto-insular vestibular cortex (PIVC; Fig. 1a), area 3aV, and area 2v in Java monkeys. The latter is most likely homologous with part of area 7ant in squirrel monkeys (Fig. 1b). The structures of this inner cortical vestibular circuit are closely interconnected with each other and receive signals from two other cortical sensory systems, a part of the primary somatosensory cortex which mediates somatosensory and proprioceptive signals and a visual association area. This area was identified in *Saimiri* and *Callithrix* as the part of area 19 located along the medial temporal sulcus (Fig. 1d), while in Java monkeys it is predominantly areas MST (the medial superior temporal area) and FST (the fundus region around the superior temporal sulcus), transmitting optokinetic responses (see also Thier and Erickson 1992) to the temporal area T3 (Figs. 1b,d). In *Saimiri* and *Macaca fascicularis*, area T3 was directly connected to the neighboring area PIVC. In addition, some of the cortical gaze motor regions, especially area 6a, are interconnected with the structures of the inner cortical vestibular circuit, and the same is true for part of the anterior cingulate cortex (see below).

7.7.4 Vestibular, Visual, and Somatosensory Interaction in Nerve Cells of the Inner Cortical Vestibular Circuit

The Parieto-Insular Vestibular Cortex. PIVC corresponds cytoarchitectonically in Java monkeys to area reitp of Pandya and Sanides (1973) and in *Saimiri* and *Callithrix* to area Ri and an occipital part of the area Ig of Jones and Burton (1976; Fig. 1b). Fairly similar neuronal responses have been obtained in the PIVC of *Saimiri* and Java monkeys. PIVC is perhaps the cortical region which in man exhibits signs of increased activity during vestibular stimulation (regional cerebral blood flow measurements; Friberg et al. 1981, 1985). All neurons exhibited a directional preponderance for semicircular canal stimulation when the rotation occurred in a plane located within a certain spatial response sector. To characterize this directional preponderance, the classification scheme of Duensing and Schäfer (1958) was used: type I neurons were activated during rotation toward the side ipsilateral to the recording electrode and inhibited during contralateral movement (yaw and roll), and type II neurons showed the opposite responses. With pitch stimulation type I neurons were those activated during "nose up" rotation and type II neurons during "nose down" rotation. Type III

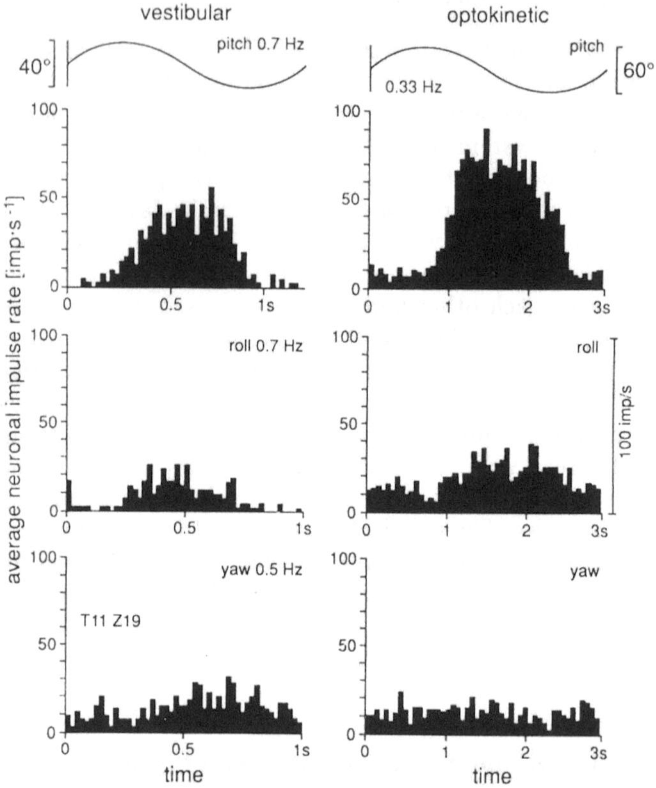

Fig. 7.7–2. Responses of a PIVC neuron of *Saimiri* to 0.7 H vestibular stimulation in pitch, roll, and yaw and to corresponding 0.33 H optokinetic stimulation (random dot planetarium projected stimuli). Note that the strongest responses were obtained in pitch

neurons were activated by acceleration in both directions of the yaw, pitch or roll plane. Fewer than 10% of the neurons recorded belonged to this class. Types I and II were found with about equal frequency. On examining the distribution of responses to sine wave rotation in different directions (Fig. 2), it was clear that the response optimum of the PIVC neurons could be related to any of the possible rotation planes. Thus all possible head-in-space rotation planes are represented by the neuronal responses of the PIVC. In addition, neighboring neurons had about the same optimum spatial response planes for vestibular stimulation and about the same response sectors. This indicated a functional organization of the PIVC in "patches" or perhaps columns of similar response preferences.

Many of the neurons of PIVC responded to mechanical stimulation of the neck muscles by either local pressure, vibration, or trunk rotation while the head remained stationary in space. Those neurons which had their preferential vestibular response sector around the horizontal plane were also activated maximally, as a rule, by horizontal trunk rotation while the head remained stationary (Fig. 3). Typically, vestibular-neck receptor interaction occurred in these neurons, whereby mutual facilitation or inhibition was found when the responses to sinusoidal head-on-trunk rotation were compared with those evoked by trunk rotation (head stationary) or rotation of the animal in yaw. Many of the vestibular PIVC neurons also responded to mechanical stimulation of a portion of the skin surface, whereby the receptive fields were always very large, extending over one arm, hand, or leg. Ipsilateral or contralateral somatosensory inputs of this type could be identified. In squirrel monkeys some of the PIVC neurons activated by vestibular stimulation in pitch also responded to mechanical stimulation of the plantar surface of the feet or deep muscle receptors of the lower limbs. Many PIVC neurons were tested by steady tilt in the dark. No indication was found that static otolith signals contributed to the activation level of PIVC nerve cells.

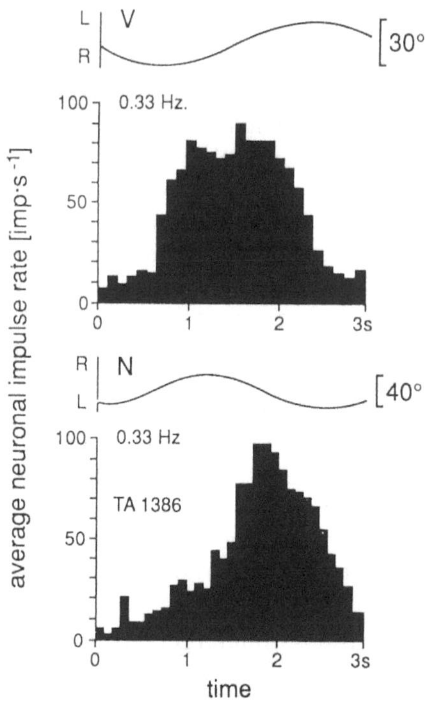

Fig. 7.7–3. Example of the responses of a PIVC neuron of a squirrel monkey to vestibular stimulation in yaw *(V)* and mechanical stimulation of the neck receptors *(N)*. For this stimulation the head was fixed in space and the trunk rotated sinusoidally

Considering psychophysical experiments (Fischer and Kornmüller 1930; Dichgans and Brandt 1978; Grüsser et al. 1992) and the findings of Waespe and Henn (1977a,b, 1978) in neurons of the brainstem vestibular nuclei, it was not surprising that nearly all PIVC neurons responding to semicircular canal stimulation were also activated by adequate optokinetic stimulation (see Sect. 7.7.2). The preferential rotation axes for optokinetic stimulation and vestibular stimulation were collinear, whereby two types of visual-vestibular interaction were found: (a) synergistic visual-vestibular interaction, a response combination in which the optimum optokinetic stimulation was obtained in the opposite direction to optimum vestibular rotation (Fig. 4); and (b) antagonistic visual-vestibular interaction, those responses in which vestibular rotation and optokinetic stimulation led to optimum activation in the same direction of rotation.

Synergistic or antagonistic visual-vestibular interaction was apparent when the animal was rotated while the optokinetic stimulus was either stationary in space or moving 180° out of phase relative to the vestibular sine wave stimulation. A synergistic visual-vestibular interaction led to an increase in neuronal impulse rate as compared to "monomodal" stimulation (rotation in the dark or optokinetic stimulation in a stationary animal). The interaction of the two modalities was rarely linear, however (Fig. 4). Antagonistic visual-vestibular interaction was characterized by a cancellation of the modulation of neuronal activity when both modalities were stimulated.

A small number of PIVC neurons also responded to small moving visual objects, whereby eye pursuit movements were not necessary for an increase in neuronal activation. Some of these neurons did not respond to pursuit of a small interesting target (e.g., a raisin) when the target was moved across an empty white background. They were activated, however, when the same target was moved in front of a patterned background. This activation was probably due mainly to the eye movement induced retinal shift of the patterned background image. A few of the PIVC neurons also increased their activity when interesting objects approached the face of the animal. No changes in neuronal activity were observed during spontaneous saccades in the dark or in the light when the animal could see a well-structured experimental room or the striped cylinder. When a PIVC neuron was activated during optokinetic stimulation, the activation continued during the first seconds of the first optokinetic after-nystagmus (OKAN I), but there was only a loose correlation between the duration of OKAN I and neuronal activation when the experimental room was suddenly darkened after optokinetic stimulation.

Auditory stimulation, either stationary or with moving stimuli, was not successful in eliciting any neuronal responses in the dark, even when tape-recorded monkey calls were used instead of sine wave or white noise auditory

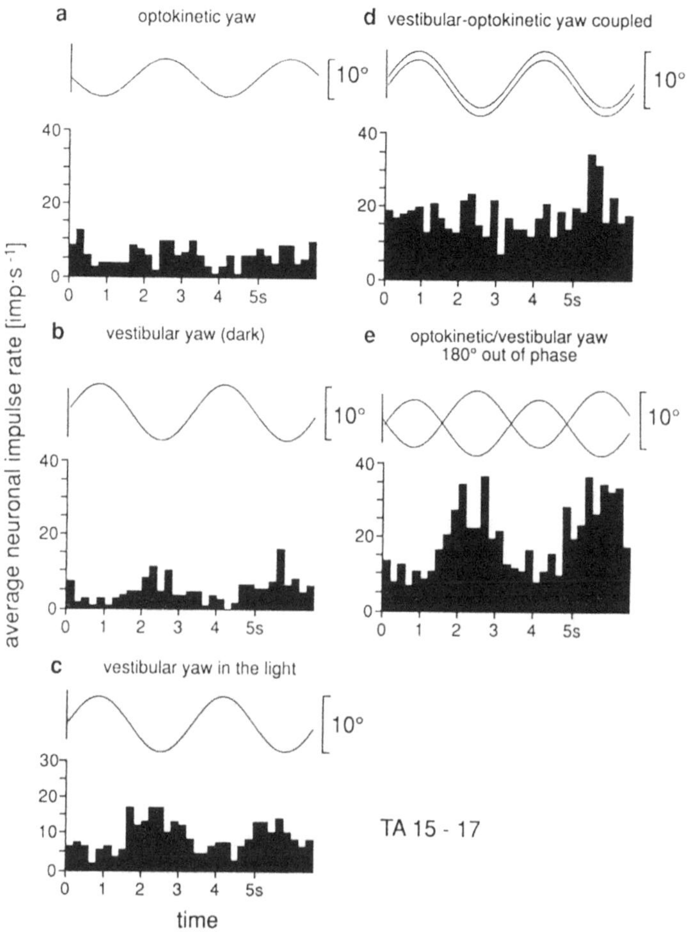

Fig. 7.7–4. Responses of a squirrel monkey PIVC neuron. Separate and combined optokinetic and vestibular sine wave stimulation in yaw. **a,b** Weak responses to monomodal stimulation. **c** When the chair was rotated in the light with the striped drum stationary, the vestibular responses increased (synergistic optokinetic-vestibular interaction). **d** When drum and chair were rotated synchronously, a general unmodulated increase in activity was recorded. **e** When drum and chair were rotated 180° out of phase, the modulation of the neuronal activity was considerably stronger than in **c**

stimuli. A few tests indicated that olfactory or gustatory stimulation did not change the activity of PIVC neurons. When the monkey was offered juice or small pieces of chocolate during the recording of the action potentials of a PIVC neuron, no change in neuronal activity were observed.

In summary, the PIVC neurons studied in Old and New World monkeys evidently monitor under natural conditions head-in-space rotation predominantly, whereby the signals contributing to this information are not only vestibular in origin but are also mediated via the visual (optokinetic) and the somatosensory modality.

Response Properties of Area 3aV Neurons. Within the somatotopic neck and upper trunk segment of area 3a, which forms a part of the anterior primary somatosensory cortex, about 10%–20% of the proprioceptive neurons also responded to stimulation of semicircular canal receptors in the dark, i.e., to sine wave stimulation in the pitch, roll, or yaw direction. As in PIVC neurons, a rather narrow response sector around the optimum rotation plane was the rule. Reconstructing the recording tracks through area 3a, we came to the conclusion that "vestibular" nerve cells were always located inside the neck or upper trunk segment of area 3a. All neurons in this region responding to vestibular stimulation also responded to proprioceptive neck stimulation (local pressure, trunk rotation with the head stationary). In contrast to PIVC neurons, none of these vestibular area 3a neurons exhibited a significant optokinetic response. We designated the part of area 3a in which vestibular-proprioceptive neck interaction occurred "area 3aV". This part of area 3a had close connections to PIVC and area 2v (or 7ant). In addition to the neck region vestibular responses were also found in the hand region of area 3a (area 3aH; Ödquist et al. 1974).

Visual and Vestibular Responses in Areas 2v and 7a. By means of microelectrode recordings Büttner and Buettner (1978) determined a small vestibular region called area 2v located at the tip of the intraparietal sulcus of macaques (see Fredrickson et al. 1966; Schwarz and Fredrickson 1971). The activity of area 2v nerve cells were modified by vestibular and optokinetic stimulation in the horizontal plane. In Java monkeys we could not find this area after a systematic search with microelectrode penetration in the region around the tip of the intraparietal sulcus. Vestibular and optokinetic responses were recorded repeatedly, however, from nerve cells located in area 7a (see also Kawano et al. 1980). The vestibular responses were similar to those described in area 2v neurons: directional selectivity for rotatory and optokinetic stimuli. Very few (presumably fewer than 2%) of the area 7a neurons showed a vestibular response during rotation in the dark. In squirrel monkeys, the homologue of area 2v is believed to be part of area 7ant, which forms histologically part of the inner cortical vestibular circuit.

7.7.5 Anatomical Connections of the Cortical Vestibular Areas

As described in Sect. 7.7.2, different anterograde and retrograde tracers were injected into small, circumscribed regions of the brainstem vestibular nuclear complex, the cortical vestibular regions PIVC and 3aV, and the "optokinetic" area T3. By these methods, thalamic nuclei with afferent connections to the three cortical areas mentioned were identified as well as thalamic structures receiving afferent connections from the vestibular nuclei (see also Liedgren et al. 1976; Büttner and Henn 1976; Buettner and Büttner 1979; Büttner and Lang 1979; Lang et al. 1979). Moreover it was found that corticofugal axons project directly to the vestibular nuclei from the following cortical areas: PIVC, 3aV, 3aH, 2v or 7ant, T3, 6a, and the cingulate regions 23c and 6c. As mentioned above, three cortical vestibular regions in the squirrel monkey were found to be fairly densely interconnected with each other, namely PIVC, 3aV, and 7ant (corresponding to area 2v in macaques). These structures were defined as the "inner cortical vestibular circuit". They receive input signals from exactly those cortical areas which project directly to the vestibular nuclei, namely area 7a, T3, 6a and 23c/6c (Akbarian et al. 1992, 1993; Guldin et al. 1992).

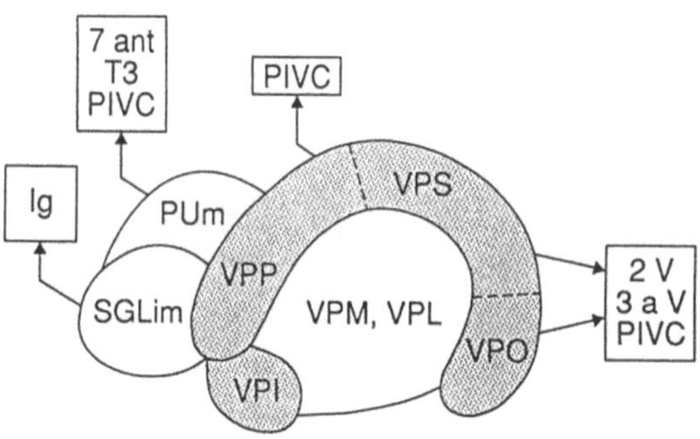

Fig. 7.7–5. Scheme of the afferent connections to cortical vestibular areas from the thalamus

The thalamic input to PIVC originates within two main regions, the oral and medial pulvinar, presumably transmitting optokinetic and space-related signals, and a complex of thalamic nuclei which also receive afferent connections from brainstem vestibular nuclei: ventroposterior nucleus pars posterior (VPP), ventroposterior lateral nucleus (VPL), ventroposterior superior nucleus (VPS; Fig. 5). The vestibular area 3aV receives its main input from the ventroposterior oral nucleus (VPO), while the areas T3 and 7ant receive quite strong projections from the medial pulvinar but only insignificant ones from the vestibular thalamic nuclei (for details see Akbarian et al. 1992). Figure 6 represents schematically the main afferent and efferent connections of the inner cortical vestibular circuit identified to date in the Java macaque and the two New World monkeys (*Saimiri, Callithrix*). Comparing the data found for *Saimiri* and Java monkeys, a fairly

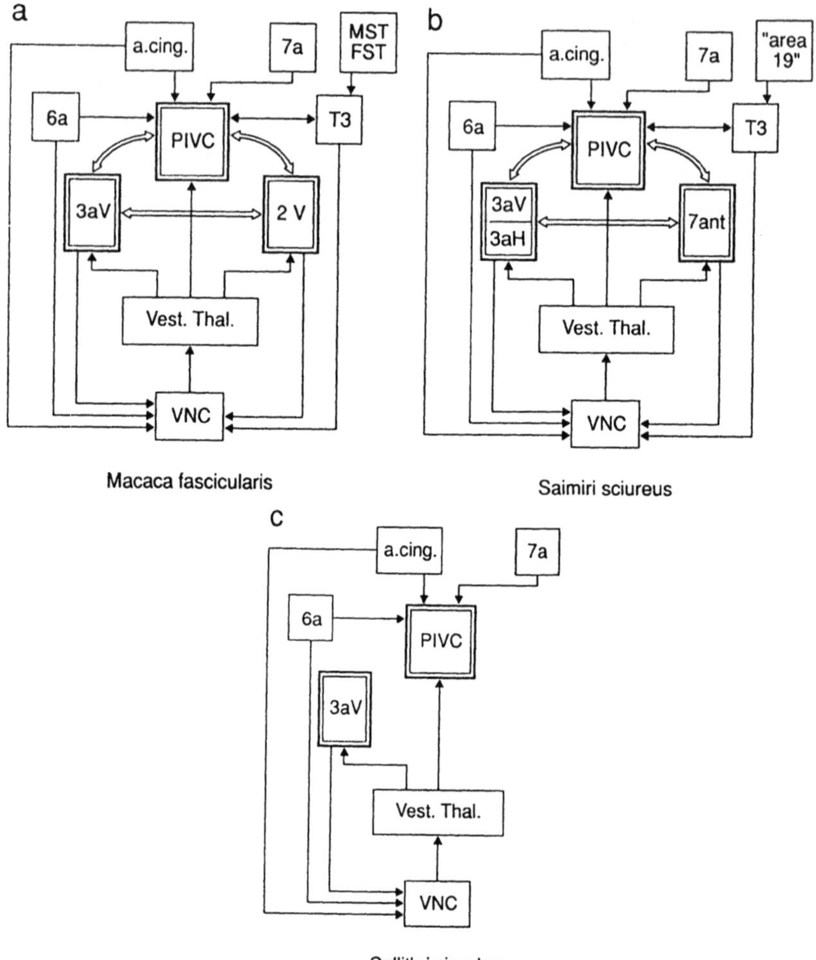

Fig. 7.7–6. **a-c** Simplified schemes of the connections of the inner cortical vestibular circuit (PIVC, 2v, 3aV) in the Old World monkey *Macaque fascicularis* (**a**), in the New World monkey *Saimiri sciureus* (**b**), and in the New World monkey *Callithrix jacchus* (**c**)

close homology was evident while some of the connections found in these two species could not as yet be identified in *Callithrix*.

The functional interpretation of the anatomical connections mentioned and depicted in Fig. 6 indicates that cortical vestibular structures, in addition to the spatiotopic, optokinetic, and vestibular thalamic inputs, receive information from cortical structures dealing with the representation of the extrapersonal space (area 7a), optokinetic stimulation (area T3), gaze motor control (area 6a), and perhaps signals related to intra-abdominal receptive mechanisms (area 23c). These connections indicate an "updating" of the afferent vestibular and optokinetic mechanisms by cortical signals.

The efferent connections from the different cortical "vestibular" structures to the brainstem vestibular nuclear complex (Akbarian et al. 1993) presumably suppress vestibular reflexes during active, goal-directed, head and body movements, which are controlled by a cortical neuronal network. Further studies are necessary, of course, to test this hypothesis on direct corticofugal inhibition modulating the sensitivity of brainstem vestibular nerve cells.

7.7.6 Summary

Cortical vestibular structures, in particular the retroinsular area PIVC, area 2v, and area 3aV, were described in Old World and New World monkeys. These structures are closely interconnected and form the inner cortical vestibular circuit in which the vestibular responses originating from the semicircular canal receptors are modified by optokinetic signals, neck receptor activation and other somatosensory inputs. The corticocortical, thalamocortical, and cortico-brainstem (vestibular nuclei) connections of these cortical vestibular structures were analyzed by anterograde and retrograde tracer techniques. It was found that area T3 bordering on the PIVC, area 6a, area 7a, and area 23c/6c are connected to the cortical vestibular structures mentioned and provide a meaningful input of spatial, optokinetic, gaze control, and somatosensory signals and perhaps signals related to mechanoreceptors of the inner organs, as well. All cortical structures mentioned project directly to the brainstem vestibular nuclear complex and are thus able to regulate the vestibular sensitivity of these structures.

Acknowledgements. This work was supported in part by a grant of the Deutsche Forschungsgemeinschaft (Gr 161). O.-J. G. was supported in 1987/1988 by an Akademie-Stipendium of the Volkswagen Foundation and more recently by a F.C. Donders Professorship at the University of Utrecht (Summer 1991). A.S.-E. was a Fellow of the "Channel-System" of the Republic of Egypt. We thank Mr. J. Lerch for his skillful assistance in constructing the equipment, Mr. P. Holzner for photographic work, Mrs. B. Krawczynski and Mrs. D. Starke for art work, Mrs. Dreykluft for technical assistance, Mrs. U. Saykam for typing the manuscript, and Mrs. J. Dames for her expert help in the English translation.

References

Akbarian S, Berndl K, Grüsser O-J, Guldin W, Pause M, Schreiter U (1988) Responses of single neurons in the parietoinsular vestibular cortex of primates. Ann New York Acad Sci 545: 187–202
Akbarian S, Grüsser O-J, Guldin W (1992) Thalamus connections of the vestibular cortical fields in the squirrel monkey (Saimiri sciureus). J Comp Neurol 326: 423–441
Akbarian S, Grüsser O-J, Guldin W (1993) Corticofugal projections to the vestibular nuclei in squirrel monkeys — further evidence of multiple cortical vestibular fields. J Comp Neurol 332: 89–104
Buettner UW, Büttner U (1979) Vestibular nuclei activity in the alert monkey during suppression of vestibular and optokinetic nystagmus. Exp Brain Res 37: 581–593
Büttner U, Buettner UW (1978) Parietal cortex (2v) neuronal activity in the alert monkey during natural vestibular and optokinetic stimulation. Brain Res 153: 392–397
Büttner U, Henn V (1976) Thalamic unit activity in the alert monkey during natural vestibular stimulation. Brain Res 103: 127–132
Büttner U, Lang W (1979) The vestibulocortical pathway: neurophysiological and anatomical studies in the monkey. In: Granit R, Pompeiano O (eds) Reflex control of posture and movement. Progress Brain Res 50: 567–579
Dichgans J, Brandt Th (1978) Visual-vestibular interaction: effects on self-motion perception and postural control. In: Held VR, Leibowitz H, Teuber H-L (eds) Handbook of sensory physiology, vol V. Springer, Berlin Heidelberg New York, pp 755–804
Duensing F, Schäfer KP (1958) Die Aktivität einzelner Neurone im Bereiche der Vestibulariskerne bei Horizontalbeschleunigung unter besonderer Berücksichtigung des vestibulären Nystagmus. Arch Psychiat u Nervenkr 198: 224–252
Fischer MH, Kornmüller AE (1930) Der Schwindel. In: Bethe A, von Bergmann G, Embden G, Ellinger A (eds) Handbuch der normalen und pathologischen Physiologie, vol XV/1. Springer, Berlin Heidelberg New York, pp 442–494
Fredrickson JM, Figge U, Scheid P, Kornhuber HH (1966) Vestibular nerve projection to the cerebral cortex of the rhesus monkey. Exp Brain Res 2: 318–327
Friberg L, Skyhoj T, Paulson OB, Lassen NA (1981) Cortical activation during vestibular stimulation and rCBF measurement. J Cerebr Blood Flow Metab 1: 473–474
Friberg L, Olsen TS, Roland PE, Paulson OB, Lassen NA (1985) Focal increase of blood flow in the cerebral cortex of man during vestibular stimulation. Brain 108: 609–623
Grüsser O-J, Pause M, Schreiter U (1982) Neuronal responses in the parieto-insular vestibular cortex of alert Java monkeys (Macaca fascicularis). In: Roucoux A, Crommelinck M (eds) Physiological and pathological aspects of eye movements. Junk W, Publishers, The Hague, pp 251–270
Grüsser O-J, Pause M, Schreiter U (1983) A new vestibular area in the primate cortex. Soc Neurosci Abstr 9, 2: 749
Grüsser O-J, Pause M, Schreiter U (1990a) Localisation and responses of neurons in the parieto-insular vestibular cortex of awake monkeys (Macaca fascicularis). J Physiol 430: 537–557
Grüsser O-J, Pause M, Schreiter U (1990b) Vestibular neurons in the parieto-insular cortex of monkeys (Macaca fascicularis): visual and neck responses. J Physiol 430: 559–583
Grüsser O-J, Guldin WO, Harris L, Lefébre C, Pause M (1992) Cortical representation of head-in-space movement and some psychophysical experiments on head movement. In: Berthoz A, Graf W, Vidal PP (eds) The head-neck sensory motor system. Oxford University Press, Oxford New York, pp 497–509
Guldin WO, Grüsser O-J (1987) Single unit responses in the vestibular cortex of squirrel monkeys. Neurosci Abstr 13: 1224
Guldin WO, Akbarian S, Grüsser O-J (1989) The interconnections of the parieto-insular vestibular cortex (PIVC) in the squirrel monkey. Soc Neurosci Abstr 15: 519
Guldin WO, Akbarian S, Grüsser O-J (1992) Cortico-cortical connections and cytoarchitectonics of the primate vestibular cortex. A study in squirrel monkeys (Saimiri sciureus). J Comp Neurol 326: 375–401
Jones EG, Burton H (1976) Areal differences in the laminar distribution of thalamic afferents in cortical fields of the insular, parietal and temporal opercular regions of primates. J Comp Neurol 168: 197–247
Kawano K, Sasaki M, Yamashita M (1980) Vestibular input to visual tracking neurons in the posterior parietal association cortex of the monkey. Neurosci Lett 17: 55–60

Lang W, Büttner-Ennever JA, Büttner U (1979) Vestibular projections to the monkey thalamus: An autoradiographic study. Brain Res 177: 3–17
Liedgren SRC, Milne AC, Rubin AM, Schwarz DWF, Tomlinson RD (1976) Representation of vestibular afferents in somatosensory thalamic nuclei of the squirrel monkey (Saimiri sciureus). J Neurophysiol 39: 601–612
Ödkvist L, Schwarz DWF, Fredrickson JM, Hassler R (1974) Projections of the vestibular nerve to the area 3a arm field in the squirrel monkey (Saimiri sciureus). Exp Brain Res 21: 97–103
Pandya DN, Sanides F (1973) Architectonic parcellation of the temporal operculum in rhesus monkey and its projection pattern. Z Anat u Entwicklungsgesch 139: 127–161
Schwarz DWF, Fredrickson JM (1971) Rhesus monkey vestibular cortex: a bimodal primary projection field. Science 172: 280–281
Thier P, Erickson RG (1992) Responses of visual-tracking neurons from cortical area MST-I to visual, eye and head motion. Europ J Neurosci 4: 539–553
Waespe W, Henn V (1977a) Neuronal activity in the vestibular nuclei of the alert monkey during vestibular and optokinetic stimulation. Exp Brain Res 27: 523–538
Waespe W, Henn V (1977b) Vestibular nuclei activity during optokinetic after-nystagmus (OKAN) in the alert monkey. Exp Brain Res 30: 323–330
Waespe W, Henn V (1978) Conflicting visual-vestibular stimulation and vestibular nucleus activity in the alert monkeys. Exp Brain Res 33: 203-211

Chapter 8

Functional Aspects II

8.1 Introduction

K. ALBUS
Department of Neurobiology, Max Planck Institute for biophysical Chemistry, Göttingen, Germany

This section discusses cortical mechanisms and functions related to visual perception and sensory-motor control thus continuing and extending the themes alluded to in Chap. 7.

H.-C. Nothdurft summarizes the results of psychophysical studies on phenomena attributed to preattentive vision like texture segmentation, pop out (parallel versus serial visual search) and grouping. The studies concentrate on answering the question whether feature contrast or feature properties themselves provide a cue for the task of figure-ground segmentation on which these perceptual performances depend. Findings on feature contrast versus feature identity properties will be discussed for parameters like line orientation, color, motion, luminance and disparity contrast. The results of this work suggest that visual filters at an early cortical level, i.e., in the primary visual cortex, play a role in preattentive vision, and physiological experiments supporting this notion are presented.

The talks of Hoffmann, Fischer, Berlucchi and Fromm illuminate specific aspects of sensorimotor control thus bringing into play extrastriate visual cortex, visual subcortical centers and — last but not least — motor cortex as well. K.-P. Hoffmann and collaborators present an elegant solution to the so far unexplained finding that eye movement deficits produced by lesions of posterior cortical structures or corpus callosum section are asymmetric despite the fact that more or less all stimulus directions are functionally represented in the related cortical areas. Their explanation is based on experimental evidence demonstrating functionally specified connections between the nucleus of the optic tract and the middle temporal area. The interesting and somewhat surprising consequence of their findings is that grouping of specific cortical efferents relevant for visuomotor function was achieved by selection induced in subcortical centers.

B. Fischer and H. Weber present experimental results which elucidate some of the mechanisms involved in the generation of saccadic eye movements. Their particular interest is with a type of very fast visually saccades — the so-called express saccades — which have reaction times of about 70 ms in the monkey and about 100 ms in human beings. Express saccades can be characterized as a kind of optomotor reflex which is under the inhibitory control of higher cortical functions that regulate fixation and attention. The experiments reported isolate and identify mechanisms of express saccade preparation and provide hints as to where in the cortex neuronal populations contributing to the inhibitory control function reside.

G. Berlucchi and collaborators have studied the cortical substrates of simple forms of visuomotor integration in patients with total or partial defects of the corpus callosum. As a test the crossed-uncrossed differences in speed of motor responses to lateralized light flashes were employed. Three types of motor responses were tested (distal, proximal and axial) and these responses were performed both unilaterally and bilaterally. The experiments confirm that a callosal contribution is important for the execution of fast distal and unilateral proximal responses. Additional findings in subjects with an anterior section of the corpus callosum indicate that both anterior and posterior callosal routes can subserve the integration of speeded manual responses to a visual stimulus directed to the hemisphere ipsilateral to the responding hand.

The work of C. Fromm and E. Bauswein is concerned with the sensorimotor transformations between different coordinate frames representing spatial, kinematic, and dynamic parameters that are used by the central nervous system to achieve goal-directed movements. Their experiments aimed at answering the question if and to what extent neuronal processing of muscle force and of trajectory encoding for three dimensional reaching movements is distributed over several motor cortical areas. The results of their studies in the primary motor cortex, the dorsal premotor cortex and the supplementary motor area support the evidence for parallel processing within distributed neuronal subpopulations and suggest the presence of several coexisting functional maps in each of these motor areas.

8.2 Cortical Properties of Preattentive Vision

H.-C. Nothdurft
Department of Neurobiology, Max Planck Institute for biophysical Chemistry, Göttingen, Germany

Recent neurophysiological and psychophysical experiments have provided new insight into the properties of preattentive vision and provide a better understanding of underlying neuronal processes. The experiments that started from cortex neurophysiology in the laboratory of Otto Creutzfeldt (Creutzfeldt and Nothdurft 1978; Nothdurft and Li 1985) resulted in an extensive investigation of texture segmentation and visual search in psychophysical experiments. The results are briefly summarized, and important aspects are stressed. From our data, preattentive vision appears largely based on feature contrast, not on feature properties themselves, which provides a cue for figure-ground segmentation and also may gate information for detailed analysis.

8.2.1 Preattentive and Attentive Aspects of Vision

One of the many intriguing phenomena in vision is the distinction between what are called preattentive and attentive processes (Neisser 1967; Julesz 1975; Treisman 1986). In Fig. 1, for example, segmentation of the left-hand pattern takes place spontaneously and without effort (preattentively), that of the middle pattern does not and the distinction of different elements requires directed attention. Interestingly, there is no percept of a global figure from this pattern. Similarly, from the two different elements in the right-hand figure, one is detected immediately (in parallel) and "pops out", while detection of the other one requires sequential (serial) search through all elements (see also Caelli et al. 1986). [Considerable uneasiness has developed with respect to this terminology, in particular since recent experiments have shown that even "preattentive" phenomena require attention to be seen (Mack et al. 1992; Ben-Ave et al. 1992), and processes presumed serial in visual search may become parallel under prolonged training (Von der Heydt and Dürsteler 1993; Sireteanu and Rettenbach 1993).]

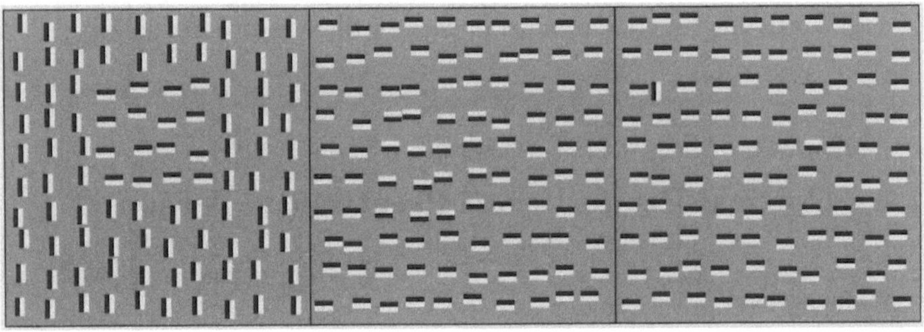

Fig. 8.2–1. Segmentation and pop-out as examples of preattentive visual processes. Certain stimulus properties provide spontaneous segregation and the percept of distinct texture areas (*left-hand pattern*), other properties do not segregate and can only be distinguished from attentive visual analysis ("by scrutiny", see Julesz 1975; *middle pattern*). In a similar fashion, some visual properties are immediately ("preattentively") detected even from a large sample of displayed items whereas others require focal attention to be shifted serially across the whole pattern (*right-hand display*)

Similar phenomena are obtained for a variety of, but not for all, visual properties. For example, segmentation and pop-out result from differences in luminance, line orientation, spatial frequency, direction of motion, disparity, or color, a list that immediately reminds us of cortical cell properties. In the cat or monkey, cortical cells are known to be sensitive to line orientation, spatial frequency, and motion direction, some cells are sensitive to color or luminance differences, and others even prefer a certain stimulus disparity (Van Essen 1985; Von der Heydt 1987). Thus, it is reasonable to assume that the perceptual phenomena described above are based on response properties of visual filters at an early cortical level. Because this assumption has been gaining further support from recent experiments, I draw attention here to some important aspects of this early cortical analysis.

8.2.2 Feature Contrast: A Prediction from Psychophysics

The left-hand column in Fig. 2 shows two squares, each of which segregates well from the background and is instantaneously seen. However, there is an important difference between these patterns. While the pattern in A may easily segment from the analysis of feature properties and the combination of similar features across the pattern (see the example plotted as "feature map activity"), it is difficult to see how such a process could help us to segment the pattern in B. The distribution of responses in a certain orientation channel correlates with the extent of texture areas in A, but there is no channel in which activity distribution would reflect the percept of pattern B. What is seen instead, and what leads to the percept of distinct areas in this pattern, is the locally increased texture gradient (Nothdurft 1985; Julesz 1986; Sagi and Julesz 1987), i.e., the orientation contrast at the texture border.

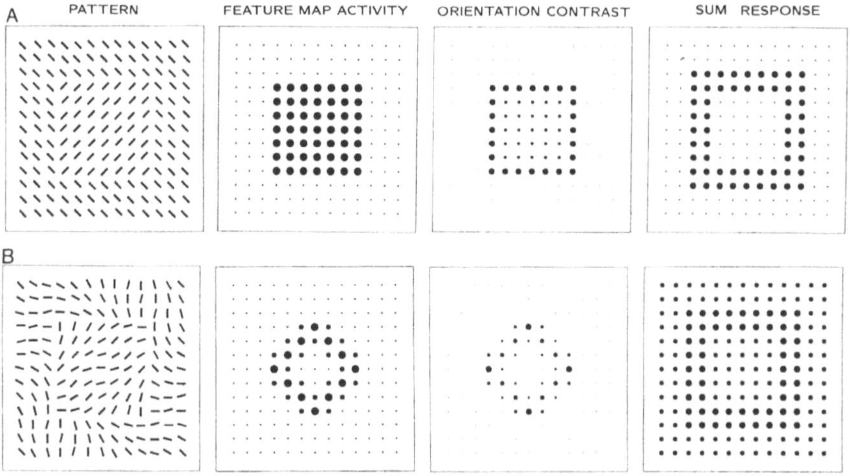

Fig. 8.2–2. Schematic response distribution for two patterns (**A** and **B**; *first column*). Whereas the energy distribution within certain feature maps (e.g., the representation of oblique orientations tilted to the right; *second column*) correlates with the percept seen in A, no such correlate is found in B. Only if responses are modulated by local feature (orientation) contrast (*third column*; same orientation filters as before) and responses from cells with different orientation preferences are pooled (*fourth column*), resembles the activity distribution the texture borders seen even for pattern B. Notice that, according to this model, noise increases with increasing feature variation throughout the pattern ("background orientation shift") and signal-to-noise ratio in texture border detection is reduced (*right-hand column*; A versus B)

Fortunately, there is a series of physiological experiments both in cat and monkey which has provided evidence that cells are sensitive to local orientation contrast. Knierim and Van Essen (1992) have shown that neurons in V1 respond, on average, more strongly to line elements surrounded by orthogonal lines than to line elements surrounded by identical lines. In many cells, lines presented outside the "classical receptive field" modulate the response to an optimal stimulus in the receptive field center, and responses are on average larger when local orientation contrast is increased, for example, near a texture border (Nothdurft et al. 1992). Response modulation by oriented surrounds was also reported for the cat (Blakemore and Tobin 1972; Maffei and Fiorentini 1976; Fries et al. 1977), although not consistently found (Bishop et al. 1973; Albus and Fries 1980). If response modulation by feature contrast ("orientation contrast") is taken into account in the response distributions of Fig. 2, the *summed responses of all cells* (right-hand column) can represent the texture border. There are two requirements here in order to obtain such a result. One is that local orientation differences must be detected by some cells at some level, and, secondly, responses from differently oriented cells must be pooled in a non-specific way in order to encode the texture border. Although there is not yet any direct physiological evidence of how such a pooling may be achieved, some cells at higher areas seem to respond in a generalized way to the movement of texture borders (Albright 1992; Olavarria et al. 1992).

8.2.3 Pop-out Under Noise

Several predictions can be made from this model. One is that segmentation, and similarly pop-out, should be strongly affected by an increased feature contrast among background elements, i.e., by an overall variation of feature properties in the background of the pattern. This has indeed been observed in psychophysical experiments (Nothdurft 1991, 1992). When subjects were asked to detect a single line element that popped out from homogeneous or inhomogeneous background patterns (where neighboring lines were either the same or slightly different), the required orientation contrast for a target to pop out was found to increase with increasing background variation. Already for an orientation shift of 30° between neighboring elements, a target at maximal orientation contrast (90°) could not reliably be detected from preattentive vision (Fig. 3, bottom left).

Similar properties were obtained for pop-out of motion where, again, the local differences in motion direction that were required for a target to pop out increased with increasing background variation. In fact, the curves obtained for motion targets look very similar to those obtained for orientation differences thus suggesting a similar mechanism underlying pop-out of orientation and pop-out of motion (Fig. 3, bottom right).

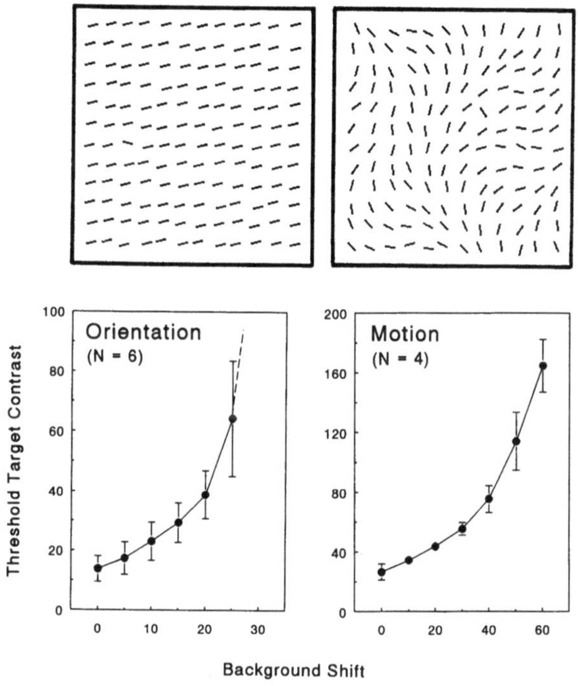

Fig. 8.2–3. Pop-out of orientation or motion from local feature contrast. Subjects looked for a single conspicuous target occurring on either the left or the right half of each pattern. Two parameters were varied in experiment, the overall feature variation (measured as the shift in orientation, or direction of motion, from one element to the next; *background shift*) and the local orientation, or direction of motion, contrast at the target's position (*target contrast*). For each background condition, the target contrast producing 75% correct responses was taken as the threshold value. *Top*, examples of stimulus patterns used in the orientation task; background orientation shift 0° in the left-hand pattern, and 20° in the right-hand one; target orientation contrast 20° and 90°, respectively. Lines were 20' by 5' in size. For motion experiments, the lines were replaced by dots (7.2' diameter) moving back and forth around their central location, amplitude 7–9', speed 1.5–3°/s. Feature variation was in the direction of motion. Patterns were shown for 100–200 ms. *Bottom*, threshold target contrast as a function of background orientation shift (left) or direction of motion shift (right). Pooled data from different subjects. The local feature contrast required to make a target pop out increases with increasing background variation. For medium variations (30° for orientation, 60°–70° for motion), even targets at maximal feature contrast (orthogonal orientation, movement in opposite direction) could not reliably be detected. The two visual dimensions showed similar properties in this test

8.2.4 Saliency and Feature Identity

Whereas the role of local feature contrast in preattentive vision seems to be well established, the second aspect of the model outlined above has not yet been tested. Are feature identities processed in preattentive vision, or are they simply ignored? I studied this question in two series of psychophysical experiments. One of them was related to visual search. I wondered whether the fast search properties found with some targets really depend on the target properties themselves, or whether they might not be due to the locally increased feature contrast which makes the target more salient than other elements in the pattern. In a series of experiments it was found that only salient targets can be detected fast (Fig. 4), irrespective of which visual property caused their salience (Nothdurft 1992, 1993). Non-salient targets not only required more time to be detected but, for these search time also increased with the number of displayed items, suggesting a serial search process. Among different salient elements, however, search was serial again (Fig. 4; see Nothdurft 1992). These data suggest parallel and non-specific analysis of preattentive aspects of the stimulus (salience), and serial analysis of specific properties for target identification, as is also concluded from other studies (Sagi and Julesz 1985; Braun and Sagi 1990).

The other series of experiments investigated the feature specificity of grouping processes in vision. Are targets grouped for feature properties or only for salience from local feature contrast? In other words, is the information from different orientation or motion detectors indeed pooled in preattentive vision, or are figures seen from the linking of certain feature properties (feature binding)?

Grouping was studied using texture displays with three or four pop-out elements that formed various global figures such as differently oriented triangles or a square. Local feature contrast of these elements was varied systematically but was the same for all pop-out elements in a given pattern. Triangles were made from identical or non-identical elements, as shown in Fig. 5. In square configurations, three of the four elements were identical (forming a triangle with similar feature properties), the fourth target was plotted at a different orientation. If figures were seen on the basis of feature identity, one would expect subjects to detect triangles of identical elements easier (i.e., at lower local orientation contrast) than triangles of non-identical elements. They also should easily detect the similar-target triangles in the four-element presentations. This was not the case, however, and both in the orientation and in the motion test subjects could group dissimilar targets as well as similar ones. They also could not detect the triangle of similar elements from the square figures.

8.2.5 Generalized Feature Contrast

The experiments reported here were repeated in three other stimulus dimensions with patterns displaying color, luminance, or depth differences. In all cases the threshold feature contrast of a single pop-out target increased with increasing background variation, suggesting that feature contrast may play

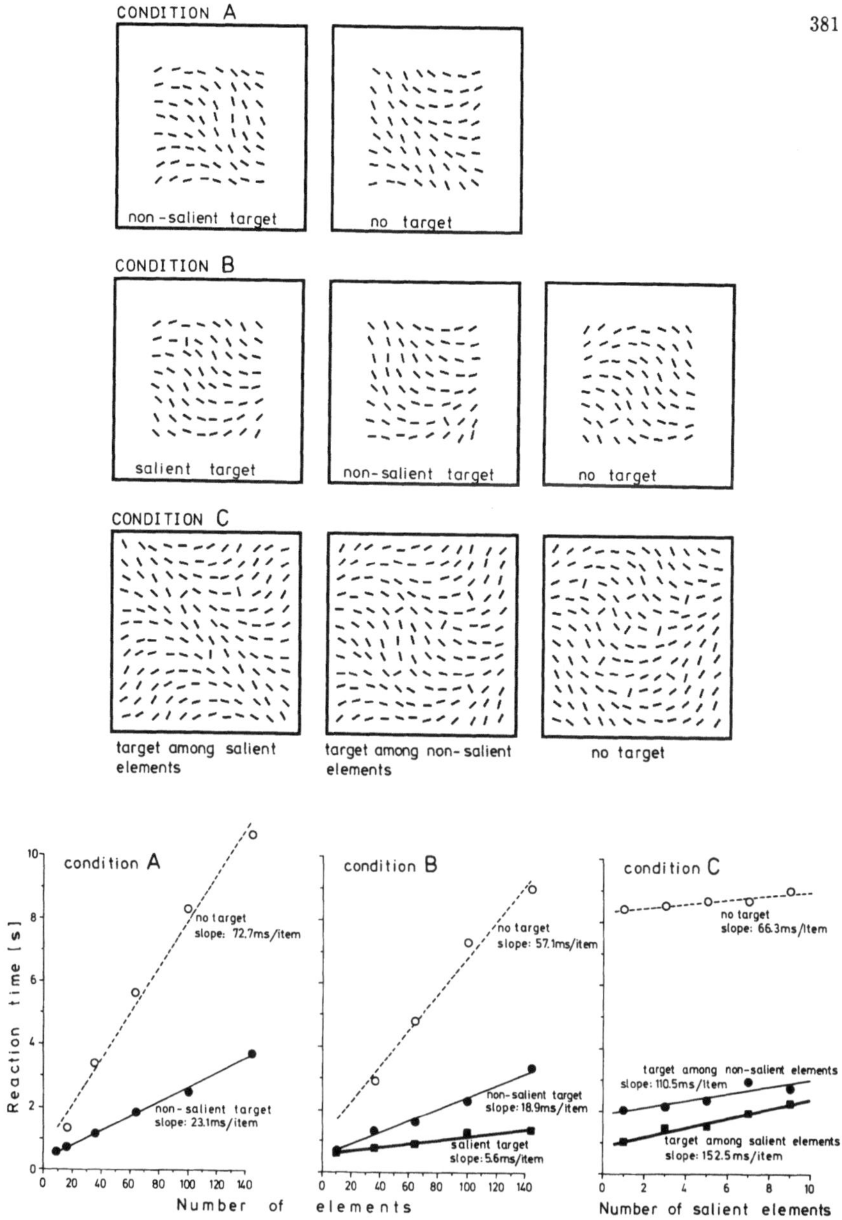

Fig. 8.2–4. Fast search of a vertical line when the target is *salient* (from locally increased feature contrast) but slower search when the target displayed the same local feature contrast as the other elements in the pattern (*non-salient targets*). Search speed is also reduced (*serial search*) when the target is not the only salient element in the display (condition C). The different test conditions are depicted *above*, reaction times for target detection or pattern rejection (*no target*) are plotted *below*. Pooled data from three subjects. In *conditions A and B*, the number of displayed elements was varied, in *condition C*, all patterns displayed 144 elements but a variable number of salient ones. The data indicate that fast (and "parallel") visual search is not obtained from feature properties themselves but depends on target salience obtained from increased feature contrast. (Patterns at top reproduced from Nothdurft 1992)

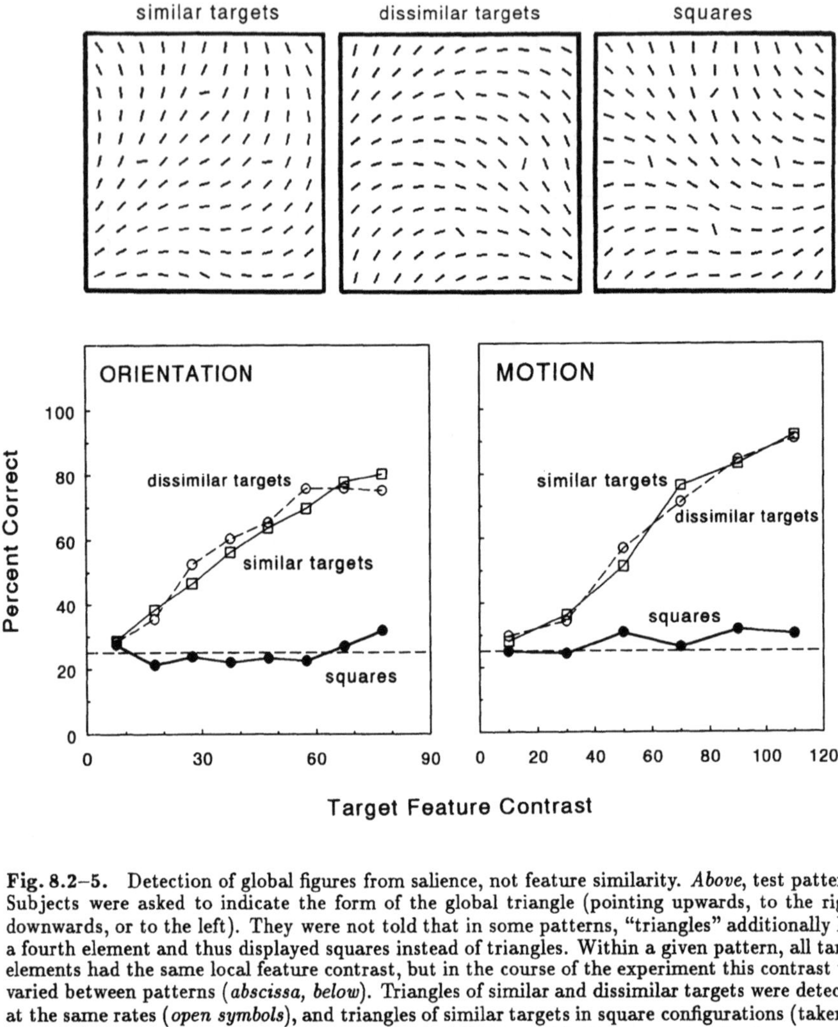

Fig. 8.2–5. Detection of global figures from salience, not feature similarity. *Above*, test patterns. Subjects were asked to indicate the form of the global triangle (pointing upwards, to the right, downwards, or to the left). They were not told that in some patterns, "triangles" additionally had a fourth element and thus displayed squares instead of triangles. Within a given pattern, all target elements had the same local feature contrast, but in the course of the experiment this contrast was varied between patterns (*abscissa, below*). Triangles of similar and dissimilar targets were detected at the same rates (*open symbols*), and triangles of similar targets in square configurations (taken as correct responses) were seen only at chance level (*filled symbols*). Thus, preattentive figure-ground discrimination in these patterns is achieved from target saliency and not from linking processes between identical targets. This holds for both orientation and motion properties. Pooled data from altogether four subjects, two in each test. Orientation patterns were presented for 160 ms and then masked; motion patterns were shown for 120 ms without a mask. Background shifts were 10° (orientation) and 20° (motion)

a general role for the evaluation of target salience and for pattern segmentation. In fact, in none of these tests, could pop-out be explained by the detection of *certain* stimulus *properties*, since such properties did not pop out when present elsewhere in the pattern at lower feature contrast. For grouping, however, the feature non-specificity was less general, and figures from targets of the same color, showing the same luminance polarity, or at the same relative disparity categories (near–far) were usually seen better than figures from dissimilar targets in these properties.

8.2.6 Conclusions

Segmentation, pop-out, and figure-ground discrimination in preattentive vision are not obtained from complicated feature processing in the visual cortex but display characteristics as would be obtained from the evaluation of local feature contrast (which, of course, also requires feature processing, but to a limited extent). Response variation with local feature contrast can be seen already in area V1, at least for orientation (Van Essen et al. 1989; Knierim and Van Essen 1992) and motion (Allman et al. 1991) and probably to some extent also for color (e.g., the "double opponent cells"; Livingstone and Hubel 1984; but see Creutzfeldt et al. 1987; Ts'o and Gilbert 1988). As far as typical preattentive visual phenomena are concerned, the feature information is not specifically retained, but feature contrast from different orientation or motion detectors is pooled. Less complete pooling has, however, been seen for color, luminance and disparity properties. For orientation, grouping was also affected by configurational aspects of line elements (Beck et al. 1989; Nothdurft 1992; Field et al. 1993; Polat and Sagi 1993). These additional findings, not reported here, indicate that further mechanims, beyond the non-specific evaluation of local feature contrast, are involved in preattentive vision.

References

Albright TD (1992) Form-cue invariant motion processing in primate visual cortex. Science 255: 1141–1143

Albus K, Fries W (1980) Inhibitory sidebands of complex receptive fields in the cat's striate cortex. Vision Res 20: 369–372

Allman J, Miezin F, McGuinness EL (1991) Effects of background motion on the responses of neurons in the first and second cortical visual areas. In: Edelman GM, Gall WE, Cowan MW (eds) Signal and sense: local and global order in perceptual maps. Wiley-Liss, New York, pp 131–142

Beck J, Rosenfeld A, Ivry R (1989) Line segregation. Spatial Vision 4: 75–101

Ben-Ave MB, Sagi D, Braun J (1992) Visual attention and perceptual grouping. Percept Psychophys 52: 277–294

Bishop PO, Coombs JS, Henry GH (1973) Receptive fields of simple cells in the cat striate cortex. J Physiol (Lond) 231: 31–60

Blakemore C, Tobin EA (1972) Lateral inhibition between orientation detectors in the cat's visual cortex. Exp Brain Res 15: 439–440

Braun J, Sagi D (1990) Vision outside the focus of attention. Percept Psychophys 48: 45–58

Caelli T, Hübner M, Rentschler I (1986) On the discrimination of micropatterns and textures. Human Neurobiol 5: 129–136

Creutzfeldt OD, Nothdurft HC (1978) Representation of complex visual stimuli in the brain. Naturwissenschaften 65: 307–318

Creutzfeldt OD, Weber H, Tanaka M, Lee BB (1987) Neuronal representation of spectral and spatial stimulus aspects in foveal and parafoveal area 17 of the awake monkey. Exp Brain Res 68: 541–564
Field DJ, Hayes A, Hess RF (1993) Contour integration by the human visual system: evidence for a local "association field". Vision Res 33: 173–193
Fries W, Albus K, Creutzfeldt OD (1977) Effects of interacting visual patterns on single cell responses in cat's striate cortex. Vision Res 17: 1001–1008
Julesz B (1975) Experiments in the visual perception of texture. Sci Am 232(4): 34–43
Julesz B (1986) Texton gradients: the texton theory revisited. Biol Cybern 54: 245–251
Knierim JJ, Van Essen DC (1992) Neuronal responses to static texture patterns in area V1 of the alert macaque monkey. J Neurophysiol 67: 961–980
Livingstone MS, Hubel DH (1984) Anatomy and physiology of a color system in the primate visual cortex. J Neurosci 4: 309–356
Mack A, Tang B, Tuma R, Kahn S (1992) Perceptual organization and attention. Cognitive Psychol 24: 475–501
Maffei L, Fiorentini A (1976) The unresponsive regions of visual cortical receptive fields. Vision Res 16: 1131–1139
Neisser U (1967) Cognitive Psychology. Appleton-Century-Crofts, New York
Nothdurft HC (1985) Sensitivity for structure gradient in texture discrimination tasks. Vision Res 25: 1957–1968
Nothdurft HC (1991) Texture segmentation and pop-out from orientation contrast. Vision Res 31: 1073–1078
Nothdurft HC (1992) Feature analysis and the role of similarity in pre-attentive vision. Percept Psychophys 52: 355–375
Nothdurft HC (1993) Saliency effects across dimensions in visual search. Vision Res 33: 839–844
Nothdurft HC, Li CY (1985) Texture discrimination: representation of orientation and luminance differences in cells of the cat striate cortex. Vision Res 25: 99–113
Nothdurft HC, Gallant JL, Van Essen DC (1992) Neural responses to texture borders in macaque area V1. Soc Neurosci Abstr 18: 1275
Olavarria JF, DeYoe EA, Knierim JJ, Fox JM, Van Essen DC (1992) Neural responses to visual texture patterns in middle temporal area of the macaque monkey. J Neurophysiol 68: 164–181
Polat U, Sagi D (1993) Lateral interactions between spatial channels: suppression and facilitation revealed by lateral masking experiments. Vision Res 33: 993–999
Sagi D, Julesz B (1985) "Where" and "what" in vision. Science 228: 1217–1219
Sagi D, Julesz B (1987) Short-range limitations on detection of feature differences. Spatial Vision 2: 39–49
Sireteanu R, Rettenbach R (1993) Practice makes the difference: learning and memory in visual search. In: Elsner N, Heisenberg M (eds) Gene – Brain – Behaviour. Thieme, Stuttgart New York, p 31
Treisman A (1986) Features and objects in visual processing. Sci Am 255(11): 106–115
Ts'o DY, Gilbert CD (1988) The organization of chromatic and spatial interactions in the primate striate cortex. J Neurosci 8: 1712–1727
Van Essen DC (1985) Functional organization of primate visual cortex. In: Jones EG, Peters A (eds) Cerebral cortex. Plenum, New York, pp 259–329
Van Essen DC, DeYoe EA, Olavarria J, Knierim J, Fox J, Sagi D, Julesz B (1989) Neural responses to static and moving texture patterns in visual cortex of the macaque monkey. In: Lam DMK, Gilbert CD (eds) Neural mechanisms of visual perception. Portfolio Publishing, The Woodlands Texas, pp 137–154
Von der Heydt R (1987) Approaches to visual cortical function. Rev Physiol Biochem Pharmacol 108: 69–150
Von der Heydt R, Dürsteler M (1993) Visual search: monkeys detect conjunctions as fast as features. Invest Ophthalmol Vis Sci 34: 1288

8.3 Structural and Functional Organization of Connections Between Neocortex and Midbrain

K.-P. HOFFMANN, U. ILG, AND C. DISTLER
Lehrstuhl für Allgemeine Zoologie und Neurobiologie, Ruhr-Universität Bochum, Germany

The ascendency of the cerebral cortex in mammals naturally raises questions about the role of the architypal subcortical centers that we share with other classes. Here we report a situation in which an ancient oculomotor control center, the nucleus of the optic tract, is less dominated by the cerebral cortex than served by it. We suggest that the organization of cortical output to subcortical centers may be helpful in understanding the function of the cerebral cortex.

Studies in rhesus monkeys and in man demonstrate a directional deficit in optokinetic and smooth pursuit-eye movements for tracking targets that move toward the lesion site following unilateral lesions of the posterior cortex (Bogousslavsky and Regli 1986; Leigh and Thurston 1986; Dürsteler et al. 1987; Thurston et al. 1988; Morrow and Sharpe 1990). This directional deficit is difficult to explain with current knowledge: there is no anisometry in the sensitivity for different directions of stimulus movement in the tuning of cortical neurons (Wurtz et al. 1990) to explain the selective impairments of a cortical injury. In contrast to cortical areas, however, a nucleus of the pretectum, the nucleus of the optic tract, in conjunction with the dorsal terminal nucleus of the accessory optic system (NOT-DTN) consists exclusively of neurons that are activated during ipsiversive stimulus movement, i.e., stimulus movement toward the recorded hemisphere (Hoffmann et al. 1988; Mustari and Fuchs 1990). A lesion of this nucleus produces directionally selective deficits in optokinetic nystagmus (Kato et al. 1986, 1988; Schiff et al. 1990; Ilg et al. 1993).

We propose a solution to the problem of directional deficits following lesions of the posterior cortex by demonstrating that cortical neurons with ipsiversive preferred directions provide exclusive direct cortical input to the (NOT-DTN). In this report we describe the cortical analysis of moving visual stimuli in the middle temporal area (MT) and demonstrate presence and functional characteristics of strong projections from this area to the (NOT-DTN) in macaque monkeys. In addition to neurons in the striate cortex (V1) only neurons located in the lower bank and fundus of the superior temporal sulcus (STS), presumably in a subregion of MT, are shown to project to the (NOT-DTN).

Cortical lesions damage neurons representing all directions, but since only neurons with an ipsiversive component provide an input to the ipsilateral optokinetic system, the lesion eliminates this input, and the resulting oculomotor deficit is asymmetric.

8.3.1 Methods

Experiments were performed in awake and trained or anesthetized and paralyzed macaque monkeys. The procedures for surgery, electrical and visual stimulation have been described elsewhere (Hoffmann et al. 1988, 1991, 1992; Ilg and Hoffmann 1991). Orthodromic activation refers to an electrical transsynaptic activation of a neuron via its input fibers. This shows the source of input to the neuron. Antidromic activation refers to the electrical activation of a neuron via its own axon. This involves no synapse and shows where the neuron under study sends its output.

8.3.2 Responses of Neurons in Extrastriate Area MT During Smooth-Pursuit and Optokinetic Nystagmus

First we present data concerning the dynamic role of MT neurons during slow eye movements. We recorded 239 single MT units during smooth pursuit and optokinetic nystagmus in awake and trained monkeys. These neurons respond strongly and always direction specifically to the retinal slip of small targets within their restricted receptive fields. Movement of the entire visual scene as used to elicit optokinetic nystagmus leads to similar direction specific responses and neuronal modulations as single targets. Presenting both stimuli simultaneously, there is, however, only a weak influence of passive background movement during pursuit of a single target. In all MT neurons tested the modulation of the neuronal activity was entirely dependent on the direction of target motion if the monkey's eyes followed the target. The presence of the background movement across the retina due to the eye movements never led to a change in the preferred direction of the neurons (see Fig. 1).

In 148 neurons complete directional tuning curves were accumulated while the animal was fixating. The distribution of preferred directions did not show any bias. All directions were evenly represented by MT neurons. During optokinetic after-nystagmus (OKAN) the neuronal activity returns to the spontaneous level. MT neurons thus seem to code the velocity error between eye and the visual target and not the eye velocity per se. What happens when the two stimuli moving in opposite directions are of equal quality, e.g., two random dot stimuli? Our preliminary results on MT neurons agree completely with those recently published by Snowden et al.

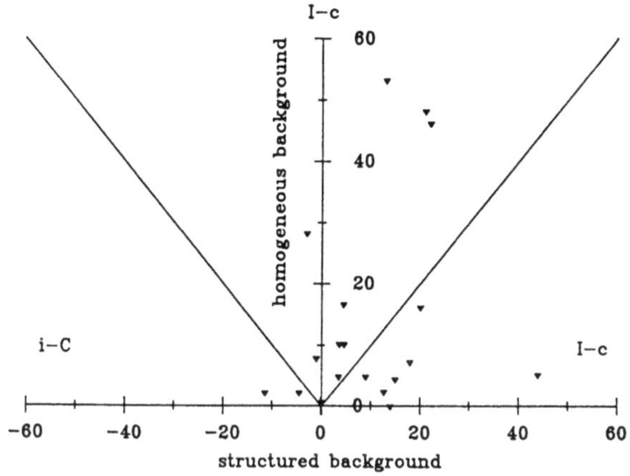

Fig. 8.3–1. Neuronal modulation of 19 MT neurons during smooth pursuit over homogenous (*ordinate*) and structured background (*abscissa*). *i-C, I-c*, neuronal modulation as difference between the neuronal response to ipsi- and contraversive target movement. Positive values indicate preference for ipsiversive, negative values preference for contraversive target movement. Background structure has no significant influence on the preferred direction of MT neurons

(1991). Simultaneous presentation of two random dot patterns, one moving in the preferred and the other in the opposite direction (transparency paradigm), invariably leads to a decrease in response strength when compared to the preferred stimulus only condition.

8.3.3 Cortical Areas Projecting to the Direction-Specific Neurons of the NOT-DTN

In a first approach in anesthetized and paralyzed monkeys NOT-DTN neurons were identified physiologically and tested for cortical input by orthodromic electrical stimulation in various cortical areas. Successful stimulation sites to activate NOT-DTN neurons lie in the primary visual cortex (V1) and in the motion processing areas in the superior temporal sulcus (STS). In contrast, electrical stimulation in area V4 and in parietal areas never yielded any spike discharges in the NOT-DTN.

Striate Cortex (V1). Central and peripheral field representations in area V1 were electrically stimulated in two experiments (Fig. 2B). All but one NOT-DTN unit tested could be orthodromically activated from V1. The latency distribution is presented in Fig. 3A.

Superior Temporal Sulcus (STS). For the stimulation of extrastriate cortex we used two different approaches. In the first two animals we recorded

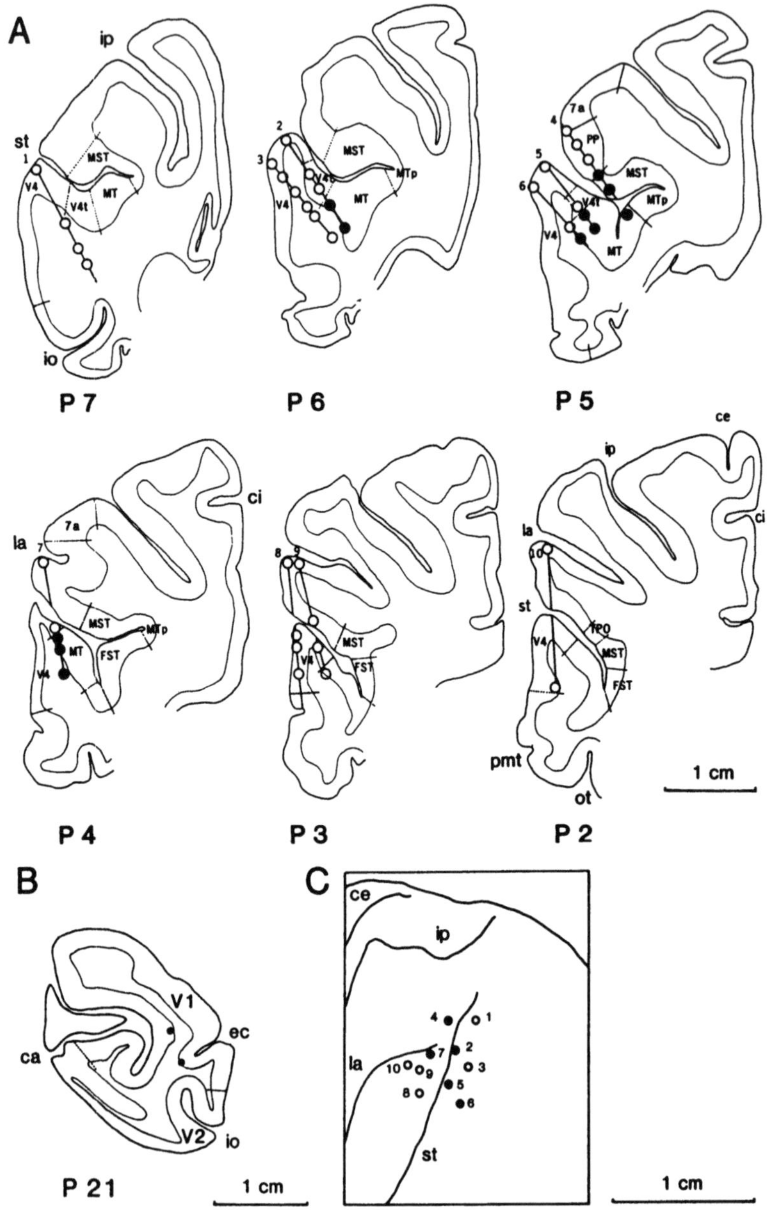

Fig. 8.3–2. **A** Reconstruction of frontal sections through the left hemisphere of monkey 4. The sections are presented from posterior (*top left*) to anterior (*bottom right*); the stereotaxic levels are indicated. *1–10*, penetration tracks; *filled circles*, effective stimulation sites; *open circles*, ineffective stimulation sites; *dotted lines*, borders between visual areas. **B** Section through the right hemisphere of monkey 4 showing the location of two stimulation electrodes in area V1 between which electrical stimuli were delivered effectively. Conventions as in **A**. **C** Reconstruction of the lateral view of the left hemisphere of monkey 4 showing the location of penetration tracks (numbered as in **A**) along which electrical stimulation was effective (*filled circles*) or non-effective (*open circles*) in orthodromically activating NOT-DTN cells. Note that effective stimulation sites never occurred in superficial cortex. For abbreviations see text

from cortical visual areas in the STS to identify the location of direction-selective cells and the representation of the fovea presumably in area MT. The physiological response properties and receptive field locations of cortical neurons recorded at certain sites which later were electrically stimulated were also used for identifying the location of these cortical stimulation sites. The recording electrode was then replaced by stimulating electrodes which were cemented into place. This procedure allowed us to sample orthodromic latencies from a large number of subcortical cells in response to electrical stimulation of one particular part of the extrastriate visual cortex (Fig. 3B). According to the physiological data, the two stimulation sites in the first animal correspond to the region spanning the border between foveal MT and fundal superior temporal area (FST). Shocks delivered between these electrodes were indeed successful in eliciting action potentials in NOT-DTN cells.

The second monkey had a total of seven stimulation sites. Three effective ones were in the left STS corresponding to two foveal MT locations and one FST location. Of the four stimulating electrodes placed in the right hemisphere three were in the STS. Only one was effective and corresponded to the foveal MT. The other two were ineffective and were in the peripheral MT and the white matter underlying the medial superior temporal area (MST). The fourth site was also ineffective and lay within area 7a on the inferioparietal gyrus.

In the remaining two monkeys (cases 3 and 4) we used another procedure. In these animals, three and ten penetrations, respectively, were made in the temporal and parietal cortex to map cortical areas from which single NOT-DTN neurons could be orthodromically activated. Furthermore, this approach enabled us to compare the orthodromic latencies and thresholds of a limited number of individual NOT-DTN cells in response to electrical stimulation of various cortical areas.

Case 3 had three parallel cortical penetrations. It is interesting to note that the stimulating electrode went through posterior parietal areas, areas within the lateral sulcus and MST, or areas V4 and V4t without activating NOT-DTN neurons. Only with stimulation in areas MT and FST were clear orthodromic responses obtained in NOT-DTN. Their latencies are presented in Fig. 3B.

In case 4 (Fig. 2A,C) cortical penetrations 1–6 went parallel to each other and to the banks of the STS, whereas penetrations 7–10 went almost in the dorsoventral direction and reached mostly superficial cortical areas. Again, only stimulation sites in MT, and this time additionally in MST, were effective. Stimulation in areas V4, V4t, posterior parietal area, and at the lower lip of the lateral sulcus were ineffective even at the highest stimulus strengths.

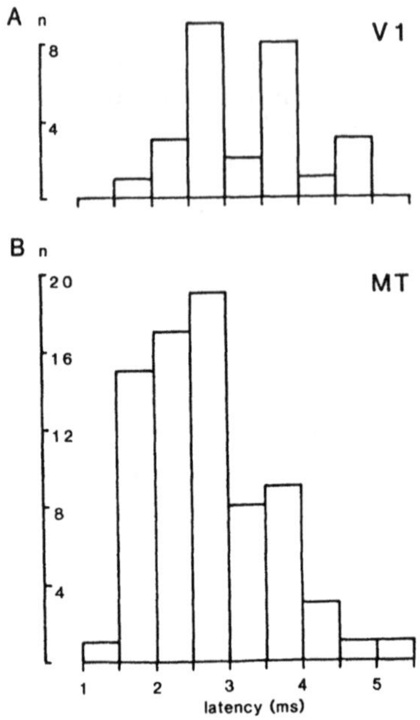

Fig. 8.3–3. A,B Frequency distributions of orthodromic latencies measured at NOT-DTN neurons after electrical stimulation in area V1 (**A**) and in area MT (**B**). *Abscissa*, latency in ms; *ordinate*, number of cells

Differences in Cortical Projections. The efficacy of electrical stimulation was very similar for the positive sites in different cortical areas. Single shocks always elicited at least two action potentials and often a burst of up to five to eight spikes with intervals of 2–4 ms. Applying two electrical stimuli at V1 or the optic chiasm and the STS sites with variable intershock delays (5–100 ms) never demonstrated inhibitory interactions. Therefore, the cortical projection to NOT-DTN seems to be exclusively excitatory in action. The latencies of the first spikes elicited from either V1 or STS stimulation sites were not identical, however. Latencies from V1 were on average 0.5 ms longer than those from the STS (3.51 ± 0.81 ms and 2.99 ± 0.85 ms, respectively). This latency difference was statistically significant ($p<0.01$, *t*-test). In one animal (case 1) we placed stimulating electrodes both in V1 and MT. For 22 NOT-DTN neurons we obtained latencies from both sites. The latency differences between V1 and MT varied between 0 and 2.5 ms, with V1 always having the longer latency. Latency differences of 1 ms or less make a relay of V1 activity in V2 and/or in MT en route to the NOT-DTN very unlikely. This was the case in 16 of the 22 cells. We take this as evidence for a direct projection of V1 to the NOT-DTN.

Receptive Fields of Cortical Neurons Projecting to the NOT-DTN. In a second approach we tried to identify the cortical neurons giving rise to the projection to the NOT-DTN. Electrical stimulation was applied to the NOT-DTN and antidromically activated single units in extrastriate cortex were analyzed. Here we present only the characteristics relevant for extent

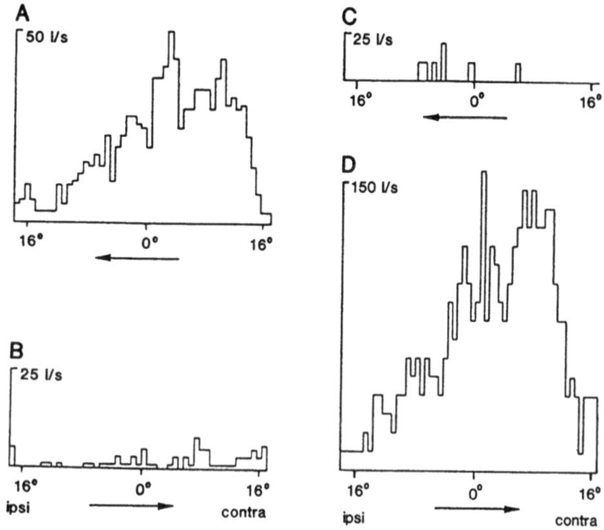

Fig. 8.3–4. Population responses from the left hemisphere of cortical neurons projecting (**A**,**B**) and not projecting (**C**,**D**) to the NOT-DTN during horizontal movement of a random dot pattern. **A**,**C** Ipsiversive stimulus movement, **B**,**D** contraversive stimulus movement. *Ordinate*, neuronal activity in impulses per second; *abscissa*, position in the visual field; 0°, vertical 0° meridian; *ipsi*, ipsilateral visual field; *contra*, contralateral visual field

and position of receptive fields as well as coding of ipsiversive stimulus movement in the NOT-DTN.

The data presented in Fig. 4A and B stem from such antidromically identified cortical neurons recorded in the floor of the left STS of two normal monkeys. The receptive fields include in addition to the contralateral hemifield and the fovea also up to about 15° of the ipsilateral hemifield. These data from extrastriate cortical cells projecting to the NOT-DTN are very similar to receptive field characteristics and visual field covering of neuronal populations in the NOT-DTN of normal monkeys. In addition, these cortical cells resemble those in the NOT-DTN very closely with respect to their direction selectivity and their direction tuning. All 41 neurons antidromically identified as cortex-to-NOT-DTN projection neurons preferred ipsiversive stimulus movement (Fig. 4A) whereas 43 not antidromically driven neurons recorded in the same penetration and interleaved with the antidromic ones preferred other directions of stimulus movements (Fig. 4C,D). Therefore, it seems very likely that this subpopulation of cortical cells constructs its large central receptive fields via callosal connections and then provides the NOT-DTN with the very specific and unmistakable visual information about ipsiversive stimulus movements in the central visual field on both sides of the vertical 0° meridian.

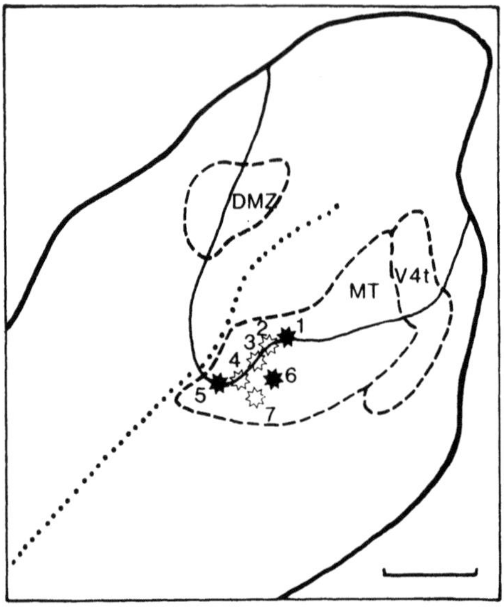

Fig. 8.3–5. Two-dimensional map of the superior temporal sulcus (STS) derived from sagittal sections. Myeloarchitectonic borders are given by *broken lines*, the fundus of the sulcus by the *dotted line*. The sagittal section through STS containing five microlesions (1–5) is represented by a *solid line*. *Solid asterisks* (1,5,6) indicate locations of microlesions at recording sites of cells antidromically activated from the NOT-DTN; *open asterisks* (2,3,4,7) indicate sites where cells could not be antidromically activated. *Scale,* 5 mm. *DMZ*, densely myelinated zone of medial superior temporal area (MST); *MT*, middle temporal area; *V4t*, transitional zone of area V4

This type of neuron was recorded exclusively in the floor of STS, presumably in area MT. In the two-dimensional reconstruction of the left STS of one monkey (Fig. 5) lesions marking cortical recording sites are indicated by asterisks. Area MT and the densely myelinated zone (DMZ) within area MST are marked by broken outlines. Lesions at recording sites of antidromically activated neurons (solid asterisks) were always in infragranular layers, whereas at least some of the lesions at other cortical sites (open asterisks) were located in supragranular layers.

8.3.4 Discussion

Taken together, our results indicate a strong cortical input from striate as well as extrastriate visual areas to the NOT-DTN in the macaque. The input from extrastriate areas, mainly MT, was found to provide the NOT-DTN with visual information about ipsiversively moving visual stimuli from the contra- as well as ipsilateral hemifield and thus promote the NOT-DTN role in gaze stabilizing and pursuit eye movements.

With orthodromic electrical stimulation we demonstrated activation of the NOT-DTN cells from a number of cortical areas (Hoffmann et al. 1991). The broad latency range of orthodromic potentials in NOT-DTN cells would allow a number of cortical relays before an excitatory corticofugal fiber projects to the NOT-DTN. So far we have found antidromically activated cells only in the subregion of MT, but negative results are not very convincing. Further anatomical work must show whether the cortical extrastriate motion analyzing pathway converges onto the corticopretectal cells shown in this study before the information is transmitted to the NOT-DTN. These cells would then be the only source from STS which provides the NOT-DTN with information about ipsiversive stimulus movement across large central receptive fields reaching far into the ipsilateral visual field.

In conclusion, the described projection of a specific group of cells in the fundus of the STS to the NOT-DTN would be the solution to two so far unexplained findings concerning eye movements after lesions of posterior cortical structures: (a) ipsiversive directional deficits in smooth-pursuit eye movements after lesions of the posterior cortex despite the fact that the lesioned areas contain neurons selective for all stimulus directions (Dürsteler and Wurtz 1988; Komatsu and Wurtz 1988a,b, 1989; Newsome et al. 1988), and (b) asymmetry of optokinetic nystagmus (OKN) with hemifield stimulation after corpus callosum transection (Pasik and Pasik 1964; Mehdorn 1984).

Figure 6 displays the wiring diagram for the cortical influence on the subcortical pathway for the generation of slow eye movements. Neurons in each NOT-DTN sharply increase their firing rate during ipsiversive pursuit if eye velocity is lower than stimulus velocity. They do so for stimuli presented to either eye and in both hemifields because of their cortically generated binocular and bilateral receptive fields. Let us construct, for example, the cortical input to the left NOT-DTN. An ipsiversive movement (to the left) across the visual field is split at the vertical 0° meridian. As long as the stimulus moves in the right (contralateral) hemifield, it is transmitted through the left geniculocortical system (solid arrows and lines). When the stimulus moves in the left (ipsilateral) hemifield it is transmitted through the right geniculocortical system (open arrows and dotted lines). However, to get to the left NOT-DTN all axons must leave from the left cortex. The information from the right cortex (left hemifield) must cross via the corpus callosum. Then in the left STS the whole field information is reconstituted and sent to the NOT-DTN (solid line with dashes). The defect in ipsiversive tracking eye movements to stimuli moving on both sides of the fovea after MT and MST lesions is fully explained by the destruction of these neurons or the input to these neurons in the STS projecting to the NOT-DTN. The remaining direct retinal input to the NOT-DTN is too weak to maintain high gain OKN. Why smooth-pursuit also suffers the same direction specific deficit must be further investigated. One possibility is that directionally biased cortical output through the NOT-DTN has an influence on the velocity of smooth pursuit as well (Ilg et al. 1993).

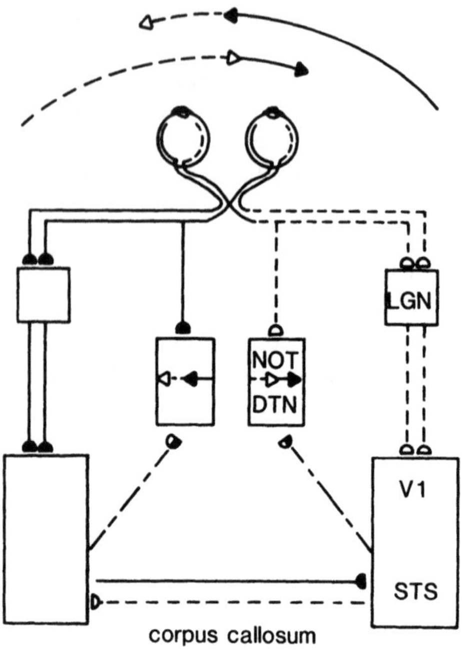

Fig. 8.3–6. Wiring diagram of the subcortical and cortical components of the pathway subserving the optokinetic reflex in normal monkeys. The flow of visual information from the left hemifield is indicated by *broken lines*, that from the right hemifield by *solid lines*. With intact corpus callosum the corticopretectal projection carries information about ipsiversive movement from the entire visual field, and NOT-DTN receptive fields extend far into the ipsilateral hemifield. *LGN*, lateral geniculate nucleus; *V1*, primary visual area; *STS*, visual areas in the superior temporal sulcus

The general implication of our findings is that grouping of cortical activity for certain visuomotor functions could be achieved by subcortical selection of a specific cortical output. In our example only cortical neurons with specificity for ipsiversive stimulus movement are connected with the NOT-DTN (Hoffmann 1987).

Neuroscientists have come to look upon the cerebral cortex as an evolutionarily new structure dominating older brain centers with its many subcortical projections. For the NOT-DTN, it seems as if a subcortical nucleus with its innate connections and preference for ipsiversive movement has commissioned the cortex to create a subcenter for providing information integrated from the two visual hemifields about ipsiversive movements in the visual world.

References

Bogousslavsky J, Regli F (1986) Pursuit gaze defects in acute and chronic unilateral parieto-occipital lesions. Eur Neurol 25: 10–18

Dürsteler MR, Wurtz RH (1988) Pursuit and optokinetic deficits following chemical lesions of cortical areas MT and MST. J Neurophysiol 60: 940–965

Dürsteler MR, Wurtz RH, Newsome WT (1987) Directional pursuit deficits following lesions of the foveal representation within the superior temporal sulcus of the macaque monkey. J Neurophysiol 57: 1262–1287

Hoffmann K-P (1987) Influence of visual experience on ontogeny of the optokinetic reflex in mammals. In: Rauschecker J, Marler P (eds) Imprinting and Cortical Plasticity. Wiley, New York, pp 267–286

Hoffmann K-P, Distler C, Erickson RG, Mader W (1988) Physiological and anatomical identification of the nucleus of the optic tract and dorsal terminal nucleus of the accessory optic tract in monkeys. Exp Brain Res 69: 635–644

Hoffmann K-P, Distler C, Erickson R (1991) Functional projections from striate cortex and superior temporal sulcus to the nucleus of the optic tract (NOT) and dorsal terminal nucleus of the accessory optic tract (DTN) of macaque monkeys. J Comp Neurol 313: 707–724

Hoffmann K-P, Distler C, Ilg U (1992) Callosal and superior temporal sulcus contributions to receptive field properties in the macaque monkey's nucleus of the optic tract and dorsal terminal nucleus of the accessory optic tract. J Comp Neurol 321: 150–162

Ilg UJ, Hoffmann K-P (1991) Responses of monkey nucleus of the optic tract neurons during pursuit and fixation. Neurosci Res 12: 101–110

Ilg UJ, Bremmer F, Hoffmann KP (1993) Optokinetic and pursuit system: a case report. Behav Brain Res 57: 21–29

Kato I, Harada K, Hasegawa T, Igarashi T, Koike Y, Kawasaki T (1986) Role of the nucleus of the optic tract in monkeys in relation to optokinetic nystagmus. Brain Res 364: 12–22

Kato I, Harada K, Hasegawa T, Igarashi T (1988) Role of the nucleus of the optic tract in monkeys in relation to optokinetic nystagmus optokinetic afternystagmus. Brain Res 474: 16–26

Komatsu H, Wurtz RH (1988a) Relation of cortical areas MT and MST to pursuit eye movements. I. Localization and visual properties of neurons. J Neurophysiol 60: 580–603

Komatsu H, Wurtz RH (1988b) Relation of cortical areas MT and MST to pursuit eye movements. III. Interaction with full-field visual stimulation. J Neurophysiol 60: 621–644

Leigh RJ, Thurston SE (1986) Recovery from ocular motor function in humans with cerebral lesions. In: Keller EL, Zee DS (eds) Adaptive processes in visual and oculomotor systems. Pergamon Press, Oxford, pp 231–238

Mehdorn E (1984) The importance of the corpus callosum for the optokinetic nystagmus in man. Fortschr Ophthalmol 81: 157–160

Morrow MJ, Sharpe JA (1990) Cerebral hemispheric localization of smooth pursuit asymmetry. Neurol 40: 284–292

Mustari MJ Fuchs AF (1990) Discharge patterns of neurons in the pretectal nucleus of the optic tract (NOT) in the behaving primate. J Neurophysiol 64: 77–90

Newsome WT, Wurtz RH, Komatsu H (1988) Relation of cortical areas MT and MST to pursuit eye movements. II. Differentiation of retinal from extraretinal inputs. J Neurophysiol 60: 604–620

Pasik T, Pasik P (1964) Optokinetic nystagmus: an unlearned response altered by section of chiasm and corpus callosum in monkeys. Nature 203: 609–611

Schiff D, Cohem B, Büttner-Ennever JA, Matsuo V (1990) Effects of lesions of the nucleus of the optic tract on optokinetic nystagmus and after-nystagmus in the monkey. Exp Brain Res 79: 225–239

Snowden RJ, Treue S, Erickson RG, Andersen RA (1991) The response of area MT and V1 neurons to transparent motion. J Neurosci 11: 2768–2785

Thurston SE, Leigh JL, Crawford T, Thompson A, Kennard C (1988) Two distinct deficits of visual tracking caused by unilateral lesions of cerebral cortex in humans. Ann Neurol 23: 266–273

Wurtz RH, Komatsu UR, Dürsteler MR, Yamasaki DS (1990) Motion to movement: Cerebral cortical visual processing for pursuit movements. In: Edelman G, Gall WE, Cowan WM (eds) Signal and sense: local and global order in perceptual maps. Wiley, New York, pp 233–260

8.4 The Visual Guidance of Saccades

B. Fischer and H. Weber
Department of Neurophysiology, University of Freiburg, Germany

This contribution collects converging pieces of evidence for the existence of two reflexlike mechanisms that control the cooperation between vision and saccadic eye movements. Starting with the modulation of the neural responses of visual cells prior to saccades directed to the receptive field stimulus, the significance of the absence or presence of a fixation point for these responses is discussed. Attempts to find a behavioral indication that differentiates between an attentively fixating eye and a nonmoving eye have led to the discovery of the express saccade in monkey and in man. The investigation of the conditions under which these reflexlike eye movements occur or are suppressed, suggest that attentive fixation of foveal stimuli or their appearence just before a saccade inhibits the saccade generating neural structures in an almost reflexlike manner. The activation of the fixation system, however, depends on the exact visual conditions. The existence of an additional system for the control of fixation and its inhibitory action on the saccade system is further supported by the observation of subjects who are unable to suppress saccades without showing any other deficit in their optomotor system.

8.4.1 Introduction

During the last century it was suggested from electrical stimulation studies that there exists a motor cortex for eye movements as for skeletal movements. Later, however, it was found that eye movements could be elicited from frontal as well as from occipital cortical structures, and it remained unclear which of these areas should be regarded as the oculomotor cortex. When microelectrodes became available and were applied for electrophysiological investigations in monkeys, an area — called the frontal eye fields — was found where saccadic eye movements could be elicited by electrical stimulation (Robinson and Fuchs 1969). However, the neurons in this area showed increased activation not before but only after spontaneous saccades in the dark (Bizzi 1968). This finding was of course incompatible with the idea of an oculomotor cortex. In the late 1960s the method of microelectrode recording was applied in awake behaving monkeys for studying the visual and oculomotor system; predominantly in the work of Wurtz and Goldberg some light was then shed on these questions: it was found that cells in the frontal eye fields have visual "ON" responses that are enhanced when the receptive field stimulus is made the target for a saccade; moreover, in this case the cells were activated before the saccade (Bushnell and Goldberg 1979). Such an enhancement effect related to visually guided saccades was also observed in the parietal cortex (Robinson et al. 1978), superior colliculus (SC; Wurtz and Goldberg 1972), pulvinar (Petersen et al. 1985), and prelunate visual association cortex, area V4 (Fischer and Boch 1981). Nevertheless, some characteristic differences were found for the individual structures. In the frontal eye fields and the SC the neuronal activity was closely related to the saccade as a motor event, whereas in the parietal cortex the enhancement occured with saccades as well as with arm movements directed toward a given target (Goldberg and Bruce 1985). In the prelunate cortex it was found that the cells' activity could be modulated by switching the monkeys' fixation target on and off, while a visual stimulus was constantly presented in the cells' receptive field (Fischer and Boch 1985). Thus it became clear that the occipital areas were not involved directly in the motor processes of the saccade system, and it was suggested that they may rather participate in the processes of visual attention so intimately related to changes in the direction of gaze. Applying electrical microstimulation in animals trained to fixate on a small spot of light, it was found that saccades could be readily elicited only when the animal was not actively fixating; threshold currents were increased in the frontal eye field (Goldberg et al. 1986), parietal cortex (Shibutani et al. 1984), and superior colliculus when a fixation target was presented in the fovea. These results suggest that the saccade generating neural structures must be under the inhibitory control of a system that is involved in the processes of visual attention and/or active fixation. Still missing, however, was a behavioral indication of whether the monkey was indeed actively fixating or merely not moving his eyes.

More insight into this problem derived from a rather simple experiment. Monkeys were trained to make saccades in two tasks: the gap and the overlap paradigm. In the gap task the fixation point was turned off some time (gap) before the saccade target was presented, in the overlap task the fixation point remained on. It was found that the introduction of the gap not only decreased the time needed to generate a saccade (saccadic reaction time; SRT), but — as a new and striking result — the SRT distribution exhibited a clearly separated peak at about 70 ms. The saccades contributing to this peak were called express saccades (Fischer and Boch 1983). It was known in man that the introduction of a gap reduces the saccadic reaction times (Saslow 1967), but it took 17 years until the express saccade was first reported for human observers.

8.4.2 Brain Structures Involved in the Generation and Control of Express Saccades

The extremely short latency of express saccades indicates that only those visual structures can be involved in their direct generation, the cells of which have visual "ON" responses of correspondingly short latencies. Therefore the first structure to consider is the SC. Unilateral SC lesions in fact eliminate express saccades, whereas unilateral lesions of the frontal eye fields leave them intact (Schiller et al. 1987). Similarly, after small lesions in striate cortex monkeys can no longer make express saccades into the affected region of the visual field (Boch 1989). Interestingly, during the days following the SC lesion, when saccades to the affected side are completely eliminated, the saccades to the "unaffected" side are all of the express type. Only after recovery of the regular saccades to the affected side was the percentage of express saccades to the other side reduced to the preoperative level. Another indirect control over the express saccades is evident from local chemical lesions in the prelunate cortex, area V4, where the number of express saccades to the lesioned side was increased to 100% following the lesion (Weber and Fischer 1990). Figure 1 shows data from one monkey. The intact (left) side served as control, the lesioned (right) side for both pre- and postoperative testing.

From these observations it can be concluded that the SC and striate cortex participate directly in the generation of express saccades, whereas higher cortical structures have an indirect control function leading to suppression of express saccades.

8.4.3 The Express Saccade: An Optomotor "Go" Reflex

While the existence of express saccades in monkeys was readily accepted and replicated by other investigators (Schiller et al. 1987), the observation of human express saccades remained a contradictory phenomenon for many years, as other investigators regarded them as an artifact or as anticipatory reactions. However, after the first report (Fischer and Ramsberger 1984) a number of additional experiments were made, and the conditions under which express saccades occur in humans were investigated. The arguments against and in favor of human express saccades have recently been summarized (Fischer and Weber 1993).

One of the experiments that proved human saccades of extremely short reaction times (100 ms) to be under visual control, was an antisaccade task in conjunction with the gap task (Fischer and Weber 1992). In the antisaccade task the subjects are instructed to make saccades to the side opposite to that on which a visual stimulus appears. Figure 2 shows the main re-

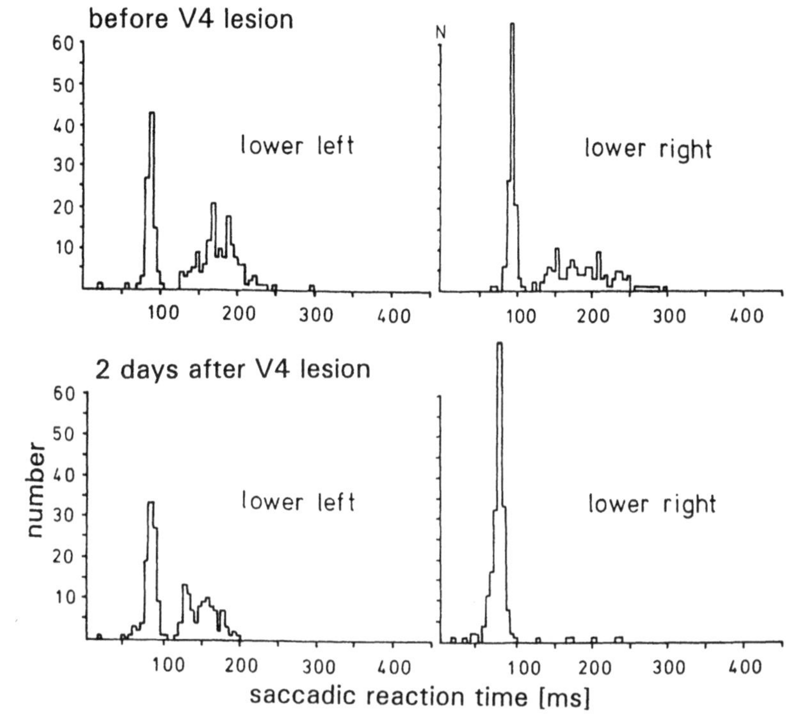

Fig. 8.4–1. Distributions of saccadic reaction times before and after a chemical lesion (ibotenic acid) in area V4. Saccades to the left are unaffected (control); those to the right exhibit a single express peak resulting from the lesion

sults. The distribution in Fig. 1A was obtained in a control experiment and shows the SRTs of normal target directed saccades (prosaccades), the bottom distribution gives the SRTs of the antisaccades. The antisaccades have significantly longer latencies even though they were executed under the same physical experimental conditions. The middle distribution depicts direction errors in the sense of the antitask, i.e., data from trials where the subjects (during the antitask) made saccades to the target. The clear express peak present in this distribution suggests that express saccades are not only under visual control but also need a visual stimulus as trigger and as target for the saccade.

As these direction errors occured involuntarily, one can consider the express saccades to represent a reflexlike optomotor response to the sudden occurence of a visual stimulus. (We see below that there are subjects who have little or no control over their express saccades, an observation which supports the idea of an optomotor reflex.)

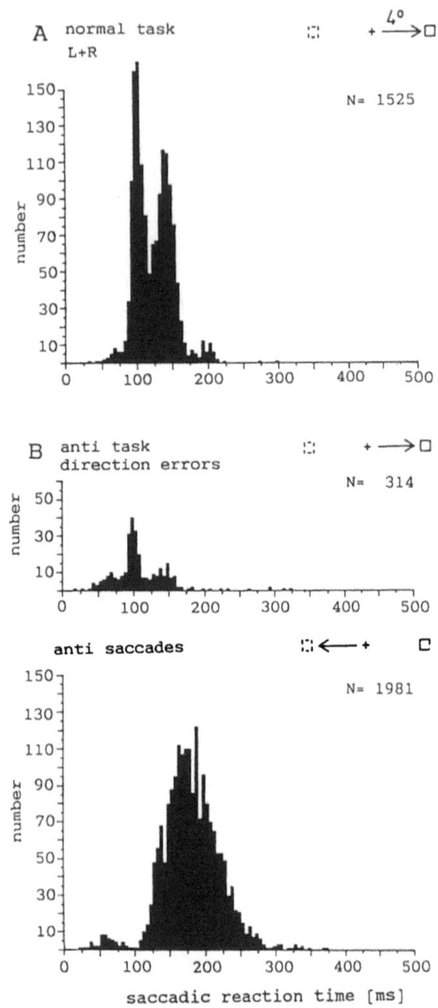

Fig. 8.4–2. Human saccadic reaction times of pro- (**A**) and antisaccades (**B**, *below*). Direction errors from the antitask are depicted in **B**, *above*. Note the absence of the 100 ms peak in the antisaccade data and its presence with the direction errors. The targets were presented randomly at the right or left side of the fixation point. A small number of anticipations is present in all three conditions

8.4.4 Suppression of Express Saccades

The question arises why saccades are not all of the express type? Several attempts have been made to find the physical, psychological, and context conditions in which express saccades are absent in subjects who can make them in the normal gap task. An effective method to reduce the number of express saccades is to leave the fixation point on while the target occurs, i.e., to use the overlap task. In most subjects the number of express saccades is drastically decreased in this condition, and even the number of fast regular saccades is reduced in favor of slow regular saccades with SRTs of 200 ms or

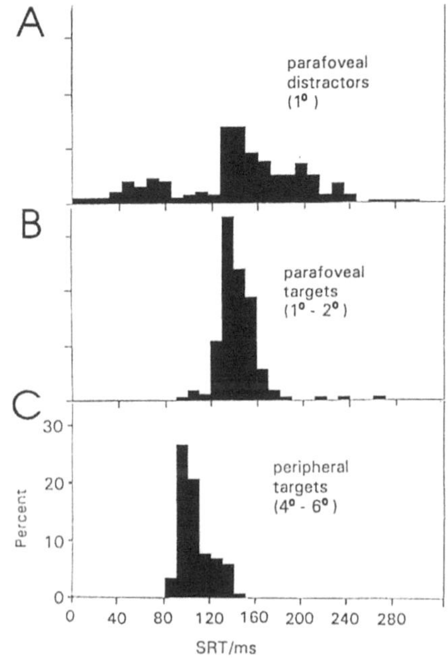

Fig. 8.4–3. The influence of parafoveal nontarget and target stimuli on saccadic reaction times. **A** Small parafoveal (presented at ±1° eccentricity) distractors presented simultaneously with the saccade target (at 4° eccentricity) suppress the express peak, whereas anticipations and regular saccades are present. **B** Targets presented within 1°–2° eccentricity cannot be reached by express saccades. **C** The same targets presented between 4° and 6o eccentricity are reached almost exclusively by express saccades

more. This change in the SRT distribution when switching from the overlap to the gap task has been called the gap effect. Nevertheless, it has been demonstrated that the permanent physical presence of a foveal stimulus per se is not the reason why express saccades tend to be absent (Mayfrank et al. 1986). A crucial point is the way in which the subject deals with it. It is the instruction to fixate on the foveal target actively and attentively that leads to a complete absence of all fast reactions, whereas the instruction not to attend to the foveal stimulus allows at least some express saccades and many fast regular saccades to occur. Daily practice using the overlap task leads to express saccades in monkeys (Fischer et al. 1984) and in man (Fischer 1987). This observation lead to the hypothesis that active fixation has an inhibitory effect on the neural structures responsible for express saccades and perhaps on the saccade system altogether (Fischer 1987).

With this idea in mind a number of experiments were carried out, the results of which were the following: (a) Dead zone for express saccades: if saccade targets are presented close to the fovea, they can be reached by fast or slow regular saccades but not by express saccades (Weber et al. 1992). There exists a dead zone for express saccades around the fovea. This phenomenon is demonstrated by the data shown in Fig. 3, comparing the reaction times of small (Fig. 3B) and larger (Fig. 3C) saccades. The complete absence of small express saccades is evident. (b) Distractor effects: if one presents non-target distracting stimuli (which have nothing to do with the subject's task) simultaneously with the saccade target, express saccades are largely reduced in number and may be suppressed altogether depending upon the contrast, size, number, and position of the distractors. Contralaterally presented distractors are more effective than those presented ipsilaterally (Weber and Fischer 1994). Of special interest are small and relatively weak distractors presented close to the fovea, i.e., within the dead zone. An example is shown in Fig. 3A. One sees some anticipatory reactions with SRTs below 80 ms and a second peak around 140 ms. Almost no data points are present in the region of the express saccades (for control see Fig. 3C).

8.4.5 Fixation: A "Stop" Reflex

The conditions under which express saccades are absent lead to the suspicion that a foveal stimulus has the tendency to arrest saccades, regardless of whether it represents a target to be foveated or not. One may even speculate that there exists a fixation reflex which almost automatically inhibits fast movements of the eyes. Further support for this hypothesis comes from an analysis of secondary saccades. If a saccade does not bring the fovea right onto the target a second (corrective) saccade is usually generated to account for the remaining retinal error. Because of the existence of the dead zone the corrective saccade can be made almost instantaneously only if the target is still outside the dead zone after the primary saccade. If the fovea is brought closer to the target, the secondary saccade is delayed depending upon the size of the error. With this idea in mind we analyzed – in a single target gap task with left/right randomization of the saccade target – those trials in which subjects made larger or smaller errors. Figure 4 shows the main result. Panel A depicts all primary saccades. One clearly sees the two modes of express and regular saccades. Also included are some anticipatory reactions (SRT below 80 ms, see Wenban-Smith and Findlay 1991) in the direction of the target and in the opposite direction. The latter represent direction errors from trials where the subjects anticipated target occurence at the wrong side. The distributions in Fig. 4B depict the reaction times of the secondary saccades (measured from target onset, not from the end of the primary saccade). The upper graph shows data from trials in which the second saccade was preceded by anticipatory saccades in either direction. Clear peaks of express and fast regular saccades can be seen. The lower graph represents data from trials with primary express or regular saccades. Fast secondary reactions are completely absent.

Further analysis shows that the express and fast regular secondary saccades occur almost exclusively in trials with direction errors. In this case the dead zone remains empty at the end of the primary saccade; in the case of anticipatory reactions in direction of the target or in the case of primary

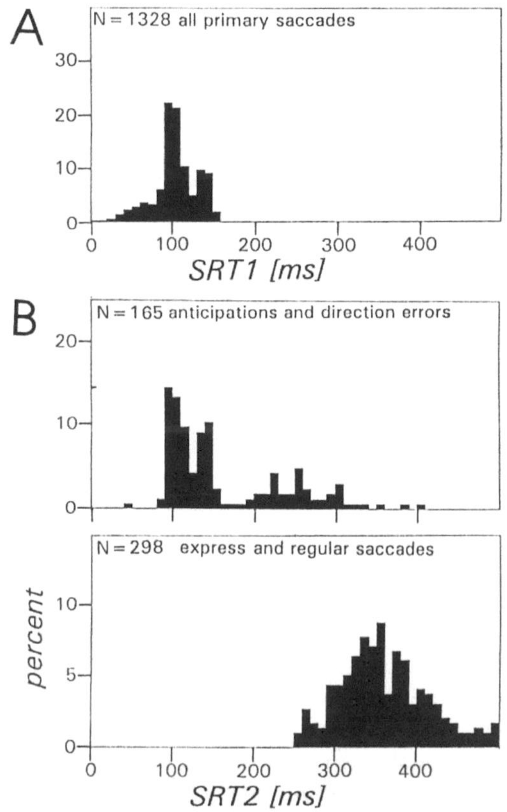

Fig. 8.4–4. Reaction times from a single target gap task. **A** Reaction times of all primary saccades containing anticipations, express, and fast regular saccades. **B** Reaction times of secondary saccades (measured from target onset) following primary saccades that did not reach the target at once; *above*, secondary saccades that were preceded by anticipations in the direction or against the direction of the target; *below*, secondary saccades that were preceded by express or regular saccades

express or regular saccades the dead zone is invaded by the target, and this leads to a delay of the next saccade. If an anticipatory saccade in the direction of the target also leaves a sufficiently large error, it may be also corrected at about 100 ms after the onset of the target. Further details of this study are presented elsewhere (Fischer et al. 1994).

The other line of evidence for a fixation reflex comes from double step experiments in which the target was switched off at the beginning of the primary saccade. After a variable delay (gap) the target reappeared at a position farther away or closer to the fovea, requiring a secondary saccade to reach it. It was found that these secondary saccades could be of the express type with gap durations only half as long as the optimal gap duration for primary express saccades. This indicates that at least one preparatory process is not necessary for the initiation of these secondary saccades. Switching the primary target off only 50 ms after the primary saccade reduced the number of second express saccades drastically; the fixation reflex arrested the eyes and it took more preparatory time to generate the next saccade (Kustov and Fischer submitted).

So far we have evidence supporting the idea that there exists a system involved in the control of fixation. When activated it arrests saccades, presumably by inhibiting the saccade generating neural structures. As mentioned above, directed attention may reduce the number of express saccades, and the question arises of whether such effects are restricted to foveal attention (allocated at the point of fixation), or whether they are also observed when attention is directed to a peripheral target. The results of Braun and Breitmeyer (1988, 1990) support the latter notion and confirm the idea of an inhibitory influence of attention on the saccade system (Mayfrank et al. 1986).

8.4.6 Deficits in Fixation

If there indeed exists a separate neural system that controls fixation by arresting saccades, one may have the chance to see subjects who have somehow lost or never developed complete control over their saccade system, without exhibiting any other deficit. In fact we have seen a few subjects of this kind. Figure 5 shows the case of a subject who was unable to suppress saccades to a suddenly appearing stimulus. In all conditions tested (gap, overlap, antitask, remembered or delayed saccade tasks) he made a saccade upon the onset of any visual stimulus, and, moreover, these were almost exclusively express saccades. Such "express makers", as one might call them, seem to be rare among normal adult subjects. More frequently they may be found among dyslexic children, but certainly not all dyslexics are express makers (Fischer and Weber 1990; Biscaldi and Fischer 1993), and not all express makers are dyslexics. Neurological patients with lesions in the frontal cortex involving the frontal eye fields also produce extremely high numbers of express saccades in the gap task, and they cannot suppress reflexlike target directed express saccades in an antitask (Guitton et al. 1985; Braun et al. 1992); however in the overlap task their SRT distributions resemble those of normal subjects.

8.4.7 Conclusions

From the experimental results briefly summarized above we conclude that the natural sequence of saccades alternating with periods of no eye move-

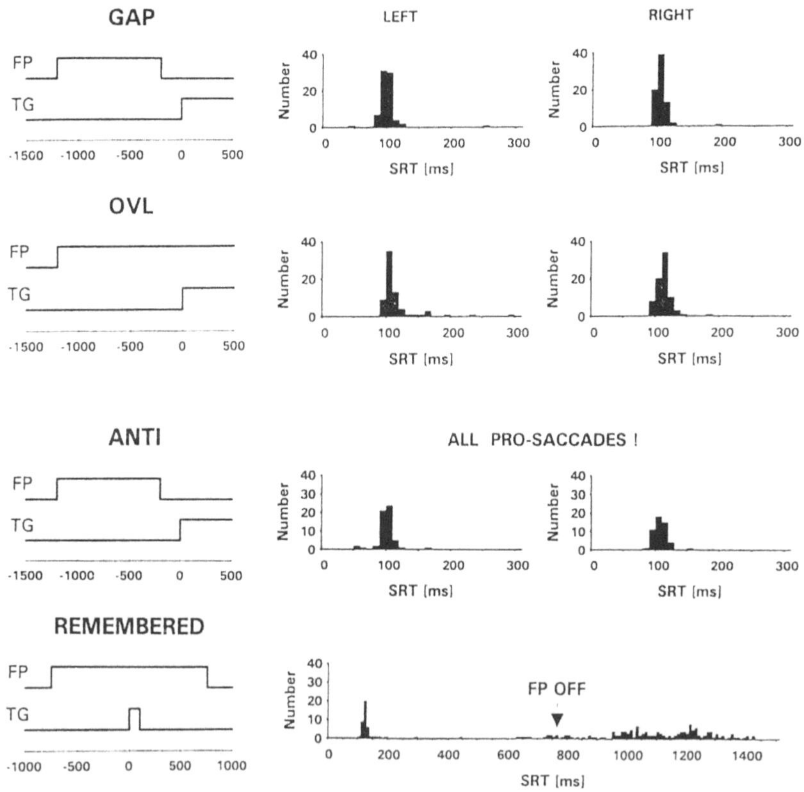

Fig. 8.4–5. Saccadic reaction times of a single subject who was unable to suppress saccades to suddenly appearing targets regardless of the task used. In the gap and in the overlap task (*OVL*) he produced single express peaks to both sides. In the antitask he made almost exclusively pro-saccades, which were all of the express type. In the remembered saccade task he looked at the cue in most of the trials instead of maintaining fixation until the offset of the fixation point (*FP*). All these saccades were of the express type

ments (often called fixations) is controlled by two reflexlike mechanisms. The first can be regarded as a "go" reflex taking the fovea to the next target; the other can be interpreted as a "stop" reflex that prevents saccades when a pattern has entered the dead zone at the end of a saccade. What remains open is the question of target selection: what makes a visual stimulus a target for a saccade? A related but even more difficult problem is how volition is implicated, by which we can make voluntary saccades, or by which we can decide to make no saccade, even though we might have selected a particular target. To what extent is the volitional control of our saccades dependent upon conscious perception of the possible targets? In any instance, to fully understand the optomotor system of primates one must at the same time understand these higher aspects of brain functions.

Acknowledgements. This work was supported by the Deutsche Forschungsgemeinschaft, Sonderforschungsbereich SFB 325, TpC5.

References

Biscaldi M, Fischer B (1993) Saccadic eye movements of dyslexic children in non-cognitive tasks. In: Wright SF, Groner R (eds) Facets of dyslexia and its remediation. Elsevier North Holland, pp 217–233

Bizzi E (1968) Discharge of frontal eye field neurons during saccadic and following eye movements in unanaesthetized monkeys. Exp Brain Res 6: 69–80

Boch R (1989) Saccadic reaction times after chemical lesions in striate and prestriate cortex of the rhesus monkey. Invest Ophthal Vis Sci 30: 184

Braun D, Breitmeyer BG (1988) Relationship between directed visual attention and saccadic reaction times. Exp Brain Res 73: 546–552

Braun D, Breitmeyer BG (1990) Effects of reappearance of fixated and attended stimuli upon saccadic reaction time. Exp Brain Res 81: 318–324

Braun D, Weber H, Mergner T, Schulte-Mönting J (1992) Saccadic reaction times in patients with frontal and parietal lesions. Brain 115: 1359–1386

Bushnell MC, Goldberg ME (1979) The monkey frontal eye fields have a neuronal signal that precedes visually guided saccades. Soc Neurosci Abstr 5: 779

Fischer B (1987) The preparation of visually guided saccades. Rev Physiol Biochem Pharmacol 106: 1–35

Fischer B, Boch R (1981) Enhanced activation of neurons in prelunate cortex before visually guided saccades of trained rhesus monkeys. Exp Brain Res 44: 129–137

Fischer B, Boch R (1983) Saccadic eye movements after extremely short reaction times in the monkey. Brain Res 260: 21–26

Fischer B, Boch R (1985) Peripheral attention versus central fixation: modulation of the visual activity of prelunate cortical cells of the rhesus monkey. Brain Res 345: 111–123

Fischer B, Ramsperger E (1984) Human express saccades: extremely short reaction times of goal directed eye movements. Exp Brain Res 57: 191–195

Fischer B, Weber H (1990) Saccadic reaction times of dyslexic and age-matched normal subjects. Perception 19: 805–818

Fischer B, Weber H (1992) Characteristics of "anti" saccades in man. Exp Brain Res 89: 415–424

Fischer B, Weber H (1993) Express saccades and visual attention. Behav Brain Sci 16:3:

Fischer B, Boch R, Ramsperger E (1984) Express-saccades of the monkey: effect of daily training on probability of occurrence and reaction time. Exp Brain Res 55: 232–242

Fischer B, Weber H, Biscaldi M (1994) The time of secondary saccades to primary targets. Exp Brain Res (in press)

Goldberg ME, Bruce CJ (1985) Cerebral cortical activity associated with the orientation of visual attention in the rhesus monkey. Vision Res 25: 471–481

Goldberg ME, Bushnell MC, Bruce (1986) The effect of attentive fixation on eye movements evoked by electrical stimulation of the frontal eye fields. Exp Brain Res: 61: 579–584

Guitton D, Buchtel HA, Douglas RM (1985) Frontal lobe lesions in man cause difficulties in suppressing reflexive glances and in generating goal-directed saccades. Exp Brain Res 58: 455–472

Mayfrank L, Mobashery M, Kimmig H, Fischer B (1986) The role of fixation and visual attention in the occurrence of express saccades in man. Eur Arch Psychiatry Neurol Sci 235: 269–275

Petersen SE, Robinson DL, Keys W (1985) Pulvinar nuclei of the behaving rhesus monkey: visual responses and their modulation. J Neurophysiol 54: 867–886

Robinson DA, Fuchs AF (1969) Eye movements evoked by stimulation of frontal eye fields. J Neurophysiol 32: 637–648

Robinson DL, Goldberg ME, Stanton GB (1978) Parietal association cortex in the primate: sensory mechanisms and behavioral modulations. J Neurophysiol 41: 910–932

Saslow MG (1967) Effects of components of displacement-step stimuli upon latency of saccadic eye movements. J Opt Soc Am 57: 1024–1029

Schiller PH, Sandell JH, Maunsell JH (1987) The effect of frontal eye field and superior colliculus lesions on saccadic latencies in the rhesus monkey. J Neurophysiol 57: 1033–1049

Shibutani H, Sakata H, Hyvärinen J (1984) Saccade and blinking evoked by microstimulation of of the posterior parietal association cortex. Exp Brain Res 55: 1–8

Weber H, Fischer B (1990) Effect of a local ibotenic acid lesion in the visual association area on the prelunate gyrus (area V4) on saccadic reaction times in trained rhesus monkeys. Exp Brain Res 81: 134–139
Weber H, Fischer B (1994) Differential effects of non-target stimuli on the occurrence of express saccades in man. Vision Res (in press)
Weber H, Aiple F, Fischer B, Latanov A (1992) Dead zone for express saccades. Exp Brain Res 89: 214–222
Wenban-Smith MG, Findlay JM (1991) Express saccades: is there a separate population in humans? Exp Brain Res 87: 218–222
Wurtz RH, Goldberg ME (1972) The role of the superior colliculus in visually-evoked eye movements. Bibl Ophthal 82: 149–158

8.5 Callosal Pathways for Simple Visuomotor Control in Man

G. Berlucchi, G. Tassinari, and S. Aglioti
Istituto di Fisiologia umana, Università di Verona, Italy

8.5.1 Simple Visuomotor Control

In ordinary life conditions, many human movements occur under visual control and therefore require a functional coordination between visual and motor centers in the cerebral cortex as well as in subcortical centers. Complex sequences of activation of visual and motor centers have been postulated to underlie those motor responses which must be shaped and/or continuously guided by the visual stimulus as actually perceived or remembered (e.g., Glickstein 1990). In more elementary forms of visuomotor integration visual stimuli may act merely as triggers for simple motor responses which, once initiated, can proceed unaided by visual perception or memory. At least the fastest of these responses can be emitted in a reflexlike fashion, such that the underlying neural circuits are likely to involve relatively fixed and straightforward connections between visual and motor cortical areas.

In 1912 Poffenberger applied chronometric analysis to the dissection of the central neural pathways subserving the execution of a fast manual or digital movement (such as pressing a key) in reaction to a light flash presented in the right or left hemifield, that is, to the visual cortex of the opposite hemisphere. The motor reaction of each hand is under the control of the contralateral hemisphere due to the crossing of the major motor pathways.

Uncrossed reactions (i.e., reactions of each hand to stimuli in the ipsilateral hemifield) can thus be integrated within a hemisphere, whereas crossed reactions (i.e., reactions of each hand to contralateral hemifield stimuli) require an interaction between the hemisphere receiving the visual stimulus and that emitting the response, probably through the corpus callosum. On this basis Poffenberger (1912) argued that the reaction time (RT) of crossed responses should be longer than that of uncrossed responses, and that the crossed-uncrossed time difference (CUD) should correspond to the additional time needed for interhemispheric communication. He found a CUD of a few milliseconds in the expected direction, and his finding has been repeatedly confirmed by modern studies which have demonstrated CUDs of 2–3 ms, a difference which can be accounted for by the conduction time along the largest fibers of the corpus callosum (Bashore 1981; Aglioti et al. 1993; Tassinari et al. 1993).

8.5.2 Crossed Visuomotor Control in Patients with Partial or Complete Callosal Defects

If the corpus callosum is crucially involved in the fast integration of crossed responses, the execution of these responses should be impossible or very slow in acallosal subjects. To test these possibilities we measured the CUD in six subjects with an callosal section sparing the splenium, one subject with total section of the corpus callosum, two subjects with complete callosal agenesis, and one subject with an agenetic absence of the splenium due to a cerebrovascular malformation. The callosotomy patients had been suffering for several years from drug-resistant forms of epilepsy and had been submitted to section of corpus callosum at the Institute of Neurosurgery of the Catholic University in Rome in an effort to reduce the severity of their condition. Objective assessments based on magnetic resonance imaging (MRI) showed the completeness of the callosal section in one case whereas in the remaining cases the extent of the callosal section varied from the anterior one-third to the anterior four-fifths of the corpus callosum, the splenium being consistently spared. In the three subjects with dysgenetic callosal defects MRI showed a complete absence of the corpus callosum in two cases, whereas in the third only the posterior third of the callosum with the splenium was lacking, most probably because its development was prevented by an arterovenous malformation attached to the great cerebral vein of Galen. The rest of the corpus callosum is normal in the latter patient. Detailed descriptions of these cases can be found in Aglioti et al. (1993) and Tassinari et al. (1993).

Table 1 shows that CUDs of the subjects with partial callosal defects, whether surgical or congenital, fell by and large within the normal range. By

Normal subjects (n=12) (Di Stefano et al. 1980)	2.2 ± 0.4
Normal subjects (n=8) (Tassinari et al. 1983)	2.7 ± 1.3
Normal subjects (n=48) (Aglioti et al. 1991)	7.4 ± 1.9
Anterior callosotomy patients (n=6) (Tassinari et al. 1993)	3.3 ± 2.8
Complete callosotomy patient ME (Aglioti et al. 1993; Tassinari et al. 1993)	83.2
Complete callosal agenesis patient RB (Aglioti et al. 1993; Tassinari et al. 1993)	22.5
Complete callosal agenesis patient PM (Aglioti et al. 1993; Tassinari et al. 1993)	25.4
Posterior callosal agenesis patient AZ (Tassinari et al. 1993)	11.8

Table 8.5–1. Crossed-uncrossed differences in reaction time (ms) Data for groups are means ± standard errors of the mean. Note the abnormally prolonged CUDs of patients with complete callosal defects (ME, RB, PM). The CUD of patient AZ with a posterior callosal defect, although slower than the normal mean CUDs, is still within the normal range.

contrast, the CUDs exhibited by the two subjects with a complete callosal agenesis and by the subject with a complete callosal section were at least an order of magnitude greater than the typical 2–3 ms CUDs of normals. These findings confirm previous reports of abnormally prolonged CUDs values in subjects with a congenital absence of the corpus callosum (e.g., Milner et al. 1985; Di Stefano et al. 1992) as well as in split-brain patients (Sergent and Myers 1985; Clarke and Zaidel 1989; Di Stefano et al. 1992). In addition, the contrast between the presence of an abnormal prolongation of CUDs in the completely acallosal subjects and its absence in the subjects with partial callosal defects, whether anterior or posterior, suggests that both anterior and posterior callosal routes can subserve the integration of speeded crossed responses.

8.5.3 Callosal and Non-Callosal Mechanisms of Crossed Visuomotor Control

In principle, the corpus callosum may subserve the integration of crossed visuomotor reactions by transferring the visual input across the midline or by transmitting a "go" signal to premotor and motor areas of the hemisphere which emits the response (Berlucchi 1978). The fact that CUD remains invariant across major changes in intensity and eccentricity of visual stimuli strongly suggests that the callosal transfer mediating crossed responses is not a replica of the visual input but rather a trigger for the response (Berlucchi 1978; Milner and Lines 1982; Milner et al. 1985). This implies that the transfer normally occurs by way of non-visual callosal routes, perhaps through anterior and/or middle callosal portions interconnecting premotor and motor areas of the frontal lobes, a possibility which has been supported indirectly by the finding that the CUD is matched by interhemispheric differences in latencies of potentials evoked by lateralized visual stimuli at central, but not occipital sites (Rugg et al. 1984). However interhemispheric routes for the initiation of motor responses to visual stimuli may also run in the splenium of the corpus callosum along with purely visual interhemispheric pathways. Some evidence on callosal topography in man indicates that the splenium contains not only the callosal connections of visual areas in the occipital lobes, but also those of the posterior-inferior parietal cortex (De Lacoste et al. 1985). This part of the parietal lobe includes cortical areas which appear to be important for the initiation of oculomotor and skeletomotor responses to visual stimuli (Andersen 1987), and very long CUDs have been reported in patients with parietal lesions (Anzola and Vignolo 1992). Conceivably, parietal regions with callosal connections running in the splenium may have a role in coordinating motor initiation functions across the hemispheres (Tassinari et al. 1993).

More direct evidence for the normal involvement of the corpus callosum in the fast integration of crossed manual responses has been obtained by studying CUDs for responses to unilateral flashes performed bilaterally and/or with effectors other than the hands. Studies in normals (Di Stefano et al. 1980) have borne out a clearcut distinction between crossed responses which presumably utilize interhemispheric transfer from those which presumably do not. The first set of responses includes unilateral and bilateral distal responses (e.g., flexion of the thumb) and unilateral proximal responses of the upper limb (e.g., flexion of the elbow), all associated with significant CUDs reflecting dependence on interhemispheric transfer. The second set includes bilateral proximal responses and unilateral and bilateral axial responses of the upper limb (e.g., elevation of the shoulder), all associated with null CUDs reflecting independence from interhemispheric transfer. Logically this distinction leads one to predict that impairment of interhemispheric transfer by callosal defects should alter the CUDs associated with the first set of responses, but not those associated with the second set. The CUD pattern that we found in our three subjects with complete callosal defects, either surgical or dysgenetic, fitted this prediction very well (Aglioti et al. 1993). As with normal subjects, these acallosal subjects showed insignificant CUDs for bilateral proximal responses and for unilateral and bilateral axial responses, in sharp contrast with their greatly prolonged CUDs for distal responses, either unilateral or bilateral, and for unilateral proximal responses. These results confirm that a callosal contribution is important for the execution of fast distal and unilateral proximal responses to a visual stimulus directed to the hemisphere ipsilateral to the responding hand. The importance of the corpus callosum derives from the fact that motor pathways for manual or digital movements are completely crossed. By contrast, other upper limb movements can be directly initiated by either hemisphere through bilaterally distributed motor pathways. Bilaterally distributed motor systems originating from each hemisphere are indeed available for the activation of axial and proximal limb muscles involved in global body movements, general postural adjustments, and integrated synergistic limb-body movements. Their existence has been demonstrated anatomically and physiologically in nonhuman primates (Kuypers 1987; 1989) and confirmed by clinical and experimental evidence in man (e.g., Freund 1987; Colebatch and Gandevia 1989; Müller et al. 1991; Colebatch et al. 1991; Benecke et al. 1991). Thus it is possible for a visual input channeled into a single hemisphere to directly initiate and guide axial and proximal limb movements on both sides of the body, and this is why CUDs are absent in normal and acallosal subjects when crossed as well as uncrossed visuomotor responses can be initiated by the hemisphere receiving the flash without the aid of the corpus callosum.

It must be emphasized that our completely acallosal subjects were fully capable of making all types of speeded responses both ipsilaterally and contralaterally to the visual stimulus on both right and left sides. Their strikingly long CUDs depended on a selective retardation of crossed responses relative to uncrossed responses. Crossed responses appear to be mediated in these subjects by an extracallosal interhemispheric transfer via relatively long cross-midline pathways which are able to transmit at least crude visual information. Several considerations suggest that this transfer occurs by way of subcortical commissures and other brainstem routes for indirect communication between the cortices of the two sides (Aglioti et al. 1993), as has been postulated to account for residual abilities for interhemispheric interaction in visually guided behavior in callosal agenesis and after forebrain commissurotomy (Trevarthen and Sperry 1973; Holtzmann 1984; Milner 1982; Milner et al. 1985; Myers and Sperry 1985; Sergent 1986, 1987).

8.5.4 Extracallosal Interhemispheric Transfer of Simple Visual Information

We obtained internal additional evidence for an abnormally slow interhemispheric transmission of crude visual information in our subject with a complete callosotomy. We used a two-flash paradigm in which subjects must detect a first extrafoveal flash of light (or prime) without overtly reacting to it, and then press a key as fast as possible upon seeing a second extrafoveal

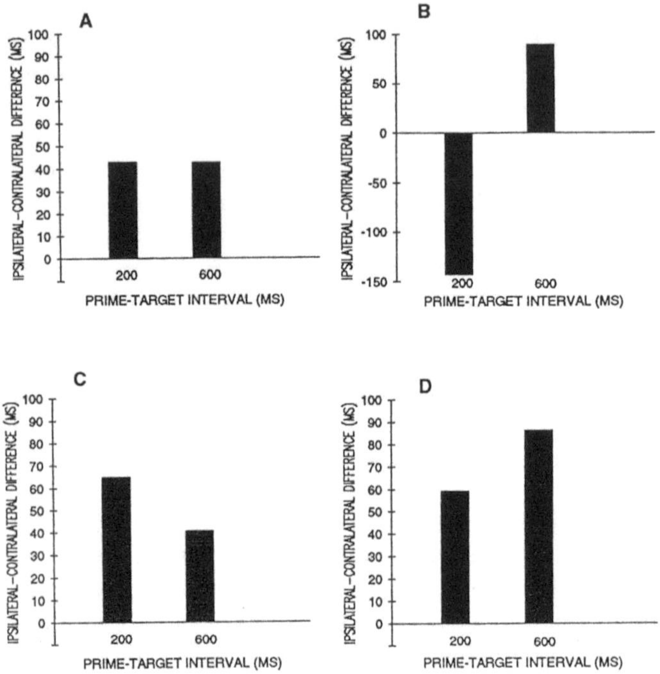

Fig. 8.5–1. Differences between RT for ipsilateral prime-target combinations and RT for contralateral prime-target combinations as a function of prime-target interval in 6 normal subjects (**A**,**C**; from Tassinari et al. 1987) and in the completely callosotomized patient ME (**B**,**D**). The stimulus array was aligned with the horizontal meridian in **A**,**B** and with the vertical meridian in **C**,**D**

flash (target). During the presentation of both prime and target as well as in the prime-target interval the eyes must remain fixated on a mark in front of the subject. Normal subjects are faster in reacting to targets contralateral than to targets ipsilateral to the prime (Fig. 1A). When prime and target are presented in the same visual hemifield they can be processed by the same hemisphere, whereas with contralateral combinations the prime and the target are processed by different hemispheres, and therefore an interhemispheric integration is required for the execution of the task. Nevertheless, the response speed for ipsilateral prime-target combinations is subject to an inhibition which annuls and overrides the advantage of intra- over interhemispheric integration (Posner and Cohen 1984; Tassinari et al. 1987; Berlucchi et al. 1989). But if the interhemispheric exchange of visual information subserving the response to contralateral prime-target combinations is slowed down by the absence of the corpus callosum, then one might expect RT for contralateral combinations to be longer than RT for ipsilateral combinations, in spite of ipsilateral inhibition. As shown in Fig. 1B, this is precisely what we found in our completely callosotomized subject. In support of the hypothesis of a slow interhemispheric transmission of visual information via an extracallosal route in this subject, the relative contralateral advantage was absent at short but present at long intervals between prime and target. At short intervals, the impairment of interhemispheric communication overrode ipsilateral inhibition. By contrast, ipsilateral inhibition was evident in normals at both short and long intervals because of the rapid interhemispheric transmission. Ipsilateral inhibition can also be demonstrated in normal subjects with primes and targets presented along the vertical meridian (Fig. 1C). In this case ipsilateral inhibition occurs with combinations involving primes and targets occurring on the same side of (i.e., both below or both above) the horizontal meridian (Berlucchi et al. 1989). However, primes and targets are projected to both hemispheres with both ipsilateral and contralateral combinations, and therefore the latter combinations do not require interhemispheric cooperation. Accordingly, Fig. 1D shows a completely normal ipsilateral inhibition in the completely callosotomized subject tested with primes and targets presented along the vertical meridian. This result provides further evidence for a causal role of the callosal disconnection in the abnormal response to horizontal stimuli at short prime-target intervals.

Acknowledgments. This research was supported by grants from the Consiglio Nazionale delle Ricerche, the Ministero dell'Università e della Ricerca Scientifica e Tecnologica, and the Human Frontier Science Programme Organization.

References

Aglioti S, Dall'Agnola R, Girelli M, Marzi CA (1991) Bilateral hemispheric control of foot distal movements: evidence from normal subjects. Cortex 27: 571–581

Aglioti S, Berlucchi G, Pallini R, Rossi GF, Tassinari G (1993) Hemispheric control of unilateral and bilateral responses to lateralized light stimuli after callosotomy and in callosal agenesis. Exp Brain Res 95: 151–165

Andersen RA (1987) Inferior parietal lobule function in spatial perception and visuomotor integration. In: Mountcastle VB, Plum F, Geiger SR (eds) Handbook of physiology. American Physiological Society, Bethesda, Vol V, Part 2, pp 483–518

Anzola GP, Vignolo LA (1992) Interhemispheric communication following unilateral cerebrovascular lesions. Ital J Neurol Sci 13: 649–655

Bashore TR (1981) Vocal and manual reaction time estimates of interhemispheric transmission time. Psychol Bull 89: 352–368

Benecke R, Meyer B-U, Freund H-J (1991) Reorganisation of descending motor pathways in patients after hemispherectomy and severe hemispheric lesions demonstrated by magnetic brain stimulation. Exp Brain Res 83: 419–426

Berlucchi G (1978) Interhemispheric integration of simple visuomotor responses. In: Buser PA, Rougeul-Buser A (eds) Cerebral correlates of conscious experience. North-Holland, Amsterdam, pp 83–94

Berlucchi G, Tassinari G, Marzi CA, Di Stefano M (1989) Spatial distribution of the inhibitory effect of peripheral non-informative cues on simple reaction time to non-fixated visual targets. Neuropsychologia 27: 201–221

Clarke JM, Zaidel E (1989) Simple reaction times to lateralized light flashes: varieties of interhemispheric communication routes. Brain 112: 871–894

Colebatch JG, Gandevia SC (1989) The distribution of muscular weakness in upper motor neuron lesions affecting the arm. Brain 112: 749–763

Colebatch JG, Deiber M-P, Passingham RE, Friston KJ, Frackowiak RSJ (1991) Regional cerebral blood flow during voluntary arm and hand movements in human subjects. J Neurophysiol 65: 1392–1401

De Lacoste MC, Kirkpatrick JB, Ross ED(1985) Topography of the human corpus callosum. J Neuropathol Exp Neurol 44: 578–591

Di Stefano M, Morelli M, Marzi CA, Berlucchi G (1980) Hemispheric control of unilateral and bilateral movements of proximal and distal parts of the arms as inferred from simple reaction time to lateralized light stimuli in man. Exp Brain Res 38: 197–204

Di Stefano M, Sauerwein HC, Lassonde M (1992) Influence of anatomical factors and spatial compatibility on the stimulus-response relationship in the absence of the corpus callosum. Neuropsychologia 30: 177–185

Freund H-J (1987) Abnormalities of motor behavior after cortical lesions in humans. In: Mountcastle VB, Plum F, Geiger SR (eds) Handbook of Physiology, American Physiological Society, vol V, Part 2. Bethesda, pp 763–810

Glickstein M (1990) Brain pathways in the visual guidance of movement and the behavioural function of the cerebellum. In Trevarthen CB (ed), Brain circuits and functions of the mind. Cambridge University Press, Cambridge, pp 157–167

Holtzman JD (1984) Interactions between cortical and subcortical visual areas: evidence from human commissurotomy patients. Vision Res 8: 801–813

Kuypers HGJM (1987) Some aspects of the organization of the output of the motor cortex. In Ciba Foundation Symposium 132: Motor Areas of the Cerebral Cortex. Wiley, Chichester, pp 63–820

Kuypers HGJM (1989) Motor system organization. In: Adelman G (ed) Encyclopedia of neuroscience, Suppl 1. Birkhäuser, Boston, pp 107–110

Milner AD (1982) Simple reaction times to lateralized visual stimuli in a case of callosal agenesis. Neuropsychologia 20: 411–419

Milner AD, Lines, CR (1982) Interhemispheric pathways in simple reaction time to lateralized light flash. Neuropsychologia 20: 171–179

Milner AD, Jeeves MA, Silver PH, Lines CR, Wilson JG (1985) Reaction times to lateralized visual stimuli in callosal agenesis: stimulus and response factors. Neuropsychologia 23: 323–331

Müller F, Kunesch E, Binkofski F, Freund H-J (1991) Residual sensorimotor functions in a patient after right-sided hemispherectomy. Neuropsychologia 29: 125–145

Myers JJ, Sperry RW (1985) Interhemispheric communication after section of the forebrain commissures. Cortex 21: 249–260

Poffenberger AT (1912) Reaction time to retinal stimulation with special reference to the time lost in conduction through nervous centers. Arch Psychol 23: 1–73

Posner MI, Cohen Y (1984) Components of visual orienting. In: Bouma H, Bowhuis DG (eds) Attention and Performance X. Erlbaum, Hillsdale, NJ, pp 531–556

Rugg MD, Lines CR, Milner AD (1984) Visual evoked potentials to lateralized visual stimuli and the measurement of interhemispheric transmission time. Neuropsychologia 22: 215–225

Sergent J (1986) Subcortical coordination of hemisphere activity in commissurotomized patients. Brain 109: 357–369

Sergent J (1987) A new look at the human split brain. Brain 110: 1375–1392

Sergent J, Myers JJ (1985) Manual, blowing, and verbal simple reactions to lateralized flashes of light in commissurotomized patients. Percept Psychophys 37: 571–578

Tassinari G, Morelli M, Berlucchi G (1983) Interhemispheric ransmission of information in manual and verbal reaction-time tasks. Hum Neurobiol, 2: 77–85

Tassinari G, Aglioti S, Chelazzi L, Marzi CA, Berlucchi G (1987) Distribution in the visual field of the costs of voluntarily allocated attention and of the inhibitory after-effects of covert orienting. Neuropsychologia 25: 55–71

Tassinari G, Aglioti S, Pallini R, Berlucchi G, Rossi GF (1993) Interhemispheric integration of simple visuomotor responses in patients with partial callosal defects. Behav Brain Res (in press)

Trevarthen C, Sperry RW (1973) Perceptual unity of the ambient visual field in human commissurotomy patients. Brain 96: 547–570

8.6 Distributed Processing of Movement Dynamics and Kinematics in the Three Motor Cortical Areas of the Rhesus Monkey

C. FROMM AND E. BAUSWEIN
Department of Neurobiology, Max Planck Institute for biophysical Chemistry, Göttingen, Germany

Contemporary motor control theories are concerned with the sensorimotor transformations between different coordinate frames and the processing of spatial, kinematic, and dynamic parameters that are used and implemented by the CNS to achieve goal-directed movements (for reviews see e.g., Alexander and Crutcher 1990a; Kalaska and Crammond 1992; Soechting and Flanders 1992). *Limb kinematics* comprise the spatial position of joints, the direction, and amplitude of a movement together with the respective time derivatives. The term *movement dynamics* (synonymous with "kinetics") refers to joint torques, muscle forces and their derivatives and the underlying muscular activities which generate the movement. For the planning and execution of reaching movements to targets in space, for example, most models take a serial analytic approach. Thus planning of such movements is often thought at a higher kinematic level, either in an *extrinsic* reference frame such as the spatial hand path or an *intrinsic* one such as the joint angle motions; at a lower stage and with movement execution the required dynamics are thought to be finally determined (Georgopoulos 1991; Kalaska and Crammond 1992).

While these hierarchical models of sequential coordinate transformations have been quite stimulating and have the virtue of clarity, there are a number of problems to be addressed. For example, the motor system seems to be organized in parallel, with neuronal processing at a given hierarchical level being *distributed* over several different cortical areas. Besides other motor cortical fields, the primary motor cortex (MI), the dorsal premotor cortex (PMd), and the supplementary motor area (SMA) being reciprocally interconnected with each other have their own and equally substantial access to the spinal motor apparatus via pyramidal tract neurons (PTNs) (Dum and Strick 1991). In each of these three areas segregated groups of neurons have been found to be related either to the location of target, to the goal, spatial direction, trajectory, amplitude, and velocity of movement or to the detailed pattern of muscle activation and external loading conditions, some cells to be more involved in preparation, others in execution of movement (Schwartz et al. 1988; Alexander and Crutcher 1990a,b; Crutcher and

Alexander 1990; Bauswein et al. 1991; Caminiti et al. 1991; Georgopoulos 1991; Hocherman and Wise 1991; Lurito et al. 1991; Werner et al. 1991; Di Pellegrino and Wise 1993; Kurata 1993).

The most dominant, but not exclusive, relationship of neural activity in MI represents that to muscle force. This relationship between MI firing rates and particularly static forces and joint torques has been documented for isometric, isotonic, and auxotonic tasks around single joints (for review see Hepp-Reymond 1988; Cheney and Fetz 1980; Evarts et al. 1983; Fromm 1983; Cheney et al. 1991) and for two-dimensional tasks (Kalaska et al. 1989; Georgopoulos et al. 1992). In the first part we report our findings on the correlations of neuronal activity with static wrist torque in the other two areas, PMd and SMA. As a contrast to the processing of this basic muscle-like property we then turn to the problem of "trajectory encoding" by describing motor cortical activity with the higher-level planning stage for reaching arm movements in three-dimensional space.

8.6.1 Comparative Study of Load Relationship in the Motor Cortical Fields

The task involved positional maintenance of the wrist against graded torques loading the flexors or extensors and at three different positions of the joint. The task, the transdural recording techniques employed to isolate single unit discharges, antidromic identification procedures, and data analyses have been described previously and the data obtained from the PMd published (Werner et al. 1991). Altogether, 25% of 342 SMA and 28% of 470 PMd cells as compared to 48% of 800 neurons recorded from MI had significant relations of steady-state firing rate to static load. Firing rate-load characteristics of SMA cells resembled those found in MI. In particular, their rate-torque relationships could also vary with wrist joint position, at least qualitatively, similarly to the length-tension properties shown by the electromyogram (EMG) patterns of prime mover muscles in this task. Thus in principle, neurons in all three cortical areas can display nonlinear correlates of the length-tension relation of muscle.

The population averages of firing rate–load relationships (Fig. 1) for MI (curves A and C, filled circles) and SMA (D, crosses) essentially parallel each other (notice, however, the remarkably low mean discharge frequency of the SMA population). Both indicate an S-shaped function. In contrast, the averaged curve of the PMd cells is almost linear, with a tendency to increase with stronger torques in the preferred direction (B, open circles). For MI, mean torque sensitivity (Hz/Nm) was significantly higher than that of either SMA or PMd neurons. In the SMA, all load-related units were found in the microexcitable region of its forelimb representation. In summary, load-

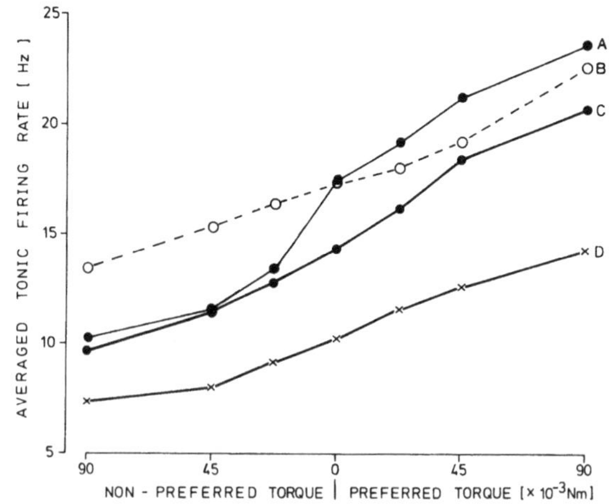

Fig. 8.6–1. Averaged firing rates with seven different static loads of monotonically related neurons from two monkeys. *Curves A, C (filled circles)*, MI population averages of 67 and 128 neurons, respectively. *Curve B (open circles, dashed line)*, PMd, n=35. *Curve D (crosses)*, SMA, n=74; *A and B* and *C and D* are from the same monkey, respectively. "Preferred torque" is the direction of load (either extensor or flexor) for which activity of any neuron was higher

sensitive neurons in the three areas share the following properties: they have tonic discharge rates, tend to be concentrated in layer V, and seem to form a continuum also with respect to their dependence on joint position, which is in full agreement with observations on MI cells in a multiarticular task (Kalaska et al. 1989). Antidromically identified PTNs possess the load sensitivity to a considerable and almost identical extent in all three areas investigated (44% in SMA, 45% in PMd, and 50% in MI).

8.6.2 Motor Cortical Activity During an Instructed Delay Period Preceding Targeted Reaching Movements

It is clear that proprioceptive information of the orientation of the arm and visual target information must be combined somewhere in a body-centered frame of reference (Georgopoulos 1991; Burnod et al. 1992; Kalaska and Crammond 1992). Soechting and Flanders (1992) have deduced from their psychophysical experiments an algorithm for the planning of the trajectory of targeted arm movements. In essence the visually derived target coordinates are transformed into shoulder-centered coordinates. Their prediction is that the intended yaw angles of the arm are linearly related to the azimuth of target location, and that the intended elevation angles of the arm are linearly related to the elevation of the target and its distance from the shoulder. In this way the target coordinates are transformed into a common frame of reference with kinesthetically derived arm coordinates in order to define the initial arm position relative to the movement endpoint.

We tested this specific hypothesis by examining neuronal activity during an instructed delay period when a target is presented, and the monkey is required to withhold the reaching movement up to a subsequent "go" signal which was a dim to bright change of the respective target spot. In two monkeys we recorded from the rostral MI and the PMd because neurons here are known to exhibit sustained activity changes during the delay period ("set" or "preparatory" activity) which has been implicated in some motor planning (e.g., Hocherman and Wise 1991; Kurata 1993). In addition, the PMd appears to be a pivotal processing stage for both visual and proprioceptive information (e.g., Bauswein et al. 1991; Wise et al. 1992). In our task, which has been fully described elsewhere (Bauswein and Fromm 1992), the two spatial variables determining different trajectories of the impending reaching movement could be controlled separately; target location was varied with arm position held constant, and the starting position of the arm was changed via a motor-driven forearm support before movements to the same target. The influence of changing the initial arm position could also be compared before and after the target instruction. The passive arm displacement in the horizontal plane over an angle of 36° moved the forearm between a left and right position, involving mainly a rotation of the shoulder joint. Two targets were projected at the same horizontal level on a screen 30 cm in front of the monkey's eyes. Thus, we changed chiefly the yaw/azimuth parameters; the straight distances of hand movements along the three different trajectories performed in this task ranged from 36 to 28 cm.

We found that the post-instruction activity depended not only on the target location but also on the initial arm position. Neuronal modulation with the passive arm shift could be contingent upon prior target instruction or it was different from the effects of the same arm displacement when applied before target appearance. This is shown by the neuron illustrated in Figs. 2 and 3. Notice also that the cell's discharge rate attained a comparable tonic level of set-activity whenever both parameters (target instruction and actual arm position) indicated that the next trajectory to be made is from left starting position of the hand to the right target (LR, cf. Fig. 3 bottom, Fig. 2 left part of bottom display and the middle display). Of 100 set-related neurons in PMd, 28 varied their tonic activity only with target location, in 23 units it depended solely on the initial arm position (like the neuron shown in Figs. 2 and 3), and for 34 neurons the set-activity varied with all three trajectories (i.e., dependent on both parameters).

Phasic responses (see Fig. 3) to presentation of targets for reaching movements also seem to play a role in visuomotor transformation (e.g., Wise et al. 1992). These phasic responses of 165 PMd cells depended in 36% only

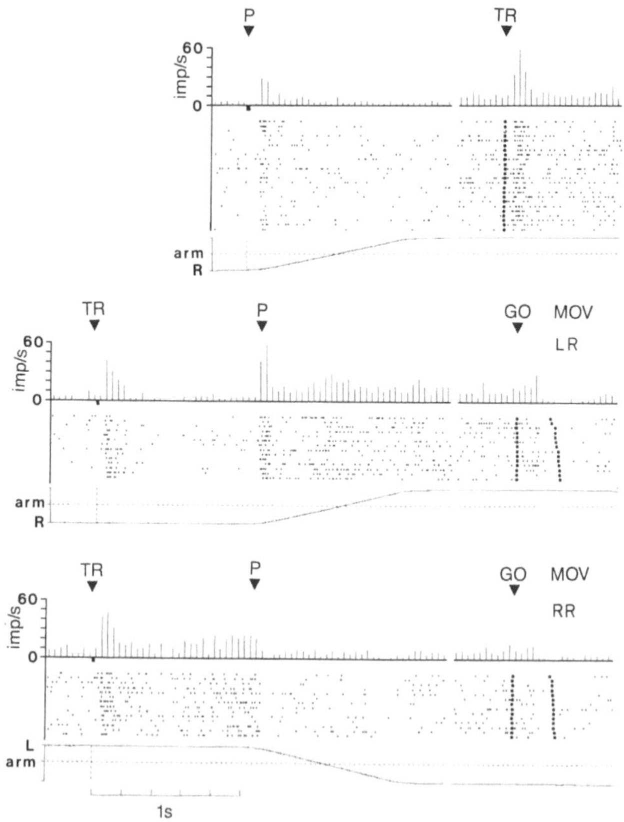

Fig. 8.6–2. Change in the activity of a PMd cell with shift of arm posture is contingent upon target instruction. *Small raster dots*, single discharges; *large dots* mark the "go" signal, onset of reaching movements (*MOV*) or appearance of the right target (*TR*), respectively (also denoted by an arrow). The ramp-shaped traces below the rasters represent the motor-driven arm displacement, its onset is marked by *P*; upward deflection indicates arm shift to the left. Sustained activity occurred with that arm displacement after TR presentation (*middle display*) but not before it (*top display*). Post-instruction activity was reciprocally influenced by a passive ramp displacement of the arm to the right position (*bottom display*). Notice that this cell shows also a phasic component of response to target illumination and ramp onset

on target location, in 20% exclusively on the initial arm posture, and in 29% on both target and initial arm position. In addition to having a cue signal related response, 83 PMd neurons showed an early activity increase after the "go" signal (i.e., during the reaction time). In the majority of these neurons (57/83), both the post-cue and premovement activations had the same directional preference. Thus in these cells both activation phases changed in parallel with the three movement trajectories performed. We suggest that such visuomotor responses in PMd reflect some selection of movement

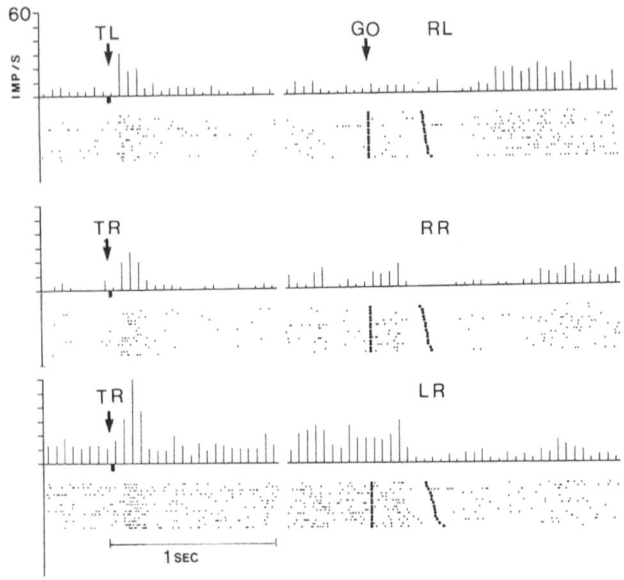

Fig. 8.6–3. The different phasic as well as tonic responses of the neuron (same as in Fig. 2) to presentation of the same target (*right, TR*) depend on the starting position of the arm which is at right in the middle display and at left in the bottom display. In this neuron, the post-cue activity is quite similar for different spatial locations of the target if the initial position of the arm is the same (*right*, cf. top and middle displays; *TL*, left target). After the "go" signal the monkey lifted his forearm from the support in the left position and reached forward to touch the right target (*LR*, *bottom*); the other trajectories were performed from the right starting position of the hand to right target (*RR*), or to the left target (*RL*). Trials are rank-ordered according to increasing intervals between "go" and movement onset (*two large raster dots*). The trials were presented in randomized order and with varying delay periods (notice interrupted sections, also in Fig. 2)

trajectories or the association between spatial cue and starting point of movement. The target instruction seems to "prime" these cells conditional on an intrinsic parameter, i.e., the initial position of the hand/arm.

8.6.3 Concluding Remarks

Altogether, our findings presented in the second part are consistent with the notion that kinesthetically and visually derived information interact in motor cortical cells in an appropriate manner for trajectory formation of the ensuing movement. They are compatible with but do not prove the above hypothesis of Soechting and Flanders (1992) and a related network model which, by a bilinear operation, projects the visual information about the target on a reference frame that rotates with the arm (Burnod et al. 1992). In general, one can question whether it may be possible at all to define specific variables encoded explicitly and independently by single neurons and any formal reference frame signalled by motor cortical activity. One of several problems is that spatial, intrinsic kinematic, and dynamic variables are often closely coupled and interdependent (e.g., Mussa-Ivaldi 1988; Kalaska and Crammond 1992). Important differences exist between the representational features of single cells and neuronal populations (Caminiti et al. 1991). Thus, a vectorial representation of the movement trajectory can emerge as an invariant from the whole population of motor cortical neurons while the single "synergy" output units can give commands in terms of movement dynamics (Burnod et al. 1992).

We would like to point out that the terms "neuronal processing" during a "planning stage" are potentially misleading where learned reaching movements are concerned. Alternatively, preparatory or set-activity whose occurrence during the delay period is known to depend on the intent to move may reflect the retrieval from memory of similar sets of initial conditions that establish and select a particular movement (see discussion e.g., in Hocherman and Wise 1991; Mitz et al. 1991). If instructions about movement amplitude and direction are given sequentially (in any order during the delay period), full set-activity in most PMd neurons develops only after complete information (Kurata 1993). The same is true for the effects of change of the arm's starting posture being conditional on target information (Fig. 2). In this connection, the terms "kinesthetic" or "proprioceptive" may be too suggestive of a direct somatosensory influence. Alternatively, the variations of motor cortical activity with maintained joint positions as demonstrated for single joint (Cheney and Fetz 1980; Evarts et al. 1983; Fromm 1983; Werner et al. 1991), two-dimensional (Kalaska et al. 1989) and three-dimensional tasks (Caminiti et al. 1991) could represent the outflow from a "motor" source which adjusts the amount of muscle activity to different joint postures in order to produce the necessary forces. Transferred to our observations on the activity during an instructed delay period, it would mean that the change of the arm's initial geometry could elicit a motor set appropriate for the biomechanical requirements of the forthcoming movement. At present, this possibility cannot be ruled out.

To understand the meaning of the remarkably different representational properties coexisting even in a single motor area, as exemplified by our findings summarized here, will be the challenge to future neurophysiological research and modelling approaches. An important next step will be to elucidate the role of corticospinal neurons in spatiomotor transformations in order to reconcile conflicting views about the causal relationship to movement (Cheney et al. 1991; Georgopoulos et al. 1992).

References

Alexander GE, Crutcher MD (1990a) Preparation for movement: neural representations of intended direction in three motor areas of the monkey. J Neurophysiol 64: 133–150

Alexander GE, Crutcher MD (1990b) Neural representations of the target (goal) of visually guided arm movements in three motor areas of the monkey. J Neurophysiol 64: 164–178

Bauswein E, Fromm C (1992) Activity in the precentral motor areas after presentation of targets for delayed reaching movements varies with the initial arm position. Eur J Neurosci 4: 1407–1410

Bauswein E, Fromm C, Werner W, Ziemann U (1991) Phasic and tonic responses of premotor and primary motor cortex neurons to torque changes. Exp Brain Res 86: 303–310

Burnod Y, Grandguillaume P, Otto I, Ferraina S, Johnson PB, Caminiti R (1992) Visuomotor transformations underlying arm movements toward visual targets: a neural network model of cerebral cortical operations. J Neurosci 12: 1435–1453

Caminiti R, Johnson PB, Galli C, Ferraino S, Burnod Y (1991) Making arm movements within different parts of space: the premotor and motor cortical representation of a coordinate system for reaching tò visual targets. J Neurosci 11: 1182–1197

Cheney PD, Fetz EE (1980) Functional classes of primate corticomotoneuronal cells and their relation to active force. J Neurophysiol 44: 773–791

Cheney PD, Fetz EE, Mewes K (1991) Neural mechanisms underlying corticospinal and rubrospinal control of limb movements. Progress Brain Res 87: 213–252

Crutcher MD, Alexander GE (1990) Movement-related neuronal activity selectively coding either direction or muscle pattern in three motor areas of the monkey. J Neurophysiol 64: 151–163

Di Pellegrino G, Wise SP (1993) Visuospatial versus visuomotor activity in the premotor and prefrontal cortex of a primate. J Neurosci 13: 1227–1243

Dum RP, Strick PL (1991) The origin of corticospinal projections from the premotor areas in the frontal lobe. J Neurosci 11: 667–689

Evarts EV, Fromm C, Kröller J, Jennings VA (1983) Motor cortex control of finely graded forces. J Neurophysiol 49: 1199–1215

Fromm C (1983) Changes of steady state activity in motor cortex consistent with the length-tension relation of muscle. Pflügers Arch 398: 318–323

Georgopoulos A.P. (1991) Higher order motor control. Ann Rev Neurosci 14: 361–377

Georgopoulos AP, Ashe J, Smyrnis N, Taira M (1992) The motor cortex and the coding of force. Science 256: 1692–1695

Hepp-Reymond MC (1988) Functional organization of motor cortex and its participation in voluntary movements. In: Steklis HD, Erwin J (eds) Comparative primate biology, vol 4. Alan R Liss, pp 501–624

Hocherman S, Wise SP (1991) Effects of hand movement path on motor cortical activity in awake, behaving rhesus monkeys. Exp Brain Res 83: 285–302

Kalaska JF, Crammond DJ (1992) Cerebral cortical mechanisms of reaching movements. Science 55: 1517–1523

Kalaska JF, Cohen DAD, Hyde ML, Prud'homme M (1989) A comparison of movement direction-related versus load direction-related activity in primate motor cortex, using a two-dimensional reaching task. J Neurosci 9: 2080–2102

Kurata K (1993) Premotor cortex of monkeys: set- and movement-related activity reflecting amplitude and direction of wrist movements. J Neurophysiol 69: 187–200

Lurito JT, Georgakopoulos T, Georgopoulos AP (1991) Cognitive spatial-motor processes. VII. The making of movements at an angle from a stimulus direction: studies of motor cortical activity at the single cell and population levels. Exp Brain Res 87: 562–580

Mitz AR, Godschalk M, Wise SP (1991) Learning-dependent neuronal activity in the premotor cortex: activity during the acquisition of conditional motor associations. J Neurosci 11: 1855–1872

Mussa-Ivaldi FA (1988) Do neurons in the motor cortex encode movement direction? An alternative hypothesis. Neurosci Lett 91: 106–111

Schwartz AB, Kettner RE, Georgopoulos A (1988) Primate motor cortex and free arm movements to visual targets in three-dimensional space. I. Relations between single cell discharge and direction of movement. J Neurosci 8: 2913–2927

Soechting JF, Flanders M (1992) Moving in three-dimensional space: frames of reference, vectors, and coordinate systems. Ann Rev Neurosci 15: 167–191

Werner W, Bauswein E, Fromm C (1991) Static firing rates of premotor and primary motor cortical neurons associated with torque and joint position. Exp Brain Res 86: 293–302

Wise SP, Di Pellegrino G, Boussaoud D (1992) Primate premotor cortex: dissociation of visuomotor from sensory signals. J Neurophysiol 68: 969–972

Chapter 9

Human Cerebral Cortex

9.1 Introduction

H.H. Kornhuber
Abteilung für Neurologie, Universitätsklinik, Ulm, Germany

In this section, which deals with the cerebral cortex of man, we come to a field which was of particular interest to Otto Creutzfeldt — not surprisingly, as he was a doctor of medicine and his father was a great neurologist. The brain/mind problem preoccupied Otto throughout his life; a tragic preoccupation because he believed it was unsolvable. He supported clinically oriented research, although he was busy to do basic research. For instance, applying his knowledge on the visual system, he was interested in the development of better tests for the diagnosis of visual impairments. It was because of his interest in man and his mind that he kept up to date on a wide range of subjects despite his specialized research in fields as different as recording methods, synaptology, and electroencephalography (EEG). Otto Creutzfeldt was a warm-hearted physician, who showed enthusiasm and support for the research of many young colleagues.

9.2 Cerebral Subsystems Within Language

H.J. NEVILLE[1,2] AND C.M. WEBER-FOX[1]
[1]The Salk Institute, La Jolla, California
[2]Department of Cognitive Science, University of California, San Diego, California

Professor Creutzfeldt, in addition to his classic research on the neuroanatomy and neurophysiology of the visual system, performed research that was pivotal in the development of the electroencephalogram (EEG) and the event-related brain potential (ERP) as research tools in the study of the human brain. He employed this technique in studies of humans that encompassed the development of the brain, variations in cerebral organization with fluctuations in hormones, and language processing (Creutzfeldt and Kuhnt 1967; Creutzfeldt et al. 1976, 1985, 1989a,b). Creutzfeldt also considered the neurophysiological mechanisms of consciousness (Creutzfeldt 1979). The emerging discipline of cognitive neuroscience strives to emulate Professor Creutzfeldt's depth of commitment to a multileveled, convergent approach to the study of the brain and the study of the mind.

In this chapter we summarize research on the effects of the nature and the timing of early language experience on the development of neural systems specialized for particular aspects of language processing. Additionally, we report results from a study in which we manipulated conscious awareness of language material in an effort to test the hypothesis that there exist subsystems within language that are automatically activated and separate from those that require conscious awareness in their operation.

9.2.1 Effects of Experience on Neural Development

Most research on the effects of experience on neural development has been conducted within the realm of simple sensory processing, but nonetheless provides general guidelines concerning the formation of systems specialized for cognitive processing. For example, deprivation of input within a sensory modality results in marked reorganization both of the cortical areas that normally process the lost input and of remaining sensory systems (Garraghty and Kaas 1992; Neville 1992; Sur et al. 1990). Many such effects of experience only occur at particular time points in the immature brain and thus define critical periods, i.e., times in development when experience can influence neural development. Moreover, even within a modality, different neuronal response properties and different visual capabilities can display different critical periods (Mitchell 1978; Harwerth et al. 1986). This is likely to be in part a consequence of dependence on distinct neural systems with different developmental time courses.

Several studies of humans have attempted to verify experimentally the considerable anecdotal evidence that suggests deprivation of input in one sensory modality leads to compensatory increases in the functioning of remaining modalities. The reports present a confusing array of results, due in part to variability in the age of onset and completeness of sensory deprivation of the populations studied. It is very likely that in humans, as in animals, there are different time periods in development when different types of experience exert their major effect, but very little evidence exists on this matter. In studies of adults who have been profoundly deaf since birth due to a genetic etiology, we have reported marked changes in the organization of the visual system (Neville et al. 1983, Neville and Lawson 1987a,b,c; Neville 1990). However, these effects are not uniform on different attentional and behavioral aspects of visual processing. The neurophysiological and behavioral responses to peripheral visual stimuli are significantly enhanced in deaf subjects compared to normal hearing individuals. By contrast, responses to foveal visual stimuli do not differ in the two groups. Recent and ongoing studies demonstrate that the visual periphery is largely represented along the dorsal processing pathway, while central space is represented along the ventral visual pathway (Baizer et al. 1991). Our results imply different developmental vulnerabilities for these putative subsystems within vision.

This same developmental approach can be powerful in identifying subsystems within language. An enduring proposal in theoretical linguistics and in psycholinguistic research has been that language processing is decomposable into separate subsystems, including semantic processing (e.g., as occurs in processing the meaning of "open class" words such as nouns and verbs that make reference to specific objects and events) and syntactic processing (e.g., as occurs in processing the structural or relational information provided in English primarily by "closed class" words such as articles and auxiliaries; Chomsky 1965; Garrett 1980). Empirical studies have provided support for this idea and have led many investigators to propose that open and closed class words may be represented in distinct "lexicons" that may not be subject to the same processing constraints. On the other hand, other investigators argue against the idea that separate systems process meaning and syntax. Instead, it is proposed that the empirical findings distinguishing these word classes are due to differences between them in characteristics such as frequency, length, and imageability.

One approach we have taken to this controversy has been to compare the timing and the distribution of ERPs elicited by open and closed class words in sentences and to assess whether any observed differences can be accounted for in terms of general processing differences, i.e., of frequency, length, and imageability, rather than to the different functions these words perform in language. In addition, we compared results from normal hearing subjects to those from a group of deaf subjects whose early experience with English was abnormal, i.e., American Sign Language was their first language, and English was the second language and was learned late. This comparison permitted us to assess the hypothesis that the systems that mediate semantic and syntactic processing are differentially vulnerable to altered early language experience (for a complete description of these studies, see Neville et al. 1992).

In normal hearing adults, the ERP response to meaning-bearing, open-class words was characterized by a prominent negative component maximal around 350 ms after word onset. This component was large over posterior brain regions of both hemispheres. In previous studies we and others have linked this component to the process of "look-up" or search through the "lexicon" (Holcomb 1988; Holcomb and Neville 1990; Rugg et al. 1988, Rugg 1990). In contrast, ERPs to closed-class words displayed a prominent negative potential at 280 ms that was localized to anterior and temporal regions of the left hemisphere. An assessment of the effects of frequency and length revealed that these variables did not account for the different results for the two word classes. Current source density (CSD) analyses of the N280 peak to closed-class words indicate it is strongly localized to anterior and temporal regions of the left hemisphere and is smaller or absent to open-class words, even to very short, highly frequent open-class words. Thus, these results are consistent with the idea that nonidentical neural mechanisms mediate search and/or retrieval of the open and closed-class vocabularies that are important in processing semantic and syntactic aspects of language, respectively.

Further support for this hypothesis came from the results of congenitally deaf individuals. These subjects displayed ERP responses to the open-class words that were virtually identical to those observed in the normal hearing subjects, suggesting that aspects of semantic processing were not affected by delayed exposure to English. Further evidence on this point came from a comparison of the ERP responses to semantic anomalies in sentences in deaf and hearing subjects. The responses of the two groups were highly similar. In contrast to these results, however, responses from the deaf subjects to the closed-class words were markedly different from those of the hearing subjects. The deaf subjects' ERPs to closed-class words lacked the negative potential (N280) over anterior regions of the left hemisphere and also any other evidence of asymmetry between the hemispheres. Figure 1 shows the CSD analyses of the N280 peak to closed-class words in hearing subjects (top row) and deaf subjects (bottom row). The activity patterns within the right hemisphere are similar in both groups. By contrast, over the left hemisphere, only the hearing subjects display the sharply focused N280 current sink over anterior temporal regions. These results suggest that the neural systems that mediate syntactic processing are more vulnerable to alterations in the nature and timing of early language experience than are those that mediate semantic processing. More generally, these results are in accord with the idea that different brain systems with different developmental time courses and associated vulnerabilities mediate semantic and syntactic aspects of language processing and that it is biologically meaningful to consider these as separate subsystems within language.

Further studies of normally developing children are in agreement with the hypothesis that the asymmetrical systems, important in mediating grammatical processing, develop along a more protracted time course than do those linked with semantic processing (Neville 1994). Additionally, studies of language-impaired children suggest the systems linked to grammatical processing are more vulnerable in developmental disorders (Neville et al. 1993).

We sought further evidence for this hypothesis by employing a different paradigm designed to manipulate semantic and grammatical processing in studies of bilingual subjects whose exposure to English had been delayed until different and specifiable times in development.

9.2.2 Effects of Delayed Acquisition on Language Relevant Subsystems

Results from a variety of behavioral studies have consistently indicated that for primary and secondary language learning, age of exposure is the best predictive variable for ultimate linguistic performance (Johnson and Newport 1989; Mayberry and Eichen 1991; Newport 1988; Oyama 1982). The consistent findings of a relationship between language learning capabilities and age of acquisition are compatible with a critical period hypothesis for language learning and suggest that, similar to other neurophysiological pro-

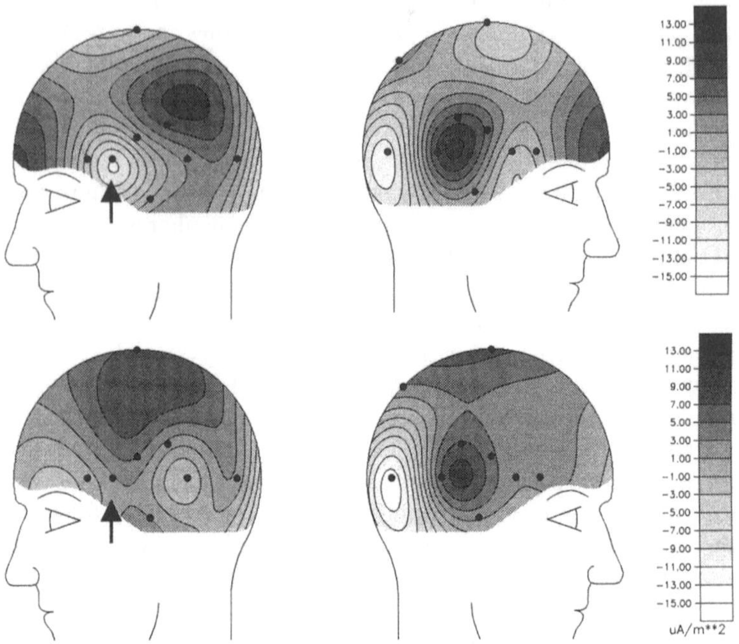

Fig. 9.2–1. Scalp topography of current source densities ($\mu A/m^2$) calculated for the N280 (most negative peak in the window 235–400 ms) in response to closed-class words. Maps of *top row* are averages across 17 hearing subjects; *below* are maps from ten congenitally deaf subjects. The longer wave lengths represent current sources (current flowing out of the head), and the shorter wave lengths represent current sinks. Each map utilizes the same scale. The maps of the hearing and deaf subjects were similar over the right hemisphere. Over the left hemisphere, the maps of hearing subjects display a prominent sink over the anterior temporal region that is absent in the maps from the deaf subjects (marked by the *arrows*). (From Neville et al. 1992)

cesses (e.g., Freeman and Thibos 1973; Knudsen and Knudsen 1990; Wiesel and Hubel 1963), language learning may be constrained by postnatal maturational mechanisms. Further, as is the case for neurophysiological processes within vision, where different functions such as stereopsis, monocular spatial resolution, and spectral sensitivity display unique vulnerabilities to altered timing of input (Harwerth et al. 1986), the acquisition of different linguistic processes also appears to be differentially affected by delays in exposure (Johnson and Newport 1989; Mayberry and Eichen 1991; Newport 1988). Aspects of grammatical processing have been found to be most profoundly impacted by alterations in timing of input, while acquisition of vocabulary appears to be quite resistant. Moreover, different grammatical skills have been shown to be affected to varying degrees by delays in exposure (Johnson and Newport 1989; Newport 1988). The ERP findings described above (Neville et al. 1992) provide evidence that the relevant functional cerebral subsystems associated with semantic and grammatical processing are also differentially affected by altered timing of language experience. Given that language proficiency is impacted differentially, even within the grammatical/syntactic domains, it could be hypothesized that specialized subsystems associated with processing specific types of linguistic constructs display nonuniform maturational constraints. The following sections describe the findings of two relevant studies examining the nature of cerebral organization for specialized linguistic processing and the effects of altered timing of language experience on these cerebral organizations.

The cerebral organizations for different types of syntactic processing, as well as semantic processing, were investigated in an ERP study of a large group of monolinguals who made grammaticality judgements on visually presented sentences (Neville et al. 1991). Half of the stimuli contained sentence-medial words that violated semantic expectations (e.g., "The scientist criticized Max's event of the theorem") or one of three types of syntactic rules: (a) phrase structure (e.g., "The scientist criticized Max's of proof the theorem"), (b) specificity constraint (e.g., "What did the scientist criticize Max's proof of?"), or (c) subjacency constraint (e.g., "What was a proof of criticized by the scientist?"). The remaining sentences were semantically and syntactically appropriate controls for these sentences. The ERPs to comparison points in the violation and control sentences were at the point of linguistic deviation in the anomalous sentences (underlined in example sentences above).

As in previous studies, the ERP responses to the words which violated semantic expectations were characterized by increased negativity between 300 and 500 ms post-stimulus onset with a posterior distribution over both hemispheres (Fig. 2, top left). In contrast, violation of phrase structure resulted in ERP changes lateralized to the left hemisphere. The response was characterized by increased negativity between 300–500 ms which was maximal over left temporal and parietal sites (Fig. 2, top right). Violations of specificity constraint rules also elicited changes in ERP lateralized over the left hemisphere; however, the response pattern was characterized by an early increased and sustained negativity over left frontal and anterior temporal regions (Fig. 2, bottom left). The ERP responses associated with violations of subjacency constraints were characterized by increased positivity at approximately 250 ms and again later between 500 and 700 ms (Fig. 2, bottom right).

In summary, ERPs elicited by semantic violations were found to be distinct in timing and distribution from ERPs elicited by syntactic violations. Moreover, different aspects of syntactic processing were associated with unique brain subsystems.

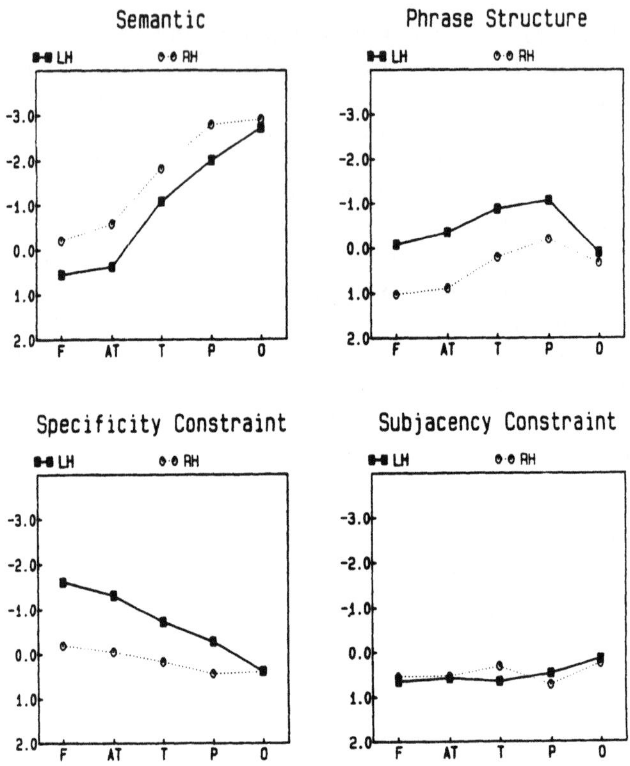

Fig. 9.2–2. Effects on mean area (μV) from 300 to 500 ms (n=40 subjects) in the difference in ERP formed by subtracting ERP elicited by comparison words in the control sentences from those at the anomaly points in sentences that violated semantics, phrase structure rules, specificity constraints, and subjacency constraints. Values from frontal (*F*), anterior temporal (*AT*), temporal (*T*), parietal (*P*) and occipital (*O*) sites over the left (*LH, solid lines*) and right (*RH, dotted lines*) hemispheres. (Reprinted from Neville et al. 1991; copyright by the Massachusetts Institute of Technology)

One model for investigating the effects of altered timing on language processing subsystems is to examine the brain organization of bilingual speakers who were exposed to their second language at different and specifiable ages. Our goal in a recent study was to investigate, using a bilingual model, the hypothesis that distinct functional specializations of language subsystems within and between the hemispheres may be affected by maturational constraints and that these effects are not uniform for different subsystems (Weber-Fox and Neville, 1992 and in preparation). The language stimuli containing semantic and syntactic violations were identical to those described above (Neville et al. 1991). The subjects were adult

Chinese–English bilinguals who were grouped according to their age of exposure to English: 1–3, 4–6, 7–10, 11–13, or greater than 16 years of age. The bilingual subjects had been immersed in English for at least 5 years at the time of testing.

Consistent with previous behavioral studies, the several behavioral measures we utilized, including self-rated proficiency, standardized tests of knowledge of English grammar, and grammaticality judgements of the experimental sentences, indicated that age of acquisition is a crucial variable for ultimate proficiency attained in a second language. Further, the effects of maturational constraints were not uniform on the accuracy for grammaticality judgements of the different types of linguistic constructs presented in the experimental sentence stimuli. Consistent with previous behavioral research, language proficiency for syntactic processing was more vulnerable to delays in language exposure compared to semantic processing. Much shorter delays affected proficiency for syntactic accuracy (< 7 years), and only delays greater than 16 years affected semantic proficiency. The findings suggest that earlier exposure to a language may be necessary for individuals to achieve native-like proficiency in grammar as compared to semantic knowledge. Further, linguistic proficiency for different types of syntactic processing were uniquely affected by delays in exposure. For example, shorter delays impacted on accuracy in judgement of subjacency constraint rules compared to phrase structure rules.

The associated ERPs for semantic and syntactic processing also showed differential vulnerabilities to alterations in the timing of language experience. The amplitude and distribution of the N400 response to semantic anomalies were not affected by delays in the age of exposure. The latency of the N400, however, was longer (approximately 20 ms) for delays in exposure greater than 11 years, suggesting a slight slowing in processing. This finding in bilingual speakers and the evidence described above for deaf individuals processing open-class words indicate that the brain systems that mediate semantic processing of this type are very robust.

In contrast, ERP responses to each of the syntactic violations resulted in changes in amplitude and distribution with increases in age of exposure. ERP responses to violations of phrase structure differed from monolinguals' responses with delays of only 4–6 years. The ERP responses in these bilinguals did not display the same degree of asymmetry evident in monolinguals and bilinguals exposed prior to 4 years of age. With longer delays (11–13 years), broader, bilateral distributions in ERP responses were observed. For delays greater than 16 years, the increased negativity associated with processing phrase structure violations was symmetrical or larger over the right hemisphere. Figure 3 shows the progression of a decrease in left lateralization as age of exposure increases.

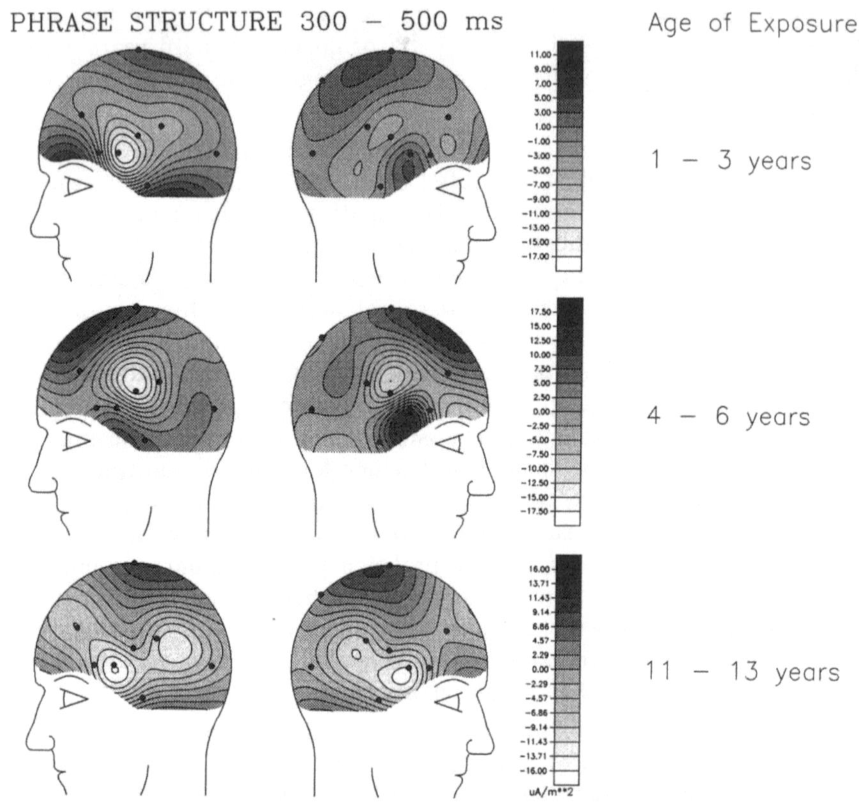

Fig. 9.2–3. Scalp topography of current source densities ($\mu A/m^2$) calculated for the mean amplitude from 300–500 ms post stimulus onset in the difference event-related brain potentials (ERP) formed by subtracting ERP elicited by the comparison words in the phrase structure control sentences from those at the deviance points in the sentences which violated phrase structure rules. Maps are averages from three of the groups of bilinguals. The groups shown were exposed to English between 1–3 years (*top row*, n=15), 4–6 years (middle row, n=13), and 11–13 years of age (bottom row, n=13). The current sink elicited by phrase structure violation was strongly lateralized in bilinguals exposed to English prior to 4 years of age, was less lateralized in those exposed between 4-6 years, and was symmetrical in subjects exposed between 11–13 years of age

In summary, language proficiency and cerebral organization for language processing are altered by delays in exposure to language. Our results are consistent with a critical period hypothesis for language learning. Performance and cerebral specialization for syntactic processing appear more vulnerable to delays in language exposure compared to proficiency and brain organization for semantic processing. Further, the type of changes occurring in ERP with delays in exposure were qualitatively different for semantic and syntactic processing. For semantic processing, changes were primarily a slight shift in latency of the response. For syntactic processing, ERP differences were seen in the amplitudes and distributions of responses. Different types of syntactic processing were associated with unique age of exposure effects, suggesting that critical period effects are not uniform across various linguistic constructs even within grammatical aspects of language. Our findings are consistent with the view that maturational changes significantly determine the capabilities of humans to acquire language and constrain the development of language-specific cerebral subsystems.

9.2.3 Automatic and Conscious Processing of Language

In discussions of the organization of language, considerable debate centers on the claim that there exist systems specialized for linguistic analysis that are activated automatically, without conscious attention, and that these are autonomous from nonlinguistic information and support rapid rates of language processing. Such domain-specific, "modular" systems are contrasted with more general systems in which words are represented together with general world knowledge. These less specialized systems are activated in conjunction with conscious, strategic, controlled processing. In this study, we investigated the hypothesis that specialized, automatically activated representations of words exist and can be distinguished from more general representations of words activated in conjunction with conscious awareness.

Adults rapidly decided whether a string of (uppercase) letters ("target") was or was not a word. On one half of the trials, the targets were preceded by identity (lowercase) "primes", e.g., sun – SUN, gif – GIF, or by semantically related primes, e.g., cat – DOG. ERP and behavior were examined for evidence that distinct patterns of priming occurred when subjects were aware of the primes, as contrasted to when the duration of the prime was very short and a preceding pattern mask rendered subjects unaware of the primes. The sequence of stimuli and their durations are shown in Fig. 4. Three groups of young adults participated in the three experiments in which the stimulus sequence and duration of the primes varied. In experiment 1, subjects were aware of both the prime and target words. In experiment 2, subjects reported seeing only the target word. In experiment 3, subjects could report both prime and target words.

The results for the (control) unmasked presentation displayed typical behavioral and ERP effects for semantic priming and for word and nonword identity priming. Subjects made lexical decisions more quickly when targets were preceded by identical or semantically related items. The ERP effect consisted of a negative peak maximal at 400 ms (N400) that onset at 300 ms, was 300 ms in duration, and displayed a centroposterior distribution (Fig. 5a, left). The difference waves, formed by subtracting ERPs to primed

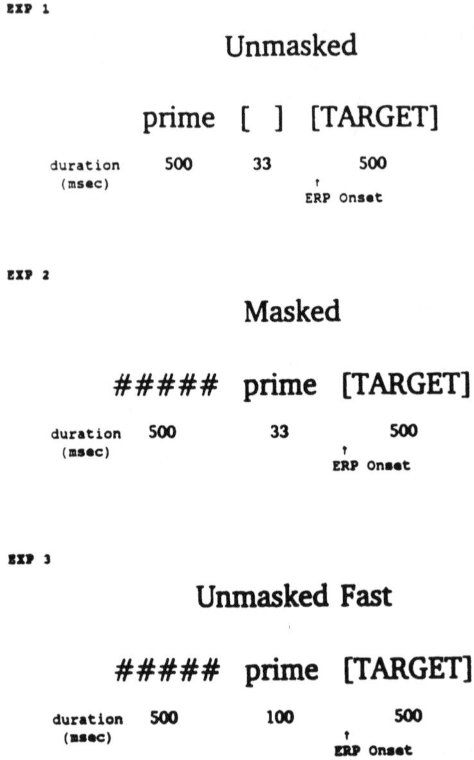

Fig. 9.2–4. Stimulus durations and sequences in the three experiments. In experiments 1 and 3, subjects were aware of (i.e., able to report) both the prime and the target. In experiment 2, subjects only reported being aware of the target

targets from ERPs to unprimed targets, isolate the effect of semantic priming (Fig. 5b, left). Both words and nonwords display a similar N400 effect of repetition priming. Remarkably, under masked conditions (i.e., when subjects were unaware of the primes), there was also a consistent ERP effect of semantic priming (Fig. 5a, right). However, this effect onset at 200 msec, was approximately 120 ms in duration, and displayed a centroanterior distribution. This effect is clearly seen in the difference waves (Ne; Fig. 5b, right). The different timing and distribution of the masked and unmasked effects suggest that nonidentical systems are activated under the two conditions. Moreover, whereas nonwords displayed clear evidence of repetition priming in the unmasked experiment, little if any priming of nonwords was apparent under the masking paradigm. This result is compatible with the hypothesis that the priming activity seen under masked conditions indexes activity within a system that is more specialized (i.e., does not include

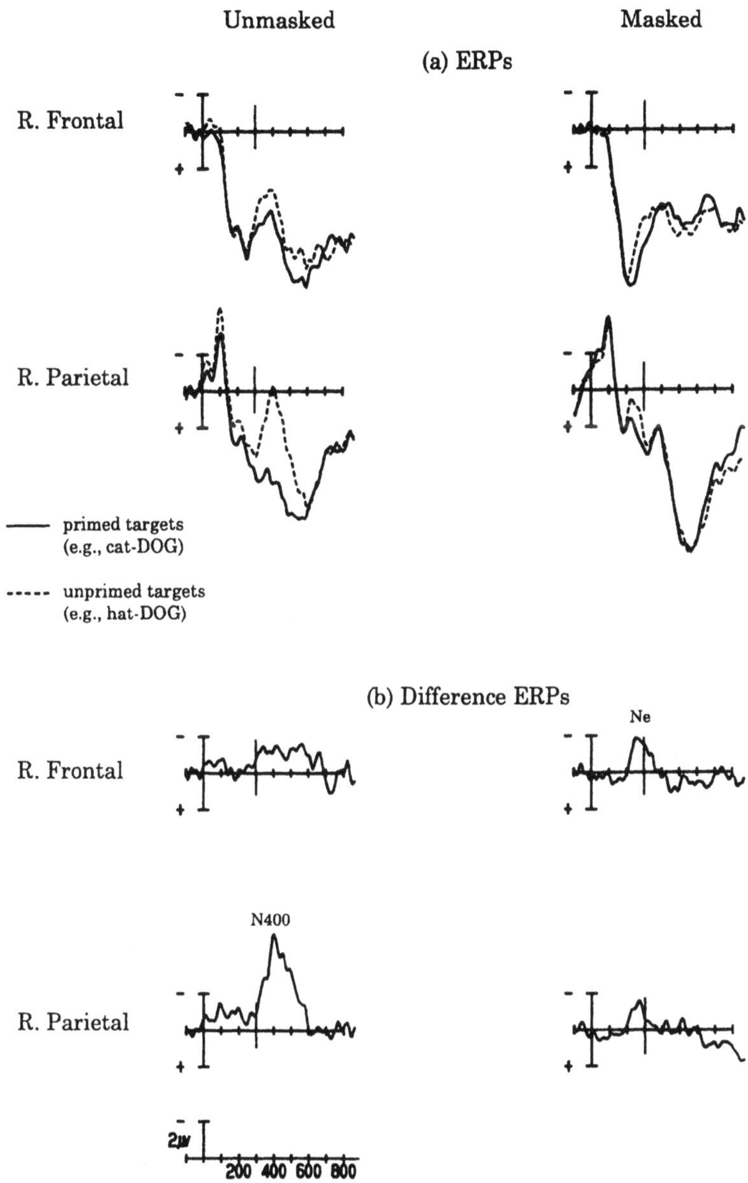

Fig. 9.2–5. Effects of semantic priming when subjects were able (unmasked; *left*) and were not able (masked; *right*) to report the prime. a Event-related brain potentials (ERP) elicited by primed and unprimed targets. b Difference ERPs formed by subtracting ERPs to primed targets from those to unprimed targets

nonwords) than that observed under unmasked conditions. These results raise the hypothesis that the later, posterior N400 priming effect indexes conscious, controlled access to or decisions about general representations of words. In contrast, the earlier, anterior Ne priming effect may index access to more specialized representations of words in a system that does not include nonwords. The earlier onset and shorter duration of Ne are compatible with the operation of an automatically activated system. These factors together with the different scalp distributions of N400 and Ne suggest they may be indexing distinct processes within the language comprehension system.

The third experiment was designed to determine whether we could elicit both Ne and N400 in the same experiment. We hypothesized that the early, automatic system (indexed by Ne) may require high rates of stimulus presentation to be activated and that it may be highly refractory (i.e., very short duration priming effects). One or both of these factors may have resulted in the absence of Ne in the first, unmasked experiment. In experiment 3, we presented the prime temporally adjacent to the target and briefly, thus achieving a higher rate of word presentation, but subjects were still able to report the prime and so presumably the controlled representation was also activated. Under these conditions, we observed both an early, anterior (Ne) priming effect and a later, posterior (N400) effect. These data provide support for the hypothesis that words are represented both within automatically activated systems that can be dissociated from conscious awareness and within less specialized systems that are activated under controlled, consciously mediated processing. More generally, the results suggest that this type of paradigm can be useful in suppressing the operation of consciously controlled, strategic processes that dominate most experimental paradigms and that it can help to reveal the operation of automatically activated systems. Further studies will elucidate the organization of these different systems, the extent to which they communicate with each other, and the conditions necessary for their development.

In summary, results from these different studies suggest that the combined ERP/behavioral approach employed in the study of populations that vary in specific aspects of early experience can contribute to the identification of different functional systems within a cognitive domain and can help to determine the nature and the timing of the effects of environmental input on their differentiation.

Acknowledgments. This research was supported by grants DC00128 and DC00054. We thank M. Mitchell for manuscript preparation and S. Coffey-Corina for help with the figures.

References

Baizer JS, Ungerleider LG, Desimone R (1991) Organization of visual inputs to the inferior temporal and posterior parietal cortex in macaques. J Neurosci 11: 168–190

Chomsky N (1965) Aspects of the theory of syntax. MIT Press, Cambridge

Creutzfeldt OD (1979) Neurophysiological mechanisms and consciousness. Ciba Found Symp 69: 217–233

Creutzfeldt OD Kuhnt U (1967) The visual evoked potential: physiological, developmental and clinical aspects. Electroenceph Clin Neurophysiol Suppl 26: 29–41

Creutzfeldt OD, Arnold PM, Becker D, Langenstein S, Trisch W, Wilhelm H, Wuttke W (1976) EEG changes during spontaneous and controlled menstrual cycles and their correlation with psychological performance. Electroenceph Clin Neurophysiol 40: 113–131

Creutzfeldt OD, Bodenstein G, Barlow JS (1985) Computerized EEG pattern classification by adaptive segmentation and probability density function classification. Clinical evaluation. Electroenceph Clin Neurophysiol 60: 373–393

Creutzfeldt O, Ojemann G, Lettich E (1989a) Neuronal activity in the human lateral temporal lobe. I. Responses to speech. Exp Brain Res 77: 451–475

Creutzfeldt O, Ojemann G, Lettich E (1989b) Neuronal activity in the human lateral temporal lobe. II. Responses to the subjects own voice. Exp Brain Res 77: 476–489

Freeman RD, Thibos LN (1973) Electrophysiological evidence that abnormal early visual experience can modify the human brain. Science 180: 876–878

Garraghty PE, Kaas JH (1992) Dynamic features of sensory and motor maps. Curr Opin Neurobiol 2: 522–527

Garrett M (1980) Levels of processing in sentence production. In: Butterworth B (ed) Language production. 1. Speech and talk. Academic, New York, pp 177–220

Harwerth RS, Smith III EL, Duncan GC, Crawford MLJ, von Noorden GK (1986) Multiple sensitive periods in the development of the primate visual system. Science 232: 235-238

Holcomb PJ (1988) Automatic and attentional processing: an event-related brain potential analysis of semantic priming. Brain Language 35: 66–85

Holcomb PJ, Neville HJ (1990) Visual semantic priming in lexical decision: a comparison using event-related brain potentials. Language Cognitive Processes 5: 281–312

Johnson J, Newport E (1989) Critical period effects in second language learning: The influence of maturational state on the acquisition of English as a second language. Cognitive Psychology 21: 60–99

Knudsen EI, Knudsen PF (1990) Sensitive and critical periods for visual calibration of sound localization by barn owls. J Neurosci 10: 222–232

Mayberry R, Eichen E (1991) The long-lasting advantage of learning sign language in childhood: another look at the critical period for language acquisition. J Memory Language 30: 486–512

Mitchell DE (1978) Effect of early visual experience on the development of certain perceptual abilities in animals and man. In: Walk RD, Pick HL (eds) Perception and experience. Plenum, New York, pp 37–75

Neville HJ (1990) Intermodal competition and compensation in development: evidence from studies of the visual system in congenitally deaf adults. In: Diamond A (ed) The Development and Neural Bases of Higher Cognitive Function. NY Acad Sci, New York, pp 71–91

Neville HJ (1992) Cross Modal Plasticity. Neuroscience Facts 3 (5): 2

Neville HJ (1994) Developmental specificity in neurocognitive development in humans. In: Gazzaniga M (ed) The cognitive neurosciences. MIT Press, New York (in press)

Neville HJ, Lawson D (1987a) Attention to central and peripheral visual space in a movement detection task: an event-related potential and behavioral study. I. Normal hearing adults. Brain Res 405: 253–267

Neville HJ, Lawson D (1987b) Attention to central and peripheral visual space in a movement detection task: an event-related potential and behavioral study. II. Congenitally deaf adults. Brain Res 405: 268–283

Neville HJ, Lawson D (1987c) Attention to central and peripheral visual space in a movement detection task. III. Separate effects of auditory deprivation and acquisition of a visual language. Brain Res 405: 284–294

Neville HJ, Schmidt A, Kutas M (1983) Altered visual evoked potentials in congenitally deaf adults. Brain Res 266: 127–132

Neville HJ, Nicol J, Barss A, Forster K, Garrett M (1991) Syntactically based sentence processing classes: evidence from event-related brain potentials. J Cognit Neurosci 3: 155–170

Neville HJ, Mills DL, Lawson DS (1992) Fractionating language: different neural subsystems with different sensitive periods. Cerebral Cortex 2: 244–258
Neville HJ, Coffey SA, Holcomb PJ, Tallal P (1993) The neurobiology of sensory and language processing in language impaired children. J Cogn Neurosci 5: 235–253
Newport E (1988) Constraints on learning and their role in language acquisition: studies of the acquisition of American Sign Language. Lang Sci 10: 147–172
Oyama S (1982) The sensitive period and comprehension of speech. In: Krashen SD, Scarcella RC, Long MH (eds) Child-adult differences in second language acquisition. Newbury House, Rowley, pp 39–51
Rugg MD (1990) Event-related brain potentials dissociate repetition effects of high- and low-frequency words. Memory Cognition 18: 367–379
Rugg M, Furda J, Lorist M (1988) The effects of task on the modulation of event-related potentials by word repetition. Psychophysiology 25: 55–63
Sur M, Pallas S, Roe A (1990) Cross-modal plasticity in cortical development: differentiation and specification of sensory neocortex. Trends Neurosci 13: 227–233
Weber-Fox CM, Neville HJ (1992) Maturational constraints on cerebral specializations for language processing: ERP and behavioral evidence in bilingual speakers. Soc Neurosci Abstr 18: 335
Wiesel T, Hubel D (1963) Effects of visual deprivation on morphology and physiology of cells in the cat's lateral geniculate body. J Neurophysiol 26: 978–993

9.3 Human Left Temporal Neural Mechanisms for Recent Verbal Memory

G.A. Ojemann
Department of Neurological Surgery, University of Washington, Seattle, Washington

Prof. Otto Creutzfeldt spent nine months of the 1984–1985 year on sabbatical at the University of Washington in Seattle. A primary goal of that sabbatical was microelectrode recording from humans during awake craniotomies. Prof. Arthur Ward Jr. had initiated human microelectrode recording in that context at the University of Washington in the late 1950s (Ward and Schmidt 1961). Prof. Ward's primary interest was in identifying changes in neuronal activity associated with epilepsy, especially patterns of burst firing (Ward 1969; Calvin et al. 1973; Wyler et al. 1982). There had been little investigation of changes in activity of neurons with normal nonburst firing pattern or of any changes during measures of human higher functions. That was what Prof. Creutzfeldt and I planned to investigate during his sabbatical, particularly changes in activity of seemingly normal neurons during measures of language and recent memory. We were not sure that we would ever record a neuron that changed activity with language or memory, so Prof. Creutzfeldt also planned a second project, an extension of his interest in visual responses of neurons in monkey temporal cortex. He never really had time to devote to this second project during the sabbatical. Instead, we found that many of the neurons that we recorded in human lateral temporal cortex changed activity during various measures of language, memory, and other higher functions, often in unexpected ways. Those investigations, initiated during that sabbatical, resulted in a series of publications (Creutzfeldt et al. 1989a,b; Creutzfeldt and Ojemann 1989, 1990; Ojemann et al. 1988; Ojemann and Creutzfeldt 1987).

This paper summarizes the present findings of one aspect of the investigations begun during Prof. Creutzfeldt's sabbatical, the changes in human left dominant hemisphere temporal cortical neuronal activity during recent verbal memory. Although memory mechanisms in temporal lobe are usually related to function of mesial structures, especially hippocampus, there is a body of evidence indicating that lateral cortex is also important to this function. In an early study of the effects of temporal resections on recent verbal memory, Milner noted that extending the resection laterally had almost as much deleterious effect on memory as extending the resection medially (Milner 1967). Subsequently I reported that lateral temporal cortical stimulation altered performance on a recent verbal memory measure (Ojemann 1978), an effect that was independent of effects on language. In addition, in studies with Dodrill, I found that recent verbal memory deficits after left temporal resections were in part related to the lateral extent of the resection, were more likely when the resection included lateral cortical sites related by stimulation to memory, and that in patients at risk for a memory deficit after a temporal resection based on performance on the intracarotid amobarbital perfusion test, the memory deficit could be avoided by preserving lateral cortical memory sites (Ojemann and Dodrill 1985, 1987). In monkey, cooling of inferior temporal cortex interferes with recent memory performance (Fuster et al. 1981), and changes in neuronal activity have been recorded there during memory measures (Fuster and Jervey 1982).

Two separate studies of changes in the frequency of temporal cortical neuronal activity during recent verbal memory are summarized and integrated here. The first study was begun during Prof. Creutzfeldt's sabbatical (Ojemann et al. 1988); these investigations have been continued at the University of Washington, and the second study is one of the results of that (Haglund et al. 1993a). Many features of the two studies were similar. They were undertaken during left temporal craniotomies under local anesthesia for medically intractible epilepsy in patients known to be left brain dominant for language based on a preoperative intracarotid amobarbital perfusion test, although each study was derived from a different series of patients. The University of Washington Institutional Review Board approved the studies and patients gave specific, informed consent for them.

Microelectrode recording used tungsten electrodes, with tip diameters of 1–5 μm and impedences of 3–7 mΩ advanced by a Trent-Wells micromanipulator through a plexiglas footplate, to obtain extracellular recordings of neuronal activity free of injury discharges and largely free of burst activity. Recordings were made from the least epileptogenic portion of cortex that was subsequently resected, in lateral cortex of middle or superior temporal gyrus. Memory measures were presented during these recordings as series of slides. The recent verbal memory measures used in both studies were based on the Peterson and Peterson (1959) input-distractor-retrieval paradigm and used an object name as the material to be remembered and retrieval of that name at a cue, following storage in memory for the 8 s (study 1) or 12 s (study 2) occupied by distracting linguistic tasks. However, different distracting tasks were used in the two studies. Control measures were established in the same way in the two studies, by presenting the same slides that constituted the input and part of the distracting phases of the memory measure in two other sets of tasks: language measures, where the linguistic information did not have to be retained in verbal memory and spatial tasks involving matching a feature on those slides. The two studies used different techniques to separate records into populations reflecting activity of one or a few neurons and different techniques of analysis and statistics.

In the first study, recordings were obtained from 13 patients. The distracting task involved word and sentence reading. Activity was separated into 17 populations based on action potential amplitude as determined by manual counting from Grass C4K camera films made from the frequency modulated (FM) tapes used to archive the operating room recordings. A relation to memory input or storage phases was established when activity then differed significantly from both the same visual stimuli used in the analogous language task and spatial task. A relation to memory retrieval was established if activity then differed significantly from overt naming when the same word was said aloud but in response to the object picture rather than retrieved from memory. A relation to language was established when activity during both the memory measure and the language task differed significantly from that during the spatial task. In the second study, recordings were obtained from 11 other patients. The distracting linguistic task was naming of object pictures other than the one to be retained in memory. Activity was separated into 20 neuronal populations by digitizing the FM tape record at 5 kHz on a MacIIx computer and then separating activity with multiple window discriminators. A population was related to memory based on significant differences in two comparisons: (1) between the same slides used in three conditions — as the input to memory, the cue for the language task, naming, or as the cue for a spatial matching task — and (2) between the input, distracting, and retrieval phases of the memory measure, when the same language dimension, naming, was used as the input to memory or the distractor during which the memory had to be stored and an object name as the word to be produced at memory retrieval.

Despite these technical differences between the two studies, the major results are quite similar:

1. A large proportion of the neuronal populations sampled in either study changed activity with the memory measure. Comparison 1 of the second study and the method used in the first study to establish a relation to memory are quite similar, and in those analyses, 71% of the neuronal populations of the first study and 65% of the second had statistically significant changes during memory. By contrast, a much smaller proportion of the neuronal populations had statistically significant changes during naming: 12% in the first study and 20% in the second.

2. Most of the changes related to memory represent increased activity at memory input or retrieval. Both studies were designed to detect either increases or decreases in activity during a particular behavior. Combining both studies, 72% of the changes related to memory were increases in activity. Significant changes in the much fewer populations related to naming were about equally divided between increased or decreased activity.

   In each study, changes occurred predominately during the input and retrieval phases of the memory measure. No neuronal population was identified in either study that increased activity only during storage. In the first study, 50% of the 12 populations related to memory demonstrated changes with input, 67% with retrieval, and 25% with storage, in all cases an extension of the change in activity during input into the next phase. In the second study, statistically significant numbers of populations increased activity only during memory input and retrieval, with 75% of the 16 populations related to memory showing changes with input and 50% with retrieval. Most populations changed activity with only input or retrieval and not both. Changes with both (in the same direction) were identified in only 8% of the populations related to memory in study 1 and 31% in study 2.

   The duration of the changes during memory input or retrieval differed some between the two studies. In study 1, the increased activity at memory input or retrieval lasted longer than 2.6 s in half the patients. In study 2, significant effects occurred in only the first 2 s of memory input or the first second of retrieval. Study 2 also examined the effect of subsequent retrieval of the same item after additional distractions on the increased activity at initial retrieval. Activity began to diminish by the second retrieval and was further diminished at the third retrieval of the same item. There was no evidence of any overall decline in activity or habituation of activity to items that only required naming, rather than retrieval from memory.

3. Neuronal activity related to memory usually differs from that related to naming. Electrical stimulation mapping studies had often shown separation of the sites where memory was altered by stimulation from those where changes in naming were evoked (Ojemann 1978, 1983; Ojemann and Dodrill 1985). That separation seemed less evident at a neuronal level in our previously published analysis of study 1 (Ojemann et al. 1988). However, in reviewing the results of that study and adding the results of study 2, it is apparent that if neuronal populations change activity with both naming or memory, it is usually in opposite directions. In study 1, although both populations with changes during naming also showed changes with memory, those changes with memory input or storage were always opposite to those for naming. In study 2, three of the five populations with changes in naming also showed changes with memory input, but again always in the opposite direction. Changes at the time of retrieval might be expected to be more often similar to those during naming, reflecting activity common to overt speech. However, in the combined studies, only three of 18 populations with any significant changes at memory retrieval showed changes during naming that were at all similar.

Between the two studies, eight sets of activity of nearby neurons recorded through the same microelectrode are available. In two sets, three different neuronal populations were identified, in the others, only two. In three of these eight sets, changes in activity of these nearby neurons were either to no measured behaviors (one set) or different behaviors (two sets, both with three separate neuronal populations and both with one population related to memory, one to naming, and one to no measured behavior). The remaining five sets showed changes related to memory in both populations, but in all cases the changes were slightly different, usually (four sets) one population changing activity with input and retrieval phases of the memory measure and the other only input or retrieval.

These studies indicate that when a new item of information enters recent verbal memory, there is an increase in activity of many temporal neocortical neurons. That increase lasts a few seconds, somewhat longer than the time required to execute any linguistic aspects of the information. This increase is not due to novelty of the information per se, for it is not seen when naming novel pictures. Rather it is related to retention of this new item in recent verbal memory. Most of the neurons showing this effect respond differently during the linguistic task of naming. During the time the information is stored in recent verbal memory while other linguistic functions are occurring, lateral temporal cortical neurons show few changes in activity. However, when that item of information is first retrieved from memory, there is another period of increased activity. This increase at first retrieval seems to be slightly shorter than the increase at memory entry. It appears in some of the same neurons that were active with memory entry, but also in neurons that did not change activity then. Often these are nearby neurons. However, these, too, are usually neurons showing a different change in activity with the linguistic task. With each subsequent retrieval, this increase becomes less evident, but activity again increases when another new item of information enters memory (for only that pattern would achieve statistical significance for changes in activity with different memory phases in the two studies reviewed here).

These findings are similar to those reported for recordings of neuronal activity in monkey inferior temporal gyrus during delayed match to sample paradigms when there is a distracting task (Fuster 1990), but differ from the patterns of activity recorded in the absence of a distracting task (Fuster and Jervey 1982). In that setting, sustained activity during the time the memory is stored has been identified. In our human recordings, the neurons showing increased activity with memory often show decreased activity with naming when it occurs in the absence of any active storage of information in recent memory. Perhaps the increased activity at memory entry would be sustained if those neurons were not inhibited as part of the changes associated with the linguistic task used as a distractor during memory storage. If recent memory storage involves sustained increased activity, as has been suggested in invertebrate studies of conditioning (Hawkins et al. 1983), it must depend on cells other than those sampled in our studies.

The increased activity in human temporal cortical neurons with memory entry and initial retrieval is perhaps best viewed as a promoter for memory retention. It is apparently essential, however, since altering temporal cortical function during memory input, by electrical stimulation, interferes with retention of that incoming information (Ojemann 1978, 1983). This need for increased activity in so many temporal cortical neurons may also be the "weak link" in recent memory, explaining why it is so vulnerable to head injury, dementia, aging, and surgical removals of the left temporal lobe, as it would not require destruction of very many temporal cortical neurons to reduce the available neuronal population below the 65%–70% of neurons in our relatively young patients that were active with memory entry or the 50%–60% active with initial retrieval. If substantially fewer cortical neurons were active, this aspect of memory might well fail, while retrieval of already stored information would be successful, since many fewer temporal cortical neurons need to be active with later retrievals from memory.

Intraoperative microelectrode studies also provide some insights into the organization of human association neocortex (Ojemann 1991a). As is apparent in the studies reviewed here, there is little evidence for multipurpose neurons. Many recordings obtained during multiple behavioral measures either demonstrate changes in activity to only one of the behaviors or changes in opposite directions to multiple behaviors, as demonstrated here for memory compared to naming and in other studies for naming in different languages (Ojemann 1991b), including naming in oral or sign language (Haglund et al. 1993b), naming, reading and spatial matching (Schwartz and Ojemann, unpublished data), word production or perception (Creutzfeldt et al. 1989a), and several spatial functions (Ojemann et al. 1992). Thus, individual neurons are often related to a system for one specific behavior. Nearby neurons recorded through the same microelectrode not infrequently are related to different systems, a finding somewhat less evident in the studies reviewed here than in other investigations of changes in frequency of neuronal activity of human neocortex during linguistic and spatial measures (Ojemann et al. 1992).

References

Calvin WH, Ojemann GA, Ward AA Jr (1973) Human cortical neurons in epileptogenic foci: comparison of interictal firing patterns to those of "epileptic" neurons in animals. EEG Clin Neurophysiol 34: 337–351

Creutzfeldt O, Ojemann G (1989) Neuronal activity in human lateral temporal lobe. III. Activity changes during music. Exp Brain Res 77: 490–498

Creutzfeldt O, Ojemann G (1990) Neuronal responses to the human lateral temporal lobe to speech. In: Eccles J, Creutzfeldt O (eds) Study week on the principles of design and operation of the brain. Vatican, Pontificae Academiae Scientarium, Scripta Varia 78, pp 501–518

Creutzfeldt O, Ojemann G, Lettich E (1989a) Neuronal activity in human lateral temporal lobe. I. Responses to speech. Exp Brain Res 77: 451–475

Creutzfeldt O, Ojemann G, Lettich E (1989b) Neuronal activity in human lateral temporal lobe. II. Responses to the subject's own voice. Exp Brain Res 77: 476–489

Fuster JM (1990) Neuronal discrimination and short-term memory in association cortex. In: Scheibel A, Wechsler A (eds) Neurobiology of higher cognitive function. Guilford, New York, pp 85–101

Fuster JM, Jervey JP (1982) Neuronal firing in the inferotemporal cortex of the monkey in a visual memory task. J Neurosci 2: 361–375

Fuster JM, Bauer RH, Jervey JP (1981) Effects of cooling inferotemporal cortex on performance of visual memory tasks. Exp Neurol 71: 398–409

Haglund MM, Ojemann GA, Schwartz TW, Lettich E (1993a) Neuronal activity in human lateral temporal cortex during serial retrieval from short-term memory. J Neurosci (in press)

Haglund MM, Ojemann GA, Lettich E, Bellugi U, Corina D (1993b) Dissociation of cortical and single unit activity in spoken and signed languages. Brain Lang 44: 19–27

Hawkins R, Abrams T, Carew T, Kandell E (1983) A cellular mechanism of classical conditioning in aplysia: activity dependent amplification of presynaptic facilitation. Science 219: 400–405

Milner B (1967) Basic mechanisms suggested by studies of temporal lobes. In: Millikan C, Darley F (eds) Basic mechanisms underlying speech and language. Grune and Stratton, New York, pp 122–131

Ojemann GA (1978) Organization of short-term verbal memory in language areas of human cortex: evidence from electrical stimulation. Brain Lang 5: 331–340

Ojemann GA (1983) Brain organization for language from the perspective of electrical stimulation mapping. Behav Brain Sci 6: 189–230

Ojemann G (1991a) Cortical organization of language. J Neurosci 11: 2281–2287

Ojemann G (1991b) Cortical organization of language and verbal memory based on intraoperative investigations. Prog Sens Physiol 12: 192–230

Ojemann G, Creutzfeldt O (1987) Language in humans and animals: contribution of brain stimulation and recording. In: Mountcastle V, Plum F, Geiger S (eds) The nervous system. 5. Higher functions of the brain. American Physiological Society, Bethesda, pp 675–700 (Handbook of Physiology, Sect 1)

Ojemann GA, Dodrill CB (1985) Verbal memory deficit after left temporal lobectomy for epilepsy: Mechanism and intraoperative prediction. J Neurosurg 62: 101–107

Ojemann GA, Dodrill CB (1987) Intraoperative techniques for reducing language and memory deficits with left temporal lobectomy. In: Wolf P, Dam M, Janz D, Dreifuss F (eds) Adv Epileptol 16: pp 327–330

Ojemann GA, Creutzfeldt OD, Lettich E, Haglund MM (1988) Neuronal activity in human lateral temporal cortex related to short-term verbal memory, naming and reading. Brain 111: 1383–1403

Ojemann J, Ojemann G, Lettich E (1992) Neuronal activity related to faces and matching in human right nondominant temporal cortex. Brain 115: 1–13

Peterson L, Peterson M (1959) Short-term retention of individual verbal items. J Exp Psychol 58: 193–198

Ward AA Jr (1969) The epileptic neuron: chronic foci in animals and man. In: Jasper H et al (ed) Basic mechanisms of the epilepsies. Little Brown, Boston, pp 263–288

Ward AA Jr, Schmidt RP (1961) Some properties of single epileptic neurons. Arch Neurol (Chic) 5: 308–313

Wyler AR, Ojemann GA, Ward AA Jr (1982) Neurons in human epileptic cortex: correlation between unit and EEG activity. Ann Neurol 11: 301–308

9.4 Verbal and Visual Thinking in the EEG

H. Petsche, S.C. Etlinger, P. Richter, and O. Filz
Institut für Neurophysiologie der Universität, Wien, Austria

When electric aspects of the brain are to be studied, either of two approaches may be chosen. One follows a bottom-up strategy and attempts to understand how membranes and nerve cells behave and why cells, conceived of as ensembles, show ordered behavior and exhibit such phenomena as Abeles' "synfiring chains" (1982). In other words, bottom-up strategies of brain research start with the behavior of its smallest parts to draw conclusions on the cooperativity and functioning of the nervous system as a whole. The other way starts with the macrophenomena of the system, in our case, with the electroencephalogram (EEG). Such a strategy asks whether any features accompanying information processing by the brain may be detected in the EEG; in other words; are mental processes in any way reflected by the ongoing EEG? For more than a decade we have been investigating this question by making use of the parameters power and coherence obtained by fast Fourier transform (FFT); it was particularly the parameter of coherence that proved useful for the detection of traces of thinking in the EEG. Based on today's knowledge of the anatomy of the cortex this is not surprising. According to recent neuroanatomical studies (Braitenberg and Schüz 1991) the main function of the cortex must be to ensure highest possible convergence and divergence of signals. The structural prerequisite for this function is the extremely extended horizontal fiber network that mutually connects all cortical areas. This horizontal feltwork seems to be a conditio sine qua non for higher mental processes to arise. As a consequence, brain research in this field ought ultimately to aim at elucidating the processes contingent upon this anatomical particularity. Thus, the proper functions of the cortex seem to be based on its high cooperativity. And it is just this cooperativity which may be elucidated by coherence estimates.

A second point worth mentioning in this context concerns the EEG. The results presented by Nunez (1989) suggested that the horizontal feltwork of fibers is a main condition for potential oscillations to arise in the neuronal structure of neocortex. The postulated range of their frequencies roughly corresponds to those observed in the human EEG. Therefore it seemed reasonable to study the EEG by means of the parameter of coherence in order to elucidate electric relationships between different areas and their possible changes accompanying mental work. Coherence seems an appropriate parameter to study cooperativity: it is the normalized cross-spectrum of two signals and provides information about the linear correlation between these two signals with respect to frequency. More casually expressed, coherence can be considered as a measure of the degree of similarity of two EEG signals or of the electric kinship of two brain regions with respect to frequency. It can take any value between 1 (synchrony) and 0 (no correlation between the two signals).

Coherence depends on several conditions: the number of fiber connections between the two respective recording areas, their distance, and the functional relationships between these two areas. Only this latter is of interest for our purposes since functional relationships between areas change with the respective tasks performed by the brain. Studying such changes of coherence caused by distinct mental tasks by looking for statistically significant differences with respect to the EEG at rest was at the core of our approaches. Methodological details are described by Rappelsberger and Petsche (1988). In this way we found more or less characteristic topographic patterns of significant coherence changes by subjecting groups of individuals to the same tasks, such as listening to a certain piece of music (Petsche et al. 1988), silent reading, spatial imagination (Rappelsberger et al. 1987), mental arithmetic, and visual imagery (Petsche et al. 1992). The results are presented as probability maps of coherence changes. Based on these results our method was extended to single cases. We demonstrate here in two subjects how efficacious the parameter coherence can be for the elucidation of brain processes involved in verbal and in visual thinking.

The methodological procedure is simple. EEG is recorded from 19 electrodes (the Jasper-Penfield system) with averaged ear electrodes taken as reference. EEG epochs of at least 1 min duration are recorded while certain mental tasks are be performed. Between all these tasks epochs of EEG at rest are recorded, again for at least 1 min. All EEG recordings are Fourier transformed in sections of 2 s. The spectrum is subdivided into six frequency bands between 1 and 32 Hz; after elimination of artifacts significant differences of coherence are computed for each frequency band betweeen all 19 electrodes, which yields 171 values per frequency band. Only those site-pairs with significant differences in the confidence intervals ($p < 0.05$, two-tailed) for the coherence measures for each task compared to the EEG averaged from the between-task epochs (the rest periods) are taken into account. To represent these findings graphically different kinds of representation were developed, of which only the topographical ones are shown here.

The first subject was a 26-year-old man, right-handed interpreter of English and Russian, with German as native language. He was asked to simultaneously interpret texts, presented binaurally, from English to German and from Russian to German in both ways (E > G, G > E, R > G and G > R). The EEG was recorded with closed eyes. Each trial comprised 5 min interpreting. The interpretation had to be done mentally, without articulation, in order to avoid interference from muscle artifacts. After each trial the interpreter was interviewed about his feelings during the task. The texts, unknown to the interpreter, were presented stereophonically. They had been chosen from a conference on politics and economics, a common topic which the interpreter was accustomed to dealing with in his job.

In Fig. 1 significant increases and decreases caused by these four interpretation tasks are shown for the frequency range beta 3 (24–32 Hz). For each trial, significant changes in coherence are entered between the respective electrodes. Probabilities ($p \leq 0.01$, 0.02, and 0.05) are marked by different widths of the lines. Increases are full, decreases broken lines. The results are represented separately for each hemisphere and for interhemispheric coherence changes. To the left of the figure are interpretations from (a) and into (c) English; the right half of the figure concerns Russian (b and d). The upper half represents interpretations from the foreign into the native language, the lower one interpretations into the respective foreign languages.

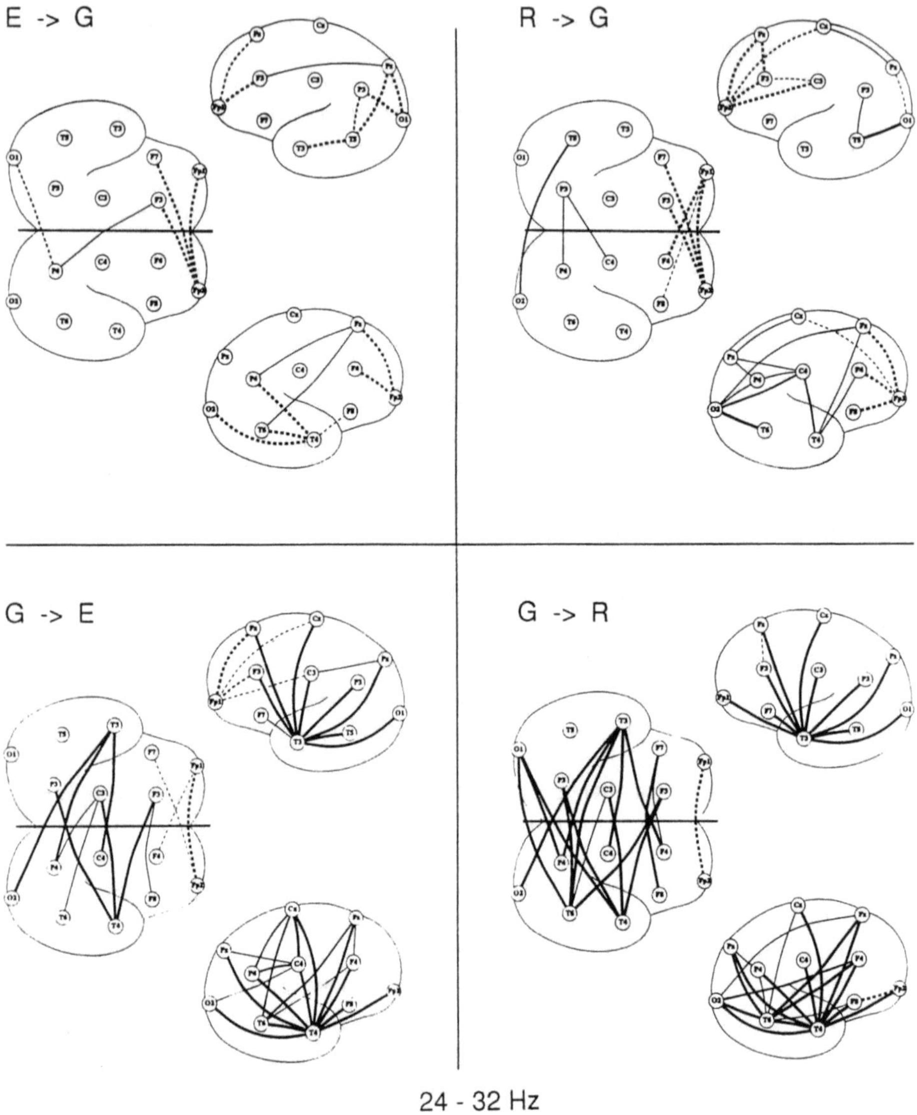

Fig. 9.4–1. Significant coherence changes with respect to the averaged EEG at rest while mentally interpreting texts for 5 min from English to German (**a**), from Russian to German (**b**), and vice versa (**c** and **d**, respectively) in a 26-year-old man, right-handed professional interpreter with German as native language. *Solid lines*, increases in coherence; *broken lines*, decreases. The width of the lines corresponds to $p \leq 0.01$, 0.02, and 0.05, respectively. View of the brain from each side (for intrahemispheric) and from top (for interhemispheric connections). Particularly while interpreting from native into foreign languages, coherence significantly increases between numerous brain regions, and the number of interhemispheric coherence increases is larger, and even more so while interpreting into Russian than into English

Most striking is the contrast between interpreting from and into the foreign language. In the former only few coherence increases appear, mainly in the right hemisphere for R > G. But in the reverse case, while interpreting from German into the foreign language, a great number of coherence increases appears, with an accentuation of the temporal regions, and more so for G > R than for G > E. Moreover, there is a great number of interhemispheric coherence increases, and even more for G > R than for G > E. Apparently interpreting from one language into another activates numerous connections centered around both temporal areas and extending to locations in both the same and the contralateral hemisphere. In addition (and this can also be concluded from studying mental events other than verbal ones), the number of coherence increases in some way reflects the engagement of the brain in the respective task, which is obviously greater while interpreting into the foreign than into the native language. According to this presentation, the most important nodal points for language interpretation in this person seem to be the mid temporal areas where strong information transfer seems to take place with many regions of both hemispheres. Figure 1, however, represents only the findings in the frequency band between 24 and 32 Hz. In our studies on the reflections of mental processes by the EEG it has always turned out that not all frequency bands are equally concerned in mental tasks, even if the left temporal region in most verbal tasks acts as a kind of center of gravity around which coherence increases are concentrated.

Studying these patterns of coherence changes, one may notice that the areas in which coherence changes seem to take their origin are by no means randomly distributed but have preferred locations, as in this interpreter the temporal regions. Our experience with more than 400 subjects who have performed different mental tasks show that these places evidently have functional significance for the respective mental process and can be conceived of as kinds of nodal points. For this reason they were plotted topographically, as shown in Fig. 2.

These diagrams summarize the above findings. They are entered into five schemata of the brain, one below the other, for the five frequency bands, with some electrodes marked by squares. For each task, the sizes of these squares are proportional to the number of coherence increases (black squares) and decreases (empty squares) that are found at these areas. The first four columns are for the four verbal tasks just dealt with, the fifth and sixth concern other tasks (doing mental arithmetic and listening to music, each for 1 min) for comparison. Figure 2 demonstrates that the four verbal tasks are fairly uniformly in this respect. Apart from the above mentioned accentuation of the temporal regions one may notice that the number of coherence changes emerging when interpreting into the two foreign languages

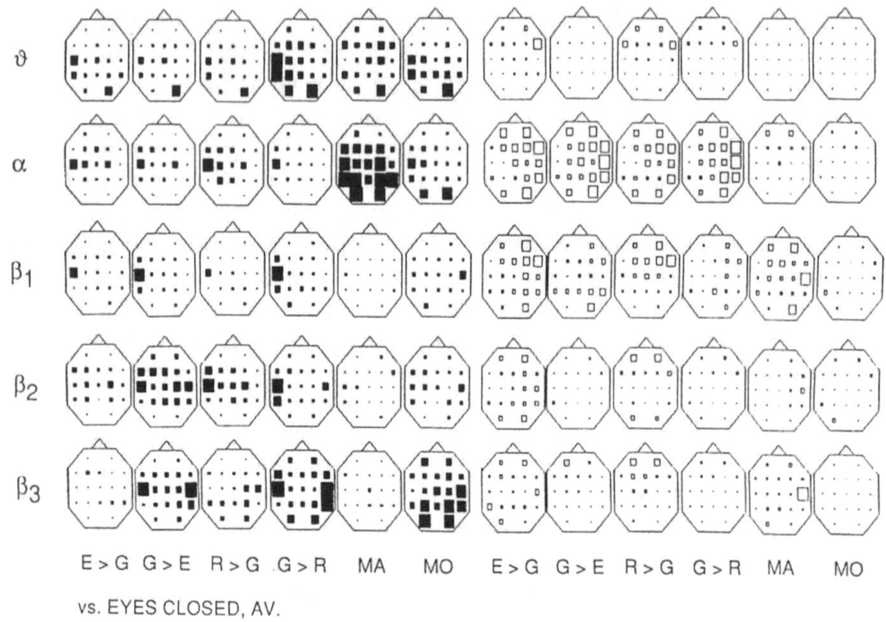

Fig. 9.4–2. Same case as in Fig. 1. Topographic charts of the weightings for the number of significant ($p \leq 0.05$) increases (*black squares*) and decreases (*empty squares*) observed at each electrode obtained from the data of Fig. 1. For further comparison, the results while doing mental arithmetic (*MA*) and listening to a Mozart quartett (*MO*), for 1 min each, are shown. Mainly in the beta bands, the highest incidence of focal areas of coherence increases is found at the left midtemporal region, the right midtemporal region becoming more involved in beta 3 while interpreting into the foreign than into the native language. During MA and MO, additional focal areas are also observed in the right hemisphere; with both these tasks, the distribution of the focal areas is clearly different from that found with the verbal tasks

is larger than vice versa, and generally also for Russian larger than for English. This representation also clearly demonstrates that the involvement of the contralateral, right hemisphere is greater when interpreting from the native into the foreign language than vice versa (the person was right-handed). On the other hand, the patterns for doing mental arithmetic and listening to music are distinctly different from those obtained while interpreting; this holds true for both the number of coherence increases and the number of decreases. In the left hemisphere, the decreases of coherence in the right, non-dominant hemisphere could indicate that a great number of the signals whisking incessantly through the brain are withdrawn from the hemisphere less engaged in the act of language interpretation in order to support the speech-dominant hemisphere.

The second example concerns visual thinking in a 36-year-old, right-handed female painter. We were interested in learning whether characteristic coherence patterns are also found when visual processing takes place in the brain, and, in addition, whether any differences are found between actual visual perception and visual imagining. Therefore we devised a paradigm in which four pictures were presented to subjects. They were asked to contemplate them carefully in order to keep each of them in memory. The pictures were projected onto the wall in a frame. Immediately after the presentation, and for distraction for another 2 min, the persons had to read a text that was also projected onto the wall. Then they were asked to memorize each picture as vividly as possible. Before, between, and after each trial, EEG at rest was recorded for 2 min each. This experiment is from a group study on female painters whom we compared with women not educated in the fine arts. The pictures were paintings by Jordaens, Holbein, Kandinsky, and an etching by Rembrandt. Only the differences between the coherence patterns elicited by contemplating and by memorizing the pictures are discussed here.

Again, the patterns found for contemplating were different from those for memorizing and also different in the different frequency bands. Most differences were found in the beta range. For this reason, only that frequency band in which most significant changes were found (between 13 and 18 Hz) is presented in Fig. 3. In this figure, the width of the lines joining the electrodes corresponds to the number of significant ($p < 0.05$) coherence increases. If, for instance, coherence was increased by each of the four pictures while either contemplating or memorizing them, the thickest line was drawn; if a significant coherence change was found only in three out of the four pictures, the line is thinner, the thinnest lines being drawn, if only two of the four pictures produced significant coherence changes. Thus, the strength of the connecting lines is proportional to the incidence of occurrence of significant coherence increases for all four pictures together. The analysis comprises time periods of 256 s for contemplating and 328 s for memorizing with respect to 746 s EEG at rest. The largest number of significant increases for both tasks was found in the frequency range between 13 and 18 Hz. In both tasks the right hemisphere turned out to be more involved, and especially while contemplating the pictures, whereas while memorizing them, the number of coherence increases consistent for all four pictures became smaller. For memorizing there is greater interhemispheric cross-talk and stronger (contralateral) engagement of frontopolar regions than for contemplating. These are the very general findings. It should be added that such consistent results were found only for coherence and not for power changes of the EEG during mental tasks.

Due to the constraints of space it is not possible to discuss the full range of possible interpretations of these findings. Only a few remarks are

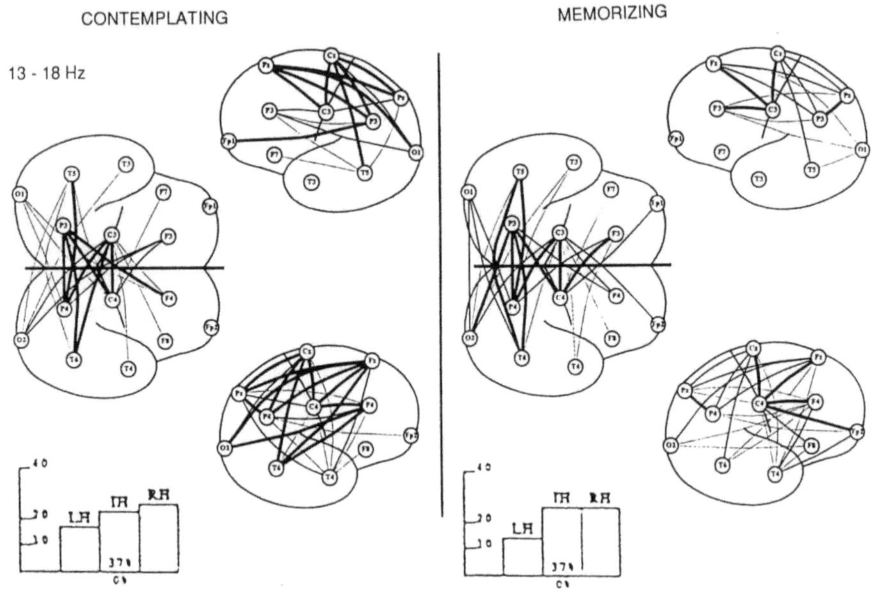

Fig. 9.4–3. Significant coherence changes during visual thinking with respect to the EEG at rest in a 36-year-old female painter while contemplating (*left*) and memorizing (*right*) four paintings (recordings 2 min each). The width of the lines indicates the incidence of significant ($p <= 0.05$) coherence changes (*solid* for increase, *broken* for decrease). *Widest lines*, coherence is significantly changed by each of the four tasks; *narrowest*, by only two of them. 13–18 Hz range. The histograms indicate the numbers of intra- (*LH, RH*) and interhemispheric (*IH*) connections with increased coherence. Note the different involvement of cortical regions while contemplating and memorising pictures

added about our ideas of their possible significance. First, these findings clearly demonstrate that mental phenomena represent themselves in the parameter of coherence. Changes in this parameter differentiate distinctly between verbal and visual tasks. Moreover, they are fairly constant as they reflect averages over several minutes of EEG recording. Therefore, they can probably be taken as indicators of functional changes of brain processes due to mental acts. Most likely these changes are the manifestations of changes of attention, of receptiveness, which, in contrast to the general opinion, turns out to entail most intricate brain processes that involve many brain areas in different ways, according to the demands of the situation at the moment. If we think that, by estimating coherence, we may be able to comprehend some basic mental processes, then it seems to be but a representation of the electric preparation for the performance of those mental processes that we are getting hold of rather than their realization in the cortex.

References

Abeles M (1982) Local cortical circuits. Springer, Berlin Heidelberg New York

Braitenberg V, Schüz A (1991) Anatomy of the cortex: statistics and geometry. Springer, Berlin Heidelberg New York

Nunez PL (1989) Generation of human EEG by a combination of long- and short-range neocortical interactions. Brain Topography 1: 199–215

Petsche H, Lindner K, Rappelsberger P, Gruber G (1988) The EEG: an adequate method to concretize brain processes elicited by music. Music Perception 6: 133–160

Petsche H, Lacroix D, Lindner K, Rappelsberger P, Schmidt-Henrich E (1992) Thinking with images or thinking with language: a pilot EEG probability study. Int J Psychophysiol 12: 31–39

Rappelsberger P, Petsche H (1988) Probability mapping: power and coherence analysis of cognitive processes. Brain Topography 1: 46–54

Rappelsberger P, Krieglsteiner S, Mayerweg M, Petsche H, Pockberger P (1987) Probability mapping of electroencephalogram changes: application to spatial imagination studies. J Clin Monitoring 3: 320–322

9.5 Magnetoencephalography: A New Window into the Functional Organization of the Cerebral Cortex

M. Hoke, T. Elbert, and C. Pantev

Institute of Experimental Audiology, University of Münster, Germany

Physiological investigations are not performed as an end in themselves to satisfy the scientist's thirst for knowledge, although many studies may give that impression. Eventually, a new technique is also going to be evaluated in a clinical environment. Electric phenomena are the first physical quantities used for studying physiological processes, and innumerable studies, especially microelectrode studies on the cellular level, have unveiled the function and functional organization of the central nervous system. However sophisticated a technique may be, its applicability to studies in humans is limited by its invasiveness. In contrast to animal experiments, the use of electrical potentials for diagnostic purposes must be restricted — with few exceptions — to potentials recorded from the surface of the body. Their interpretation has for decades been a merely pragmatic attribution of distinct features of the recorded waveforms to specified generator structures.

The development of several imaging techniques — especially magnetic resonance imaging (MRI) and positron emission tomography (PET) — has opened previously unimagined possibilities of visualizing not only structures but also functional processes especially of the brain. The temporal resolution of these techniques, however, is unsatisfactory, creating the need to develop techniques capable of visualizing or more precisely localizing the sources which generate the surface-recorded electrical potentials. Potential differences recordable at the surface are, however, a rather indirect outcome of the underlying processes: intracellular currents of high-density flowing in the excited portions of dendrites, whose spread in the volume conductor body is substantially distorted owing to tissue layers of different conductivity and anisotropy. In contrast to this, an immediate outcome of the intracellular currents are magnetic signals whose spread is undistorted by the body's physical properties. With the advent of the supraconducting quantum interference device (SQUID) these extremely weak biomagnetic signals became detectable.

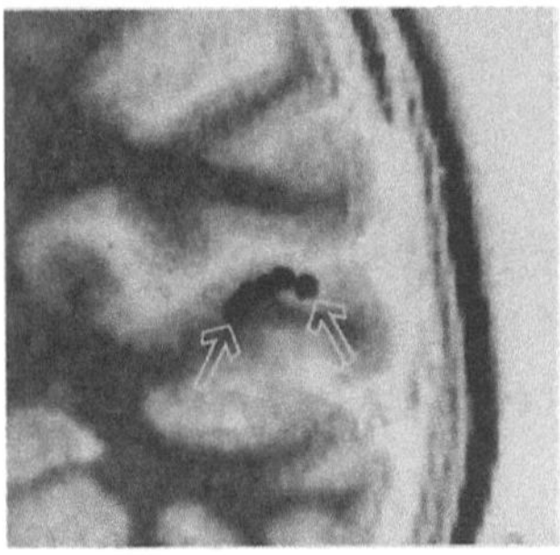

Fig. 9.5–1. Example of the high spatiotemporal resolution of neuromagnetic source localization. *White dots*, location estimates for successive peaks of the human gamma-band field, forming an arcing trajectory which follows the surface of the transverse temporal gyrus. (From Pantev et al. 1991)

Only true imaging techniques such as computed tomography (CT), MRI, and PET offer the reconstruction of arbitrary slices — tomograms — of the investigated structures. Although often referred to as source imaging, the recording of neuromagnetic — as well as neuroelectric — data from the surface of the head is not sufficient for the spatial imaging of the active generator structures. It is rather the attempt to determine a solution of a set of differential (Maxwell's) equations which, by their nature, is not unique. This "inverse problem", which was first formulated by Helmholtz in the middle of the last century, results from the fact that an infinite number of source configurations within the body can produce exactly the same distribution of electric and magnetic activity on its surface. In many cases, however, it is possible to construct constraints such that only one solution remains which satisfies the equations. Hence, the spatiotemporal distribution of electromagnetic generators becomes unique. The simplest constraint is the assumption of a single focal source which can be modeled with good approximation by a current dipole — regardless of the real shape of neuronal arrays. More sophisticated approaches utilize the knowledge of anatomical structures. It is obvious, for instance, that areas covered by ventricles or white matter do not incorporate active structures. More explicitly, it is reasonable to assume that only current dipoles with an orientation perpendicular to the surface of gray matter contribute to electromagnetic activity on a macroscopic scale. The individual anatomical information is necessary for three different reasons: to construct a realistic body model, to retrieve constraints for a realistic source configuration, and to project the computed source location(s) onto the corresponding anatomical structures. This information is available only from three-dimensional reconstructions of MRI tomograms. Therefore, neuromagnetic (or neuroelectric) source localization must rely on its being combined with a real imaging technique. If the sources can be restricted to cortical structures (which is generally the case for potentials or fields related to cognitive processes), and if both electroencephalogram (EEG) and magnetoencephalogram (MEG), measured simultaneously from many different sites, can be integrated into one unitary model for the solution of Maxwell's equations, then this solution becomes unique. The resulting cortical map, although similar to a PET scan in its appearance, has the unique advantage of not displaying a slowly changing cerebral blood flow but of imaging the neuronal activation changing within a fraction of a second. The image of static activity, averaged across minutes in PET, is replaced by a moving image, unique in respect to temporal and spatial resolution and allowing the detection of the functional sequence in patterns of neural mass activation (Fig. 1).

Neuromagnetic fields are extremely weak. With approximately 1–5 pT (10^{-12} Tesla) for the MEG and some 10 to some 100 fT (10^{-15} Tesla) for cortical evoked magnetic fields, the amplitudes of neuromagnetic signals are not only far below that of the earth magnetic field (seven to nine orders of magnitude) but also considerably below that of environmental noise. This implies that not only an extremely sensitive instrument — as given by the SQUID — is required, but also an effective shielding — a magnetically shielded room — to protect against strong environmental noise.

The distribution pattern, on which even the simplest source localization is based, requires measuring of the magnetic signal from multiple positions. Measurement of both MEG and EEG over the whole neurocranium facilitates arriving at a unique solution. Multi-sensor instruments have been developed during the past decade allowing the measurement of the magnetic field simultaneously from up to 37 different positions (Fig. 2). The latest developments are whole-head systems allowing a simultaneous scanning of the entire neurocranium. These powerful neuromagnetic instruments and the localization techniques associated with them offer the neuroscientist a novel and unique view of cerebral processes and functions in humans.

What other advantages of the MEG justify the use of such an expensive instrument? What are the disadvantages? Distinct advantages are the

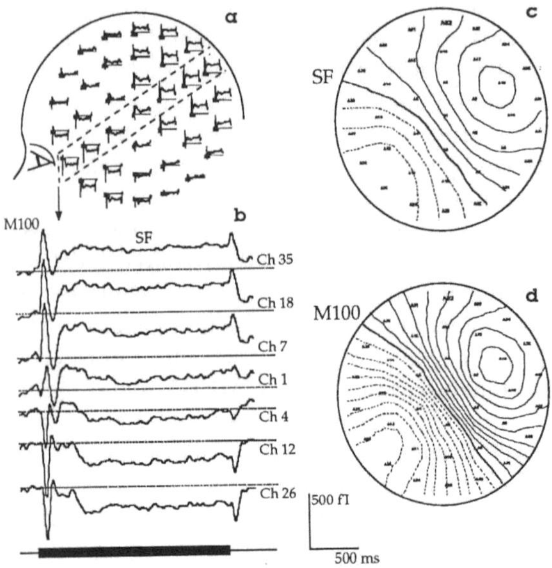

Fig. 9.5–2. a Projections onto the outline of the head of a set of 37 averaged waveforms (length 2400 ms), recorded over the left hemisphere in response to a 2s, 1000 Hz tone burst as indicated at the bottom of **b**. **b** Waveforms recorded along a line connecting the minimum and the maximum of the sustained field pattern (enclosed in *dotted lines* in **a**) are represented in magnification, clearly showing polarity reversal (or, more precisely, reversal of field direction) of both the transient on- and off-responses and the sustained field (SF). **c** The surface distribution of the magnetic field is shown in terms of isofield contour maps for selected instants of time (sustained field at a latency of 1300 ms). **d** M100 at a latency of 94 ms. Outwardly going flux marked by *solid lines*, inwardly going flux by *dotted lines*. Spacing between two adjacent isocontour lines is 50 fT. (From Pantev et al. 1993b)

easy modeling for the magnetic signal (the body is transparent for such low-frequency magnetic fields), the higher relative precision of localization, and the faster placement of the neuromagnetometer system without many of the problems of artifact contamination inherent in electrode-based techniques. The major restriction of neuromagnetic measurements is that current dipoles oriented normally to the surface do not produce a magnetic field outside of the body. This restriction is not that crucial, however, in the case of the cerebral cortex. Firstly, approximately two thirds of the cortical tissue is located in sulci. Secondly, the major direction of cortical current flow is, owing to the cortical cytoarchitecture, normal to the cortical surface. Thirdly, the tangential component of an arbitrarily oriented dipole corresponds to the total dipole vector weighted by the sine of the angle by which it is tipped from the normal to the surface. This implies that magnetically detectable cortical sources encompass not only current flows directed tangentially to the surface of the head, but even these sources contribute 50% or more of the total dipole vector whose deviation from the normal to the surface is not less than 30°. Hence, cortical activity is a principal target for tracking functional activation with neuromagnetic measurements.

It goes without saying that MEG measurements cannot compete with microelectrode studies, which certainly provide a main route to the functional organization of cortical structures. However, with microelectrodes it is possible only to monitor a small fraction of actual ongoing activity. Furthermore, knowledge of functional properties of a few single neurons is not sufficient to understand neural nets, and the study of neural nets is insufficient to grasp the dynamics of neural mass action. The magnetic sensors, with a diameter of approximately 20–30 mm, pick up activity from a very large number of active neurons. This activity is not restricted to that of the investigated structures but also incorporates spontaneous activity from surrounding areas. With appropriate experimental paradigms, however, a simultaneous activation of a circumscribed neuronal population can be achieved. The activity of a larger number of similarly oriented, simultaneously active individual neuronal current dipoles can be modeled after the activity of a single equivalent current dipole (ECD), which corresponds to the vectorial integral of all individual dipole vectors. Regarding the complexity of cerebral processes initiated by external stimuli, especially of cognitive processes, it is obvious that a single, or prevailing, source is rather an exception than the rule. Modeling multiple dipoles cannot be a solution unless, as previously mentioned, sufficient restrictions for source configurations, retrieved from individual MRI tomograms and physiological knowledge can be implemented in the model (Lütkenhöner et al. 1992). In addition, the use of distributed source models is another way of treating multiple sources (Grunwald et al. 1992).

Neuromagnetic research originated in the exploration of the spontaneous MEG, followed by the investigation of auditory evoked magnetic fields. Since then magnetic fields elicited by almost all major stimulus modalities have been investigated (visual, somatosensory, olfactory, pain). Magnetic fields preceding and accompanying movements have been studied as well (see Hoke et al. 1992, for a survey of current research).

Where neuroelectric techniques have failed, neuromagnetic measurements have succeeded in demonstrating the cortical representation of specified stimulus features in humans, as well as in separating cortical sources for different components of the evoked fields. These achievements are mainly due to the high precision in relative localization, which is of the order of few millimeters. One of the most impressive examples is the study of the tonotopic organization of the primary auditory cortex. After several futile efforts to reveal tonotopicity in humans by means of neuroelectric measurements or PET (for more information see Pantev et al. 1988), Romani et al. (1982) and Elberling et al. (1982) succeeded with neuromagnetic measurements in demonstrating a tonotopic organization of the human auditory cortex: the depth of the ECD increases with the logarithm of stimulus frequency. A decisive step forward was taken by Pantev et al. (1989a). According to the place principle of the classical hearing theory, the physical entity frequency is encoded in the auditory periphery as place information which is decoded in more central parts of the auditory system to form the perceptual entity pitch. This relation turned out to be true only for pure-tone signals (spectral pitch); it can be quite different in the case of complex auditory stimuli (virtual pitch), thus requiring a multistage process for pitch formation. Neuromagnetic measurements have shown that the tonotopic organization of the primary auditory cortex reflects the pitch rather than the frequency of the stimulus, implying that the pitch formation process takes place in subcortical regions.

Another discovery also concerning functional organization first achieved with neuromagnetic measurements was that of an amplitopic organization of the human primary auditory cortex (Pantev et al. 1989b). In contrast to the place coding of intensity in the auditory periphery, where increasing intensity shifts the place of maximal excitation into the same direction as does increasing frequency, the ECD depth of wave N_1m is shifted for increasing intensity in opposite direction as for increasing frequency. This implies that the unidimensionality of frequency and intensity coding in the auditory periphery is, at cortex, abandoned in favor of a bidimensional representation of the two stimulus features.

Coherent oscillatory brain activity in the 20–90 Hz frequency (gamma) band is now being regarded as a possible fundamental physiological mechanism which may bind features to form objects. An emerging consensus maintains that oscillatory rhythms play an important role in perception and attention and may support the encoding of spatial and temporal order during behavioral adaptation. Such activity was first described for the olfactory (Adrian 1942) and visual (Gray et al. 1989) systems. It was again neuromagnetic measurements which first allowed the noninvasive detection of the transient auditory evoked gamma-band response (Pantev et al. 1991), consisting of four or more cycles phase-locked to stimulus onset in the 20–130 ms poststimulus interval. The source resides in the primary auditory cortex, clearly separated from that of the auditory wave N_1m, and moves in a posterior arcing trajectory 1 cm or more in length. Neuromagnetic source localization demonstrated that the gamma-band response originates from a source location which is also clearly separated from that of the steady-state response of basically the same frequency. The steady-state response, in contrast, originates from the same location as the middle-latency responses, indicating that the former is a mere superimposition of middle-latency responses (Pantev et al. 1993a). A clearly distinct source location has also been demonstrated for the sustained field, which also resides in the primary auditory cortex and also exhibits tonotopicity (Pantev et al. 1993b).

A final example for the localization of spatially distinct sources is the auditory mismatch field in response to minor pitch changes (Csépe et al. 1992). The mismatch field, which corresponds to the so-called mismatch negativity, occurs in response to a deviant stimulus presented with low probability in an otherwise homogeneous sequence of standard stimuli. Results of Csépe et al.'s (1992) neuromagnetic study support Näätänen's (1990) proposal on the basic processing of auditory stimuli. According to his proposal the "task-independent" sensory analysis is composed of a transient detector system, which activates the N_1m generator process and has a permanent feature detector system. Whereas slight deviations in frequency do not activate different N_1m generators, they are obviously sufficient to trigger a spatially distinct mismatch generator.

These few examples of the high precision of relative localization, i.e., the identification of spatially distinct sources in the same experiment, demonstrate that if there is one major source active at a time, i.e., if the requirements for the application of a single moving dipole model are met, neuromagnetic source localization is a powerful, noninvasive tool for the investigation of cerebral organization and cortical processes in humans.

The situation is quite different, however, if the spontaneous MEG is to be investigated. Spontaneously occurring EEG and MEG fluctuations reflect the dynamics of neural mass excitation. Alterations in attention as well as gross changes in cognitive states are related to changes in macroscopic patterns of excitation and hence should result in changes of dynamical attributes such as the largest Lyapunov exponent (LLE) or the Kolmogorov entropy. Together with the fractal dimension (commonly calculated as correlation dimension), such variables characterize the current brain dynamical state, particularly its complexity (Elbert et al. 1993). Mapping the LLE of magnetic activity measured over the auditory regions reveals deviances in specified disorders or during specified symptoms. For instance, the average LLE is larger in tinnitus sufferers than in controls (Kowalik et al. 1993).

Generally these dynamic variables are useful in tracing alterations in the brain's state. Prior to a phase transition, the system exhibits a loss in its global stability, well described by the LLE, a measure of predictability. Consequently, a critical transition can be found by tracing the LLE across time. For instance, an epileptic attack constitutes such a phase transition and can be observed in both the electrocorticogram (ECoG)- and the MEG-derived LLE (Elbert et al. 1993). Distinct local variations in the LLE suggest that under resting conditions the brain dynamics vary considerably between regions, meaning that there is only limited or no coupling among different brain regions. Owing to volume conduction, such a separation is difficult to observe in the EEG. Nearby MEG channels, on the other hand, may often display equal values in their dynamical variables. This cannot be explained solely on the basis of magnetic field propagation but demonstrates how LLE maps indicate the task-dependent coupling among various brain regions.

The closest that we come to an actual clinical application of the MEG is the source localization for the detection of epileptogenic zones. This was previously considered too difficult because, on the one hand, recordings of ictal activity are often contaminated with movement artifacts, and, on the other hand, interictal phenomena seem to be of little validity due to the possible mismatch of the pacemaker zone and the irritative zone. We have explored a unique way of circumventing the necessity of recording ictal activity: the epileptic focus is activated by the administration of methohexital, a short-term anesthetic which provokes epileptic activity while the patient is immobilized (Hufnagel et al. 1992). The results of neuromagnetic localization of epileptic foci (Fig. 3) obtained in patients suffering from temporal lobe epilepsy (Wienbruch et al. 1993) were verified by intracranial recordings. Even at the early stages of development of this technique it has become obvious that — at least in selected cases — such information can provide valuable aid for the presurgical diagnosis and, in addition, is gathered easily by noninvasive measurement.

A final example may highlight perspectives offered through mapping cortical electromagnetic activity. It has been assumed that chronic stimulation leads to changes in the cortical representation of pain, specifically to increased receptive fields for the painful area. Consequently, cortical evoked magnetic activity originates from more extended neural assemblies when evoked from the affected body part of a patient. We have indeed been able to show that pain patients who experience chronic noxious stimulation over extended periods of time produce stronger magnetic fields when phasic somatosensory stimuli are applied to the location of their pain, and that the magnitude of activation is significantly related to the chronicity, i.e., the duration of illness (Flor et al. 1993). These results are in accordance with observations from animal experiments (e.g., Merzenich et al. 1983, Milliken et al. 1992) demonstrating the plasticity of receptive fields subsequent to long-term alterations of afferent input. Magnetoencephalography offers new possibilities of monitoring such alterations in humans.

These examples of ongoing research on spontaneous MEG activity should be sufficient to demonstrate that MEG has indeed opened up a new and unique window into the functional organization and functional processes of the human cerebral cortex.

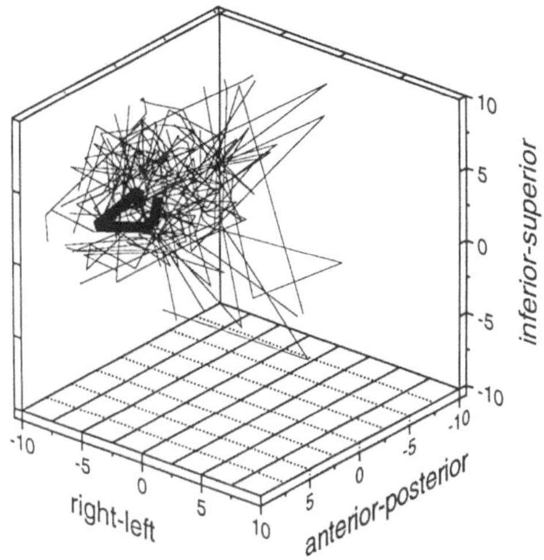

Fig. 9.5–3. Aspects of the spatiotemporal pattern of spontaneous MEG activity as demonstrated by the temporal sequence of ECD locations calculated for every sampling instant (3 ms). *Thin line*, trajectory obtained from the MEG of an epileptic patient measured over the right hemisphere during methohexital induction. To determine whether the trajectory moves in arbitrary directions, or whether it follows at times some main stream, i.e., whether it frequently runs through the same sequence of adjacent voxels, for every volume element the net flow of all trajectories running through it was evaluated (according to Mühlnickel et al. 1993). *Thick*, stereotypic pattern obtained, which in this case corresponds to methohexital-induced epileptic spikes

References

Adrian ED (1942) Olfactory reactions in the brain of the hedgehog. J Physiol 100: 459–473

Csépe V, Pantev C, Hoke M, Hampson S, Ross B (1992) Evoked magnetic responses of the human auditory cortex to minor pitch changes: localization of the mismatch field. Electroenceph clin Neurophysiol 84: 538–548

Elberling C, Bak C, Kofoed B, Lebech J, Saermark K (1982) Auditory magnetic fields. Source localization and "tonotopical organization" in the right hemisphere of the human brain. Scand Audiol 11: 61–65

Elbert T, Rockstroh B, Kowalik ZBJ, Hoke M, Molnar M, Skinner JE, Birbaumer N (1993) Chaotic brain activity. Electroenceph clin Neurophysiol, Suppl (in press)

Flor H, Braun C, Birbaumer N, Elbert T, Ross B, Hoke M (1993) Chronic pain enhances the magnitude of the magnetic field evoked at the site of pain. Soc Neurosci Abstr 19: 233

Gray CM, König P, Engel AK, Singer, W (1989) Oscillatory response in the cat visual cortex exhibit intercolumnar synchronization which reflects global stimulus properties. Nature 338: 334–337

Grunwald A, Lütkenhöner B, Pantev C, Hoke M (1992) Distributed-source analyses of the auditory evoked field compared with single-dipole analyses. Proc IEEE Satellite Symposium on Neuroscience and Technology, pp 68–72

Hufnagel A, Burr W, Elger CE, Nadstawek J, Hefner G (1992) Localization of the epileptic focus during Methohexital-induced anesthesia. Epilepsia 33 (2), 271–284

Kowalik ZBJ, Elbert T, Hoke M (1993) Mapping dynamic brain functions: the largest Lyapunov exponents derived from multichannel magnetoencephalography. In: Janssen BH, Brandt ME (eds) Nonlinear dynamical analysis of the EEG. World Scientific, Singapore, pp 156–164

Hoke M, Erné SN, Okada Y, Romani GL (eds) (1992) Biomagnetism: clinical aspects. Excerpta Medica, Amsterdam

Lütkenhöner B, Pantev C, Hoke M (1992) Source analysis of the auditory evoked field: bias introduced by the choice of the model. Proc IEEE Satellite Symposium on Neuroscience and Technology, pp 112–118

Merzenich MM, Recanzone CE, Jenkins WM, Allard TT, Nudo RJ (1983) Cortical representation plasticity. In: Rakic P, Singer W (eds) Neurobiology of the cortex. Wiley, New York, pp 41–67

Milliken GW, Nudo RJ, Grenda R, Jenkins WM, Merzenich MM (1992) Expansion of distal forelimb representations in primary motor cortex of adult squirrel monkeys following motor training. Soc Neurosci Abstr 18:506

Mühlnickel W, Wienbruch C, Kowalik ZJK, Hoke M, Elbert T, Elger CE (1993) Automatic detection and localization of focal MEG activity. 9th Internat. Conference on Biomagnetism, Vienna

Näätänen R (1990) The role of attention in auditory information processing as revealed by event-related potentials and other brain measures of cognitive function. Behav Brain Sci 13: 201–288

Pantev C, Hoke M, Lehnertz K, Lütkenhöner B, Anogianakis G, Wittkowski W (1988) Tonotopic organization of the human auditory cortex revealed by transient auditory evoked magnetic fields. Electroenceph clin Neurophysiol 69:160–170

Pantev C, Hoke M, Lütkenhöner B, Lehnertz K (1989a) Tonotopic organization of the auditory cortex: Pitch versus frequency representation. Science 246: 486–488

Pantev C, Hoke M, Lehnertz K, Lütkenhöner B (1989b) Neuromagnetic evidence of an amplitopic organization of the human cortex. Electroenceph clin Neurophysiol 72:225–231

Pantev C, Makeig S, Hoke M, Galambos R, Hampson S, Gallen C (1991) Human auditory evoked gamma-band magnetic fields. Proc Natl Acad Sci USA 88: 8896–9000

Pantev C, Elbert T, Makeig S, Hampson S, Eulitz C, Hoke M (1993a) Relationship of transient and steady-state auditory evoked fields. Electroenceph Clin Neurophysiol 88: 389–396

Pantev C, Eulitz C, Elbert T, Hoke M (1993b) The auditory evoked sustained field: origin and frequency dependence. Electroenceph Clin Neurophysiol (in press)

Romani GL, Williamson SJ, Kaufman L (1982) Tonotopic organization of the human auditory cortex. Science 216: 1339–1340

Wienbruch C, Lehnertz K, Elger CE, Elbert T, Hoke M (1993) Non-invasive localization of methohexital-induced epileptic activity, 9th International Conference on Biomagnetism, Vienna

9.6 Electroencephalography in Psychiatry: From Hans Berger to the "Decade of the Brain"

T. Dierks and K. Maurer

Zentrum für Psychiatrie, Universität Frankfurt am Main, Germany

From the discovery of the human electroencephalogram(EEG) by Hans Berger (1929) to the "Decade of the Brain", as proclaimed by the US government, more than 60 years have passed, and knowledge about the human neurophysiology has increased greatly in that time. New hardware technologies, but also in particular the development of new methods for analysis of electromagnetic fields in the brain, have contributed to progress in the field. There has been a continuous improvement of the methods, and Hans Berger's dream recorded in his diary in 1924, may soon be fulfilled: "I hope that I may realize a 20-year-old plan, to create a kind of brain mirror, the 'elektenkephalogram'." However, at the beginning of the EEG era, the main application of EEG was in the field of neurology and not so much in psychiatry. Focal graphoelements such as spikes and slow waves helped to localize cerebral pathologies, before other imaging methods such as computed tomography (CT) were available. Today structural and functional imaging methods of brain function such as magnetic resonance imaging (MRI) and positron emission tomography (PET) seem to put the clinical importance of EEG in the background. However, the quantitative neurophysiology has several advantages as a functional method that are not achievable using other functional and structural methods, especially the very high time resolution (in the millisecond range) and the noninvasiveness, which allows numerous follow-up investigations without problem. Until recently, most psychiatrists did not expect paper EEG to be used other than to exclude organic brain disease and epilepsy. However, with new possibilities to digitize EEG, new fields of application can be seen. Particularly noteworthy was the introduction of evoked potentials (EP) in the middle of the seventies, i.e., visual (VEP), somatosensory (SEP), and auditory evoked potentials (AEP). The exogenous components of the EP showed a higher usefulness in neurology than in psychiatry. Promising exceptions for psychiatric application are some endogenous components of the EP, such as the P300 wave (positive inflection around 300 ms after stimulus presentation) and the contingent negative variation (CNV).

In the frequency domain of EEG analysis, the fast Fourier transform (FFT; Cooley and Tukey 1965) promised to enhance the usefulness of EEG in psychiatry; however, except for some diseases, no specific frequency patterns could be discerned, although in the field of psychopharmacology characteristic patterns could be observed for different substances. Relative specific frequency patterns have been observed for neuroleptics, antidepressants, tranquilizers, psychostimulants, and cognitive enhancers (Itil 1982). It is, however, important to remember that EEG shows drug effects and does not indicate whether a drug is efficacious.

With the increasing number of electrodes, made possible by increasingly better hardware, the amount of information which confronts the neurophysiologist is becoming overwhelming. Consequently, many clinicians lost interest in EEG recordings. This gap was filled by mapping; at the beginning of the seventies, Lehmann (1971) started to use maps of the topographic distribution of electrical field potentials, and at the end of the seventies, Duffy and coworkers (1979) developed a system for the clinical application of mapping.

Mapping of EEG and EP does, of course, not hold more information than conventional EEG and EP, but it presents the data in a well-ordered, clear, and informative manner if used in a responsible way. Mapping allows us to see details which could not be detected before due to the extreme complexity and large amount of data. Its clinical use will be the topic of this paper.

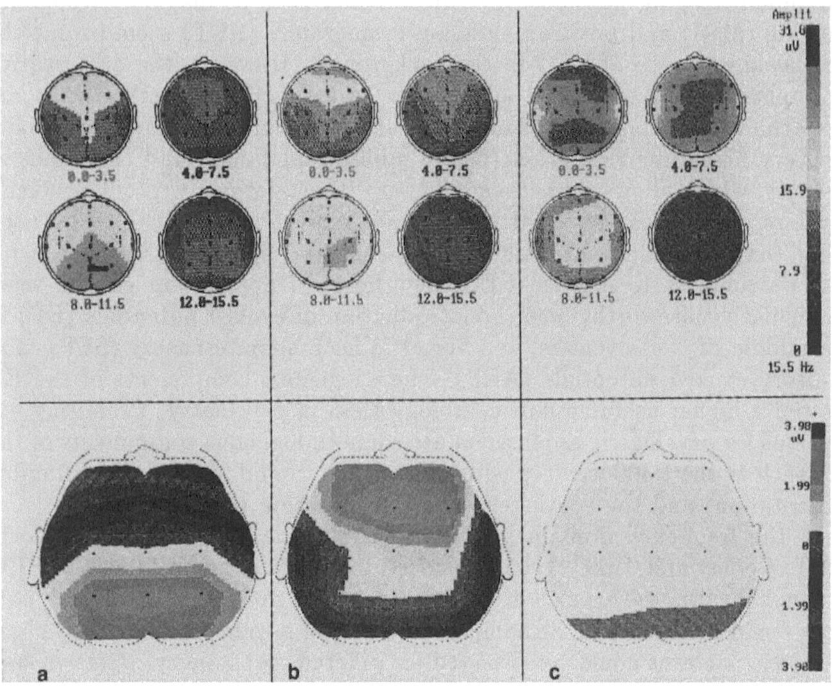

Fig. 9.6–1. Topographical display of stage dependent alterations of EEG activity and P300 fields. **a** Normal properties. **b** EEG and P300 topography of moderately demented patients. **c** Topograhical features of severely demented patients

9.6.1 Dementia

Due to the increase in life expectancy, the number of elderly people in our society is steadily increasing and also the number of subjects suffering from age-related disorders. The most common psychiatric disorder of advanced age is dementia and especially the dementia of Alzheimer type (DAT; Jorm et al. 1987). As early as 1932, Hans Berger observed in patients with impaired cognitive performance a slowing of the EEG. With the help of mapping and topographical analysis of the EEG, it was possible to elucidate that with increasing severity of dementive symptoms, slow-wave activity, especially theta activity, was increasing and alpha and fast-wave activity was decreasing (Giaquinto and Nolfe 1986; Brenner et al. 1986; Coben et al. 1990; Dierks et al. 1991). Furthermore, with increasing degrees of cognitive impairment, alpha and beta activity was more anteriorly distributed. Regarding the endogenous EP component P300, an anteriorization of the wave was also observed with increasing severity (Dierks and Maurer 1990; Fig. 1); however, when patients reached a stage of more than moderate dementia, P300 paradigms (selective attention to certain stimuli) were mostly not performed correctly anymore (Maurer et al. 1988). Concerning the differential diagnosis between different kinds of dementia, e.g., DAT versus multi-infarct dementia (MID), the EEG was not very helpful. Except for more focal abnormalities in MID patients, no clear differences in the spontaneous electrical activity of the brain could be observed. The P300, on the other hand, provided better discrimination of DAT and MID patients, especially in slightly and moderately demented patients. DAT patients had a more anteriorly localized maximal P300 amplitude, whereas in MID patients the topographic pattern of the P300 was similar to the one in healthy control subjects (Fig. 2). It seems as if the P300 allows quite a reliable differentiation between dementia and pseudodementia and depression. In clinical practice, differentiation between these disease entities is often difficult, especially if depressive patients show attention and memory disorders. The P300 of a depressive patient is not topographically dislocated, but might show a decreased amplitude.

Differential diagnosis DAT-MID			
	DAT	MID	Statistics
Delta	++	+	Trend
Theta	++	++	n.s.
Alpha	--	--	n.s.
Beta	---	--	p<0.05
P300	+++	+++	p<0.05

Fig. 9.6–2. Neurophysiological differences between patients with *DAT* (dementia of Alzheimer type) and *MID* (multi-infarct dementia). *n.s.*, not significant

9.6.2 Major Psychoses

In mapping of schizophrenic patients, it was possible to make productive symptoms such as hallucinations visible. Figure 3 displays a map of beta activity in a 30-year-old male schizophrenic patient. He suffered from delusions, auditory hallucinations, illusions, and ideas of reference. During an acute psychotic episode, a digital EEG was recorded, and in the map of the beta activity increased temporal amplitudes were observed, which probably were not due to muscle activity. As a trait marker, the EEG regarding schizophrenia was rather disappointing, and apart from a well-known alpha activity reduction, no clear brain-mapping patterns were observed. However, studies using cognitive loading and activation procedures may to a higher degree elucidate brain dysfunction in schizophrenic patients (Kokkou 1982, Westphal et al. 1992). More promising was also the result concerning the P300. We investigated 18 schizophrenic patients (according to DSM-III-R; mean age 31 years). The data were compared to age- and sex-matched healthy control subjects. For assessment of the psychopathological state, the scale for assessment of negative symptoms (SANS) and the schedule of affective disorders and schizophrenia (SADS) were applied to the patients. When correlating these two scales with the P300 amplitude, a negative correlation between SANS and P300 amplitude was revealed, meaning that with increasing amounts of negative symptoms in a patient, the amplitude of P300 was reduced. Also SADS correlated negatively with P300 amplitude; schizophrenic patients with more pronounced impaired overall functioning showed lower P300 amplitudes. Consequently, P300 amplitude could be seen more as a state marker and not so much as a trait marker in schizophrenic disorders (Strik et al. 1993).

Mapping of EEG and EP led to new insights into psychiatric diseases, but perhaps not to the high degree as was expected at the beginning. Other functional imaging methods have similar problems to brain mapping, but are more invasive.

9.6.3 Beyond Mapping

Mapping of EEG and EP has to be seen as one step in the continuous process to understand human brain function. New concepts and methods of analysis of electrical brain currents have been developed and applied to healthy subjects as well to patients suffering from various diseases.

Since mapping results gained in the frequency domain are highly dependent on the choice of reference, the results are quite ambiguous and badly suited for a physiological interpretation. Furthermore, mapping still confronts the user with a huge amount of data and dependent variables, making a confirmative statistical analysis difficult.

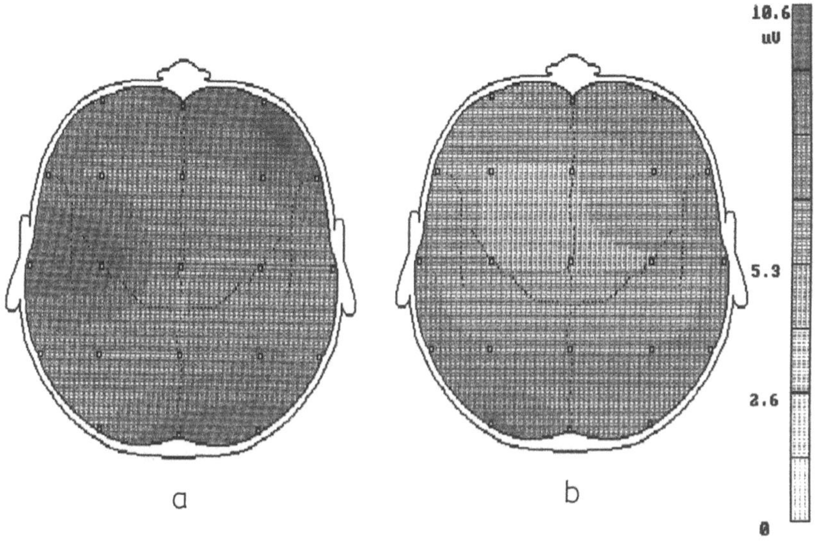

Fig. 9.6–3. Topographic display of EEG activity in a patient during a psychotic episode. **a** During the psychotic episode, an increased beta activity at electrode sites T3 and C3 occurred. **b** Attenuation of the formerly described increased beta activity

Various reference-free methods have lately been developed. One which combines reference independency with substantial data reduction is the "FFT approximation". The method developed by Lehmann and Michel (1989, 1990) allows an estimation of equivalent dipoles in the frequency domain. Figure 4 shows the result of a comparison between 35 healthy geriatric controls and 35 patients suffering from DAT. The figure allows lateral two-dimensional insight into the model brain, and one can clearly observe a more anterior located EEG generator in DAT patients in alpha and beta bands. The correlation between the psychopathological state of schizophrenic patients and equivalent dipole data gained by FFT approximation revealed a correlation between brief psychiatric rating scale (BPRS) and depth of the theta generator in the sense that theta activity appeared more deeply with increasing severity of psychiatric symptoms. In the case of beta activity a correlation was found in the sense that with increasing BPRS score, beta activity appeared more anteriorly. Similar results have been reported by Koukkou et al. (1991).

9.6.4 Conclusions

More than 60 years have passed since the detection of the human EEG by Hans Berger and 3 years have passed since the start of the "decade of the

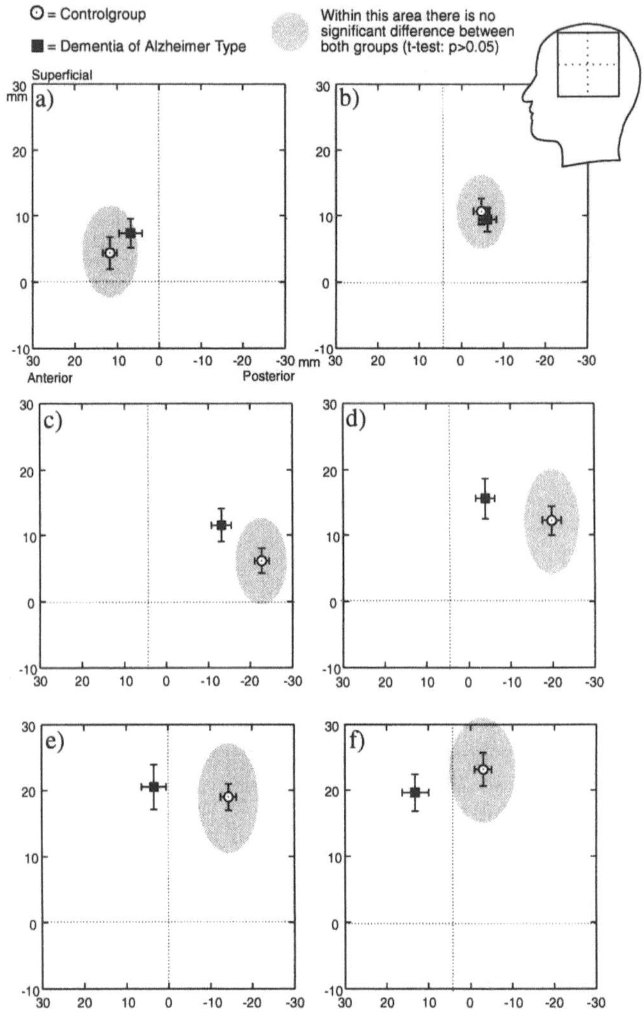

Fig. 9.6–4. Two-dimensional plot of mean values and standard errors of equivalent dipole localizations in anterior-posterior direction and for the depth in the **a** delta (1.0–3.5 Hz), **b** theta (4.0–7.5 Hz), **c** alpha (8.0–11.5 Hz), **d** beta-1 (12.0–15.5 Hz), **e** beta-2 (16.0–19.5 Hz), and **f** beta-3 band (20.0–23.5 Hz) for the DAT (demetia of Alzheimer type) group (*black square;* n=35) and the healthy age- and sex-matched control group (*white circle;* n=35). Zero value is middlepoint of a spherical head model. Within the hatched areas shown there is no significant difference between the two groups (t-test: $p>0.05$)

brain". Concerning neurology, Berger's dream of a brain mirror has been realized to a great extent; in psychiatry, however, we must admit that only a few steps have been made. It is not so much the weakness of the methods applied to the electrical brain activity that is the main reason for this, but rather the complexity of the subject. The ancient wisdom of Goethe is still largely true, that "we cannot fathom nature with mechanical devices". Theoretical models based on classification systems such as ICD-10 or DSM-III are also probably not so suited for neurophysiological investigations of brain function in psychiatric disorders, since these classification systems can change and have a strong conventional aspect. Consequently, diagnostic classification of patients using neurophysiological parameters into disease entities based on such classification systems is more or less fruitless. It would probably be more helpful to use a symptom-oriented method, i.e., positive or negative symptoms, which we also call "state variables". Besides these clinically important state phenomena, a clinician needs help in making prognostic decisions. The search for state overlasting measurements is indicated, i.e., the accepted P300 amplitude reduction in schizophrenia which points to a structural defect within the P300 loop. If a patient shows an amplitude reduction during remission, one could assume that a trait phenomenon which outlasts fluctuations is present. Furthermore, quantitative EEG combined with topography will give a better indication if a so-called central nervous system (CNS)-active substance shows effects within the CNS or not. The combination of basic science, as in the outstanding scientific work performed by Otto Creutzfeldt, and clinical research will help us to understand the organization of the human cortex and its normal and pathological function.

References

Berger H (1929) On the electroencephalogram of man. Archiv Psychiatrie Nervenkrankheiten 87: 527–570

Brenner RP, Ulrich RF, Spiker DG, Sclabassi RJ, Reynolds CF, Marin RS, Boller F (1986). Computerized EEG spectral analysis in elderly normal, demented and depressed subjects. Electroencephalography Clin Neurophysiol 64: 483–492

Coben LA, Chi D, Snyder AZ, Storandt M (1990) Replication of a study of frequency analysis of the resting awake EEG in mild probable Alzheimer's disease. Electroencephalography Clin Neurophysiol 75: 148–154

Cooley JW and Tukey JW (1965) An algorithm for the machine calculation of complex fourier series. Mathematics comput 19: 297–301

Dierks T, Maurer K (1990) Reference-free evaluation of auditory evoked potentials – P300 in aging and dementia. In: Dostert P, Riederer P, Strolin Benedetti M, Roncucci R (eds) Early Markers in Parkinson's and Alzheimer's disease. Springer, Vienna, pp 197–208

Dierks T, Perisic I, Froelich L, Ihl R, Maurer K (1991) Topography of quantitative EEG in dementia of Alzheimer type: relation to severity of dementia. Psychiatry Res Neuroimaging 40(3): 181–194

Duffy FH, Burchfiel JL, Lombroso CT (1979) Brain electrical activity mapping (BEAM): A method for extending the clinical utility of EEG and evoked potential data. Ann Neurol 5: 309–321

Giaquinto S, Nolfe G (1986) The EEG in the normal elderly: a contribution to the interpretation of aging and dementia. Electroencephalography Clin Neurophysiol 63: 540–546

Jorm AF, Korten AE, Henderson AS (1987) The prevalence of dementia: a quantitative integration of the literature. Acta Psychiatrica Scand 76: 465–479

Koukkou M, Michel CM, Tremel E (1991) FFT dipole sources and schizophrenic symptomatology. Brain Topography 3: 466–467

Koukkou M (1982) EEG states of the brain, information processing, and schizophrenic primary symptoms. Psychiatry Res 6: 235–244
Lehmann D (1971) Multichannel topography of human alpha EEG fields. Electroencephalography Clin Neurophysiol 31: 439–449
Lehmann D, Michel CM (1989) Intracerebral dipole sources of EEG FFT power maps. Brain Topography 2(1/2): 155–164
Lehmann D, Michel CM (1990) Intracerebral dipole source localization for FFT power maps. Electroencephalography Clin Neurophysiol 76: 271–276
Maurer K, Dierks T, Ihl R (1988) Quantitative P300 data and their topography in dementia. In Samson-Dolfuss (ed) Statistics and topography in quantitative EEG. Elsevier, Paris, pp 243–250
Strik WK, Dierks T, Maurer K (1993) Amplitudes of auditory P300 in remitted and residual schizophrenics: correlations with clinical features. Neuropsychobiology 27: 54–60
Westphal KP, Grözinger B, Becker W, Diekmann V, Scherb W, Reess J, Leibing U, Kornhuber HH (1992) Spectral analysis of EEG during self-paced movements: differences between untreated schizophrenics and normal controls. Biol Psychiatry 31: 1020–1037

9.7 The Frontal Lobe and the Essentials of Homo Sapiens

H.H. Kornhuber
Abteilung für Neurologie, Universitätsklinik, Ulm, Germany

There is no principal difference between the columnal organization of the cerebral cortex of man and that of an ape. The main difference is that there are more columns in man (Sawaguchi and Kubota 1986) and, correspondingly, more fiber connections. The essential characteristic of man has often been thought to be language. However, language was already represented by areas in the lower parietal and the upper temporal lobe and lower frontal lobe in the brain of *Homo habilis* 2 million years ago and certainly in *Homo erectus* 1.5 million years ago (Holloway 1983; Tobias 1983, 1987). Both *Homo habilis* and *Homo erectus* also had large brains (Eccles 1989), used stone tools and fire, provided long family care for children, and were cooperative hunters. An asymmetry of the cerebral hemispheres similar to that in *Homo sapiens* may have existed a million years ago and earlier (LeMay 1976). The language areas have, of course, increased since that time and perhaps even somewhat more than the rest of the neocortex; however this, it seems to me, was not the origin of the human mind. Language is a means for communication, and thinking assisted by conceptual language is more efficient, but language is not the fountain of truth. As a matter of fact, language is often used for lying.

Although *Homo habilis* and *Homo erectus* possessed language, there was, as far as we know, for over a million years little creativity in these forerunners of man. Even *Homo sapiens neandertalensis* — who lived from about 200 000 until about 40 000 years ago and had on average 100 g brain more than the recent *Homo sapiens*, the Cro-Magnon man, now called *Homo sapiens sapiens* — made little progress in making tools, in contrast to *Homo sapiens sapiens*, who had new types of tools at least every 2000 years. Furthermore, the Neandertals had more primitive cooking techniques than present-day man and, correspondingly, stronger teeth, although they had fire, were able to hunt large animals (which certainly needed elaborate cooperation including verbal communication), and even buried their dead. Neandertal man (Bosinski 1985) used the same stone tools for over 100 000 years and, as far as we know, developed no arts, whereas *Homo sapiens sapiens* soon produced works of art, such as the statues of Vogelherd, the paintings of Lascaux, and many others.

Because of the belief that language is the essence of man and because of the slowness of cultural progress over 2 million years the theory has been advanced that despite the existence of the cerebral substrates there was no use of language (Jaynes 1976; Lieberman 1975; Marshack 1976). For a neurologist, it seems unlikely that all the cerebral substrates subserving language have not been used over hundreds of thousands of years without undergoing atrophy.

The obvious difference between Neandertal man and *Homo sapiens sapiens* — corresponding to the creativity of the latter — is, however, the large frontal lobe of *Homo sapiens sapiens*. Typical of Neandertal man was the low and narrow forehead (Fig. 1) indicating a small frontal lobe, while the parietal, temporal, and occipital lobes were highly developed, so that the total brain size exceeded that of present-day *Homo sapiens* on average by 100 g.

The smallness of the frontal lobe of Neandertal man was noticed by nineteenth century scientists, but it was suppressed later in our century. A German encyclopedia published in 1909 (Meyers Großes Konversations-Lexikon, 6th edn. Leipzig, Vol. 13, p. 606) remarks (in English translation): "The skull of Neanderthal man was long but low; the anterior part was narrow while the posterior broad. From this one would conclude that the frontal lobe was little developed". The author of the article inferred from the low narrow frontal lobe a lower level of intelligence as compared to present-day man. Intelligence, however, is a cooperative function of the brain which does not depend critically on the frontal lobe. We know from cerebral lesions in children that a diminution of intelligence depends only on the total size of the defect, not on the lobe involved (Kornhuber et al. 1985). Consequently, Neandertal man, who had on average a slightly

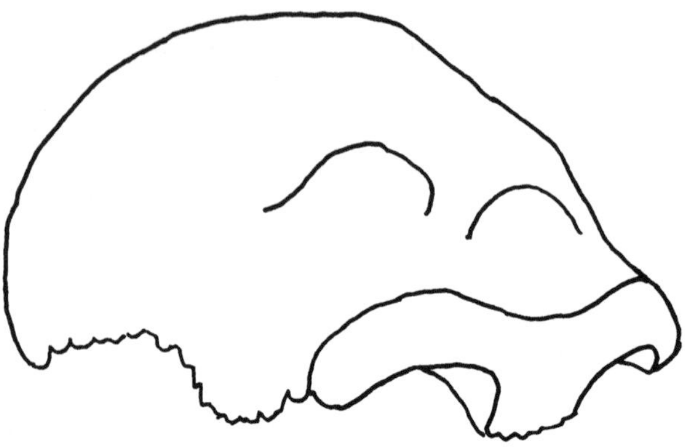

Fig. 9.7–1. Skull of Neandertal (drawn by the author from the original skull in the Bonn museum)

larger brain than present-day *Homo sapiens*, probably had a similar level of intelligence when tested by present-day tests which measure principally speed of information processing and error detection rather than creativity.

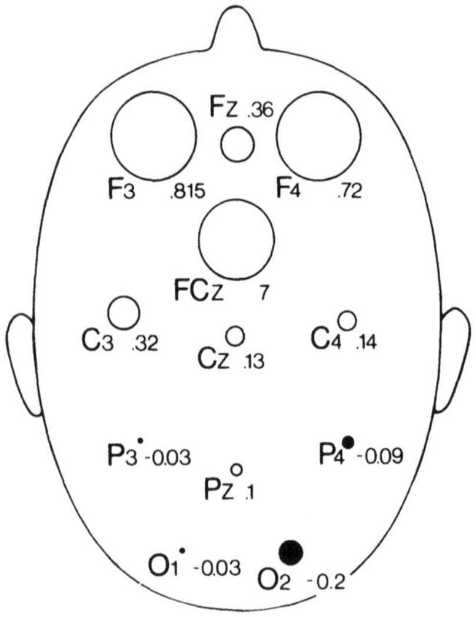

Fig. 9.7–2. Hand-tracking task with inverted (mirror) feedback; because of the concepts right and left, language is involved in this task in addition to the spatial problem. Circles (diameter) give the correlation beween the human cerebral potentials (enhancement of cortical negativity) and visuomotor learning (reduction of error in motor performance between the beginning and the end of the experiment). Significant positive correlations are found only over the frontal lobe. Recording electrode positions according to the 10–20 system. *F*, Frontal; *C*, central; *P*, parietal; *O*, occipital. Odd numbers refer to the left hemisphere, even numbers to the right hemisphere. *Z* are electrodes in the midline. (Modified from W. Lang et al. 1983)

What was added by means of a large frontal lobe was more initiative, leading to less dependency on external stimuli, including tradition. To respond to challenges with novel solutions is a matter of creativity. To search for information and to develop methods require will. It is known from neurophysiological experiments and from clinical neuropsychology that the frontal lobe is a center of volition (Kornhuber 1984, 1987; Kornhuber et al. 1989). The *Bereitschaftspotential* preceding each volitional action is localized midfrontally (Deecke and Kornhuber 1978; Kornhuber and Deecke 1965). A significant positive correlation between both motor and cognitive learning (which requires an effort of will) and cortical potentials in man has been found only for the frontal lobe (W. Lang et al. 1983; Fig. 2). There is even a hemispheric frontal specialization of learning motivation depending on the task which results in such correlations only over the right or left frontal lobe (M. Lang et al. 1987, 1989; W. Lang et al. 1983, 1986a,b, 1987, 1988; Figs. 3,4). Kleist (1934) described two different types of disturbances of the will as a consequence of bilateral frontal lesions: lack of initiative after prefronto-lateral lesions and lack of self-control after fronto-orbital lesions. The former have been confirmed after prefrontal lobotomy (Freeman and Watts 1942).

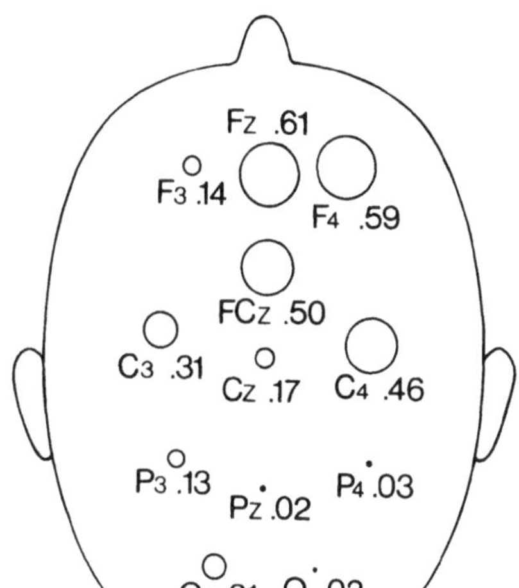

Fig. 9.7–3. Correlations of human cerebral potentials (additional cortical negativity accompanying the learning effort) and learning performance (reduction of motor error) in a hand tracking task with spatial distortion of the visual feedback. In contrast to Fig. 2, this was a purely visuomotor task without language involvement. The significant positive correlations are in this case only over the right frontal lobe. Letters and numbers for electrode positions as in Fig. 2. (Modified from W. Lang et al. 1986b)

It is not stubbornness but rational will which is the speciality of the frontal lobe, a will which initiates planning, makes purposeful use of imagination, and solves problems by reasoning. Deficits in the Wisconsin card-sorting test after prefrontal lesions (Milner 1964) are a sign of defective initiative in developing new strategies. Furthermore, the failure in the verbal fluency task after left frontal lesions (Milner 1964) corresponds to a defect in imagination; it is not caused by a lesion to Broca's area. The importance of Broca's area has been overestimated in the past. Lesions confined to this area cause only transient aphasia or even no obvious language disturbance at all; longer-lasting signs of aphasia in cases of lesions

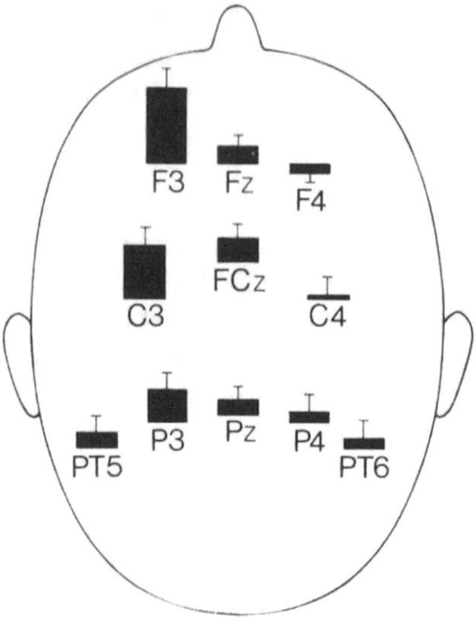

Fig. 9.7–4. Topography of increased cortical negative potentials during the encoding in an associative verbal learning task. The significant increase is only over the left frontal lobe. *PT*, Posterior temporal; otherwise labeling of electrode positions as in Fig. 2. (Modified from M. Lang et al. 1989)

in Broca's area are usually caused by the combination with lesions of the basal ganglia (Brunner et al. 1982). This was also true in Broca's original case.

Apes are able to learn sign language, and chimpanzees even have self-consciousness, but only human children, even deaf mutes, are able to create sign language on their own initiative (Goldin-Meadow and Feldman 1977). It is not language per se but creativity that constitutes the main difference between apes and man. Binocular vision with depth perception, bipedal walking, social cooperation, family life with both father and mother and long-term education of children, rich communication by language, hunting in groups, tool making, use of fire, self-consciousness, etc. — all these were of course prerequisites on the path of evolution to *Homo sapiens sapiens*, but the large frontal lobe brought the breakthrough to creativity.

Creativity, which is a basis of human freedom (Kornhuber 1978, 1984, 1987, 1992), may, however, result not only in progress, but also in such problems as atomic bombs and overpopulation. Thus, the large frontal lobe of *Homo sapiens sapiens*, which added creativity to cooperativity, eventually became a problem for man and for life on earth, leading to the development of morals and ethical reflection.

Individual self-conscious will and creativity are active factors in the development of personality in the human child and teenager. It is not only by genes and environment, but also because of their own personality that humans develop in different ways, for example, some self-disciplined and others pleasure seeking. Self-education acts over tens of years in *Homo sapiens sapiens*, in addition to genes and environment, and results in differences between individuals beyond those found in any other species. This is not a philosophy of elitism; on the contrary, everyone has a chance. There is, on each level of intelligence, a human nobility which results from long-term good will.

In summary, while bipedal walking, cooperation in families, long education, self-consciousness, language, etc. were prerequisites, the essential characteristic of *Homo sapiens sapiens* — creativity — was added by the intelligent will of a large frontal lobe. This resulted, among other things, in arts and in ethical rules.

References

Bosinski G (1985) Der Neandertaler und seine Zeit. Rheinland, Cologne

Brunner RJ, Kornhuber HH, Seemüller E, Suger G, Wallesch CW (1982) Basal ganglia participation in language pathology. Brain and Language 16: 281–299

Deecke L, Kornhuber HH (1978) An electrical sign of participation of the mesial "supplementary" motor cortex in human voluntary finger movement. Brain Res 159: 473–476

Eccles JC (1989) Evolution of the brain: creation of the self. Routledge, London New York

Freeman W, Watts JW (1942) Psychosurgery. Thomas, Springfield Thomas

Goldin-Meadow S, Feldman H (1977) The development of language-like communication without a language model. Science 197: 401–403

Holloway RL (1983) Human paleontological evidence relevant to language behavior. Hum Neurobiol 2: 105–114

Jaynes J (1976) The evolution of language in the late Pleistocene. Ann NY Acad Sci 280: 312–325

Kleist K (1934) Gehirnpathologie. Johann Ambrosius Barth, Leipzig

Kornhuber HH (1978) Geist und Freiheit als biologische Probleme. In: Die Psychologie des 20. Jahrhunderts, vol 6. Kindler, Zürich, pp 1122–1130

Kornhuber HH (1984) Attention, readiness for action, and the stages of voluntary decision - some electrophysiological correlates in man. Exp Brain Res Suppl 9: 420–429

Kornhuber HH (1984) Von der Freiheit. In: Lindauer M, Schöpf A (eds) Wie erkennt der Mensch die Welt? Klett, Stuttgart, pp 83–112

Kornhuber HH (1987) Handlungsentschluß, Aufmerksamkeit und Lernmotivation im Spiegel menschlicher Hirnpotentiale. Mit Bemerkungen zu Wille und Freiheit. Heckhausen et al. (eds) Jenseits des Rubikon; Der Wille in den Humanwissenschaften. Springer, Berlin Heidelberg New York, pp 377–407

Kornhuber HH (1992) Gehirn, Wille, Freiheit. Rev Metaphysique Morale 2: 203–223

Kornhuber HH, Deecke L (1965) Hirnpotentialänderungen bei Willkürbewegungen und passiven Bewegungen des Menschen: Bereitschaftspotential und reafferente Potentiale. Pflügers Arch Ges Physiol 284: 1–17

Kornhuber HH, Bechinger D, Jung H, Sauer E (1985) A quantitative relationship between the extent of localized cerebral lesions and the intellectual and behavioural deficiency in children. Eur Arch Psychiatr Neurol Sci 235: 129–133

Kornhuber HH, Deecke L, Lang W, Lang M, Kornhuber A (1989) Will, volitional action, attention and cerebral potentials in man: Bereitschaftspotential, performance-related potentials, directed attention potential, EEG spectrum changes. In: Hershberger WA (ed) Volitional action. Elsevier, Amsterdam, pp 107–168

Lang M, Lang W, Uhl F, Kornhuber A (1987) Slow negative potential shifts in a verbal concept formation task. In: Johnson R, Rohrbaugh JW, Parasuraman R (eds) Current trends in event-related potential research (EEG Suppl 40). Elsevier, Amsterdam, pp 335–340

Lang M, Lang W, Uhl F, Kornhuber A (1989) Patterns of eventrelated brain potentials in paired assoviative learning tasks: learning and directed attention. In: Topographic Brain Mapping of EEG and Evoked Potentials. Maurer K (ed). Springer Berlin Heidelberg New York, pp 323–325

Lang W, Lang M, Kornhuber A, Deecke L, Kornhuber HH (1983) Human cerebral potentials and visuomotor learning. Pflügers Arch Eur J Physiol 399: 342–344
Lang W, Lang M, Kornhuber A, Kornhuber HH (1986a) Cerebral potentials during hand tracking with inverted feedback: frontal lobe dominance in human visuo-motor learning. In: McCallum WC, Zappoli R, Denoth F (eds). Cerebral psychophysiology: studies in event-related potentials (EEG Suppl 38). Elsevier, Amsterdam, pp 255–257
Lang W, Lang M, Kornhuber A, Kornhuber HH (1986b) Electrophysiological evidence for right frontal lobe dominance in spatial visuomotor learning. Arch Ital Biol 124: 1–13
Lang W, Lang M, Kornhuber A, Deecke L, Kornhuber HH (1987) Slow negative potential shifts in conflicting response paradigms. In: Johnson R, Rohrbaugh JW, Parasuraman R (eds) Current trends in event-related potential research (EEG Suppl 40). Elsevier, Amsterdam, pp 53–58
Lang W, Lang M, Uhl F, Kornhuber A, Deecke L, Kornhuber HH (1988) Left frontal lobe in verbal associative learning: a slow potential study. Exp Brain Res 70: 99–108
LeMay M (1976) Morphological cerebral asymmetries of modern man, fossil man, and nonhuman primate. Ann NY Acad Sci 280: 349–366
Lieberman P (1975) On the origins of language. Macmillan New York
Marshack A (1976) Some implications of the paleolithic symbolic evidence for the origin of language. Ann NY Acad Sci 280: 312–325
Milner B (1964) Some effects of frontal lobectomy in man. In: Akert K, Warren JM (eds) The frontal granular cortex and behavior. MacGraw-Hill, New York, pp 313–331
Sawaguchi T, Kubota K (1986) A hypothesis on the primate neocortex evolution: column-multiplication hypothesis. Int J Neurosci 30: 57–64
Tobias PV (1983) Recent advances in the evolution of the hominids with special reference to brain and speech. In: Chagas C (ed) Recent advances in the evolution of primates. Pontificiae Academiae Scientiarum Scripta Varia 50: 85–140
Tobias PV (1987) The brain of Homo habilis: a new level of organization in cerebral evolution. J Hum Evolut 16: 741–761

MIX
Papier aus verantwortungsvollen Quellen
Paper from responsible sources
FSC® C105338

If you have any concerns about our products,
you can contact us on
ProductSafety@springernature.com

In case Publisher is established outside the EU,
the EU authorized representative is:
Springer Nature Customer Service Center GmbH
Europaplatz 3, 69115 Heidelberg, Germany

Printed by Libri Plureos GmbH
in Hamburg, Germany